Sustainability Education, Youth and Innovation

Jacinta Dsilva
Editor

Sustainability Education, Youth and Innovation

Engaging the Next Generation in Innovating the Future—Concise Contributions

Springer

Editor
Jacinta Dsilva
SEE Institute
SEE Global Research & Partnership Center
Dubai, United Arab Emirates

ISBN 978-3-032-21947-3 ISBN 978-3-032-21948-0 (eBook)
https://doi.org/10.1007/978-3-032-21948-0

This Springer imprint is published by the registered company Springer Nature Switzerland AG
The registered company address is: Gewerbestrasse 11, 6330 Cham, Switzerland

Scientific and Reviewing Committee Members

Foreword

It gives me great pleasure to welcome you to the proceedings of the SEE Global Sustainability Summit (SGSS) 2025 organized by SEE Institute, The Sustainable City, Dubai, UAE. I would like to congratulate the committee members of SGSS 2025 for hosting a timely Summit on *Sustainability Education, Youth and Innovation: Engaging the Next Generation in Innovating the Future.*

As a hub for sustainability research, education and innovation, SEE Institute works on leveraging on all possible opportunities and outputs to develop knowledge and upskill the next generation to contribute meaningfully to the green economy. Through Summits such as the SEE Global Sustainability Summit, research driven innovative outputs can always be presented and later be further communicated through publications such as this to enrich the existing body of knowledge in the field of sustainability, youth development and education.

I am pleased to acknowledge the UAE Ministry of Education and United Nations Principles for Responsible Management Education, as the official patronage that enhanced the quality and reach of the Summit. I would also like to acknowledge University of Birmingham, Dubai and Canadian University in Dubai and Prince Sultan University, Saudi Arabia for their collaboration and support for the summit.

It is gratifying that the number of papers and ideas presented continues to increase with each summit, and this trend was maintained with the present Summit as well. This Summit has brought together experts and practitioners from across the globe to share ideas, research, and practical solutions. This gathering of thought leaders, innovators, visionaries and changemakers reflected our shared commitment to building a more resilient, inclusive, and sustainable world. Each paper, discussion, and presentation contained within this proceeding is a valuable contribution to the global dialogue on sustainability. They are not just academic reflections but examples of how knowledge can inspire action and how innovation can transform communities.

As Chairman of SEE Institute, I am humbled by the passion and dedication of everyone who has contributed to this Summit. I extend my heartfelt appreciation to the researchers, delegates, and partners who have made this event possible. Your insights and commitment remind us that sustainability is a collective journey, one that requires unity, courage, and perseverance.

Together, we hold the power to shape a brighter future, one where economic growth, social equity, and environmental stewardship go hand in hand. Let us rise to this challenge with optimism and determination.

April 2025

Engr. Fares Saeed
Founder and Chairman
SEE Institute
The Sustainable City, Dubai, United Arab Emirates
President—SEE Global Sustainability Summit
Dubai, United Arab Emirates

Preface

It is with great pride, I present the proceedings of the SEE Global Sustainability Summit (SGSS) 2025, held at SEE Institute, The Sustainable City, Dubai, on April 24–25, 2025, under the theme: *Sustainability Education, Youth and Innovation: Engaging the Next Generation in Innovating the Future*. This year the Summit was organized in collaboration with Ministry of Education—UAE.

The Summit was conceived as a platform to bring together academics, researchers, practitioners, policymakers, and students to collectively advance dialogue and action around the challenge of creating a sustainable future for all. Each year, the Summit seeks to shed light on a theme that resonates with local and global sustainability imperatives and transformative practices. In 2025, our focus on sustainability education, youth, and innovation reflecting on the crucial need to empower younger generation with the skills, knowledge, and creativity to address complex environmental and social challenges.

As the UAE celebrates the Year of Community, the choice of this theme was deliberate and significant. Across the globe, youth are emerging as passionate advocates, entrepreneurs, and change-makers in the sustainability movement. They bring a unique combination of energy, digital fluency, and openness to experimentation. However, to realize their full potential, they must be equipped with education systems that go beyond traditional learning, systems that embed sustainability principles, foster interdisciplinary and critical thinking, and cultivate innovation as a way of life. This Summit provided a necessary platform to explore how higher education institutions, research organizations, governments, and industry can work together to reimagine learning and empower youth to lead in the transition toward net zero and circular economies.

We were delighted by the diversity and quality of contributions received this year. Submissions came from across the world, reflecting a wide range of cultural, academic, and professional perspectives. Papers and presentations covered multiple dimensions of the central theme, including:

- Innovative pedagogical approaches to embed sustainability into curricula at all levels of education.

- Case studies showcasing youth-led projects, entrepreneurial ventures, climate resilience, inclusive innovation and community-based sustainability initiatives.
- Research on circular economy models, climate policy, sustainable impact investment and technological enablers for sustainable futures.
- The role of leadership and governance in nurturing innovation ecosystems that prioritize environmental and social well-being.

The best part of this Summit's proceedings is that readers will find both theoretical insights and practical applications. The proceedings capture academic rigor, policy relevance, and real-world experiences, all of which are essential for advancing sustainability discussion and practice.

Beyond the academic content, the Summit served as a dynamic forum for dialogue and collaboration through keynote addresses, panel discussions, and fireside chats. The energy of student participants was particularly noteworthy, as their voices infused the Summit with creativity, critical reflection, and bold visions for the future.

The SEE Global Sustainability Summit is more than an academic gathering and through this proceeding, we intend to share the knowledge generated during the Summit with a wider audience and provide a resource that can guide future research, teaching, and practice. Our hope is that this volume will not only document the intellectual contributions of the Summit but also inspire new collaborations, innovations, and initiatives in the field of sustainability education and youth development.

I would like to extend my heartfelt gratitude to all contributors, reviewers, speakers, and participants who made the 2025 Summit a great success. I would like to thank our Chairman, Engr. Fares Saeed, for his visionary direction and acknowledge the dedicated efforts of the SEE Institute team, whose commitment continue to drive this important initiative forward.

We at SEE Institute believe that the journey toward sustainability always requires dialogue, bold innovation, and unwavering commitment between all the generations. By engaging the next generation, we are not only preparing them to inherit the future but also empowering them to reshape it. May this volume serve as both a record of the conversations we began at SGSS 2024 and a catalyst for the transformative actions that must follow in the future.

Dubai, UAE

Dr. Jacinta Dsilva
Director of Research
Living Lab and International Partnerships
Conference Chair

About SEE Institute

SEE Institute is a hub for research, education and innovation in the areas of sustainability, seeking to empower the growth of sustainable practices and transition to a net zero world. The institute is committed to supporting the demand for sustainability experts contributing to the green economy. UAE and the world need the right human capital base that meets current and future sustainability needs and demands. Through our professional education and research programs, we prepare existing and future industry professionals and companies for diversified employment opportunities in the sphere of sustainability. SEE Research Center seeks to advance sustainable practices through research initiatives conducted in collaboration with industry, governments, and academia. Our emphasis is upon projects that focus on the three pillars of sustainability (social, environmental and economic), sustainability leadership, sustainable behaviour, sustainability in the built environment and net zero initiatives, and it is done through peer reviewed journal papers, white papers and recommendation papers as well as conducting impact assessment and benchmark studies for several projects in the UAE. SEE Institute also supports startups and sustainability businesses to test and validate their innovative sustainable products and services through the Living Lab program.

Contents

About the Editor

Dr. Jacinta Dsilva is a renowned sustainability researcher and strategist with over 25 years of academic and industry experience. She holds the position of Director of Research, Living Lab and International Partnerships at SEE Institute. Her expertise lies in the intersection of sustainability and consumer behavior, and has worked at prestigious institutions in UAE, India, and China. With a strong focus on higher education, strategic partnership and research, Dr. Dsilva has published numerous books, book chapters, and research papers on topics such as Sustainability, Green Technologies, Circular Economy, Consumer Behaviour and Strategic Corporate Governance.

She has led research teams, established collaborative partnerships with high-level stakeholders, and developed frameworks for conducting research and education in circular economy and climate change among other topics. Driven by her commitment to sustainability, she has established a research center and living lab at SEE Institute to facilitate impactful research and knowledge dissemination. Dr. Dsilva's strategic insights and market analysis skills have supported businesses in reducing plastic waste and developing sustainable products and services. With her leadership and expertise, she continues to make significant contributions to academia, industry, and the pursuit of a sustainable future.

Recognized for her thought leadership, Dr. Dsilva has published numerous academic papers in reputable Scopus-based, UGC-Care and ABDC journals as well as presented her research findings at prestigious local and international conferences. She has advised universities and schools on embedding sustainability in their curricula, demonstrating her commitment to shaping future leaders with a sustainability mindset. She has led two successful SEE Global Sustainability Summits attended by over 1200 academics, start-ups, Industry and government officials from across the world. She also worked with National Dialogues along with the Ministry of Climate Change and Environment, UAE and played a significant role in organising Climate Talk Series during COP28 alongside the Ministry of Education. Recently, she has started her venture called InLabs Group (Institute for Learning and Breakthrough Systems), focusing on sustainability, gamification, and emerging technologies.

Part I
Circular Economy and Resource Management

Chapter 1
Advancing Circular Economy Through Higher Education: Opportunities and Challenges in UAE Universities

Afseer Majeed

Abstract This paper aims at exploring the role of universities in promoting the Circular Economy (CE) in the United Arab Emirates focusing on five universities: Khalifa University, University of Dubai, American University of Sharjah, Zayed University and University of Sharjah. Using a qualitative case study design, interviews of stakeholders and documents were analyzed to determine how these universities implement the CE ideas in their teaching, research, and campus activities. The results of the study show that although there is progress, for example, through the provision of specialized courses, sustainable campus projects, and cooperation with industry, there are still numerous problems, including the underrepresentation of CE in interdisciplinary curricula, lack of experienced teaching staff, and financial restrictions. The potential for improvement is seen in the enhancement of interdisciplinary cooperation, the use of partnership with industry, and the integration with the UAE 2021 sustainable development objectives. As such, the universities in the UAE are well-placed to lead sustainably in higher education regionally, despite challenges such as institutional stiffness. These barriers include the need for increased faculty training, more funding, and a more integrated approach to operationalizing CE principles. Thus, through the removal of the identified barriers, the UAE universities can contribute to the global and national sustainability agendas by moving from the linear economic model to the circular economic model.

Keywords Circular economy · Higher education · Sustainability · UAE universities · Curriculum integration

A. Majeed (✉)
College of Business, Jumeira University, Dubai, United Arab Emirates
e-mail: afseer.majeed@ju.ac.ae

J. Dsilva (ed.), *Sustainability Education, Youth and Innovation*,
https://doi.org/10.1007/978-3-032-21948-0_1

Introduction

The world is facing severe challenges of depletion of natural resources, climate change and environmental degradation. The traditional linear economic models of production and consumption, which embrace the 'take make dispose' approach, are now known to be unsustainable. In response, the Circular Economy (CE) has emerged as a transformative model that aims to redesign industrial systems to reduce waste, achieve maximum resource utilization and environmental and economic performance. The linear economy is a one-way system that involves the extraction of raw materials, their use in the production of goods and services and then their disposal. The Circular Economy, however, has its focus on the recycling and reuse of materials, extending the product's lifespan through repair and reuse (Geissdoerfer et al., 2017).

As learning institutions and engines of change, universities are well-placed to advance the Circular Economy agenda. Universities can achieve this by incorporating the CE into their courses, encouraging research on circular practices, and exercising leadership through sustainable campus management. Nonetheless, the current state of the CE discourse in the UAE, a rapidly developing country with distinct sustainability targets, and the extent to which its universities are implementing the CE are not fully understood. Hence, this paper aims at examining how universities in the UAE are implementing the Circular Economy strategies in their teaching, research, and operations and the challenges and opportunities that arise from them.

This research consists of a case study of five universities in the UAE: Khalifa University, University of Dubai, American University of Sharjah, Zayed University, and the University of Sharjah. It examines their strategies for implementing Circular Economy principles and assesses the effect of these strategies on sustainability in the context of the UAE.

Background and Literature

The Circular Economy paradigm has been widely discussed in academic literature, and scholars have identified its potential to solve global environmental problems while driving economic development. The core principles of CE are, reducing the intake of resources, using products and materials again, and recycling waste (Bocken et al., 2016). The Ellen MacArthur Foundation (2019) argued that the implementation of CE can minimize waste and carbon emissions, which can be used as a foundation for the future.

The Circular Economy cannot be achieved without the involvement of higher education institutions. They contribute by preparing the future generations for sustainable leadership, by researching and developing new sustainable solutions and by implementing sustainability actions on the campus. Some of the universities leading in the integration of CE principles into their curricula and campus activities are the University of Cambridge in the UK and Stanford University in the United States (Bocken et al., 2016).

Sustainable development is an important part of the national policy in the UAE, which is witnessed by the UAE Vision 2021 and the National Agenda that aim to shift to a green economy (UAE Vision 2021). This vision includes the integration of sustainability and Circular Economy principles in the educational systems, and therefore, some universities in the UAE have incorporated these principles in their teaching and learning processes. Still, there is limited research on how these universities are incorporating and promoting CE and the gaps are even more evident when the universities' distinctness and obstacles are considered.

Methodology

This study employs a qualitative case study design to explore how five universities in the UAE, namely Khalifa University, University of Dubai, American University of Sharjah, Zayed University, and University of Sharjah, are integrating Circular Economy principles into their curricula, research and operations. The research uses interviews with key stakeholders, such as faculty members, sustainability officers, and administrators, to obtain an understanding of the practices, challenges, and opportunities of CE. Moreover, institutional documents like sustainability reports, curricula, and research publications are reviewed to determine the level of CE integration.

The data is analyzed thematically to identify patterns and trends in the implementation of the Circular Economy practices in these universities. The findings are placed in the context of national and international sustainability strategies and tactics, as well as the problems of higher education institutions in the UAE.

Results and Discussion (Summarized)

This paper investigates the adoption of Circular Economy principles in higher education institutions in the UAE, and the implications for sustainability. This paper presents a case study of five universities in the UAE to examine their adoption of CE principles and the sustainability impact of these universities.

University	CE integration in curriculum	Research focus	Campus sustainability measures	Partnerships and industry linkages	Key challenges
University of Dubai (UD)	Courses on sustainable business; CE themes like material/energy productivity, sustainable design, waste minimization	Research on sustainable business models and technologies; real-world solutions for waste/resource recovery	Green campus: solar energy, water conservation, waste management	Strong industry links	Limited CE integration across all departments; need for curriculum development
Khalifa University (KU)	CE integrated into engineering programmes (e.g. sustainable energy, materials)	Center for Sustainable Energy and Environment: energy conservation, waste-to-energy	Waste separation and recycling systems; energy-efficient buildings	Limited cross-department collaboration	CE confined mainly to engineering; lack of interdisciplinary application
American University of Sharjah (AUS)	Courses in sustainability and CE in business and environmental science	Research on urban sustainability and CE; waste management systems	Campus farm; waste reduction initiatives	Partnership with Sharjah Municipality for circular waste management	CE underrepresented in non-science subjects; financial limitations
Zayed University (ZU)	CE covered in environmental sciences, business, and engineering	Research on waste-to-resource strategies, plastic recycling, green materials	Energy-efficient buildings; water conservation; recycling; awareness campaigns	Collaboration with government entities on sustainability	Needs stronger infrastructure and system-wide CE adoption
University of Sharjah (UOS)	Courses in CE and sustainable development in engineering, business, environmental science	Green building technologies; sustainable construction research	Renewable energy usage; water saving; waste management	Campus sustainability practices underway	Weak leadership and coordination; financial constraints; partial CE operational adoption

Barriers to Advancing Circular Economy in UAE Universities

The adoption of Circular Economy principles in UAE universities encounters multiple obstacles which prevent their implementation in academic and management systems. The main obstacle stems from the restricted implementation of CE principles throughout different fields of study. The majority of CE-related courses exist only in engineering and environmental sciences departments according to Al Nuaimi & Al Zaabi (2020). The creation of disciplinary silos prevents the implementation of necessary interdisciplinary approaches needed for effective CE education. According to Geissdoerfer et al. (2017) a complete disciplinary structure is necessary for the implementation of Circular Economy instead of fragmented approaches.

The lack of faculty expertise combined with low awareness about Circular Economy stands as a major obstacle. The insufficient exposure of instructors to CE concepts prevents them from effectively incorporating these ideas into their teaching practice (Ellen MacArthur Foundation, 2019). Bocken et al. (2016) stress that faculty development represents a crucial requirement for implementing CE, yet UAE universities do not provide structured professional training in this field.

The financial limitations create obstacles for implementing Circular Economy principles. The implementation of extensive sustainability projects including recycling systems and energy-efficient infrastructure requires significant financial resources which exceed the capabilities of traditional funding models (Kirchherr et al., 2018). The University of Sharjah Sustainability Report (2022) points out that financial constraints represent a major obstacle for Circular Economy development.

The sustainability practices implemented by Khalifa University and Zayed University exist independently of a complete CE framework (Khalifa University Sustainability Report, 2022; Zayed University Sustainability Report, 2022). Kirchherr et al. (2018) state that independent initiatives without systemic leadership fail to produce the essential closed-loop systems needed for Circular Economy.

Opportunities for Enhancing Circular Economy in UAE Universities

UAE universities maintain the opportunity to improve their Circular Economy implementation abilities while serving national and international sustainability goals despite existing difficulties. Academic departments should work together through cross-disciplinary partnerships as a fundamental opportunity for enhancing their operations. The incorporation of Circular Economy principles into courses beyond engineering and environmental sciences and business and social sciences and humanities will broaden student comprehension of circular systems (Geissdoerfer et al., 2017). The Ellen MacArthur Foundation (2019) indicates that interdisciplinary higher education approaches drive innovation through technical-social-economic linkages which UAE institutions could utilize effectively because of their diverse student demographics.

The partnership between universities and industry stands as an effective approach for success. Local companies in construction and energy sectors and manufacturing can create practical CE solutions through waste to resource technologies and sustainable supply chain management (Bocken et al., 2016). The American University of Sharjah and Sharjah Municipality have established a partnership for circular waste management systems which demonstrates how academic knowledge can transform into practical solutions (AUS Sustainability Report, 2022). The expansion of these partnerships would enhance resource availability while offering students hands-on experience with Circular Economy practices.

The national policy framework of UAE Vision 2021 with its green economy focus creates an opportunity for the UAE to link up with (UAE Vision 2021). Universities should utilize government support to back their CE research and campus activities because this demonstrates their contribution to national sustainability targets. The implementation of national objectives in institutional goals leads to financial backing and improved credibility for universities' sustainability initiatives according to Al Nuaimi and Al Zaabi (2020). Zayed University successfully implements this approach through government backing (Zayed University Sustainability Report, 2022).

Universities should use their position as research and innovation centers to study CE-related matters and execute campus-based actions. The Center for Sustainable Energy and Environment at Khalifa University demonstrates potential for developing waste-to-energy technologies that match circular principles through their work (Khalifa University Sustainability Report, 2022). The expansion of current sustainability initiatives together with student-led projects which are present at AUS (AUS Sustainability Report, 2022) will develop future leaders who can advocate for Circular Economy principles at local, regional and worldwide levels.

Conclusion and Way Forward

Universities in the United Arab Emirates should lead the way to accelerate Circular Economy implementation which represents an essential model for sustainable development within global challenges (Geissdoerfer et al., 2017). The CE principles get implemented through curriculum modifications and research projects and campus operations at Khalifa University and University of Dubai and American University of Sharjah and Zayed University and the University of Sharjah. Full implementation of the circular economy faces several challenges that include uneven discipline-based curriculum integration and insufficient faculty expertise and financial resource limitations (Al Nuaimi & Al Zaabi, 2020; Kirchherr et al., 2018). The substantial obstacles do not prevent UAE universities from becoming leading institutions in sustainable higher education through their potential leadership in the region.

The study indicates that engineering and environmental science programs have adopted some Circular Economy principles although there exists an extensive need to integrate these concepts throughout other academic disciplines (Ellen MacArthur Foundation, 2019). The limitation of student knowledge about CE prevents them from understanding how to develop innovative solutions for complex sustainability issues (Bocken et al., 2016). The operational initiatives such as waste management and energy conservation efforts at Khalifa University Sustainability Report (2022) and Zayed University Sustainability Report (2022) show promise but they do not meet the complete requirements of CE system implementation. A strategic plan must combine the top features of universities with outside support to resolve these deficiencies.

Several essential steps should be the focus of UAE universities to enhance CE implementation. The second step requires universities to expand CE education through subject integration across different academic fields. The University of Cambridge demonstrates how interdepartmental courses can teach CE principles in business and social science and humanitarian subjects (Bocken et al., 2016). The university needs to invest money in faculty development because their understanding of CE principles remains incomplete (Ellen MacArthur Foundation, 2019). The third requirement demands increased cooperation between academic institutions and the government as well as industry sectors through examples like AUS's partnership with Sharjah Municipality (AUS Sustainability Report, 2022). Such partnerships align with UAE Vision 2021 green economy targets which improve organizational exposure and resource accessibility (UAE Vision 2021; University of Dubai, 2022).

UAE universities need to establish research programs for CE technologies and develop sustainable campus practices as their main focus for future development (Kirchherr et al., 2018). They will effectively support UAE sustainable development goals while participating in global CE dialogues through this approach. The transformation process demands both perseverance to break traditional barriers and sustained financial support although the resulting outcomes will be significant. The new paradigm of CE requires higher education institutions to promote it as stated by Geissdoerfer et al. (2017) and UAE universities can lead this regional transformation by taking immediate action.

References

Al Nuaimi, S. A., & Al Zaabi, Z. A. (2020). Sustainability education in the UAE: An analysis of higher education initiatives. *Journal of Sustainability in Higher Education, 21*(5), 950–968. https://doi.org/10.1108/JSHE-03-2020-0085

American University of Sharjah. (2022). *AUS sustainability report 2022*. https://www.aus.edu/sustainability/our-vision/plans-and-reports

Bocken, N. M. P., de Pauw, I., Bakker, C., & van der Grinten, B. (2016). Product design and business model strategies for a circular economy. *Journal of Industrial and Production Engineering, 33*(5), 308–320. https://doi.org/10.1080/21681015.2016.1172124

Ellen MacArthur Foundation. (2019). *Circular economy in higher education: Accelerating innovation through learning*. https://www.ellenmacarthurfoundation.org
Geissdoerfer, M., Savaget, P., Bocken, N. M. P., & Hultink, E. J. (2017). The circular economy—A new sustainability paradigm? *Journal of Cleaner Production, 143*, 757–768. https://doi.org/10.1016/j.jclepro.2016.12.048
Khalifa University. (2022). *Sustainability and circular economy practices at Khalifa University*. https://www.ku.ac.ae/csr
Kirchherr, J., Piscicelli, L., Bour, R., Kostense-Smit, E., Muller, J., Huibrechtse-Truijens, A., & Hekkert, M. (2018). Barriers to the circular economy: Evidence from the European Union (EU). *Ecological Economics, 150*, 264–272. https://doi.org/10.1016/j.ecolecon.2018.04.028
UAE Vision. (2021). *The UAE Vision 2021: National agenda*. https://www.vision2021.ae
University of Dubai. (2022). *Circular economy initiatives at UD*. http://ud.ac.ae/PRME/sustainability.php
University of Sharjah. (2023). *UOS achievement and contribution in the field of environment and climate change*. Sharjah Government Media Office. https://www.sharjah.ac.ae/-/media/project/uos/sites/uos/discover-uos/annual-reports/annual_report_2023.pdf
Zayed University. (2022). *Sustainability and circular economy at ZU*. https://www.zu.ac.ae/main/en/research/_images/zu-research-strategy.pdf

Chapter 2
Strategic Approaches to Sustainable Construction Waste Management: Insights from KSA and Qatar Projects

Haitham Bataineh and Rekha Pillai

Abstract The construction industries in Qatar and Saudi Arabia are experiencing unprecedented growth driven by ambitious infrastructure projects. This rapid expansion, however, has led to significant challenges in construction waste management (CWM), contributing to environmental degradation, increased disposal costs, and inefficient resource utilization. This study critically examines current CWM practices in these regions, identifies key challenges, and explores opportunities for enhancing sustainability through the integration of circular economy principles. Employing a qualitative research approach, the study utilized semi-structured interviews with ten industry experts, including project managers, site engineers, and sustainability officers, across multiple case studies to compare various projects and identify common patterns in waste management. Findings reveal that systemic inefficiencies in project management, coupled with regulatory gaps and a pervasive lack of awareness, significantly impede sustainable waste management efforts. Furthermore, limited recycling infrastructure and inadequate workforce training exacerbate these issues. Conversely, the research identifies promising opportunities for improvement through the strategic adoption of technological advancements, such as Building Information Modeling (BIM), fostering enhanced stakeholder collaboration, and refining existing policy frameworks. The study emphasizes the pivotal role of project managers in the successful implementation of effective waste management strategies and advocates strengthening regulatory enforcement, increasing investment in recycling facilities, and promoting cross-sector collaboration. These insights offer valuable recommendations for policymakers and industry professionals, contributing to the development of more sustainable construction practices and advancing circular economy principles within Qatar and Saudi Arabia.

Keywords Strategic waste management · Circular economy · Construction waste reduction · Waste hierarchy models · Sustainable construction practices

H. Bataineh
University of West of Scotland, Paisley, UK

R. Pillai (✉)
The British University in Dubai, Dubai, UAE
e-mail: rekha.pillai@buid.ac.ae

J. Dsilva (ed.), *Sustainability Education, Youth and Innovation*,
https://doi.org/10.1007/978-3-032-21948-0_2

Introduction

The construction sector in Qatar and Saudi Arabia has undergone a period of unprecedented growth, fueled by extensive national development visions such as Saudi Vision 2030 and Qatar National Vision 2030. While this expansion has driven economic prosperity and infrastructural advancement, it has simultaneously generated substantial volumes of construction waste. The inefficient management of this waste poses significant environmental, economic, and social challenges, including increased landfill burden, depletion of natural resources, elevated project costs, and adverse ecological impacts [1]. Addressing these challenges is paramount for achieving sustainable development goals in the region.

This study aims to provide a comprehensive examination of current construction waste management (CWM) practices within Qatar and Saudi Arabia. Specifically, it seeks to identify the prevailing challenges hindering effective CWM and to explore strategic opportunities for improvement, particularly through the lens of circular economy principles. The research addresses a critical gap in existing literature, which, despite highlighting the importance of regulatory frameworks, technological advancements, and stakeholder engagement in sustainable waste management globally, offers limited region-specific insights into the unique constraints and cultural influences impacting waste practices in the Gulf Cooperation Council (GCC) countries.

By adopting a qualitative research methodology, this study employs semi-structured interviews with ten experienced industry professionals. A multiple case study approach was utilized to facilitate a comparative analysis of diverse construction projects, thereby enabling the identification of common trends and distinct variations in waste management approaches across the region. The insights derived from this research are intended to inform and enhance current waste management strategies, offering practical recommendations for policymakers and industry stakeholders committed to fostering more sustainable construction practices in Qatar and Saudi Arabia.

Methodology

This research employed a qualitative, multiple case study design to investigate construction waste management (CWM) practices in Saudi Arabia and Qatar. This approach facilitated an in-depth exploration of diverse contexts, allowing for the identification of common patterns and unique variations in waste management strategies across different construction projects [2]. The primary data collection method involved semi-structured interviews, which provided a flexible yet structured framework for gathering rich, detailed insights into the perspectives and experiences of key stakeholders [3]. This method was particularly suitable for eliciting nuanced information regarding CWM practices, associated challenges, and potential solutions within the specific regional context.

Participant Selection and Data Collection

Purposive sampling was utilized to select participants with extensive experience and expertise in construction and waste management. This non-probability sampling technique, appropriate for qualitative research aiming to target specific knowledge holders, ensured that interviewees possessed relevant insights into the study's objectives [4]. The sample comprised ten industry experts and market leaders, including project managers, construction managers, site engineers, sustainability leads, and project operations specialists, representing both Saudi Arabia and Qatar (see Table 2.1 for participant demographics). Interviews were conducted until theoretical saturation was achieved, indicating that no new significant themes or information emerged from subsequent interviews.

An interview guide, developed through a comprehensive review of existing literature and preliminary discussions with industry experts, was used to ensure consistency across interviews while allowing for emergent themes. The guide covered critical topics such as current CWM practices, regulatory frameworks, challenges encountered, and innovative solutions. Open-ended questions were predominantly used to encourage thorough responses, supplemented by follow-up questions to explore interesting or unanticipated themes [5].

Data Analysis and Ethical Considerations

Interview transcripts were analyzed using NVivo 14 software, a qualitative data analysis tool that facilitates thematic analysis. This systematic process involved familiarization with the data, generation of initial codes, searching for themes, reviewing themes, defining and naming themes, and producing the report [6]. Reliability in data analysis was supported through the use of structured interview guides and a consistent coding approach. To enhance the validity of the findings, participants were invited to review their interview transcripts for accuracy, ensuring that the research accurately reflected their views and experiences [7].

Ethical approval for the study was obtained from the University of the West of Scotland prior to the commencement of data collection. All participants provided informed consent, acknowledging their understanding of the research purpose, their right to withdraw at any time, and how their data would be utilized. To safeguard participant confidentiality and ensure anonymity, personal identifiers were removed from all transcripts, and responses were anonymized during the data analysis and reporting phases. The study strictly adhered to the ethical guidelines set forth by

Table 2.1 Interview participants' demographics

Participant ID	Country	Age group	Gender	Position	Location	Brief description
1	Qatar	20–30	Male	Project Manager	Doha	Overseeing multiple construction projects
2	Qatar	30–45	Male	Construction Manager	Al Khor	Specializes in sustainable construction
3	Qatar	45 +	Female	Sustainability Lead	Doha	Focus on environmental policies
4	Saudi Arabia	20–30	Male	Senior Project Manager	Riyadh	Manage large-scale infrastructure projects
5	Saudi Arabia	30–45	Male	Senior Site Engineer	Jeddah	Expert in waste management strategies
6	Saudi Arabia	45 +	Female	LEED Green Associate	Riyadh	Develops and implements green initiatives
7	Qatar	20–30	Male	Project Manager	Doha	Coordinates waste reduction programs
8	Qatar	30–45	Male	Site Engineer	Al Wakrah	Advocates for recycling practices on-site
9	Saudi Arabia	45 +	Male	Sustainability Officer	Riyadh	Leads corporate sustainability efforts
10	Saudi Arabia	25–30	Female	Project Operations Specialist	Jeddah	Works on innovative waste management solutions

Note: Participants were selected based on experience in sustainability and project management roles across Saudi Arabia and Qatar

the University of the West of Scotland, aligning international standards for research ethics.

Findings

The thematic analysis of the interview transcripts yielded six major themes, each with several associated sub-themes, providing a comprehensive understanding of strategic construction waste management in Qatar and Saudi Arabia. These themes align with and expand upon existing literature, offering region-specific insights into the complexities of CWM. The identified themes are visually represented in Fig. 2.1, which illustrates their interconnections and hierarchical relationships.

Key Themes and Sub-Themes Identified in the Study.

Theme 1: Project Management Challenges

Inefficient project management emerged as a primary contributor to excessive construction waste. Poor planning, material overestimation, and weak regulatory enforcement were consistently highlighted by participants. This theme underscores the critical role of project managers in mitigating waste through improved coordination and resource management. As one participant noted, "Excess materials often go to waste due to inaccurate procurement planning." This finding resonates with broader academic discourse, which consistently links inadequate project planning and management to increased waste generation in construction projects [8]. The lack of robust waste management plans at the project inception phase often leads to reactive rather than proactive waste handling, exacerbating the problem. Effective project management, therefore, requires a paradigm shift towards integrating waste minimization strategies from design to completion, emphasizing lean construction principles [9].

Inefficient Planning

Detailed planning is crucial for waste minimization. Inefficient planning often leads to material over-ordering, improper storage, and a lack of clear waste management protocols, resulting in significant waste generation.

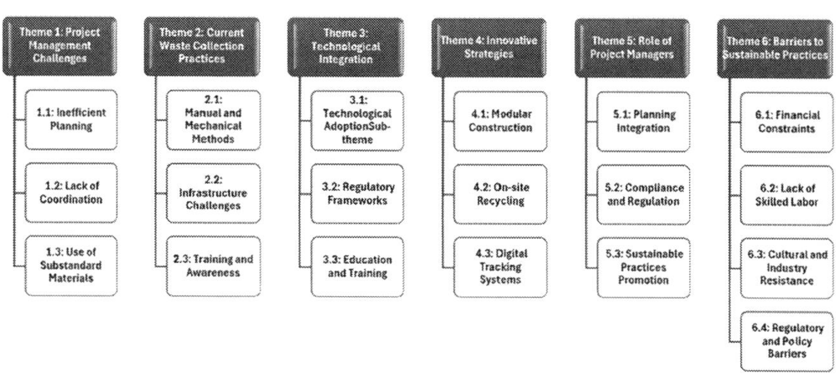

Fig. 2.1 Key themes and sub-themes identified in the study

Lack of Coordination

Poor coordination among project stakeholders, including designers, contractors, and subcontractors, contributes to material wastage and inefficient waste handling. Improved communication and integrated project delivery methods are essential.

Use of Substandard Materials

The use of substandard materials often leads to rework and increased waste. Ensuring quality control and adherence to material specifications can significantly reduce waste generation.

Theme 2: Current Waste Collection Practices

Despite growing awareness of environmental sustainability, current waste collection practices in the region face significant hurdles. Participants identified inadequate infrastructure and a lack of skilled workforce training as major impediments. The absence of readily available and accessible recycling facilities limits the diversion of construction and demolition waste from landfills. Furthermore, insufficient training in waste segregation and on-site recycling practices leads to contamination and reduced recyclability of materials. A participant emphasized the potential benefits, stating, "Using recycled aggregates could significantly reduce landfill waste." This aligns with global challenges in developing countries where informal waste management systems and limited investment in recycling infrastructure hinder circular economy initiatives [10]. Overcoming these challenges necessitates strategic investments in modern recycling technologies and comprehensive training programs for construction personnel.

Manual and Mechanical Methods

Waste collection primarily relies on manual and mechanical methods, which can be inefficient and labor-intensive. Automation and advanced equipment can improve efficiency and safety.

Infrastructure Challenges

Limited waste collection and sorting infrastructure, including insufficient recycling facilities and transfer stations, hinder effective waste management. Investment in robust infrastructure is critical.

Training and Awareness

Lack of training among construction workers on proper waste segregation and handling practices, coupled with low awareness of waste reduction benefits, contributes to inefficient waste collection.

Theme 3: Technological Integration

Technological advancements, particularly Building Information Modeling (BIM) and digital tracking systems, were identified as crucial enablers for improving material efficiency and reducing waste. However, their effective implementation requires a skilled workforce. BIM, for instance, allows for precise material estimation and clash detection, thereby minimizing material wastage during the design and construction phases. A participant highlighted this benefit: "BIM technology helps estimate material needs more accurately, reducing excess waste." This finding is consistent with research demonstrating BIM's potential to optimize resource use and enhance waste management through improved planning and visualization [11]. Beyond BIM, the adoption of digital platforms for waste tracking and management can provide real-time data, enabling better decision-making and accountability in waste reduction efforts.

Technological Adoption

The adoption of advanced technologies like BIM, prefabrication, and modular construction can significantly reduce waste by optimizing design, material use, and construction processes.

Regulatory Frameworks

Supportive regulatory frameworks are essential to encourage the adoption of new technologies and sustainable practices. Clear guidelines and incentives can accelerate technological integration.

Education and Training

Effective implementation of new technologies requires a skilled workforce. Education and training programs are necessary to equip professionals with the knowledge and skills to utilize these tools for waste management.

Theme 4: Innovative Strategies

Innovative strategies are crucial for advancing sustainable CWM practices. These include adopting modern construction techniques and implementing effective on-site waste management systems. A participant suggested an opportunity: "One opportunity is to adopt modern formwork techniques such as using plastic molds from companies like Perri *and* Doka. These molds are reusable and significantly reduce waste." This highlights the need for embracing new approaches that promote waste reduction and resource efficiency.

Modular Construction

Modular construction and prefabrication techniques can significantly reduce on-site waste generation by shifting construction activities to controlled factory environments, optimizing material use, and minimizing errors.

On-Site Recycling

Implementing on-site recycling and waste segregation facilities can reduce transportation costs and promote the reuse of materials directly at the construction site. This requires proper planning and dedicated areas for waste sorting.

Digital Tracking Systems

Digital tracking systems and waste management software can provide real-time data on waste generation, enabling better monitoring, analysis, and decision-making for waste reduction and diversion.

Theme 5: Role of Project Managers

Project managers play a pivotal role in implementing effective waste management strategies. Their leadership and commitment are crucial for integrating sustainable practices throughout the project lifecycle. This includes ensuring proper planning, compliance with regulations, and promoting a culture of sustainability. As one participant shared, "Our partnership with local recyclers has significantly reduced scrap metal waste." This underscores the importance of a multi-stakeholder approach, where shared responsibility and collective action drive sustainable outcomes [12].

Planning Integration

Project managers must integrate waste management planning into all project phases, from design and procurement to construction and demolition. This proactive approach ensures waste minimization is a core objective.

Compliance and Regulation

Ensuring compliance with environmental regulations and promoting adherence to waste management policies are key responsibilities of project managers. This includes understanding and implementing local and national guidelines.

Sustainable Practices Promotion

Project managers are instrumental in fostering a culture of sustainability on construction sites. This involves promoting waste reduction, reuse, and recycling practices among all team members and stakeholders.

Theme 6: Barriers to Sustainable Practices.

Financial and regulatory barriers significantly impede the adoption of sustainable CWM practices. Participants noted that strict enforcement of waste management laws and the provision of financial incentives for sustainable practices are urgently needed. Current penalties for non-compliance were considered insufficient to drive behavioral changes. The high cost of proper waste disposal and the lack of economic incentives for recycling often lead to illegal dumping or inefficient waste handling. This highlights the need for policy interventions that make sustainable practices economically attractive and non-sustainable practices financially punitive. Regulatory frameworks need to be robust, clearly defined, and consistently enforced to create a level playing field and encourage investment in sustainable CWM solutions [13].

Financial Constraints

High costs associated with sustainable waste management practices, including recycling and proper disposal, often deter adoption. Lack of financial incentives and subsidies further exacerbates this barrier.

Lack of Skilled Labor

Insufficient skilled labor for waste segregation, processing, and management hinders the effective implementation of sustainable practices. Training and capacity building are essential to address this gap.

Cultural and Industry Resistance

Resistance to change within the construction industry, coupled with traditional practices and a lack of awareness regarding the long-term benefits of sustainable CWM, poses a significant barrier.

Regulatory and Policy

Weak enforcement of existing regulations, coupled with a lack of comprehensive and clear policies, undermines efforts to promote sustainable waste management. Robust regulatory frameworks are essential.

This research aligns with existing literature, confirming that poor coordination contributes to construction waste [14] and that innovative methods like BIM and lean construction effectively minimize waste [15]. These findings reinfore the broader applicability of waste management strategies while providing specific context and nuances from the Gulf region. The insights gained from this study contribute to a deeper understanding of the challenges and opportunities for advancing sustainable CWM, particularly in rapidly urbanizing environments with unique socio-economic and regulatory landscapes.

Conclusions

This study has provided a comprehensive exploration of strategic approaches to construction waste management (CWM) in Saudi Arabia and Qatar, two regions experiencing unprecedented infrastructural growth. Drawing upon rich qualitative insights from ten seasoned industry professionals, the findings have illuminated systemic challenges inherent in current CWM practices. These challenges manifest as inadequate planning and project management, significant regulatory enforcement gaps, insufficient training for the workforce, and a notably limited recycling infrastructure. These barriers are further compounded by unique cultural and institutional limitations prevalent in the region, which often delay or impede the widespread adoption of sustainable construction practices.

Nonetheless, the research has successfully identified several promising pathways and critical enablers for substantial improvement in CWM. The strategic integration of digital technologies, most notably Building Information Modeling (BIM), alongside the cultivation of robust cross-sector collaborations and the targeted refinement of existing policy frameworks, emerged as pivotal elements for effective CWM. The study unequivocally underscores the central and pivotal role of project managers in aligning site-level operational practices with broader organizational and national sustainability goals. This highlights an urgent need for continuous competency development and strong leadership in implementing proactive waste reduction strategies across all project phases.

By employing a rigorous thematic analysis of the interview data, this study not only corroborates known inefficiencies in construction waste handling observed globally but also brings forward crucial region-specific nuances and contextual factors. It significantly contributes to filling a critical knowledge gap concerning CWM within the Gulf Cooperation Council (GCC) countries, a region where scholarly literature on this specific topic remains notably scarce. The findings offer a granular understanding of the unique challenges and opportunities that characterize the CWM landscape in these rapidly developing economies.

Practical Implications

The practical implications of this research are far-reaching and significant for various stakeholders. For policymakers, the results serve as a compelling call to action for the establishment of more robust enforcement mechanisms for waste management regulations, the development of enhanced waste tracking and reporting systems, and the implementation of targeted financial incentives to actively support and encourage green construction practices. For industry leaders and practitioners, the study provides actionable recommendations, emphasizing the strategic importance of investing in comprehensive workforce training programs, streamlining procurement strategies to minimize material waste, and actively leveraging local partnerships with recyclers and waste management companies. These measures can collectively lead to measurable improvements in sustainability outcomes and foster a more resource-efficient construction sector.

Limitations and Future Research

Despite its valuable contributions, this study acknowledges certain limitations. Its qualitative nature, while providing rich in-depth insights, inherently limits the generalizability of the findings to broader populations or different geographical contexts. Furthermore, the study's limited geographic scope (focused solely on Qatar and Saudi Arabia) and sample size (ten interviews) mean that the findings, while representative of the interviewed experts, may not capture the full spectrum of experiences across the entire GCC region. Future research endeavors could address these limitations by adopting mixed methods approaches, combining qualitative insights with quantitative data to enable broader statistical analysis. Longitudinal designs could also be employed to assess the long-term effectiveness and impact of policy interventions and the adoption of digital solutions over time. Moreover, future investigations into how procurement processes and contractual systems specifically influence waste generation and behavior within the construction industry would offer additional valuable insights, further advancing the development of more comprehensive and sustainable construction frameworks.

Overall, this research significantly advances both scholarly understanding and practical pathways toward achieving more circular, resource-efficient construction models in the dynamic and rapidly evolving Gulf region. It provides a foundational basis for future research and policy development aimed at mitigating the environmental impact of construction activities and promoting a more sustainable built environment.

References

1. Kingdom of Saudi Arabia. (2016). Vision 2030: Environmental and sustainability objectives. Vision 2030 Documentation. Retrieved July 23, 2024, from https://www.vision2030.gov.sa/
2. Patton, M. Q. (2002). *Qualitative research and evaluation methods* (3rd ed.). Sage Publications.
3. Kvale, S., & Brinkmann, S. (2009). *Interviews: Learning the craft of qualitative research interviewing* (2nd ed.). Sage Publications.
4. Neuman, W. L., & Robson, K. (2013). *Basics of social research: Qualitative and quantitative approaches* (3rd ed.). Pearson.
5. Bryman, A. (2016). *Social research methods* (5th ed.). Oxford University Press.
6. Braun, V., & Clarke, V. (2006). Using thematic analysis in psychology. *Qualitative Research in Psychology, 3*(2), 77–101.
7. Creswell, J. W. (2014). *Research design: Qualitative, quantitative, and mixed methods approaches* (4th ed.). Sage Publications.
8. Lingard, H., Graham, P., & Smithers, G. (2000). Employee perceptions of the solid waste management system operating in a large Australian contracting organization: Implications for company policy implementation. *Construction Management and Economics, 18*(4), 383–393.
9. Koskela, L. (1992). Application of the new production philosophy to construction. Technical Report No. 72. Stanford, CA: Stanford University, Center for Integrated Facility Engineering.
10. Poon, C. S., Yu, A. T., Wong, S. W., & Cheung, E. (2004). Management of construction waste in public housing projects in Hong Kong. *Construction Management and Economics, 22*(7), 675–689.
11. Azhar, S., Carlton, W. A., Olsen, D., & Ahmad, I. (2011). Building Information Modeling for sustainable design and LEED rating analysis. *Automation in Construction, 20*(2), 217–224.
12. Wang, J. Y., Touran, A., Christoforou, C., & Fadlalla, H. (2010). A systems analysis tool for construction and demolition wastes management. *Waste Management, 30*(11), 2223–2233.
13. Osmani, M., Glass, J., & Price, A. D. F. (2008). Architects' perspectives on construction waste reduction by design. *Waste Management, 28*(7), 1147–1158.
14. Cheng, J., Zhang, J., Wu, P., & Li, H. (2020). Enhancing construction waste management performance using stakeholder engagement approaches: A case study in Hong Kong. *Journal of Cleaner Production, 247*, Article 119593.
15. Won, J., & Cheng, J. C. (2017). Identifying potential opportunities of building information modelling for construction and demolition waste management and minimization. *Automation in Construction, 79*, 3–18.

Chapter 3
Industry 4.0 Technologies for Sustainable Plastic Waste Management: An Exploratory Study

Rita Simson Fernandez and Ahmed Mohammed

Abstract This study investigates how Industry 4.0 technologies can enhance Sustainable Plastic Waste Management (SPWM) in a Dubai manufacturing company. Encouraged by a genuine sustainability initiative from the company, the project aimed to apply research in a real-world context. Using a combination of a thorough literature review and a case study analysis with semi-structured interviews, the research identifies key enablers, such as regulatory compliance, stakeholder collaboration, and training, as well as barriers like high costs and regulatory inconsistencies. The findings from these methods provide valuable insights into the challenges and opportunities within the company's sustainability practices. Recommendations include adopting advanced technologies, implementing employee training, and engaging stakeholders. The research offers a roadmap to improve SPWM efficiency, reduce plastic waste, and support the company's sustainability goals. This applied research demonstrates the value of linking academic insights with practical solutions in a growing industry, driving meaningful sustainability outcomes.

Keywords Industry 4.0 · Sustainable plastic waste management · Enablers and barriers · Stakeholder engagement

Introduction

Since its invention, plastic's lightweight, low cost, and widespread availability have made it an indispensable material across various industries. Its resistance to decay has particularly cemented its role as the preferred choice for packaging, which accounts

R. S. Fernandez (✉)
Department of Management, Birmingham Business School, University of Birmingham, Dubai, UAE
e-mail: ritasimsonfernandez@gmail.com

A. Mohammed
University of Birmingham, Dubai, United Arab Emirates
e-mail: a.m.mohammed.2@bham.ac.uk

J. Dsilva (ed.), *Sustainability Education, Youth and Innovation*,
https://doi.org/10.1007/978-3-032-21948-0_3

for approximately 40% of global plastic production (Evode et al., 2021). Plastics are extensively used to provide hygienic and affordable packaging solutions in the food and beverage sector, pharmaceuticals, cosmetics, and other manufacturing industries. However, the increasing use of plastics, especially single-use and packaging materials, has led to significant environmental challenges due to the accumulation of waste (Duru et al., 2019). Plastics Europe reports that coastal countries discard between 4.8 million and 12.7 million metric tons of plastic waste into the oceans each year (Britannica, 2024). In 2018 alone, around 80 million metric tons of plastic waste were mishandled through open dumping, burning, and inadequate landfills (Gao et al., 2022).

Traditional plastic waste management methods, such as landfilling and recycling, have proven unsustainable. Landfilling consumes valuable space, while recycling processes are often costly, energy-intensive, and prone to emitting toxic gases (Huang et al., 2022; Pandey et al., 2023). Additionally, these methods often produce lower-quality materials and expose skilled labor to harmful emissions, making them less viable in the long term (Wen et al., 2021). As a result, inadequate plastic waste management, combined with unsustainable consumption patterns, poses serious threats to the economy, natural habitats, and public health (Prata et al., 2019). This calls for innovative solutions to manage plastic waste more effectively.

Sustainable Waste Management (SWM) aims to minimize waste generation and mitigate harmful substances while enhancing the reuse, recycling, and recovery of materials. This approach requires collaborative efforts among researchers, policymakers, and community members to balance environmental, economic, and social dimensions of sustainability (Cucchiella et al., 2017). However, current SWM practices face several challenges, including regulatory uncertainties, high implementation costs, and limited awareness of sustainable practices. The integration of advanced technologies is increasingly seen as a potential solution to these challenges.

The rise of Industry 4.0 offers promising avenues to tackle plastic pollution by leveraging advanced technologies such as the Internet of Things (IoT), Artificial Intelligence (AI), and Blockchain Technology (BCT). These technologies can significantly enhance the efficiency, traceability, and sustainability of waste management systems. For instance, IoT-enabled sensors can provide real-time data on waste location and condition, improving collection efficiency and minimizing waste (Kurniawan et al., 2022). AI-driven approaches can optimize waste collection routes and automate segregation processes, while BCT can enhance the transparency and traceability of plastic recycling across the supply chain (Ghahramani et al., 2022; Khadke et al., 2021). According to Ghobakhloo (2020), Industry 4.0 technologies can considerably enhance the economic, environmental, and social sustainability of waste management systems.

Despite their potential, these technologies' application to Sustainable Plastic Waste Management (SPWM) in the manufacturing sector remains underexplored. Key barriers include the high cost of implementation, regulatory compliance issues, and a lack of awareness among manufacturers about the benefits of Industry 4.0 solutions. Addressing these challenges requires a comprehensive understanding of both the enabling factors and the obstacles to the adoption of advanced technologies for SPWM.

This study focuses on the manufacturing stage, specifically the packaging waste generated by Company A, a Dubai-based manufacturer that relies heavily on stretching films for product protection. The increasing product demand has amplified the generation of plastic waste, emphasizing the need for responsible management practices.

This research aims to explore sustainable plastic waste management (SPWM) in the manufacturing sector using Industry 4.0 technologies. The specific objectives of this study are to:

1. Examine the current practices and challenges of plastic waste management in the manufacturing sector, focusing on packaging waste generated by Company A.
2. Identify the key enablers and barriers to the adoption of Industry 4.0 technologies for sustainable plastic waste management.
3. Analyze the potential impact of IoT, AI, and Blockchain Technology on enhancing the efficiency, traceability, and sustainability of plastic waste management.
4. Propose actionable recommendations for manufacturers and policymakers to facilitate the adoption of Industry 4.0 technologies for SPWM.

By addressing these objectives, this study seeks to contribute to the development of a strategic roadmap for effective Industry 4.0 implementation in sustainable plastic waste management, providing a sustainable solution to the growing plastic waste challenges in the manufacturing sector.

Methodology

This study adopts an interpretive research philosophy to explore how Industry 4.0 technologies influence sustainable plastic waste management (SPWM) in manufacturing. An interpretive approach emphasizes understanding context-specific factors through qualitative data, capturing the subjective experiences and meanings of stakeholders (Alharahsheh and Pius, 2020). An inductive approach was used to identify patterns and insights without seeking to confirm existing theories (Khan, 2014).

A multi-method research design was employed, combining a semi-systematic literature review, a case study, and semi-structured interviews. The literature review focused on articles published between 2014 and 2024, using keywords related to Industry 4.0 technologies, such as IoT, AI, and blockchain technology (BCT) in plastic manufacturing. Boolean operators helped refine search results for relevance. Studies focused on bioproducts, or life cycle assessments were excluded to maintain the study's focus on Industry 4.0 applications for SPWM.

The case study method enabled an in-depth exploration of SPWM practices, addressing 'how' and 'why' questions relevant to the research objectives (Ponelis,

2015). Company A, based in Dubai, UAE, was selected due to its relevance in packaging waste management within the manufacturing sector. Semi-structured interviews were conducted with a purposive sample of key stakeholders, including research architects, operations directors, and logistics specialists. This sampling strategy was chosen to ensure the data collected was aligned with the research objectives and to maximize resource efficiency (Palinkas et al., 2015).

Interviews were recorded with consent, and confidentiality was ensured through signed ethics forms and non-disclosure agreements (Shaw, 2003). The semi-structured format allowed flexibility in questioning, enabling participants to share detailed perspectives on SPWM practices and the potential of Industry 4.0 technologies.

Thematic analysis was used to analyze the interview data, focusing on identifying patterns and themes related to SPWM barriers, enablers, stakeholder involvement, and the roles of different technologies (Braun & Clarke, 2006). Key themes were derived from verbatim transcripts and were reviewed multiple times to ensure accuracy and depth. This approach provided actionable insights into how Industry 4.0 technologies can enhance SPWM practices, highlighting opportunities for integrating IoT, AI, and BCT to reduce waste and improve sustainability in the manufacturing sector.

Findings

This case study investigates sustainable plastic waste management (SPWM) practices at Company A, a Dubai-based educational furniture manufacturer, through thematic analysis of semi-structured interviews. The company is committed to sustainability, particularly in managing plastic waste, and aligns its practices with global sustainability trends to enhance its reputation as an industry leader. The study identified several key themes that have been analyzed in combination with the existing literature to gain a comprehensive understanding of SPWM and provide actionable recommendations.

i. Understanding SPWM

The study's findings revealed that Company A's approach to SPWM focuses on reducing environmental impact and ensuring economic feasibility through waste reduction and resource efficiency. These practices align with global sustainability trends, reinforcing the importance of minimizing environmental damage while supporting long-term economic benefits (Seadon, 2010). The integration of economic, environmental, and social sustainability within SPWM is underscored by Debnath et al. (2023), confirming that sustainability in waste management must account for both immediate and long-term impacts.

Government regulations, including plastic bans and recycling incentives, were highlighted as significant drivers for the company's sustainability efforts. The literature supports this, emphasizing that effective regulatory frameworks enhance the efficiency of waste management by making compliance and recycling more streamlined (Kurniawan et al., 2022).

ii. Application of SPWM to Plastic in the Manufacturing Sector

The study found that Company A applies sustainability principles to plastic waste through strategies like material substitution, waste reduction, and sustainable waste management practices. Despite these efforts, the company faces challenges such as high costs for sustainable material substitutes and the difficulty of extending these practices to suppliers and the broader network. This reflects the findings in the literature, where Horváth et al. (2018) identify the financial and operational hurdles involved in implementing sustainable practices in plastic waste management. These barriers are evident in the cost-prohibitive nature of sustainable material alternatives, presenting a challenge for manufacturing companies aiming for widespread adoption.

iii. Promoting SPWM

A key theme identified in the study was the promotion of SPWM within the manufacturing sector. This requires an integrated approach that includes enhancing employee awareness, engaging stakeholders, and leveraging government incentives. Participants emphasized the need for cultivating a sustainability culture within the organization. However, barriers such as insufficient employee awareness and compliance were noted. This is consistent with the literature, where Veleva et al. (2017) stress that employees, while crucial to sustainability efforts, often lack sufficient awareness. To address this, (Farooq & Salam, 2020) suggest that tailored communication campaigns and incentives could improve engagement and motivation, which is supported by the findings in this case study. This points to the need for ongoing training and awareness programs to cultivate a sustainability mindset across all organizational levels.

iv. Role of Industry 4.0 Technologies in Promoting SPWM

Industry 4.0 technologies such as the Internet of Things (IoT), Artificial Intelligence (AI), and Blockchain Technology (BCT) are seen as transformative tools in promoting SPWM by enhancing monitoring, waste segregation, and decision-making in manufacturing processes. The case study highlighted that Company A has already implemented tracking systems for component management and plans to integrate these technologies further into its waste management processes. However, high implementation costs and limited access to necessary knowledge remain significant barriers. The literature, including the work of Horváth et al. (2018), underscores that resource-intensive technology investments are often difficult to justify without significant financial backing. Yet, technologies like IoT, AI, and BCT have been shown to improve waste segregation efficiency and foster collaboration across supply chains, further supporting sustainability goals (Jiang et al., 2023). The study also highlights that government incentives could play a role in making the adoption of Industry 4.0

technologies more feasible, echoing suggestions from Kurniawan et al. (2022) that grants, or tax breaks could alleviate some of the financial burden of integrating these advanced technologies.

v. Waste Segregation Practices and Economic Viability

Effective plastic waste segregation emerged as a crucial factor in improving recycling efficiency. The study's findings confirm that systematic waste segregation reduces contamination and enhances recycling outcomes, aligning with existing literature that emphasizes the importance of clear sorting protocols for improving recycling efficiency. Similarly, economic viability remains a significant consideration, with the study noting the high upfront costs of implementing sustainable practices. However, the long-term benefits, including reduced disposal costs and potential revenue from recycled materials, support the economic justification of sustainable practices. This is consistent with the literature, which suggests that while initial investments are high, the long-term financial and environmental benefits outweigh these costs.

Barriers and Enablers of SPWM

The analysis identified several key enablers and barriers to effective SPWM. Enablers include stringent government regulations, the adoption of Industry 4.0 technologies, and proactive industry strategies. Barriers include high implementation costs, insufficient employee training, and limited access to technology-related knowledge. These findings contribute to the development of targeted recommendations for enhancing SPWM in manufacturing.

Conclusions

The project provided insights into the understanding of SPWM, the role of Industry 4.0 technologies in promoting it, and the barriers and enablers to practical implementation. The literature review analysis revealed that while there are developments in technology and policy, the practical implementation of SPWM is often hindered by challenges such as inadequate infrastructure, regulatory gaps, and limited awareness. Qualitative data collected via semi-structured interviews and analyzed through thematic analysis highlighted key factors influencing SPWM practices, including industry priorities and integrating Industry 4.0 technologies. The findings underscore the need for comprehensive stakeholder engagement, government support, and continuous employee education to achieve SPWM in the manufacturing sector.

Moreover, the study identified that successful SPWM initiatives are characterized by a holistic approach combining technological solutions and organizational strategies. By addressing the identified barriers and leveraging the enablers, manufacturers can better navigate the complexities of plastic waste management and contribute to a more sustainable future.

References

Alharahsheh, H. H., & Pius, A. (2020). A review of key paradigms: Positivism VS interpretivism. *Global Academic Journal of Humanities and Social Sciences, 2*(3), 39–43.

Braun, V., & Clarke, V. (2006). Using thematic analysis in psychology. *Qualitative Research in Psychology, 3*(2), 77–101. https://doi.org/10.1191/1478088706qp063oa

Cucchiella, F., D'Adamo, I., & Gastaldi, M. (2017). Sustainable waste management: Waste to energy plant as an alternative to landfill. *Energy Conversion and Management, 131*, 18–31. https://doi.org/10.1016/j.enconman.2016.11.012

Debnath, B., Bari, A. B. M. M., Ali, S. M., Ahmed, T., Ali, I., & Kabir, G. (2023). Modelling the barriers to sustainable waste management in the plastic-manufacturing industry: An emerging economy perspective. *Sustainability Analytics and Modeling, 3*, 100017. https://doi.org/10.1016/j.samod.2023.100017

Duru, R. U., Ikpeama, E. E., & Ibekwe, J. A. (2019). Challenges and prospects of plastic waste management in Nigeria. *Waste Disposal & Sustainable Energy, 1*(2), 117–126. https://doi.org/10.1007/s42768-019-00010-2.

Evode, N., et al. (2021). Plastic waste and its management strategies for environmental sustainability. *Case Studies in Chemical and Environmental Engineering, 4*, 100142. https://doi.org/10.1016/j.cscee.2021.100142

Farooq, O., & Salam, M. A. (2020). Nexus between corporate social responsibility, employee engagement, and organizational performance. *Journal of Cleaner Production, 271*, 122563. https://doi.org/10.1016/j.jclepro.2020.122563

Gao, W., et al. (2022). *Addressing the challenges of plastic waste | McKinsey*. Retrieved July 4, 2024, from https://www.mckinsey.com/industries/chemicals/our-insights/addressing-the-challenges-of-plastic-waste-circularity-and-leakage

Ghahramani, M., Zhou, M., Molter, A., & Pilla, F. (2022). IoT-based route recommendation for an intelligent waste management system. *IEEE Internet of Things Journal, 9*(14), 11883–11892. https://doi.org/10.1109/JIOT.2021.3132126

Ghobakhloo, M. (2020). Industry 4.0, digitization, and opportunities for sustainability. *Journal of Cleaner Production, 252*, 119869. https://doi.org/10.1016/j.jclepro.2019.119869

Horváth, B., Mallinguh, E., & Fogarassy, C. (2018). Designing business solutions for plastic waste management to enhance circular transitions in Kenya. *Sustainability, 10*(5), 1664. https://doi.org/10.3390/su10051664

Huang, S., et al. (2022). Plastic waste management strategies and their environmental aspects: A scientometric analysis and comprehensive review. *International Journal of Environmental Research and Public Health, 19*(8), 4556. https://doi.org/10.3390/ijerph19084556

Jiang, P., et al. (2023). Blockchain technology applications in waste management: Overview, challenges and opportunities. *Journal of Cleaner Production, 421*, 138466. https://doi.org/10.1016/j.jclepro.2023.138466

Khadke, S., Gupta, P., Rachakunta, S., Mahata, C., Dawn, S., Sharma, M., et al. (2021). Efficient plastic recycling and remolding circular economy using the technology of trust–blockchain. *Sustainability, 13*(16), 9142.

Khan, S. N. (2014). Qualitative research method: Grounded theory. *International Journal of Business and Management, 9*(11), 224–233.

Kurniawan, T. A., et al. (2022). Unlocking digital technologies for waste recycling in Industry 4.0 era: A transformation towards a digitalization-based circular economy in Indonesia. *Journal of Cleaner Production, 357*, 131911. https://doi.org/10.1016/j.jclepro.2022.131911

Lastic pollution | Definition, Sources, Effects, Solutions, & Facts | Britannica (2024) *Britannica.* Retrieved July 5, 2024, from https://www.britannica.com/science/plastic-pollution

Palinkas, L. A., Horwitz, S. M., Green, C. A., Wisdom, J. P., Duan, N., & Hoagwood, K. (2015). Purposeful sampling for qualitative data collection and analysis in mixed method implementation research. *Administration and Policy in Mental Health and Mental Health Services Research, 42*(5), 533–544. https://doi.org/10.1007/s10488-013-0528-y

Pandey, P., Dhiman, M., Kansal, A., & Subudhi, S. P. (2023). Plastic waste management for sustainable environment: Techniques and approaches. *Environmental Science and Pollution Research.* Advance online publication. https://doi.org/10.1007/s42768-023-00134-6

Ponelis, S. R. (2015). Using interpretive qualitative case studies for exploratory research in doctoral studies: A case of information systems research in small and medium enterprises. *International Journal of Doctoral Studies, 10*, 535–550. https://doi.org/10.28945/2339

Prata, J.C. et al. (2019). Solutions and integrated strategies for the control and mitigation of plastic and microplastic pollution. *International Journal of Environmental Research and Public Health, 16*(13), 2411. https://doi.org/10.3390/ijerph16132411

Seadon, J. K. (2010). Sustainable waste management systems. *Journal of Cleaner Production, 18*(16–17), 1639–1651. https://doi.org/10.1016/j.jclepro.2010.07.009

Shaw, I. F. (2003). Ethics in qualitative research and evaluation. *Journal of Social Work, 3*(1), 9–29. https://doi.org/10.1177/1468017303003001002

Veleva, V., Bodkin, G., & Todorova, S. (2017). The need for better measurement and employee engagement to advance a circular economy: Lessons from Biogen's zero waste journey. *Journal of Cleaner Production, 154*, 517–529.

Wen, Z., Liu, W., Walker, T. R., Adams, M., & Zhao, J. (2021). How does the global plastic waste trade contribute to environmental benefits: Implication for reductions of greenhouse gas emissions? *Journal of Environmental Management, 287*, 112283. https://doi.org/10.1016/j.jenvman.2021.112283

Chapter 4
Exploring Greywater Management Reuse Potential for Irrigation and Recharge—A Study in Five States of India

Uzra Sultana

Abstract Greywater management plays a crucial role in enhancing water security and sustainability, especially in semi-arid regions and agricultural zones with high water guzzler crops in India. This study examines greywater reuse and recharge potential in selected rural areas across these five states, considering policy frameworks such as the Swachh Bharat Mission-Gramin 2 (SBM-G) and Jal Jeevan Mission (JJM). Field visits to selected districts in each state provided insights into existing greywater treatment systems and its infrastructure, and community perceptions. Data analysis highlights challenges such as limited financial allocations, infrastructure gaps, and technical constraints in implementing efficient reuse systems. Findings indicate that integrating nature-based solutions can significantly enhance groundwater recharge and water reuse potential. This research underscores the need for adaptive policy measures and increased community participation to ensure long-term sustainability of greywater management practices.

Keywords Greywater reuse · Groundwater recharge · Jal Jeevan Mission · Swachh Bharat mission · Nature-based solutions

Introduction

Water security is a growing concern in India, particularly in semi-arid and water-stressed regions like Karnataka, Gujarat, Rajasthan, Punjab, and Haryana. While national programs such as the SBM-G and JJM have significantly improved water access and sanitation, greywater management remains limited, leading to environmental and public health concerns. Sustainable greywater reuse strategies can enhance groundwater recharge, reduce water pollution, and promote efficient resource use.

U. Sultana (✉)
ARGHYAM, Bengaluru, Karnataka, India
e-mail: uzrasultana@arghyam.org

J. Dsilva (ed.), *Sustainability Education, Youth and Innovation*,
https://doi.org/10.1007/978-3-032-21948-0_4

Existing studies highlight the potential of soak pits, constructed wetlands, and DEWATS for decentralized greywater treatment in rural areas. However, financial constraints, inadequate infrastructure, and limited institutional support hinder large-scale adoption. This study assesses greywater management practices in five states, analyzing technical, financial, and operational challenges. Using field visits, stakeholder interviews, and secondary data analysis, findings reveal that 65% of households have some form of greywater treatment, yet treatment infrastructure and water quality monitoring remain inadequate. The study emphasizes policy integration, community participation, and revenue models to ensure sustainable greywater reuse and recharge.

Methodology

This study adopted a mixed-method approach to evaluate greywater management practices, reuse potential, and recharge strategies in Karnataka, Gujarat, Rajasthan, Punjab, and Haryana in India. These states were selected based on criteria such as arid and semi-arid conditions, groundwater over-exploitation and the prevalence of water-intensive crops. The research framework included field observations, structured interviews, stakeholder consultations, and an in-depth analysis of secondary data.

1. *Data collection and sampling*: The selection of study locations was based on variability in water availability, Policy/Schemes implementation status and presence of decentralized treatment infrastructure. A total of 18 villages were surveyed, covering a population range of 1,500 to 12,000. Within each village, Gram Panchayats, Village Water & Sanitation Committee, Village Institution members and local implementing agencies were interviewed to assess water usage patterns, treatment infrastructure, and financial models.
2. *Qualitative and quantitative data collection*

   Field surveys, observations and stakeholder consultations

   - On-site assessment of existing greywater treatment infrastructure (soak pits, leach fields, constructed wetlands, DEWATS, etc.).
   - Visual documentation of greywater discharge and reuse patterns in agricultural fields and groundwater recharge structures.
   - Structured interviews with Gram Panchayat officials, village water committees, implementing NGOs, and local government officials to understand policy implementation and financing models.
   - Consultations with subject-matter experts and hydrogeologists to evaluate the technical feasibility of different treatment models and understand the challenges in adoption.

   Secondary data analysis

- Financial allocations from JJM, SBM-G2, MGNREGS, and the 15th Finance Commission were analyzed to understand budgetary constraints and funding patterns.
- Analysis of state-specific greywater management guidelines, policy documents, and scientific studies to contextualize the findings.
- Review of water quality reports to assess greywater treatment efficiency and reuse potential.

Data analysis and interpretation

- The collected data were processed using a comparative assessment framework to analyze the key factors affecting infrastructure adequacy (percentage of population with access to treatment systems); Financial sustainability (CapEx and OpEx requirements, revenue generation models); Community adoption (level of participation in greywater management and reuse) and Technology performance (efficiency of soak pits, constructed wetlands, and DEWATS).

Below are the details on each of the figures included below:

Figure 4.1 Stabilization Pond along with Vetivera zizanioide Plantations in Haryana: To facilitate the natural treatment of greywater through sedimentation and biological processes, while vetiver grass enhances pollutant uptake and improves water quality. They demonstrate an effective, low-cost approach to decentralized greywater management

Figure 4.2 Water storage -Maturation Pond in Punjab: The pond designed for storing partially treated water, which enables further natural treatment and improves the safety of reuse for agriculture or groundwater recharge

Figure 4.3 DEWATS and Constructed wetland in Gujarat: Shows DEWATS (Decentralized Wastewater Treatment System), an anaerobic treatment system combined with a constructed wetland that effectively removes organic matter and pathogens. This approach ensures cost-efficient, low-maintenance greywater management suitable for rural contexts

Fig. 4.1 Stabilization Pond along with Vetivera zizanioide Plantations in Haryana

Fig. 4.2 Water storage-Maturation Pond in Punjab

Fig. 4.3 Constructed wetland in Gujarat

Figure 4.4 Inline Treatment system in Karnataka: Depicts an inline greywater treatment system in Karnataka, where wastewater is treated as it flows through decentralized units embedded within the drainage channel. This approach helps reduce contamination risks, ensures partial treatment before disposal, and supports safer downstream reuse or recharge

Findings

Water supply and coverage

Villages will receive an average of 55 LPCD of water under JJM, leading to significant greywater of about 34,000 million liters per day from rural India.

Fig. 4.4 Inline treatment system in Karnataka

Population across the sampled villages ranges from 1,500 to 12,000, with water supply levels varying between 35 to 200 LPCD.
Households covered under greywater management range between 50% and 100% from the sampled villages.

Water quality and treatment

High instances of untreated black water mixing with greywater making it challenging with high organic load.
Water is not tested for BOD/COD before reusing in agriculture or groundwater recharge.

In the absence of treatment facilities, untreated greywater is discharged directly into water bodies or open lands.
Infrastructure for greywater treatment is minimal, with most households relying on direct discharge or informal soak pits.

Technology and design

Technical interventions such as decentralized treatment units, soak pits, and leach fields show promise in improving water reuse efficiency.
Nature-based solutions like constructed wetlands, DEWATS, soak pits, and kitchen gardens require minimal manual intervention.
Designing stormwater drains to divert excess water directly into wetlands can reduce the load on treatment plants.

Financial aspects

Capital expenditure required for greywater treatment ranges from Rs. 18 Lakh to 87 Lakh, with funding primarily sourced through a mix of government schemes such as MGNREGS, SBM-G2, and the 15th Finance Commission.
In Karnataka, Haryana, and Punjab, 70–80% of funding comes from MGNREGS, while in Gujarat and Rajasthan, SBM-G2 covers only 20–30% of greywater treatment costs.
Liquid waste management allocation as per the SBM-G is estimated at Rs. 280 per capita for villages with populations below 5,000 and Rs. 660 per capita for villages with populations above 5,000.
Operation and Maintenance costs range from Rs. 4,000 to Rs. 30,000 per month depending on the technology used.

Reuse and revenue generation

Reuse potential is high, with 65–70% of water being converted into greywater, of which 55% is available for reuse post-treatment.
Greywater reuse is extensively practiced for irrigation, groundwater recharge, fishing, and manure production. However, Rajasthan exhibits minimal reuse due to low water availability, high infiltration, and high evaporation rates.
NGO-supported models leveraging CSR funds have been observed in Karnataka, funding entire greywater treatment systems.
Revenue-generating activities like selling manure from treated sludge and leasing ponds for fishing are observed in Gujarat.
Revenue generation from greywater reuse ranges between Rs. 40,000 and Rs. 4.2 Lakh per annum, depending on the scale of implementation and community involvement.

Conclusions

Implementing effective greywater reuse and recharge strategies is crucial for enhancing water security in rural areas. While initiatives like SBM-G and JJM provide a strong foundation, additional investments in infrastructure and community-driven approaches are essential for long-term sustainability. Integrating decentralized treatment systems with existing rural water supply programs can optimize resource utilization. A holistic approach that combines greywater, blackwater, stormwater, and solid waste management is necessary to maximize efficiency. Empowering Gram Panchayat members to initiate revenue-generating activities, with support from NGOs or implementing partners, can ensure the financial sustainability of these systems.

Despite the potential of greywater management, several challenges persist. Funding gaps within SBM-G limit large-scale implementation, and village-level capacity for managing and scaling greywater reuse remains inadequate. While greywater reuse has been observed to contribute to groundwater recharge, there is currently no established mechanism to measure the volume of recharge accurately. Future research should focus on developing monitoring frameworks to quantify recharge impacts and refining policies to facilitate widespread adoption.

Bibliography

https://ejalshakti.gov.in/jjmreport/JJMDistrictView.aspx
https://sbm.gov.in/sbmgdashboard/StatesDashboard.aspx
Manual on Greywater Management: https://www.swachhbharatmission.ddws.gov.in/sites/default/files/Technical-Notes/Greywater_Management_Manual_English.pdf
https://jalshakti-ddws.gov.in/sites/default/files/JJM_Operational_Guidelines.pdf

Chapter 5
Sustainable Procurement in Interior Design Fit-Out Construction Projects in Qatar: Challenges and Remedies

Sudheesh Satyan and Rekha Pillai

Abstract The aim of this study is to explore and identify the challenges associated with sustainable procurement and their impact on the outcomes of interior design fit-out construction projects in Qatar. A qualitative methodology supported by Transformation-Flow-Value (TFV) theory was conducted. Utilizing thematic analysis of interviews with a purposive sample of 14 procurement professionals in Qatar, the study identified various procurement challenges such as the issue of ambiguous project scope leading to inefficiencies, unrealistic timelines causing rushed decisions, and poor cost management resulting in budget overruns. Additionally, the study reveals that while technology can streamline tasks, it does not fully resolve core procurement challenges. Effective communication, cost management, and an experienced team are crucial factors for sustainable procurement in the rapidly developing construction industry. The study highlights the importance of sustainable procurement as it can lead to cost savings, reduced waste, and more efficient use of resources, aligning with Qatar commitment to sustainable development. This research provides valuable insights for procurement professionals and stakeholders, enhancing the understanding of effective sustainable procurement practices in the context of Qatar.

Keywords Sustainable procurement · Interior design fit · Out construction · Qatar · Transformation · Flow · Value theory

S. Satyan
University of West of Scotland, Paisley, UK
e-mail: B01728411@studentmail.uws.ac.uk

R. Pillai (✉)
The British University in Dubai, Dubai, UAE
e-mail: rekha.pillai@buid.ac.ae

J. Dsilva (ed.), *Sustainability Education, Youth and Innovation*,
https://doi.org/10.1007/978-3-032-21948-0_5

Introduction

The continuous growth in the interior design fit-out industry can be ascribed to several factors, including Qatar's strategic economic planning, substantial investments in infrastructure, and the influx of multinational corporations establishing their presence in the country (International Monetary Fund, 2019). The industry's expansion is driven by ambitious infrastructure projects and the country's commitment to hosting significant global events, including the FIFA World Cup 2022. Despite the continuous growth, the interior design fit-out companies encounter many hurdles especially in the sustainable procurement process, which challenges the fulfilment of a project. Moreover, efficient procurement processes help in maintaining the quality and standards expected in high-profile projects, particularly those tied to international events and national development goals.

This research aims to identify the main procurement challenges, examine the key factors contributing to successful procurement, and evaluate the extent to which technology can address these challenges.

This study aims to answer the following research questions:

What are the challenges affecting getting materials and services for ID Fit-out projects in Qatar?

Does technology address these challenges?

What are the key factors in sustainable procurement for a successful ID Fit-out project?

The findings of this study will offer many important contributions to Qatar's interior design fit-out field. First, the research will provide a clearer view of the hurdles that companies face, by identifying and understanding the current practices and procurement challenges in the ID fit-out projects in Qatar, allowing for more specific and effective solutions.

Second, the research will highlight the key factors in sustainable procurement that contribute to successful projects, offering valuable insights into best practices and strategies that can enhance project outcomes (Schwalbe, 2013).

Third, the examination of how technology is used to address procurement challenges will shed light on the potential of innovative tools and systems to improve procurement efficiency and effectiveness (McKinsey & Company, 2022). Technologies such as Building Information Modeling (BIM), blockchain, and artificial intelligence (AI) can streamline procurement processes, enhance transparency, and improve decision-making.

Theoretical Foundations

This research draws upon the Transformation-Flow-Value (TFV) Theory of production and waste (Koskela, 2000). The TFV theory presents a complete framework that conceptualizes production as a purposeful process characterized by three fundamental dimensions: transformation, flow, and value. Transformation within the TFV theory pertains to the conversion of raw materials into finished products. Flow, another critical dimension of TFV theory, emphasizes time as a primary resource in production processes. Waste in terms of flow refers to any disruptions or delays that result in inefficiencies in the workflow. Value, the third dimension of TFV theory, focuses on the external output of the production process. Waste in terms of value occurs when products or outputs do not meet end-user requirements or fail to produce the expected benefits.

Sustainable Procurement in Construction

Sustainable procurement encompasses the essential process of obtaining goods and services that are critical to project or organizational success. Several studies have identified the procurement challenges and success factors associated with construction projects. For example, Riaz and Jaffery (2013) investigated the selection methods for construction project delivery systems (PDS) in Qatar, highlighting significant challenges such as delays and cost overruns stemming from inappropriate PDS choices. The findings underscored a substantial knowledge gap among project owners regarding available PDS options, often leading to inefficient decision-making processes characterized by reliance on broad, non-specific criteria (See also Ibrahim et al. 2014, Ahmed et al., 2023 and Bakhshi et al. 2023 for similar results). These challenges underscored the imperative for collaborative efforts among industry stakeholders, governmental entities, and educational institutions to effectively implement strategic initiatives and drive progress within the sector.

Methodology

The research employs a qualitative research design, recognizing the requirement for a deep investigation of the research phenomenon (Creswell & Creswell, 2017). The emphasis on qualitative methods is justified by the belief that they allow for a nuanced understanding of the subjective and multifaceted aspects of the research topic (Smith, 2015; Denzin & Lincoln, 2018).

Purposive sampling was used which produces a well-rounded and detailed perspective on the process under study, thereby leading to the overall validity and credibility of the research findings (Creswell & Creswell, 2017; Palinkas et al., 2015). Data collection continued until no new facts or themes emerged from additional interviews, indicating that theoretical saturation has been achieved (Guest et al., 2006). A purpose sample of 14 procurement professionals in Qatar was analyzed using thematic analysis as it offers a robust framework for uncovering and examining patterns and themes inherent in qualitative data (Braun & Clarke, 2006). To assess the data, NVivo 14 software was employed. Additionally, Descript software was used for transcription.

Analysis and Results

This section covers the result of the thematic analysis of 14 semi-structured interviews, focusing on procurement challenges in interior design (ID) fit-out construction projects in Qatar.

Thematic analysis, as outlined by Braun and Clarke (2006) begins with familiarization with the data through repeated readings, allowing the researcher to immerse themselves deeply in the content. Following this, the process involves generating initial codes that highlight interesting features or patterns within the data. These codes act as the building blocks for developing broader themes, which are then meticulously reviewed and refined iteratively to ensure they accurately represent the data. Table 5.1 lists the main themes after refining process.

As seen from Table the inductive thematic analysis identified eight emergent themes. These emergent themes not only highlight the diverse challenges faced by stakeholders but also underscore the interconnectedness of various aspects of the procurement process. They encapsulate key issues such as budget constraints,

Table 5.1 Main themes

Main themes	Description
Research Question 1—Procurement Challenges	
Budget constraints	Budget constraints pose a significant challenge in managing procurement effectively
Finding right suppliers or contractors	Identifying suitable suppliers or contractors is crucial for project success
Lack of scope or clarity	Ambiguities in project scope or clarity can lead to procurement inefficiencies
Tight project schedule	Strict project timelines often pressure procurement processes
Research Question 2—Use of technology	
Technology will ease work but doesn't solve any of the procurement challenges	While technology streamlines tasks, it does not completely address procurement challenges
Research Question 3—Key factors of sustainable procurement	
Effective communication and coordination	Effective communication and coordination are crucial for smooth procurement operations
Effective cost management	Managing costs efficiently is a key factor in successful procurement
Experienced and knowledgeable team	An experienced and knowledgeable team significantly enhances procurement outcomes

supplier reliability, regulatory compliance, communication and coordination, time management, quality control, technological integration, and stakeholder engagement. Each theme offers a lens through which to explore and examine the underlying dynamics and implications in the context of ID fit-out construction projects.

A thematic map was then created from Nvivo 14 to understand the connections between these key themes; thematic maps were created as shown below in Fig. 5.1.

Findings

Theme 1: Budget constraints

Budget constraints emerge as a significant procurement challenge. Budget constraints can lead to several complications, such as restricted access to high-quality materials, delayed payments to suppliers and contractors, and the inability to implement advanced technologies or hire skilled labor (Alsulaiti and Kerbache 2020). Stringent budgetary controls can hinder the flexibility needed to adapt to unforeseen challenges or changes in project scope, further exacerbating delays and inefficiencies. Addressing budget constraints requires careful financial planning, transparent cost management practices, and strategic allocation of resources to ensure that projects are completed within financial limitations without sacrificing quality or project outcomes.

Theme 2: Finding right suppliers or contractors

The challenge of finding the right suppliers or contractors is an important feature of procurement in ID fit-out construction projects in Qatar. Delays or disruptions in procurement can arise from mismatches in expectations between project owners and contractors, leading to issues such as project delays or disputes over contract terms and conditions. Addressing this challenge necessitates thorough vetting processes, clear communication of project requirements, and establishing robust contractual agreements to reduce risks and ensure alignment with project goals. Moreover, establishing long-term relationships with reliable suppliers and contractors can contribute to smoother project execution and enhanced project outcomes in the dynamic environment of ID fit-out construction.

Theme 3: Lack of scope or clarity

One of the emergent themes identified through the inductive thematic analysis is the lack of scope or clarity, a critical issue in the procurement process of ID fit-out construction projects in Qatar. This theme underscores significant problems arising from ambiguous or poorly defined project scopes, leading to misunderstandings and misaligned expectations among procurement professionals and stakeholders emphasizes the importance of clear scope documentation. When project specifications are vague, suppliers and contractors may interpret the requirements differently, resulting in inconsistencies in the quality of procured materials and services, and potential conflicts over deliverables.

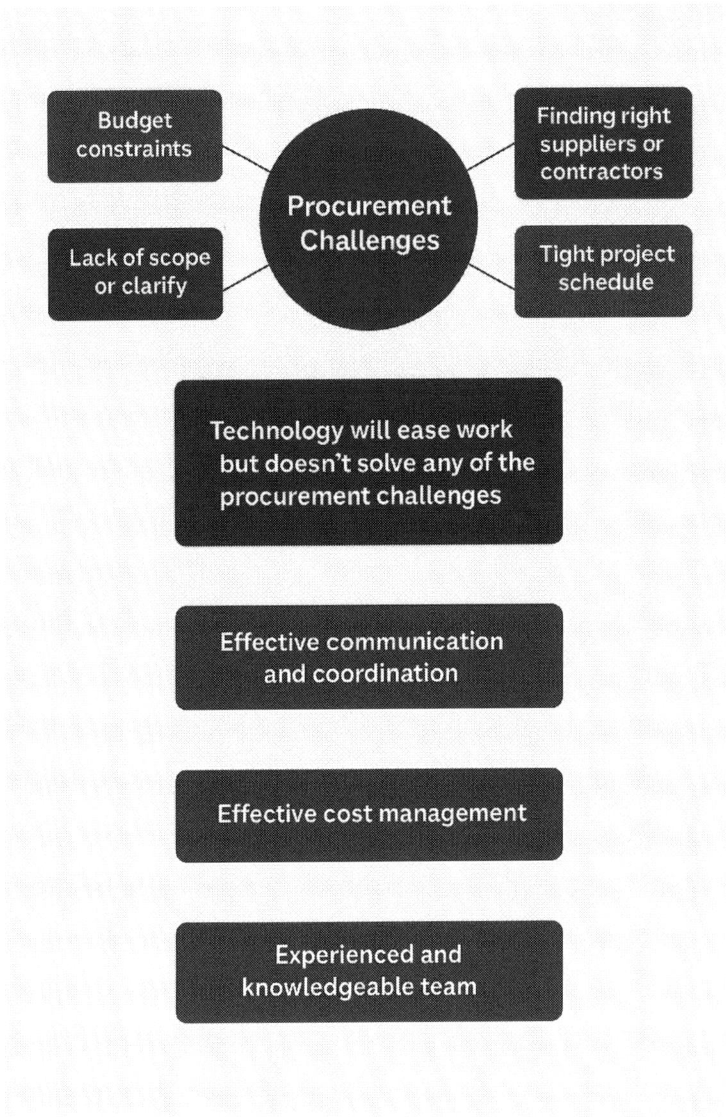

Fig. 5.1 Thematic map

Theme 4: Tight project schedule

The theme of tight project schedules in ID fit-out construction projects in Qatar underscores the significant challenges imposed by stringent timelines on procurement processes. These schedules often demand rapid decision-making and execution from procurement professionals to source materials, engage suppliers, and finalize contracts promptly. This pressure can compromise cost-effectiveness and quality assurance, as limited time frames restrict opportunities for competitive bidding and thorough supplier evaluations. To effectively manage tight project schedule proactive strategies and early involvement of procurement team is crucial.

Theme 5: Technology will ease work but doesn't solve any of the procurement challenges

The theme that technology does not significantly alleviate procurement challenges in ID fit-out construction projects in Qatar highlights a nuanced perspective. While technologies such as Enterprise Resource Planning (ERP) systems, Artificial Intelligence (AI) algorithms, and advanced analytics promise to streamline processes and improve decision-making, they do not inherently resolve core issues like ambiguous project scopes or communication gaps among stakeholders (Nizakat et al., 2022).

Theme 6: Effective communication and coordination

Effective communication and coordination in procurement extend beyond stakeholder alignment to encompass inter-supplier coordination and influence, as well as collaboration with project teams and leadership. Effective communication fosters a collaborative approach, encouraging suppliers to contribute insights and innovations that can improve project outcomes. This partnership approach not only strengthens supplier relationships but also enhances procurement effectiveness by leveraging supplier expertise and responsiveness (Ying et al., 2020).

Theme 7: Effective cost management

Effective cost management is a critical theme in the procurement of ID fit-out construction projects in Qatar, encompassing strategies to optimize spending while maintaining project quality and timelines. It begins with accurate cost estimation during the planning phase, where procurement professionals assess material costs, supplier quotations, and potential risks to develop realistic budgets. By forecasting and budgeting effectively, organizations can allocate resources efficiently, ensuring that procurement activities align with financial constraints and project goals (Ruparathna & Hewage, 2013).

Theme 8: Experienced and knowledgeable team

An experienced team brings expertise and insight honed through practical industry knowledge and previous project successes. This depth of experience allows procurement professionals to anticipate challenges, make informed decisions, and execute best practices that optimize procurement processes (Habibi et al. 2017; Gunduz & Al Naimi 2020).

In ID fit-out projects, which often involve intricate design requirements and specialized materials, having a knowledgeable team is very important. These professionals understand the nuances of procurement specifications, regulatory requirements, and industry standards specific to interior fit-out construction. Their familiarity with supplier networks and market dynamics enables them to effectively negotiate contracts, source high-quality materials, and manage vendor relationships to ensure project goals are met within budget and schedule constraints.

Conclusion

This research has explored the procurement challenges within ID fit-out construction projects in Qatar, utilizing an extensive approach that integrates theoretical frameworks, literature reviews, qualitative methodologies, and thematic analyses. The research was supported by the TFV (Transformation-Flow-Value) theory, and it integrates transformation processes, flow management, and value generation, providing a holistic framework for examining procurement optimization.

The research questions and objectives were shaped by an extensive review of literature on procurement challenges, technology adoption, and success factors in construction projects. The literature underscored prevalent issues such as budget constraints, supplier management, scope clarity, and project scheduling, highlighting their effect on procurement efficiency and project outcomes.

Eight emergent themes were revealed through thematic analysis which will answer research questions and objectives: budget constraints, finding right suppliers or contractors, lack of scope or clarity, tight project schedule, limitations of technology, effective communication and coordination, effective cost management and experienced and knowledgeable team.

The research questions were systematically answered through empirical findings and thematic analysis, providing useful cognizance into the complex nature of procurement management in interior fit-out projects. A limitation of this research is its focus on a specific geographical context (Qatar) and limited to only 14 interviews which may create bias, and future research could explore procurement challenges in ID fit-out construction projects across diverse regions to enhance generalizability and identify region-specific factors.

References

Ahmed, S., Majava, J., & Aaltonen, K. (2023). Implementation of circular economy in construction projects: A procurement strategy approach. *Construction Innovation, 20*(7), 204–226. https://doi.org/10.1108/CIN-02-2022-0227

Alsulaiti, A. A., & Kerbache, L. (2020). Analysis of critical delay factors in construction projects with a focus on Qatar. *International Journal of Business and Economics Research, 9*(3), 91–101. https://doi.org/10.11648/j.ijber.20200903.16

Bakhshi, J., Mani, S., Ahmadi Eftekhari, N., & Martek, I. (2023). Procurement practices in international development projects: Trends, networks, and performances. School of Project Management, The University of Sydney, Sydney, Australia; Tehran University of Art, Tehran, Iran; School of Architecture and Built Environment, Deakin University, Geelong, Australia.

Braun, V., & Clarke, V. (2006). Using thematic analysis in psychology. *Qualitative Research in Psychology, 3*(2), 77–101. https://doi.org/10.1191/1478088706qp063oa

Creswell, J. W., & Creswell, J. D. (2017). Research design: Qualitative, quantitative, and mixed methods approaches (4th ed.). Sage.

Denzin, N. K., & Lincoln, Y. S. (Eds.). (2018). The SAGE handbook of qualitative research (5th ed.). Sage.

Guest, G., Bunce, A., & Johnson, L. (2006). How many interviews are enough? An experiment with data saturation and variability. *Field Methods, 18*(1), 59–82. https://doi.org/10.1177/1525822X05279903

Gunduz, M., & Al-Naimi, N. H. (2022). Construction projects delay mitigation using integrated balanced scorecard and quality function deployment. *Engineering, Construction and Architectural Management, 29*(5), 2073–2105.

Habibi, M., Kermanshachi, S., & Safapour, E. (2017). Engineering, procurement, and construction cost and schedule performance leading indicators: State-of-the-art review. In *Construction Research Congress (CRC)* 2018 (pp. 378–388). https://doi.org/10.1061/9780784481271.037

Ibrahim, I. I., Githae, W., & Stephen, D. (2014). Indigenous contractors' involvement and performance in construction procurement systems in Nigeria. *Global Journal of Researches in Engineering: Journal of General Engineering, 14*(1). Online ISSN: 2249-4596 & Print ISSN: 0975-5861.

International Monetary Fund. (2019). Volume 2019: Issue 147. International Monetary Fund. https://doi.org/10.5089/9781498317863.002

Koskela, L. J. (2000). An exploration towards a production theory and its application to construction. ResearchGate. https://www.researchgate.net/publication/35018344_An_Exploration_Towards_a_Production_Theory_and_its_Application_to_Construction

McKinsey & Company. (2022). Rise of the platform era: The next chapter in construction technology. https://www.mckinsey.com/industries/private-capital/our-insights/rise-of-the-platform-era-the-next-chapter-in-construction-technology

Nizakat, R. Z., et al. (2022). E-procurement in the construction industry of Pakistan: A case study of contractors related to public sector projects. *International Research Journal of Modernization in Engineering, Technology, and Science, 4*(4). https://www.irjmets.com

Palinkas, L. A., et al. (2015). Purposeful sampling for qualitative data collection and analysis in mixed-method implementation research. *Administration and Policy in Mental Health and Mental Health Services Research, 42*(5), 533–544. https://doi.org/10.1007/s10488-013-0528-y

Riaz, M., & Jaffery, S. R. A. (2013). Investigating selection methods for construction project delivery systems based on professionals' perspective in Qatar. In *Proceedings of the CON-209-1 Conference, Montréal, Québec*, May 29 to June 1, 2013.

Ruparathna, R., & Hewage, K. (2013). Review of contemporary construction procurement practices. *Journal of Management in Engineering, 31*(3). https://doi.org/10.1061/(ASCE)ME.1943-5479.0000279

Schwalbe, K. (2013). An introduction to project management (6th ed.). Schwalbe Publishing.

Smith, J. A. (2015). Qualitative psychology: A practical guide to research methods. Sage.

Ying, F. J., Zhao, N., & Tookey, J. (2020). *Achieving construction innovation in best value procurement projects: New Zealand mega projects study*. Auckland University of Technology, Auckland, New Zealand.

Part II
Climate Action and Environmental Stewardship

Chapter 6
Beyond COP28: Path Towards UAE Climate Leadership in Driving Regional Climate Governance Through Climate Diplomacy and Trade Alignment

Nethmi Rathnayake

Abstract In the present setting, countries have agreed that "global problems require global solutions" (UNDES, 2024). As the climate challenges intensify, it requires coordinated action to ensure economic growth while considering the environmental risks that arise. From being the first Gulf nation to ratify the Paris Agreement to successfully hosting Cop 28 and building the Abu Dhabi Carbon Market, the United Arab Emirates has strategically positioned itself to strengthen regional climate governance as a leader across the Gulf Cooperation Council. Meanwhile, the EU has implemented coordinated efforts in implementing a legally binding climate framework within the EU through the EU Climate Law (2021/1119) and the Carbon Border Adjustment Mechanism (CBAM). The CBAM extends to non-EU nations for trade compliance. As per the reports, the European Union (EU) is one of the GCC's largest trading partners, with bilateral trade between the EU and GCC exceeding €175 billion annually. Therefore, it creates significant challenges to GCC states as it does not have a cohesive climate model for carbon markets. Therefore, this study revolves around how GCC could establish a climate governance model inspired by UAE leadership while drawing best practices from the EU. This research intends to explore how the UAE might lead the Gulf Cooperation Council (GCC) in building a regional climate governance model, boosting diplomatic and trade objectives, and identifying best practices that can be aligned with the EU's governance framework. Assess the possible impact of CBAM on GCC exports and how CBAM compliance could benefit Gulf nations, as well as create institutional frameworks, work on a regional carbon pricing system, and offer proposals for implementation. Following The Federal Decree-Law No. (11) of 2024 On the Reduction of Climate Change Effects and recent findings suggest the UAE is well-positioned to establish a Gulf Climate Council and a GCC-wide Emissions Trading System (Gulf ETS) and align GCC sustainability policies with international trade standards. However, further research is needed to assess long-term feasibility, economic integration challenges, and political constraints, and

N. Rathnayake (✉)
Research & Development, SEE Institute, The Sustainable City, Dubai, UAE
e-mail: nethmi.r@outlook.com

J. Dsilva (ed.), *Sustainability Education, Youth and Innovation*,
https://doi.org/10.1007/978-3-032-21948-0_6

a phased diplomatic and regulatory approach is proposed for impactful policy integration. This study contributes to the growing discourse on GCC climate governance by highlighting the UAE's pivotal role in bridging the gap between economic growth and environmental sustainability.

Keywords UAE climate diplomacy · CBAM · Gulf emissions trading system · Policy governance · EU climate law

Introduction

From shrinking glaciers and rising sea levels to shifting geographic ranges of plants and animals, climate change has significantly impacted the planet. Recurring heatwaves, wildfires, droughts, tropical cyclones, and extreme rainfall events continue to shape regions with varying levels of severity, with effects differing in intensity across the globe (NASA, n.d.). The Gulf region, in particular, faces escalating extreme weather events, including extreme heat, water scarcity, and rising sea levels, which pose a significant risk to the livelihood, ecosystems, and infrastructure with direct and indirect impacts. These challenges require collective action and cross-regional collaboration, as unified efforts enhance resilience and facilitate sustainable solutions. By encouraging cooperation that embraces diversity, regions can mitigate climate risks more effectively, ensuring equitable outcomes and shared benefits in the interest of global environmental stability.

Gulf economies face similar economic challenges as they remain heavily reliant on fossil fuels despite government efforts to diversify in line with national strategies. While Gulf nations share a common cultural background, they continue to struggle with regional integration due to diverging national interests and policy differences.

Unlike the European Union, the Gulf region lacks a unified climate governance model. The EU has established the EU Climate Law (2021/1119), trade-linked carbon regulations such as CBAM, and structured emission reduction mechanisms. In contrast, Gulf countries rely on voluntary pledges and national policies, which, in the long run, expose them to economic vulnerabilities.

As a supranational body, the EU binds its member states to a common climate goal by establishing collective targets aimed at achieving carbon neutrality by 2050, enforceable at both national and international levels. Simultaneously, it mandates at least a 55% net reduction in greenhouse gas emissions by 2030 (European Parliament & Council, 2021, p. 1). These measures apply to all sectors, including energy, industry, transport, buildings, agriculture, forestry, and waste management. Additionally, the EU encourages using carbon pricing (EU ETS), the Carbon Border Adjustment Mechanism, investment in green financing, and a Scientific and Policy Advisory Framework.

Oppositely, the EU Green Deal serves as a roadmap to achieve the EU's net-zero strategy by 2050, ensuring fairness and inclusivity during the transition (Council of the European Union, n.d.). Consequently, the EU is enacting laws and policies under the Green Deal, exemplified by the "Fit for 55" package, which targets a 55% reduction in emissions by 2030. In contrast to the Gulf Cooperation Council, the EU has established enforceability bodies, such as the European Courts, to ensure compliance. Moreover, the EU Emissions Trading System (EU ETS) and the Carbon Border Adjustment Mechanism (CBAM) provide a carbon pricing framework by imposing tariffs on high-emission imports.

This research investigates three primary questions. First, it examines how the UAE can assert climate diplomatic leadership among Gulf nations while upholding national sovereignty. Second, it explores which diplomatic instruments may be employed to align UAE diplomacy with establishing collective governance by the EU climate governance model. Finally, the study assesses the potential trade and economic benefits of adopting the EU framework.

This research intends to explore how UAE might lead the Gulf Cooperation Council (GCC) in building a regional climate governance model, boosting diplomatic and trade objectives, and identifying best practices that can be aligned with the EU's governance framework. Assess the possible impact of CBAM on GCC exports and how CBAM compliance could benefit Gulf nations, as well as create institutional frameworks, work on a regional carbon pricing system, and offer proposals for implementation.

Methodology

This research analyzes how the UAE could leverage its climate leadership among GCC nations, inspired by the EU climate governance model, to drive regional climate leadership through governance and trade alignment.

First, it conducts a comparative policy analysis, assessing the UAE's progress toward a climate governance model similar to the EU while identifying gaps that the GCC may fill. The study also uses economic modeling to examine the trade and economic impact on Gulf exports. This includes impact modeling to evaluate the implications of the EU's Carbon Border Adjustment scheme (CBAM) on GCC exports and, as a result, how the GCC could adjust by establishing a regional carbon pricing scheme.

A case study on Germany's achievement in generating €18.5 billion in revenue (Umweltbundesamt, 2024) from a carbon pricing mechanism will be discussed to explore how UAE could adopt this strategy and explore integration amongst GCC by starting a fixed price for specific sectors to attract green investments and align trade with the EU.

Finally, the research conducts a feasibility study on the UAE's soft power through climate diplomacy. It assesses its potential to negotiate and lead the Gulf region toward policy integration via the Gulf Cooperation Council (GCC).

The researcher will utilize secondary sources such as UAE and EU legislation, CBAM trade policies, UNFCCC reports, World Bank and IMF trade analyses, and case studies on climate governance in the EU.

Overall, the research framework is designed to assess how the UAE's climate diplomatic leadership could expand its influence in trade, the economy, and diplomatic metrics while evaluating the benefits of policy integration among GCC nations.

Findings/Arguments

The EU Climate Law as a Model for GCC Climate Governance

GCC participation in carbon trading has been critical to meeting the Paris Agreement's Nationally Determined Contributions (NDCs). However, the GCC lacks a unified carbon price mechanism because members make voluntary promises, resulting in weak institutional and regulatory capacities (Al-Sarihi, 2024). A legally binding method would provide legitimacy to the GCC's climate promises.

The EU Emissions Trading System (EU ETS) is the world's largest transnational compliance carbon market, and it controls through CBAM, requiring non-EU countries to have trade alignment to comply with the mechanism. As part of its climate-forward plan, the UAE has already implemented forward-thinking regional carbon pricing strategies, which might benefit from implementing a GCC ETS system (Al-Sarihi, 2024).

Establishing a Gulf Climate Council as part of the climate-forward agenda might serve as an EU model for harmonizing policy enforcement and regulatory alignment across the region while preserving country states' sovereign interests. Unlike the EU, political dynamics inside the GCC may provide a challenge because the GCC prioritizes the country's sovereignty. However, a phased approach might eventually lead to policy cooperation, and forming the Gulf Climate Council could be voluntary. Additionally, strengthening UAE-EU bilateral agreements on green finance could enhance trade compliance while making the GCC the most EU-aligned investment hub in the future.

UAE's Climate Policy and Regional Influence

The EU Climate governance model is a robust framework for GCC to ensure a centralized climate enforcement body given the increasing importance of trade alignment with the EU and as a tool of Climate Diplomacy to take its climate forward strategy above and beyond in line with the Paris agreement. UAE, as a nation, has strengthened its policies and financial initiatives by making them even stronger with the proposed penalty system for non-compliance (Tamimi & Co., n.d.) while making incentives for Research and Development, which strengthens UAE's ability to coordinate and drive climate response activities in the Gulf. However, the risks arising from CBAM tariffs for noncompliance could affect UAE and Gulf exports, making it mandatory to restrict carbon pricing systems.

With the new developments coming from the Federal Decree-Law No. (11) of 2024 On the Reduction of Climate Change Effects, the UAE strengthened institutions with advantage of creating a formal mechanism of climate efforts in local entities while creating a monitoring mechanism for emissions and reduction efforts (KPMG Lower Gulf, 2025). Additionally, the Research and Development sector has been strengthened in line with private sector collaboration to achieve Net Zero targets (KPMG Lower Gulf, 2025).

Challenges and Opportunities in GCC Climate Governance

The development of a single GCC climate framework is hampered by a number of issues, notwithstanding the UAE's leadership. Coherent climate policy implementation is difficult due to regulatory fragmentation caused by the lack of a legally binding GCC-wide policy. Climate transition measures are also politically controversial because the GCC countries continue to rely heavily on fossil fuels for their economies.

Furthermore, conflicting national objectives restrict integration. For instance, Saudi Arabia prioritizes carbon capture and oil-based strategies in its fight against climate change, whereas the United Arab Emirates adopts a different approach. The UAE backs renewable energy and global trade alignment at the same time. Balanced diplomatic actions are necessary to balance policy throughout the region.

UAE's Strategy for Leading GCC Climate Governance

Establishing a Gulf Climate Council inspired by the EU Climate commission could act as a reviewing body amongst Gulf nation while facilitating policy coordination in compliance with CBAM and intergovernmental climate negotiations.

Implementing a GCC-Wide Carbon Pricing System (Gulf ETS)

A GCC-wide carbon pricing mechanism may be established by a Gulf Emissions Trading System (Gulf ETS), which would be based on the EU ETS and guarantee adherence to global trade standards. Regional emissions could be decreased by such a system.

Strengthening UAE's Climate Diplomacy

The UAE may utilize its diplomatic influence to secure GCC-wide sustainability accords with key trading partners like the EU. Furthermore, the UAE's COP28 presidency and global leadership in green finance allow it to promote regional climate cooperation by aligning GCC policies with international climate.

Conclusion

This working paper concludes that the potential for the UAE to lead GCC climate policy through a phased approach by utilizing soft power to coordinate and negotiate amongst Gulf nations while identifying how the GCC can best unify climate governance within the region without sacrificing its national sovereignty in line with political and economic dynamics. However, this study requires further analysis to understand political dynamics and policy development, integrate the EU model into the GCC, and understand the long-term feasibility, as economic and social obstacles could arise based on country-specific strategies. As per the researcher, this study could provide insights into how GCC could integrate a GCC ETC and strengthen Gulf climate governance.

References

Al-Sarihi, A. (2024, November). *Gulf transition to low-carbon economies: The role of carbon markets*. Middle East Council on Global Affairs. https://mecouncil.org/wp-content/uploads/2024/11/CarbonMarketsIB-Aisha-WebFinal.pdf

Anwar Gargash Diplomatic Academy (AGDA). (2024, November). *Strengthening the UAE's soft power through sustainability action*. https://www.agda.ac.ae/docs/default-source/2025/strengthening-the-uaes-soft-power-through-sustainability-action_mb_7nov24-(1)-(1).pdf?sfvrsn=8154643b_1

Anwar Gargash Diplomatic Academy. (2024). *AGDA insight: Giovanna*. https://www.agda.ac.ae/docs/default-source/2024/agda-insight-giovanna.pdf?sfvrsn=2698643b_1

Bianco, C. (2021). Power play: Europe's climate diplomacy in the Gulf. European Council on Foreign Relations. https://ecfr.eu/publication/power-play-europes-climate-diplomacy-in-the-gulf/

Council of the European Union. (n.d.). *European Green Deal: A roadmap for a sustainable EU economy*. https://www.consilium.europa.eu/en/policies/green-deal/. Retrieved 17 Feb 2025

Economic Research Forum (ERF). (2024, September). *EU climate policy: Potential effects on the exports of Arab countries*. https://theforum.erf.org.eg/2024/09/10/eu-climate-policy-potential-effects-on-the-exports-of-arab-countries/

Enzymes for bioenergy application market insights | forecast. (n.d.). Allied Market Research. https://www.alliedmarketresearch.com/enzymes-for-bioenergy-application-market-A74403

European Commission. (2023). *EU climate law and emissions trading system (ETS)*. European Union. https://ec.europa.eu/climate-action/law

European Commission. (2016, June). *EU Covenant of Mayors and Compact of Mayors launched the largest global coalition of cities committed to fighting climate change.* European Commission Press Corner. https://ec.europa.eu/commission/presscorner/detail/en/ip_16_2247

European Parliament and Council. (2021). *Regulation (EU) 2021/1119 of the European Parliament and of the Council of 30 June 2021 establishing the framework for achieving climate neutrality and amending Regulations (EC) No 401/2009 and (EU) 2018/1999 ("European Climate Law"). Official Journal of the European Union, L 243/1.* https://eur-lex.europa.eu/legal-content/EN/TXT/PDF/?uri=CELEX:32021R1119&from=EN

European Parliament & Council. (2021). *Regulation (EU) 2021/1119: Establishing the framework for achieving climate neutrality.* https://eur-lex.europa.eu/legal-content/EN/TXT/PDF/?uri=CELEX:32021R1119

Ghoneim, A. F. (2024, September). *EU climate policy: Potential effects on the exports of Arab countries.* Middle East Council on Global Affairs. https://mecouncil.org/publication/gulf-transition-to-low-carbon-economies-the-role-of-carbon-markets/

Kateb, A. (2024, May). *The GCC's multipolar pivot: From shifting trade patterns to new financial and diplomatic alliances.* Carnegie Endowment for International Peace. https://carnegieendowment.org/research/2024/05/the-gccs-multipolar-pivot-from-shifting-trade-patterns-to-new-financial-and-diplomatic-alliances

KPMG Lower Gulf. (2025, March). *Understanding the UAE's Climate Change Reduction Law: A turning point for business sustainability.* KPMG. https://kpmg.com/ae/en/home/insights/2025/03/understanding-the-uae-climate-change-reduction-law.html

Lemmet, S., Ducret, P., Berthaud, C., & Evain, J. (2017). *French strategy for green finance: Executive summary.* French Ministry for the Ecological and Inclusive Transition & Ministry of Economy and Finance.

NASA. (n.d.). *The effects of climate change.* NASA Science. https://science.nasa.gov/climate-change/effects/. Retrieved 18 Feb 2025

Singh, S. (2024). Social corporate responsibility. *International Journal of Political Science and Governance.* https://doi.org/10.33545/26646021.2024.v6.i1b.309

Tamimi & Co. (n.d.). *UAE's new climate law focuses on R&D, innovation, and sustainability.* https://www.tamimi.com/news/uaes-new-climate-law-focuses-on-rd-innovation-sustainability/

UAE Government. (n.d.). *The UAE net zero 2050 strategy.* United Arab Emirates Government Portal. https://u.ae/en/about-the-uae/strategies-initiatives-and-awards/strategies-plans-and-visions/environment-and-energy/the-uae-net-zero-2050-strategy

Umweltbundesamt. (2024, March). *Revenue from emissions trading once again at record level: 2023 revenue exceeds 18 billion euros for the first time.* German Environment Agency. https://www.umweltbundesamt.de/en/press/pressinformation/revenue-from-emissions-trading-once-again-at-record

United Nations Department of Economic and Social Affairs. (2024). *Global problems need global solutions; UN officials tell ministers of development.* United Nations. https://www.un.org/en/desa/global-problems-need-global-solutions-un-officials-tell-ministers-development

Chapter 7
Fostering Inclusivity in Plazas Through Design Strategies

Jenna Harchaoui

Abstract Public plazas in hot climates, such as Dubai, often face low usability due to extreme temperatures and inadequate climate-responsive design. This study investigates City Walk Plaza to identify key urban design challenges and propose architectural design interventions. Using field observations, user surveys, and literature analysis, the research examines and addresses the challenges of outdoor plazas. The research identifies key deficiencies, including insufficient shading, ineffective cooling strategies, uncomfortable seating, and accessibility barriers related to pricing and commercial exclusivity. Findings reveal that 62.1% of surveyed users rated thermal comfort as very poor, citing extreme heat and a lack of shaded seating. The majority of respondents suggested hybrid shading systems, misting stations, and passive cooling features as necessary improvements. Additionally, affordability concerns were raised, with respondents noting that high dining costs reduce accessibility for lower-income visitors. To address these challenges, the study proposes tactical urbanism strategies such as modular shaded seating, water-integrated cooling elements, and climate-adaptive material choices to improve year- round usability. These recommendations align with Dubai's 2040 Master Plan, which emphasizes sustainable and inclusive urban spaces. The research contributes to climate-adaptive urban design, offering practical insights for architects, urban planners, and policymakers to revitalize public spaces in extreme environments and foster greater community engagement.

Keywords Urban design · Tactical urbanism · Public plazas · Climate adaptation · Inclusivity

J. Harchaoui (✉)
School of Architecture and Interior Design, Canadian University Dubai, Dubai, UAE
e-mail: jenna.harchaoui@gmail.com

J. Dsilva (ed.), *Sustainability Education, Youth and Innovation*,
https://doi.org/10.1007/978-3-032-21948-0_7

Introduction

"Space only becomes truly public when people recognize it and utilize it as such," David Adjaye Marron (2016). Imagine you are tired of virtual third place and go to a public space searching to reconnect with the community around you. However, you are in Dubai, it is mid-July, and you find the public plazas to be nearly empty. The underutilization of public spaces in arid climates in the summer is nothing new. With "sweltering" temperatures for about 5 months of the year, (*Dubai Climate, Weather by Month, Average Temperature (United Arab Emirates)—Weather Spark*, n.d.), there is little incentive for people to use these spaces.

This study aims to answer:

1. How can tactical urbanism and passive cooling strategies improve public space usability in extreme climates?
2. What design interventions can increase inclusivity and accessibility in plazas?

It examines how urban plazas, such as the City Walk Plaza in Dubai, can be adapted to improve inclusivity, comfort, and climate responsiveness through design interventions. It examines public plazas as nodes for social interaction, community engagement, and urban vitality. Culturally, it highlights architectural strategies to increase vitality and user engagement. Furthermore, it aligns with the Dubai's 2040 development goals of "developing vibrant and healthy communities," and the Dubai Social Agenda 33 (*Dubai 2040 | Dubai 2040 Urban Master Plan & Sustainable Development*, n.d.). This research provides insight on tactical urbanism, shading systems, and passive cooling techniques to revitalize the urban landscape and enhance community spaces in hot climates.

To address the main deterrent to plaza usage, the intense heat, and improve outdoor thermal comfort it is crucial to implement passive cooling strategies such as shading systems, evaporative cooling, strategic material choice, and increased vegetation. Mahmoud and Abdallah (2022) found that combining sun sail-shading, urban form planning, geometrical details, hybrid shading strategies along with planting vegetation can reduce Physiological Equivalent Temperature (PET) by up to 18.6 °C. Zhao et al. (2024) found that combining sun-sails with mist-spray systems is most effective. Through ENVI- met simulations they found that this reduced PET by up to 14.43 °C. Furthermore, visual comfort has an impact on thermal comfort. Lam et al. (2022) found that reducing illuminance by 70% using opaque materials or glass with 10% transparency improves visual and thermal comfort. As for material choice, it is crucial for creating a sustainable and comfortable environment. Doulos et al. (2004) found that materials with high albedo and high emissivity like white marble tiles significantly reduce surface temperatures and heat island effect. As for rainfall, Li et al. (2013) suggest using reflective pavements that are also permeable to help manage stormwater runoff, reduce surface temperatures and flooding. Passive cooling strategies make significant impact on plaza usage by creating more comfortable and inviting environments.

Aside from a comfortable physical environment, choosing to use a public space also relies on perceived social comfort. The urban growth in the UAE has disrupted traditional public squares and previous community bonding (Ezzeddine & Kashwani, 2019). Social inclusivity in plazas can be achieved through taking advantage of the growing emphasis on sustainable urbanism practices in the UAE by turning public spaces into hubs for social cohesion, well-being, vibrant community life, and cultural exchanges (Ezzeddine & Kashwani, 2019). Tactical urbanism projects, like the Abu Dhabi Canvas initiative that combines art and culture, help encourage social inclusion, promoting Emirati culture and enhancing the appeal of public places (Abu Dhabi Media Office, 2024). Social inclusion is heavily impacted by representation and participation. The University of Birmingham Dubai, as reported by Radfar (2024), examined key sites in the UAE, and highlighted how respondents expressed strong interest in participating in planning, designing, and developing public spaces. Implementing strategies like participatory design, and age-friendly design can enhance the vibrancy and inclusiveness of plazas like City Walk.

Tactical Urbanism is a community-driven approach to changing public areas involving quick, low-cost fixes that encourage social inclusion and safety (Lydon & Garcia, 2015) (Silva, 2016) (High Streets Task Force, 2020). These innovative short-term solutions act as a trial run for permanent measures that improve public places' usability (Davis & Buckle, 2019).

The research methods include field research, literature review, user survey, and comparative analysis. The findings of this research provide a list of potential design strategies to enhance plazas in arid climates. These strategies include providing more shading and cooling solutions, enhancing seating, enhancing accessibility, adding practical amenities, making an inclusive and affordable environment, creating opportunities for social and cultural engagement, and testing temporary solutions with tactical urbanism.

Methodology

This research includes quantitative and qualitative methods of data collection. The research is supported by a literature review including existing research and case studies of various strategies applied in plazas internationally. The research includes Field research of a focus area in the key case study site of City Walk, Dubai shown in Fig. 7.1. The observations are categorized to determine the success of the plaza. These categories include assessing the environment, society, culture and aesthetics, function, and technology integration. They are tabulated for evaluation purposes. The second research method applied is the survey of the public users in said plaza. It is intended to determine satisfaction and comfort of the outdoor plaza, and assess three main topics: Activities and Preferences, Comfort and Experience, Feedback and Suggestions. The purpose of the questionnaire is to gain insight into the experience of users in the outdoor plaza of City Walk. The objective is to identify areas of strengths and weaknesses, priorities for improvement, and preferences among users. The survey is on a voluntary basis, and responders remain anonymous. The distribution follows a convenience sampling model as this is a pilot study due to limitations.

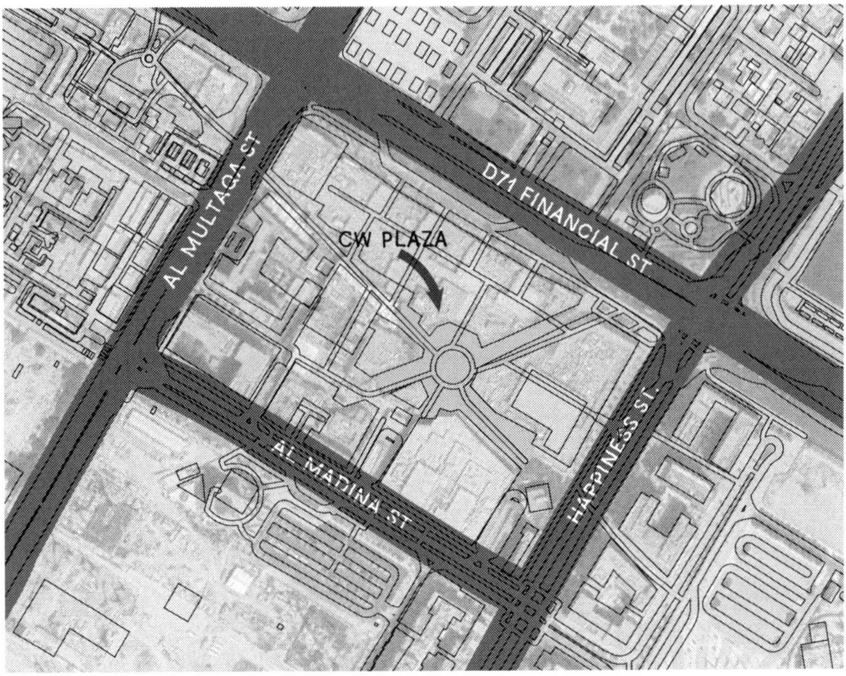

Fig. 7.1 Displays the focus area of field research highlighted in green; image developed by author

It is hosted on google forms. The data is treated and analysed with the visual aid of tables. The long-form questions are treated separately and used to draw qualitative conclusions.

Findings and Discussion

Field observations and surveys at City Walk Plaza reveal significant environmental and social design challenges:

1. Thermal Discomfort & Lack of Cooling Solutions: Field observation revealed the plaza's shading structures are insufficient, vegetation is placed sporadically, there are dark surface materials increasing heat retention, and water features are present but are inconsistently operational and not strategically placed for cooling impact. 62.1% of users rated thermal comfort as very poor, citing lack of shading and cooling options highlighted in Fig. 7.2. One respondent noted, "More shade. I know City Walk is a rush job of a venture like all locations in Dubai. But if there was genuine shade, it would be a lot more welcoming." Literature findings, such as Mahmoud & Abdallah (2022), indicate that hybrid shading systems can lower

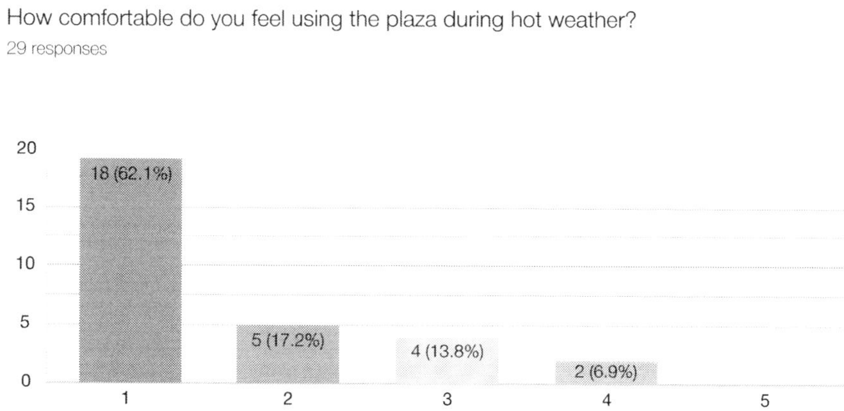

Fig. 7.2 Comfort levels of respondents using the plaza during hot weather. *1 = Very Uncomfortable; 5 = Very Comfortable*

PET by up to 18.6 °C, reinforcing the necessity of integrating dynamic shading solutions.

2. Seating & Accessibility Issues: The plaza is considered inclusive by 73% of respondents due to its wheelchair accessibility. However, others face social barriers citing lack of designated smoking areas and high-priced dining. Insufficient and uncomfortable seating arrangements were observed, with availability being poorly rated with 37.9% marking it as "fair" as shown in Fig. 7.3. Research on tactical urbanism suggests implementing modular shaded seating, integrated mist cooling, and adaptable urban furniture to enhance the plaza (Silva, 2016).

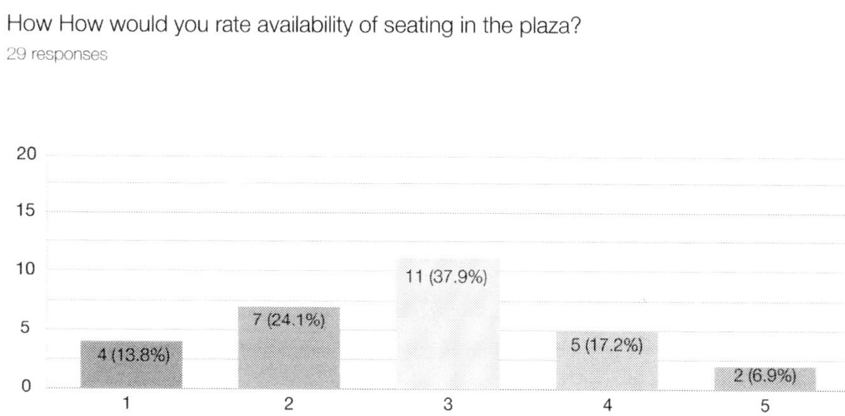

Fig. 7.3 Rating of seating availability in the plaza

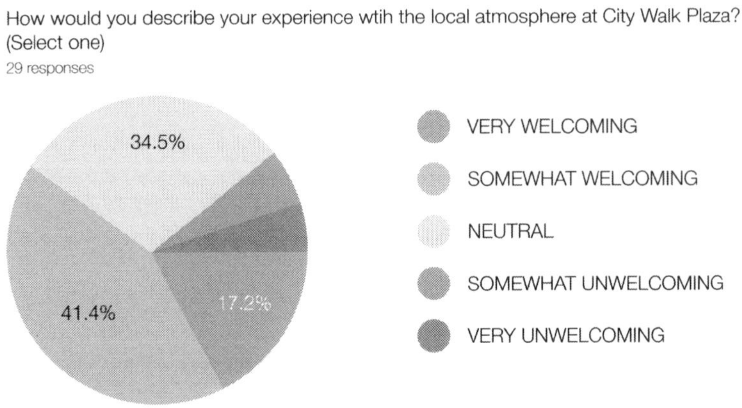

Fig. 7.4 Respondents' experience with the social atmosphere at City Walk Plaza. *1 = Very Unavailable; 5 = Extremely Available*

3. Usage & Social Engagement: 48.3% of users stay over 2 h, indicating potential for engagement, shown in Fig. 7.4, if concerns are addressed. The plaza faces challenges with valet-dominated parking and lack of affordable commercial options, while its aesthetic features are not significant.

These findings support implementing hybrid shading systems, mist cooling stations, and modular seating to enhance usability, inclusivity, and sustainability in arid climates.

Conclusion

The results of the research demonstrate the importance of addressing concerns about thermal comfort, seating, shading, affordability, and inclusivity to enhance public space usability. By introducing tactical strategies and climate-responsive design, the experience of the users within the plaza can be enhanced to fit the needs of the community, global sustainable urban planning efforts and Dubai's 2040 vision for liveable cities. While the study focuses on City Walk Plaza, similar design solutions could be applied to other arid climate cities. The limitations of the research methods include time constraints, sample size, and the broadness of the research goals. The research was conducted over a few weeks in an academic course. The sample for the surveys was limited to convenience. This research can expand to further validate the suggested strategies through pilot projects using tactical urbanism techniques. Future studies could explore long-term behavioural changes and policy integration, ensuring that public spaces remain adaptable to climate challenges and evolving community needs.

Design Recommendations

The following are recommendations for future outdoor plaza designs. These strategies can be applied to typologies that include open outdoor spaces.

Provide More Shading and Cooling Solutions

- Install hybrid shading systems to cover seating areas and walkways.
- Activate the water features consistently and add misting stations to create a cooler, more inviting environment. For example, there are misting stations designed by Mask Architects for Abu Dhabi's Heat Island. *The Oasys* is a modular cooling system that creates solar-powered, mist-spraying hubs for outdoor relaxation and community activities, adaptable for temporary or permanent use (Baldwin, 2020).
- Replace dark surface materials with lighter, reflective options to reduce heat absorption.

Enhance Seating and Accessibility

- Add shaded seating areas with comfortable designs, including benches with backrests to cater to all visitors.
- Ensure seating is evenly distributed throughout the plaza.

Make the Plaza More Affordable and Inclusive

- Prioritizing a deeper understanding of the potential users would allow for strategic choice of vendors to ensure engagement.
- Designate smoke-free zones to ensure the plaza is welcoming for everyone, including those with respiratory sensitivities like asthma.

Add Practical Amenities

- Install water fountains and refill stations to keep visitors hydrated and comfortable during their time at the plaza. This aligns with Sheikh Hamdan's "Dubai Can" initiative which involved placing 34 refilling stations around Dubai (Fletcher & Fletcher, 2022).

Create Opportunities for Social and Cultural Engagement

- Organize summer activities, such as pop-up events or cultural performances, to draw more visitors even during the hot months like the initiative by Abu Dhabi Canvas (Abu Dhabi Media Office, 2024).
- Expand the number of art installations to celebrate the local culture and make the plaza more visually appealing.

Test Temporary Solutions with Tactical Urbanism

- Use temporary shading and cooling structures during summer months to see what works best before committing to permanent changes.

- Set up pop-up seating areas or activity zones to experiment with layouts that encourage social interaction and engagement. Events like Dubai Design Week in D3 are a very successful example (Wikipedia contributors, 2024).

Appendix 1

Tabulation

Category of observations	Characteristics of observations
Environmental design & climate adaptation, sustainability features	Shading structures: Variety of shading structures over some of the alleys and only on the outer perimeter of the central area Cooling elements: There is some vegetation but not in every alley and not much in the main central area. There is a large water feature in the centre and there are water features at the bottom of the perimeter shading structure, but they are not always active Wind patterns: The courtyard configuration of the plaza along with the water feature creates a breeze in the centre and some alleys Surface materials: The surface materials of the plaza are various tints of ceramic tiles. They are mostly dark grey and light beige in color
Social interaction & inclusivity usage patterns	Seating arrangements: There are a few seats scattered in some of the alleys, but not all. The only seating in the central area belongs to businesses, like the cafes and restaurants. The public seating available is uncomfortable as some do not have backrests and are not shaded entirely Accessibility: The plaza is easily accessible by elevators from the parking lot and all changes in elevation are sloped and do not require steps, so wheelchairs are accommodated for Safety & security: The plaza has security in the shopping centre adjacent to it and within the university facilities, as well as security that circulates through the plaza Activity types: shopping, dining, children's bus tours in the plaza, shows at night with the central fountain, the cinema, and socializing Crowd density: At the observation time of 5 pm on a weekday the crowds were small with people scattered around the plaza in small groups of 2–5 people

(continued)

(continued)

Category of observations	Characteristics of observations
Aesthetic & cultural identity	Public art installations: There are some murals around the plaza, and an artistic structure Cultural landmarks: There is a view to the Burj Khalifa from the plaza. There is also a Mosque. The plaza is a landmark of the city of Dubai
Functional design	Pedestrian flow: constant in the alleys. Stagnant in front of the University. Many people stop to observe the view to the Burj Khalifa and to take pictures of it, mostly tourists. Locals walk from one commerce to another Space flexibility: The central plaza is large and open to being adapted to different events if need be. The alleys also have wide linear open spaces for this potential Facilities: There are washrooms in the various alleys, as well as access to the underground car park. There is also a valet service. There are no available water stations
Technology integration	Smart elements: The plaza has various changing screens. At night the plaza's water feature lights, up in various colors and has a show for families with projections on the walls to represent dinosaurs or animals. The shows vary. Interactive technology: N/A

Appendix2

User's Direct Feedback on Their Experience

What improvements would you suggest for making City Walk Plaza more inviting during hot months?

Of the 29 responses, 18 emphasized their desire for more shading to make the City Walk Plaza more comfortable during the hot summer months. An example of this sentiment can be seen in the following response: "More shade. I know city walk is a rush job of a venture like all locations in Dubai. But if there was genuine shade, it would be a lot more welcoming." The second most shared sentiment by respondents is the need for cooling systems. Six people mentioned it in their responses. For example: "Providing fans and AC in outdoor areas." Five answers highlighted a desire for a combined comfortable and cooled seating areas, like: "More comfortable places to sit and add something that would help people to overcome the hot weather," and "Create spaces that improve cooling around seating areas and such." Two people suggested adding water stations and water features as a potential improvement, "Water station." Two respondents did not have any suggestions. One person suggested parking improvements, “Providing more parking lots in the basement and not reserving most of it for valet parking.”

Do you feel that City Walk Plaza is inclusive for all community members? Why or why not?

Most respondents (22 of 30) felt that City Walk is inclusive with some notable responses being: “Yes, it is very friendly”, “Yes, I see people of all ethnicities having a good time”, “Yes, because I don’t struggle as a disabled person.” Four people shared that they do not find City Walk to be fully inclusive. Some mention the high pricing of dining and shopping options as a reason hindering inclusion, “I feel that City Walk is not inclusive to all parties as some of the dining options can be pricey for many individuals.” Others mention the presence of smoking and vaping as a cause for discomfort for those who have respiratory issues, “No, it’s not, because of the rampant vape and smoking problem in the area.” Two people expressed mixed feelings: “Could be, but sometimes guests want to visit, but it’s mostly uni students taking up the parking lot space.”, “I guess it is, there’s no reason why it’s not.”

What specific features or amenities do you think should be added to enhance the plaza experience?

Eight respondents highlighted the need for additional shaded areas and cooling solutions to make the plaza more comfortable, especially during hot weather: “Tools that would help reduce the hotness of the weather like umbrellas around the plaza.”, “An area where you could enjoy being outside without being fully exposed to the heat.” Seven respondents suggested that more budget-friendly or affordable dining options should be introduced, particularly to accommodate students and those seeking more accessible options: “Perhaps more seating, weather control, and budget-friendly restaurants and activities”, “More budget-friendly dining locations.” Five respondents emphasized the need for more comfortable and shaded seating areas, with specific mentions of benches with backrests and outdoor lounging spaces for relaxation: “Benches with backrests, shaded areas”, “Outdoor lounging for individuals who wish to relax rather than party.” Four respondents pointed out the need for better parking facilities, including additional parking spaces and free parking for students from nearby institutions: “Better parking situation”, “More parking spaces.” Four respondents did not have specific suggestions or were unsure of what could enhance the plaza experience.

References

Arup. (2019, October). *C40 Cities PlayBook: InclusiveCommunity Engagement*. https://www.arup.com/insights/c40-cities-playbook-inclusive-community-engagement/

Baldwin, E. (2020, October). *Mask architects design cooling stations for Abu Dhabi’s urban heat island*. ArchDaily. https://www.archdaily.com/949459/mask-architects-design-cooling-stations-for-abu-dhabis-urban-heat-island

Bedirian, R. (2024, October). Why are Abu Dhabi’s public spaces being painted with street art? *The National*. https://www.thenationalnews.com/arts-culture/2024/09/22/abu-dhabi-canvas-street-art/

Bolton, L. (2020). Tactical urbanism in low- and middle-income countries (LMICs). The Institute of Development Studies and partner organisations. Coffey, R. (2024, August). Rain making in Dubai. Was a devastating flood caused by cloud seeding? *Forbes*. https://www.forbes.com/sites/rebeccacoffey/2024/08/19/rain-making-in-dubai-was-a-devastating-flood-caused-by-cloud-seeding/

Davis, C., & Buckle, P. (2019). Successes & Challenges of Tactical Urbanism. In *Transportation Group New Zealand*. Transportation Group 2019 Conference.

Doulos, L., Santamouris, M., & Livada, I. (2004). Passive cooling of outdoor urban spaces. The role of material. *Solar Energy, 77*(2), 231–249. https://doi.org/10.1016/j.solener.2004.04.005

Dubai climate, weather by month, Average temperature (United Arab Emirates) - Weather Spark. (n.d.). Weather Spark. https://weatherspark.com/y/105470/Average-Weather-in-Dubai-United-Arab-Emirates-Year-Round#Figures-ColorTemperature

Dubai 2040. Dubai 2040 Urban Master Plan & Sustainable Development. (n.d.). http://dubai2040.ae/en/dubai-2040-urban-master-plan/

Ezzeddine, I. & Kashwani, G. (2019). Public squares in UAE sustainable urbanism: Social interaction & vibrant environment *09*, 23–32. https://doi.org/10.5923/j.arch.20190902.01. Lam, C. K. C., Weng, J., Liu, K., & Hang, J. (2022). The effects of shading devices on outdoor thermal and visual comfort in Southern China during summer. *Building and Environment, 228*, 109743. https://doi.org/10.1016/j.buildenv.2022.109743

Fletcher, L., & Fletcher, L. (2022, February). Sheikh Hamdan launches 'Dubai Can' with 34 refillable water stations. *What's On*. https://whatson.ae/2022/02/sheikh-hamdan-launches-dubai-can-with-34-refillable-water-stations/

The National. (2022, September). "Dubai Can": Sheikh Hamdan in drive to increase use of public water fountains. *The National*. https://www.thenationalnews.com/uae/2022/02/15/dubai-can-sheikh-hamdan-in-drive-to-increase-use-of-public-water-fountains/

High Streets Task Force. (2020, May). *Temporary public realm changes*. https://www.highstreetstaskforce.org.uk/reopening-high-streets-safely/temporary-public-realm-changes/. Retrieved 18 Oct 2024.

Li, H., Harvey, J. T., Holland, T. J., & Kayhanian, M. (2013). The use of reflective and permeable pavements as a potential practice for heat island mitigation and stormwater management. *Environmental Research Letters, 8*(1), Article 015023. https://doi.org/10.1088/1748-9326/8/1/015023

Lydon, M., & Garcia, A. (2015). *Tactical urbanism: short-term action for long-term change*. https://trid.trb.org/view/1360475

Mahmoud, R. M. A., & Abdallah, A. S. H. (2022). Assessment of outdoor shading strategies to improve outdoor thermal comfort in school courtyards in hot and arid climates. *Sustainable Cities and Society, 86*, Article 104147. https://doi.org/10.1016/j.scs.2022.104147

Marron, C. (2016). *City Squares: Eighteen Writers on the Spirit and Significance of Squares Around the World*. HarperCollins.

My NASA Data (n.d.). What is Albedo? | My NASA Data. MyNASA Data. https://mynasadata.larc.nasa.gov/mini-lessonactivity/what-albedo#:~:text=Albedo%20is%20the%20fraction%20of,immediately%20reflected%20back%20to%20space

Abu Dhabi Media Office. (2024, September). Department of Municipalities and Transport expanding Abu Dhabi Canvas initiative to showcase UAE's heritage and local artists in public spaces. Abu Dhabi Media Office. https://www.mediaoffice.abudhabi/en/infrastructure/department-of-municipalities-and-transport-expanding-abu-dhabi-canvas-initiative-to-showcase-uaes-heritage-and-local-artists-in-public-spaces/

Properties, T. (n.d.). *Design Quarter by Meraas at Design District (d3)—Master Plan*. Tanami Properties. https://www.tanamiproperties.com/Projects/Design-Quarter-at-D3-MasterPlan#lg=1&slide=0

Radfar, M. (2024, September). *How public spaces shape social sustainability*. Al Qasimi Foundation. https://publications.alqasimifoundation.com/blog/how-public-spaces-shape-social-sustainability. Retrieved 19 Oct 2024.

Silva, P. (2016). Tactical urbanism: Towards an evolutionary cities' approach? *Environment and Planning B Planning and Design, 43*(6), 1040–1051. https://doi.org/10.1177/0265813516657340

Wikipedia contributors. (2024, August). Dubai Design Week. In *Wikipedia, The Free Encyclopedia.* https://en.wikipedia.org/w/index.php?title=Dubai_Design_Week&oldid=1238514836. Retrieved 09:51, 21 Nov 2024.

Zhao, Y., Zhao, K., Zhang, X., Zhang, Y., & Du, Z. (2024). Assessment of combined passive cooling strategies for improving outdoor thermal comfort in a school courtyard. *Building and Environment, 252*, Article 111247. https://doi.org/10.1016/j.buildenv.2024.111247

Chapter 8
Clean Earth Movement

Jairam Chawan

Abstract The project "JaizGlobal1928 Clean Earth Movement" and its concepts are dedicated to children below 15 years of age, elderly people above 50 years of age, expecting women, and students of engineering and architecture. Debris produced by the construction industry is one of the major contributors to environmental pollution, where very valuable and expensive materials are often dumped into the ocean despite being easily recyclable, reusable, or utilizable for various social causes (Cardoso et al., Construction and Building Materials 367, 2023). If the right actions are taken and efforts are implemented to bring awareness among people, we may be able to find a fruitful solution to one of the deadliest concerns on this planet. As oceans produce almost half of the Earth's oxygen, it must be mandatory to keep them free from pollutants (NOAA, 2024; Scripps Institution of Oceanography. (2023, May). Phenomenal phytoplankton: Scientists uncover cellular process behind oxygen production. https://scripps.ucsd.edu/news/phenomenal-phytoplankton-scientists-uncover-cellular-process-behind-oxygen-production). Due to rapid deforestation, the Earth has already lost most of its green cover, reducing oxygen content, while smoke released from an increasing number of industries, hydrocarbons from petroleum companies, vehicles, and other pollutants further endanger life on the planet. Rapid urbanization results in various infrastructure developments, which generate significant construction waste that often reaches landfills and oceans, creating marine debris (Cardoso et al., Construction and Building Materials 367, 2023). Every species in this universe has equal rights to live healthily and without abuse, physical trauma, or intentional harm caused by trash, garbage, and human-created waste dumped directly into the sea or along the seashore. The goal and mission of the project is to ensure complete transformation of a 3.0 km sea shoreline from a polluted coast into one of the most hygienic places, safe for all, while creating various temporary constructions from marine debris, construction waste, and plants discarded in dustbins.

Keywords Clean earth movement · Marine debris · Construction waste · Environmental pollution · Ocean conservations

---

J. Chawan (✉)
Dubai, UAE
e-mail: jaizglobal1928@gmail.com

J. Dsilva (ed.), *Sustainability Education, Youth and Innovation*,
https://doi.org/10.1007/978-3-032-21948-0_8

Introduction

One of the major contributors to environmental pollution is the waste and debris produced by construction industries. Most of the time, very valuable and expensive materials are just dumped into the ocean which can be easily recycled, reused, or utilized for various social causes (Cardoso et al., 2023). According to the Five Element Theory in Chinese Medicine, everything in this universe and the human body is made up of five basic elements such as earth, water, fire, air, and space. Protecting these natural resources is essential to sustain life on the planet. Humans have five basic senses: vision, hearing, smell, taste, and touch, which help us to understand and perceive the world around us, and these too are deeply affected by environmental degradation. The coastline of Nurana Island is a man-made island and is private property, but open for public use. Due to construction waste dumped in the sea, waste created by fishing boats, and littering by beachgoers, the seashore has become unhygienic, creating marine debris that can cause serious health issues in children below five years and elderly people. Covering 3.0 km of the sea shoreline was not easy as it required just one high tide to bring all the marine debris to shore, and it took two months to bring the area to an acceptable condition. The project theme was divided into two schemes: the Nurana Island Clean-Up Program focusing on cleaning the seashore from marine debris and trash, and the Nurana Island Beautification Program, which involved creating artworks using lightweight construction blocks and wood materials, repairing road cavities, and developing public garden road islands for smooth traffic flow.

Methodology

The JaizGlobal1928 Clean Earth Movement was conceptualized more than 3.5 years before implementation but faced delays due to time constraints and difficulty in mobilizing volunteers. As the pandemic worsened, alternative approaches were adopted, and the project was eventually executed single-handedly. The initiative was structured under two schemes: Scheme 1 focused on cleaning the Nurana Island seashore from marine debris and trash, while Scheme 2 centered on beautification by creating artworks using lightweight construction blocks and wood, repairing graded roads, and developing public garden road islands. All activities were performed manually, with daily work sessions ranging from 6 to 14 h. Key tasks included removing non-decomposable marine debris such as plastics, metals, fishing nets, diapers, glass bottles, Styrofoam, and ropes, and repurposing recovered materials like blocks, plywood, timber, and crates for road repairs, planters, parks, and architectural islands. Gardening and landscaping transformed the site with green patches, family seating spaces, and children's play areas, while on-site educational programs promoted plantation activities and environmental awareness among children below 15 years. Water from a nearby R.O. plant was utilized for ornamental water bodies, irrigation, and gardening (Show Me Mizzou, 2023). Despite challenges posed by weather conditions, pandemic restrictions, and limited social interactions, the project steadily advanced, with occasional volunteer participation and logistical support from the local municipality and cleaning agencies, ensuring the collected debris was processed appropriately.

Background

The coastline of Nurana Island is a man-made island and private property but open for public use. Despite its abundant nature and uniqueness, the area became polluted due to construction waste dumped in the sea, waste from fishing boats, and littering by beachgoers, resulting in marine debris that, when mixed with seaweed, can cause serious health issues in children below five years and elderly people. Considering the high number of visitors for walking, swimming, and other activities, this location was chosen, and the mission began on 21st June 2020. Initially, the shoreline was filled with marine debris, and covering 3.0 km was challenging, taking two months to bring it to an acceptable condition. The project, conceptualized over 3.5 years earlier, was implemented single-handedly during the pandemic and divided into two schemes: Scheme 1 focused on cleaning marine debris, while Scheme 2 involved creating artworks from lightweight construction blocks and wood, repairing graded roads, and developing public garden road islands. Despite weather conditions, pandemic restrictions, and manual handling of heavy materials, the work progressed with occasional support from volunteers, the local municipality, and cleaning agencies. Collected debris included plastic, metal, glass, fishing nets, plywood, timber, diapers, and synthetic ropes, much of which was repurposed to repair roads, install curbs, make planters, create parks, and develop green spaces and children's play areas.

Project Objective and Motivation

The objective of the "JaizGlobal1928 Clean Earth Movement" is to add value to human life by bringing people closer to nature, making them aware of the importance of coexisting with it, and protecting and preserving flora and fauna by minimizing pollution from construction waste and debris dumped into oceans and the environment. The project spreads awareness on handling marine debris, construction waste, and discarded ornamental plants from plantations and household gardens. Rapid urbanization and high demand have caused developing nations to generate significant construction waste, but due to high recycling costs and limited opportunities for reuse, much of it is either dumped in open lands or oceans, leading to environmental pollution. Although technology exists to process and recycle waste, cost implications often prevent its effective use.

Reusing recycled materials from construction waste and marine debris can provide affordable housing for homeless populations, estimated at 150 million globally, and improve living conditions for the 1.6 billion people lacking adequate housing (UN-Habitat, 2022). Surplus materials from construction sites can also support public utilities such as parks, roads, pavements, playgrounds, and the renovation of old homes, orphanages, old-age homes, and animal rescue centers.

The COVID-19 pandemic has slowed global development and strained economies, making sustainability essential for human well-being. Through education and awareness, pollution of earth and water can be reduced, leading to cleaner air, fire, and sky and a healthier environment (Cardoso et al., 2023). As everything in the universe and human body is made of five basic elements: earth, water, fire, air, and space, a clean environment improves physical and psychological health by activating the five senses: vision, hearing, smell, taste, and touch, bringing calmness and happiness.

Front-line workers during the pandemic, including doctors, nurses, police, defense personnel, and volunteers, along with the leaders of GCC nations, were a major source of motivation. The Kingdom of Bahrain's leadership, in particular, offered exceptional support by protecting lives and providing assistance to both citizens and residents. These efforts and campaigns inspired the successful implementation of the project.

Scope and Implementation Activities

The scope of the project is to transform the 3.0 km of coastline from the contents of marine & construction debris to naturally beautiful & clean habitat for all the dependent species, spread awareness by showcasing results through effective implementation of ideas and concepts by re-using materials received from the marine and construction debris by appropriately segregating and setting up examples by the construction of art pieces and other public utilities by fixing roads and construction of the public garden, etc. Protecting source of water discharge to sea, from marine debris, construction waste & trash created by beachgoers and using the water (released from nearby R.O plant) for landscaping. Implementation of education & awareness (on-site) programs for children below 15 years to promote the importance of a clean environment, having clean oceans, and planting trees.

To abide by the social & legal requirements this facility is temporarily constructed to experiment and derive results to provide consolidated conclusions and study materials for upcoming engineers till the permanent development is carried out at this location and to ensure the coastline is free from marine & construction debris, trash formed due to public littering. Considering the uniqueness in its location the core area of the public park is dedicated to Poets, writers, thinkers, educators, health and wellbeing practitioners, musicians, songwriters, engineers, architects, & this area can be used as an open prayer hall & meditation too. Moreover, the Kingdom of Bahrain doesn't have a natural riverfront considering the topographical profile this can be visualized as an example of a stream at the base of a hill and forming a river later meeting the sea. Thus, this location may have something to learn about for everyone irrespective of their subject of interest.

Every species in this universe has equal rights to live healthily and without any abuse, physical trauma, or intentional harm and risk caused by trash, garbage, human created waste dumped directly in the sea or at the seashore.

The Activities Performed through the project are.

- Beach Clean-up: Removal of marine debris which are non-decomposable materials/items such as plastic, metal, food packaging's, empty plastic, glass, metal bottles, fishing nets, wooden pieces, children nappies/diapers, cardboard, straws, synthetic ropes, and threads, plastic and metal buckets, utensils, barrels, Wooden pallets with nails and sharp fasteners, alcohol bottles and beer cans, food leftover, carpets, mats, ropes, Styrofoam's, polystyrene, nylon and cotton ropes, clothes, shoes and other foot wares, wooden and marine plywood, wooden battens, plastic and metal pipes, seaweed, nails, ETC.
- Repair and Refurbishment of Graded Roads, installing curbs, making of planters from waste lightweight blocks, Gardening, Children play areas, Family seating areas, friends' catch-up areas, Etc.
- Educational programs for children are implemented to plant saplings and to promote clean environment habits.
- Used water released from nearby R.O plant to create artificial ornamental water body and to use for irrigation and gardening and temporary construction.
- Approaching nearby construction sites and ready-mix concrete suppliers to request for adopting the right ways of handling waste.
- Giving final rights to dead birds found on the seashore.
- Collecting saplings and other plants from trash bins and planting at the project site.
- Managing strays (Dogs and cats).
- Awareness programs to deal with migratory birds and marine life.
- Traffic management by fixing deformities and vertical cavities formed in the gradient of the graded road (alternative approach created by beachgoers) & landscape garden with curbs made from lightweight blocks pulled from marine debris.
- Implementing programs of awareness regarding allergies caused by marine debris and its harmful effects on children and elderly people.

Note all these activities are performed manually, no measuring tools, equipment or machines are used.

Innovation

Every manmade product used in industries and communities that contributes to environmental pollution can be recycled or reused through proper corrective and innovative measures. With over 18 years of experience in the construction industry across

India, Sri Lanka, Oman, and Bahrain, the project explored alternative approaches to enhance sustainable living through effective ideas, themes, and experiments.

Rapid urbanization and industrial growth have severely impacted air quality, making it essential to increase green zones, plant more trees, and restrict landfills and ocean dumping. Since half of the Earth's oxygen comes from the sea, clean oceans directly improve air quality and human health. Many countries face challenges in managing marine and construction debris; however, by applying the project's approach, items such as concrete, metals, construction wood, and paints can be segregated, recycled, and reused to create public facilities at minimal cost.

Plastics and other pollutants trapped in debris harm coastlines and marine life, but with the right strategies, waste collection and segregation can be made easier for cleaning agencies and volunteers. Water from a nearby R.O. plant, despite slightly higher salinity, was used for landscaping, ornamental plants, grass patches, washing, and even as drinking water for migratory birds and some animals. Baby plants and trimmed branches discarded from gardens, nurseries, and plantations were relocated to green barren lands or composted into natural manure, reducing dependence on chemical fertilizers.

Additionally, recycled plastic and petroleum-based waste can replace wood in construction, reducing deforestation. With accurate combinations, recycled plastic can also be used in road construction, offering greater durability than asphalt. Finally, a clean environment is vital for emotional, psychological, and physical balance. With proper awareness, legal frameworks, and discipline, governments can reduce cleanup costs and foster responsible waste management, resulting in healthier ecosystems and sustainable living.

Implementation

The total concept and implementation of the project was part of service to mankind. Due to the Covid-19 pandemic millions of people lost their lives and millions lost their jobs. The efforts taken by the government of Bahrain and hospitality received from the people of Bahrain is the prime source to implement this project. All the service has been provided free of cost to the community through the saved money and by adopting a minimal living lifestyle. To control the cost last from 26th of November project location is commuted by walk in one hour while going and another one hour while returning. Minimum service hours are 6 h a day and maximum up to 14 h.

All the materials used in the project are from marine and construction debris, plants, saplings, and grass discarded from the owners and dumped into the trash, total of 9 donors have contributed with some ornamental saplings.

The comprehensive results achieved in the past 214 days are;

- Cleaning of 3 kms of sea coastline, by handling and ensuring correct disposal for a minimum of 60,000 gallons of marine debris,

- Removal of lightweight construction blocks from the seashore, transporting and installing manually for artworks, fixing cavities in roads, installing curbs, creating planters for the landscaping, installing and pavement, etc. approximately 15 Tons.
- Removal of concrete curbs, transporting and installing manually for the landscaping boundaries, 7 tons.
- Removal of mosaic tiles, and installing manually to create a pavement for walkway, 5 tons.
- Removal of concrete construction blocks and installing for steps and hardscape boundaries, 2.0 ton.
- Filling of landscaping soil, 60 cum.
- Filling of marine sand for fixing roads and pavements, 30 cum.
- Planting of saplings 250 Nos.
- Creating a green patch with grass 60 Sq.m.
- Creating a seawater breaker with stones, 25 m.
- Surplus and waste ready mix concrete used, 20 Cum.
- Creating a green patch in a highly saline environment.
- Creating a foul odor-free environment.

Benefits of the project implementation.

- Community involvement to create a clean and safe sea coastline.
- Protecting water source from the accumulation of marine debris and risk of construction waste dump.
- Inspiring young children to learn new things and enhance their creativity.
- Creating a green patch in a highly saline environment.
- Education children to plant saplings.
- Creating a clean and safe environment for migratory birds and marine life.
- Managing strays.
- Community involvement to pick lightweight blocks and other materials to re-use for gardening and decoration works.

Appendix

Global Citizen_0000 YouTube Playlist: https://www.youtube.com/playlist?list=PL4wqbIXX_LblhArYexMeQ87OT-uZ5JNMi

Instagram Handle: https://www.instagram.com/global_citizen_0000

References

Cardoso, R., dos Santos, A. R., Lima, J. L., & Oliveira, L. A. (2023). Incorporation of plastic waste into asphalt mixtures: A sustainable solution for road construction. *Construction and Building Materials, 367*, Article 130265. https://doi.org/10.1016/j.conbuildmat.2023.130265

Dow Chemical Company. (2020). *Rethinking waste: Turning plastic into roads.* https://corporate.dow.com/en-us/news/seek-together/rethinking-waste.html

Economic Times. (2025, July). *India launches first field trial using waste plastic geocells to build roads in difficult terrains.* https://economictimes.indiatimes.com/industry/transportation/roadways/india-launches-first-field-trial-using-waste-plastic-geocells-to-build-roads-in-difficult-terrains/articleshow/122416995.cms

National Oceanic and Atmospheric Administration. (2024, June). *How much oxygen comes from the ocean?* NOAA Ocean Service. https://oceanservice.noaa.gov/facts/ocean-oxygen.html

Scripps Institution of Oceanography. (2023, May). *Phenomenal phytoplankton: Scientists uncover cellular process behind oxygen production.* https://scripps.ucsd.edu/news/phenomenal-phytoplankton-scientists-uncover-cellular-process-behind-oxygen-production

Show Me Mizzou. (2023, July). *Driving innovative solutions for plastic roads.* University of Missouri. https://showme.missouri.edu/2023/driving-innovative-solutions-for-advancing-use-of-plastic-roads/

Times of India. (2025, June). *Over 1 lakh kg plastic from MRF units used for road construction in Mangaluru.* https://timesofindia.indiatimes.com/city/mangaluru/over-1-lakh-kg-plastic-from-mrf-units-used-for-road-construction-in-mangaluru/articleshow/121710332.cms

UN-Habitat. (2022). *World Cities Report 2022: Envisaging the Future of Cities.* United Nations Human Settlements Programme. https://unhabitat.org

Chapter 9
Co-Xisting Tektons (Tectonics)

Mohammed Juned Shaikh

Abstract Co-Xisting Tektons (Tectonics), The idea embodies this vision by merging historical biogeographical studies with modern technology and architectural innovations. It questions whether future urban development should focus on a single species or foster interdependence among multiple species. The idea proposes a novel approach to urban design that considers plate tectonics and species movements, incorporating Bio-Behavioral Mechanisms to address the living requirements of all species, including navigation, habitat, geometric preferences, and resource consumption. This approach aims to recreate future tectonics by blending past species and geographic timelines with present technological advances, fostering sustainable and interdependent environments.

Keywords Interconnectedness · Terraforming · Socio-economic aspects · Geographic-geometric-reconstruction · Mixed neighborhoods

Introduction

A stable planet has permanent ice caps, flowing rivers, vast forests, healthy oceans, and active tectonics yet unchecked human expansion risks replacing this balance with the Eremocene, an era of biodiversity collapse. With over 4 million species holding unique survival solutions, we must merge natural wisdom with technology to create symbiotic systems for sustainable development or risk ecosystem and human collapse.

In just a century, humanity has challenged Earth's millions of years of stability, pursuing infinite growth on a finite planet. The Industrial Revolution spurred progress, but saturation of old ideas and poor management now drives ecological decline. Architecture, which literally reshapes Earth's surface, must shift from human-centric to nature-integrated design, preserving life-support systems.

M. J. Shaikh (✉)
Independent Researcher, Co Xisting Tektons, Dubai, UAE
e-mail: juned2502@gmail.com

J. Dsilva (ed.), *Sustainability Education, Youth and Innovation*,
https://doi.org/10.1007/978-3-032-21948-0_9

One person can meet basic needs on ~1,000 m^2 by integrating food, water, energy, and ecosystem services, yet inequality under global capitalism stalls climate action. Without structural change, both planetary stability and human survival remain at risk.

These dynamic fuels skepticism, resentment, and leads to short-sighted policies that worsen ecological damage. If this system continues to exist, then this is what our planet will face along with the failure of tipping points which include –

Risks of Inaction

2030—the amazon rainforest cut down until it no longer produces its own rain and moisture, degrades into a dry savannah bringing catastrophic species loss altering global cycle of water. At the same time the arctic becomes ice free in the summer without the white ice cap less of the sun energy reflects back and speed of global warming increases.

2040, throughout the north frozen soils thaw releasing methane a greenhouse gas as more potent than c02.

2050, As the ocean continues to heat and become more acidic coral reefs around the world die and fish population crash.

2080, Global food production enters a crisis as soil become exhausted by overuse, and the weather becomes increasingly unpredictable.

2100, Our planet becomes 4 degrees warmer, large parts of the earth are uninhabitable. Millions of people rendered homeless. A sixth mass extinction event is well underway and the collapse of the living world the very thing that gave birth to our civilization.

Most sustainability reports target governments with frameworks, not tangible, cross-border solutions. This research proposes a multidirectional "boomerang vector" using architecture to visually model development that balances people, planet, and profit cultivating resources, creating jobs, supporting businesses, and achieving net-zero, offering leaders a clear, actionable blueprint beyond text and numbers.

Bio Behavioral Mechanisms, The initial explorations of this research started with learning the core framework of our planet's ecosystem and all the biological processes and their connection to help keep this planet stable during the Pleistocene epoch Till the Holocene epoch and now the unstable epoch Anthropocene.

- Trophic Cascades & Keystone Species: Large herbivores and apex predators balance populations, shape vegetation, and maintain diverse habitats critical both in Ice Age "megafaunal" contexts and modern ecosystems.
- Connectivity & Metapopulations: Migratory routes, wildlife corridors, and "stepping- stone" habitats prevent isolation, enable gene flow, and boost long-term species survival.

- Low-Impact Adaptation: Whether through Pleistocene mobility or sustainable Holocene agriculture, communities thrive by using resources sparingly and adjusting to environmental shifts (e.g., seasonally adaptive or modular structures).
- Climate Feedback & Carbon Storage: From megafauna trampling permafrost to managed forests capturing CO_2, healthy ecosystems stabilize climate. Maintaining habitat complexity ensures resilience to environmental changes.
- Biodiversity & Resilience: Both eras show that diverse, well-connected food webs recover faster from disturbances, highlighting the need to conserve and restore ecological networks for continued stability.
- Optimal Foraging Theory: Animal generally must work harder to find their food, they must move, and this uses extra resources. Optimal foraging theory (OFT) proposes that animals try to gather resources in the most efficient way to avoid using additional energy. Habitats and migratory routes of the species are optimized for shortest walk and territorial and seasonal replenishment of resources as all the routes are carved out in a very systematic and symbiotic ways through the ecosystem with food sources.

By combining AI, eDNA analysis, and satellite data, we can map tectonic resilience and resource availability, identifying areas at risk of overshooting and those where ecosystems thrive. Integrating this ecological intelligence into architecture ensures designs respect natural boundaries, optimize resources, and foster regenerative communities that keep all nine planetary boundaries in check.

The Co-Existing Tektons vision creates hyper-local, data-driven urban forms that merge permaculture, civic governance, and new economic models. Cities become living terrariums where wildlife and humans coexist façades host nesting sites or aquatic habitats, elevated corridors allow safe animal crossings, and controlled, façade-integrated "zoo" spaces blend into commerce, housing, and infrastructure.

Nature-based valuation (Ralph Chami) and social business frameworks shown in Fig. 9.1 (Muhammad Yunus) transform animals into economic partners through services like carbon sequestration and eco-tourism. This approach redefines urban growth as a self-sustaining tapestry supporting human prosperity while safeguarding biodiversity, ensuring all life thrives together.

Methodology Phase 1: Theoretical & Historical Analysis

Archival & Field Research: Study Holocene societies (e.g., Egyptian irrigation, Mayan agroforestry) and Pleistocene proxies (e.g., Pleistocene Park, rewilding).

Comparative Framework: Develop a matrix linking architectural/urban typologies to ecological states (stable/dynamic, mono-/polyculture)

Phase 2: Parametric "Co-Existing Tektons" Model

Algorithmic Simulations: Use Maya, Houdini, Rhino, Unreal Engine, Grasshopper, Dynamo, AI to generate prototypes for rural, urban-edge, and dense-core city blocks integrating biodiversity loops.

Ecological Scenarios: Combine future climate projections with reintroduced species (herbivores, predators, pollinators).

Criterias For Anthropocene Urban Design Working	Pressures on Natural Systems	Reliance on Natural Services	Implementations From Co Xisting Tektons
Spatial Expansion Dynamics	Encroachment and habitat fragmentation from uncontrolled expansion.	Dependence on topography and terrain stability for sustainable development.	Innovative integration of ecological corridors reducing habitat loss and urban fragmentation.
Resource Cycles & Self-Sufficiency	Depletion of local resources (soil nutrients, water reserves).	Reliance on ecosystem services for replenishment of natural resources.	Implementation of localized regenerative cycles and resource banks minimizing external dependencies.
Urban Climate Footprint	Acceleration of local climate extremes (urban heat islands).	Utilization of natural climate-regulating mechanisms (winds, shade canopies)	Enhanced bioclimatic architecture reducing heat stress through integrated natural airflow systems.
Urban Pollutant Management	Urban sources contributing to cumulative ecological toxicity	Natural purification mechanisms (biotic filtration, waste assimilation)	Strategic establishment of micro-ecosystem hubs actively neutralizing pollutants at the source.
Ecological Resilience	Increased vulnerability due to invasive species and biodiversity loss.	Stability and disease control through robust biodiversity presence.	Active introduction of native species strengthening resilience against biological invasions
Bio-economic Symbiosis	Traditional economic models disconnecting urban growth from ecological health.	Ecosystem-generated livelihoods ensuring economic durability and ecological integration.	Economic diversification through regenerative industries closely linked to ecosystem functionality
Thermal Comfort & Biodiversity Enhancement	Thermal stress induced by poorly integrated natural cooling solutions.	Passive reliance on vegetation and water-based cooling solutions.	Advanced ecological cooling strategies including wildlife-assisted forest belts and aquatic systems
Sustainable Development Goals Integration	Low-level sustainability integration, relying on indirect solutions.	Active alignment with ecological-driven SDGs promoting holistic urban health.	Comprehensive direct engagement across multiple SDGs (nature-based jobs, integrated ecological agriculture)
Adaptable Ecosystem Design	Rigid urban frameworks limiting adaptation and scalability.	Dependence on flexible natural patterns and ecosystem adaptability.	Ecologically responsive designs allowing seamless adaptability to multiple geographic contexts

Fig. 9.1 (Ref—Javier García Olivares, Cristina Gómez Garcia-Reyes, Nicole Merlo, Switzerland, Nature Positive: Guidelines for the Transition in Cities, World Economic Forum, MAY 2024, 41 p.)

Architectural Forms: Minimize waste by localizing form creation using regional materials and site-specific needs.

Phase 3: Interdisciplinary Input for the final output of Co X T

Study of Zoo Typology and Architectural benchmarks: Travel to top conservation parks and zoos, visit leading studio buildings with Avant Garde architectural solutions.

Expert Inputs: Unreal Engine developer to make a configurator, Biologist, Business expert, economist, sustainability consultant, etc.

Findings

In Ecology principles competition is the driver of evolution; the need to be bigger, stronger, and better inevitably leads to adaptations that give a species an edge. When two species compete for identical resources, the one which has any advantage will outdo the other. As a result, the weaker of the two species becomes extinct. Co-Xisting Tektons Solution Mitigate this by a new framework of development for bringing back the stability of the planet using Optimization methods for developing Architecture shown in Fig. 9.2.

Conclusion

To restore planetary balance, we must adopt a Half-Earth approach allocating 50% of land for nature and 50% for humans rewriting development codes and masterplans to meet the 2030 target and keep warming below 1.5 °C. This vision merges Pleistocene rewilding with modern science, reintroducing extinct mega-flora and fauna alongside Holocene species to rebuild dense, carbon-efficient, and biodiverse ecosystems.

A hybrid master planning model between urban centers would rewild native habitats, creating a "pullback effect" toward Holocene–Pleistocene potency where global temperatures ranged from 0.4 °C to 1 °C countering today's 1.5–2 °C Anthropocene crisis. Starting with developing nations like South Africa or Pakistan, India, supported by visionary leaders such as the UAE, this bold initiative could become a global "Olympic Torch" for climate regeneration.

By creating low-toxicity, high-nutrient ecosystems, we can restore water cycles, boost food security, and transform cities into bio-synergistic systems where architecture and ecology co- evolve securing a resilient, self-sufficient future for all life (DK, 2019; Secretariat of the Convention on Biological Diversity, 2020).

Below is a 5-step category framework to achieve the impossible task of stabilizing our planet based on planetary science (Johan Rockstrom) needs.

1. Nature-Integrated Tectonics and Bio-Geometric Approaches: Toward Terraforming Futures.

Design built environments that directly incorporate flora, fauna, and local geological and Climatic conditions moving away from static, human-only buildings toward forms that are actively shaped by (and supportive of) ecological processes. (Please note that the design methodology may sound like biomimicry which is a micro level target, but Co X T is a macro level Bio Synergetic System (Oxman, 2010)).

For the purposes of achieving this, this study introduces,—'Bio-Behavioral Mechanism' in Architecture for Co Xisting Tektons based on Umwelt Terminology (Every species "reads" geometry, light, texture, sound differently) by Jakob von Uexküll.

Species rely on instinctual visual cues for survival, navigation, and habitat selection, yet human- built environments disrupt these cues, fragmenting landscapes and hindering migration. Architectural tectonics can counteract this by merging the

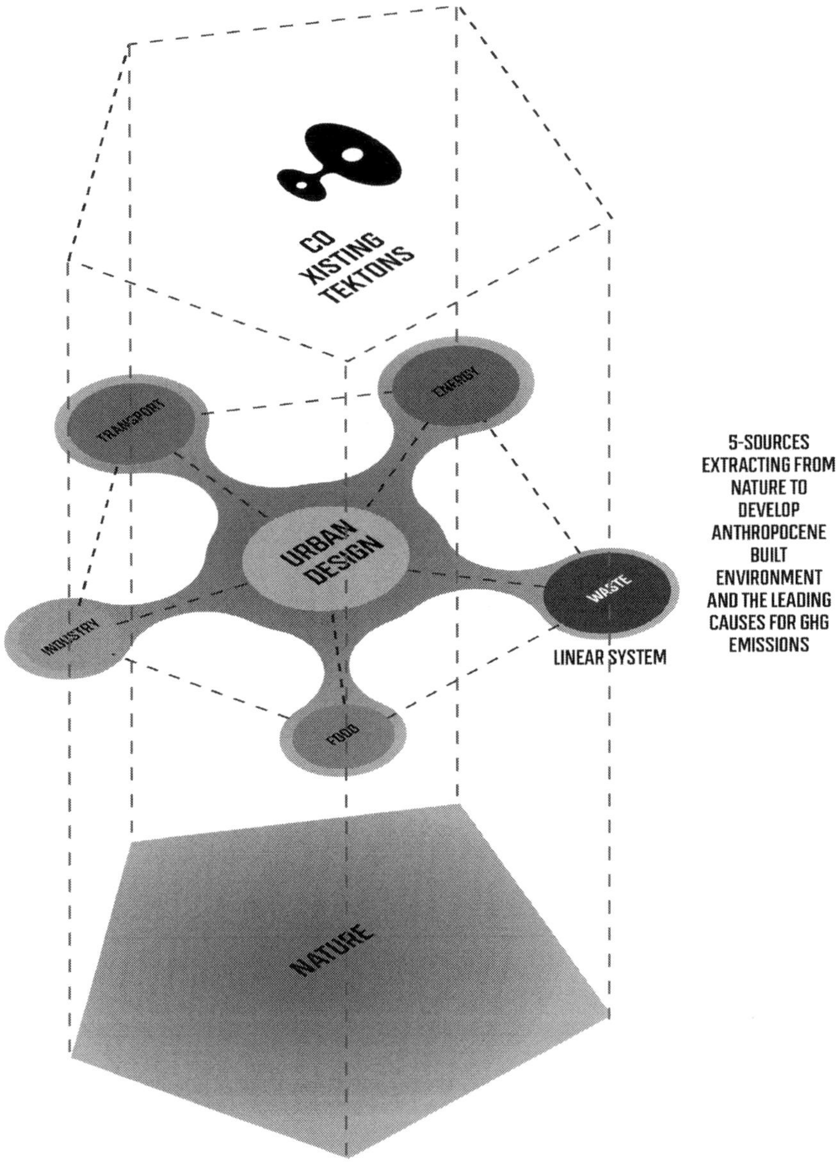

Fig. 9.2 Conceptual model of resource extraction from nature into the Anthropocene built environment, showing five key sectors (transport, energy, industry, food, waste) as leading drivers of GHG emissions

morphological language of nature, integrating forms, materials, and controlled visual connections to guide species through adaptive habitats.

Every architectural form ever designed can be repurposed for region-specific needs, ensuring the vast catalogue of geometries from past and present designers is applied to innovative, nature- positive solutions. Inspired by the "valley effect," urban structures can mimic forest tectonics curved, staggered, or layered forms resembling tree canopies and geological contours facilitating safe passage and habitat continuity aligned with the genomic memory of local fauna (AbdUllah et al., 2013).

Material selection should use biomimetic textures, colors, and reflectivity to blend with natural surroundings, while smart lighting and dynamic shading control visual permeability. Bio- responsive architecture can form functional ecotones zones where built and natural systems meet preserving instincts and biodiversity.

The Coexisting Tektons vision redefines urbanization as an extension of nature, incorporating tech-enabled habitat boundaries, shared spaces like public aviaries, and Holocene-inspired biodiversity hotspots. Access is limited, but visual connection is encouraged, fostering parallel human–fauna thriving and stimulating economies through nature-based innovation.

Key Points & References:

- Form Follows Tectonics (paradigm shift to developing box morphologies of buildings): Adapt geometry to local ecosystems (CoxT model) and reference parametric strategies (e.g., Patrick Schumacher progress of geometry) to optimize material usage and ecological performance.
- Biomorphic Networks: Use computational tools (Grasshopper, Rhino, Maya, Houdini) to generate forms that mimic or complement natural patterns based on Bio Behavioral Mechanism (e.g., Patrick Schumacher Parametric Patterns).
- Holocene vs. Anthropocene: Show the contrast between stable Holocene "tectonics" and unstable Anthropocene footprints, aiming for a "Neo Holocene" balance.
- Prehistoric + Future Forms: Factor in how the Earth might have evolved without human interference by using AI or advanced simulations to explore "what if" forms and integrate them into contemporary building design under some additional parameters considering the importance of nature's beauty for healthy human living.

2. Socio-Economic Dimensions of the Circular Economy, Resource Management, and Social Business.

Core Idea:

Shift from a linear "extract-consume-dispose" economy to holistic, closed-loop systems where resource use (including biodiversity) is valued, tracked, and replenished, benefiting both nature and local communities.

Key Points & References:

- Earth Overshoot & Resource Distribution: Address the planet's limited carrying capacity; design systems that provide for humans and wildlife.

- Multi-Species Economy: Include animals and plants as contributors, not just passive elements everything has a price tag to understand the value of that non living or a living resource of our planet (inspired by "organic robots" in Horizon Zero Dawn and Ralph Chami Nature Valuation framework.
- Social Business Models: Explore frameworks from Mohammed Yunus (Grameen Bank) and Ralph Chami (nature-backed financial instruments) to fund biodiversity conservation (Yunus, 2017). A highly effective methodology to transform our economies and support the profitably through coexisting tektons development standards and stop the trend of billionaire and wealth concentration as our planet's resources are right for every species born on this blue marble.
- One District, One Product policy (ODOP): Inspired by Uttar Pradesh's *One District One Product* policy, this anti-globalization model promotes hyper-local, sustainable industries, protecting MSMEs from large-scale corporate takeover. It fosters regional self-sufficiency, reduces carbon footprints, and ensures experts not distant capitalists benefit from local resources, preventing exploitation and enabling equitable growth for every rightful steward of Earth's assets, & it also adds the layers of tectonic preservation just like how we preserve our heritage building as a UNESCO backed conservation similarly a policy targeting regional biodiversity should be pushed.
- Life Cycle Analysis (LCA): Evaluate building materials' full carbon impact (embodied + operational) and require regenerative offsets (Secretariat of the Convention on Biological Diversity, 2020).

3. Cultural Preservation and Community Empowerment in Mixed Neighborhoods.

Honor and integrate local traditions, crafts, and knowledge systems into modern development, rather than allowing globalization to homogenize cultures, architecture, and ecosystems.

Key Points & References:

- Local Regulations & Indigenous Knowledge: Work within regional sustainability standards, reviving ancestral building techniques for resilience (e.g., using local materials) (DK, 2019).
- Tourism & Identity: Shift from profit-driven icon building (Burj Khalifa model) to culturally anchored landmarks that highlight heritage and biodiversity (e.g., Guggenheim effect reinterpreted as nature resilient forms not just a green building or bio enhanced facade systems).
- Community-Focused Infrastructure: Redesign highways, roads, and other public works to respect migratory routes and local ecosystems, transforming them into ecological corridors.
- Education & Social Reform: Embed learning about coexistence (with wildlife, plants, and culture) in schools and daily life, so future generations grow up valuing ecological interactions.

4. Advanced Technologies and Data-Driven Solutions for Geographic and Geometric Reconstruction.

Core Idea:

Leverage AI, robotics, and advanced digital tools not just for human convenience but to simulate, restore, and protect natural systems at every scale, from city blocks to entire bioregions.

Key Points & References:

- AI Modeling & Forecasting: Predict ecological outcomes under various development scenarios (e.g., "what if humans never altered Holocene ecosystems?").
- Robotics & Automation: Drone towers for resource extraction, AI-driven data analysis for biodiversity (link to IUCN metrics, Ramboll's biodiversity calculations).
- Parametric & Generative Design: Use algorithmic planning (Bartlett UCL Unit 14 research, AADRL research, etc) to create "living designs" that adapt to ecosystem feedback loops (Oxman, 2010; AbdUllah et al., 2013).
- Neuralink & Biomimicry: Speculative integration of tech into animals or flora to enhance symbiosis with human-built environments.

5. Holistic Ecosystem Restoration and Future Vision: Interconnectedness as a Core Principle.

Envision large-scale rewilding and species reintroduction, where humans live in solar-punk, eco-integrated habitats alongside prehistoric and modern wildlife. This "biological renaissance" rewrites coexistence with apex predators, blending education, architecture, and blurred human–nature boundaries, fostering interdependent, resilient communities for present and future generations to thrive in harmony with Earth's life systems.

Key Points & References:

- Rewilding & Prehistoric Species: Explore the revival of extinct fauna (e.g., woolly mammoth) and flora for enhanced biodiversity, guided by CRISPR and EDNA research (Col0ssal Company).
- Permaculture & Food Security: Emphasize integrated, community-led approaches to food production (forests, farmland, horticulture) that also feed local wildlife.
- Neo Holocene Mindset: Champion a global shift back to an epoch of ecological equilibrium, merging Holocene stability with Anthropocene tech to avert an "Eremocene" collapse (David Attenborough, Johan Rockström references).
- Global Collaboration: Create a governance network for "co-existing tektons" where countries share best practices, unify around the SDGs, and measure success by ecosystem health (Half-Earth Theory, Edward O. Wilson).

How These Categories Work Together, by involving cross disciplinary Experts with Architects and giving key Governmental and policy making positions to Architects to enable such a vision, marketing Architecture as technologist & promoting them like usual brand and products which can terraform our planet and also contribute to well being of our planet where people are investors into this grand future.

1. Nature-Integrated Tectonics forms the design backbone, ensuring buildings themselves respond to ecology and adds to planet, people and profit.

2. Circular Economy & Resource Management fuels the regenerative cycle of materials, finances, and community benefits.
3. Cultural Preservation & Community Empowerment grounds every project in the unique identity of its locale.
4. Advanced Technologies serve as enablers, gathering data and automating processes that restore and protect ecosystems.
5. Holistic Ecosystem Restoration & Future Vision unifies these efforts into a coherent, planet-wide shift toward multi-species coexistence and ecological resilience.

In the form of Deliverables after Research approval with a timeline to make the deliverables.

- Draft Policy Recommendations

Develop frameworks for urban zoning, wildlife credits, nature-backed lending, and educational reforms, beginning with the UAE (noted for experimental, forward-thinking projects).

- Configurator & Booklet

Create a digital twin for township models in both developing (South Africa) and developed (UAE) contexts, illustrating how the Co-Existing Tektons approach delivers superior economic and social benefits compared to conventional Fordist methods. Ultimately, transition into an actual developer role to implement this model.

- Exhibitions & Publications

Present and network at events like TED, COP3n, Future Summits, and global exhibitions (especially in the UAE) to share findings and garner support.

- University & Student Engagement

Act as visiting faculty, showcasing Co-Existing Tektons to inspire future generations of planners and architects to create healthier, more livable habitats for both people and wildlife.

References

AbdUllah, A. R., Said, I. B., & Ossen, D. R. (2013). Zaha Hadid: Form-making strategies for design (Unpublished doctoral thesis). Universiti Teknologi Malaysia. https://www.researchgate.net/publication/283347344_Zaha_Hadid_Form_Making_Strategies_For_Design

AbdUllah, A. R., Said, I. B., & Ossen, D. R. (2013). Zaha Hadid's techniques of architectural form-making. *Open Journal of Architectural Design, 1*(1), 1–9. https://doi.org/10.12966/ojad.11.01.2013 (available via ResearchGate)

DK. (2019). The Ecology Book: Big Ideas simply explained. DK. https://www.dk.com/us/book/9780241350386-the-ecology-book/

Oxman, N. (2010). Material-based design computation (Doctoral thesis, Massachusetts Institute of Technology). https://dspace.mit.edu/handle/1721.1/59192

Running, S. W. (n.d.). Global net primary productivity studies. University of Montana. https://www.umt.edu/numerical-terradynamic-simulation-group/project/modis/mod17.php

Secretariat of the Convention on Biological Diversity. (2020). *Global Biodiversity Outlook 5*. Montreal. https://www.cbd.int/gbo/gbo5/publication/gbo-5-en.pdf

Wilson, E. O. (2016). Half-Earth: Our planet's fight for life. W. W. Norton & Company. https://wwnorton.com/books/9781631492525

World Database on Protected Areas (WDPA)—UN Environment Programme & International Union for Conservation of Nature.

World Wide Fund for Nature. (2022). Living Planet Report 2022—Building a nature-positive society. In R. E. A. Almond, M. Grooten, D. Juffe Bignoli, & T. Petersen (Eds.), WWF. https://wwflpr.awsassets.panda.org/downloads/lpr_2022full_report_1.pdf

World Wildlife Fund. (2024, October). Living Planet Report 2024: A system in peril. WWF. https://www.worldwildlife.org/publications/2024-living-planet-report

Yunus, M. (2017). A world of three zeros: The new economics of zero poverty, zero unemployment, and zero net carbon emissions. PublicAffairs. https://www.publicaffairsbooks.com/titles/muhammad-yunus/a-world-of-three-zeros/9781610397582/

Chapter 10
Climate Action and Environment Stewardship

Priya Gunjan

Abstract The world is already witnessing the effects of climate change through rising average temperatures, shifting seasonal patterns, more frequent extreme weather events, and other slow onset environmental changes. These evolving climate conditions are affecting economies, ecosystems and communities across the globe. Since the adoption of the Paris Agreement and the release of the IPCC Special Report on Global Warming of 1.5°C, a growing number of countries have committed to net zero emissions targets (Intergovernmental Panel on Climate Change. (2018). Global Warming of 1.5 °C—An IPCC Special Report. https://www.ipcc.ch/sr15/). In the past 30 years we have emitted more than 50% of the total carbon-dioxide (CO2) humanity that has put into the atmosphere (Friedlingstein et al., 2022). Global warming is a global challenge, requiring a collaborative approach across nations to reach climate targets. Climate change is perhaps the world's most pressing concern today, with ramifications for both human and environmental systems and the potential for large changes in resource consumption, production, and economic activity. As one of the most critical issues of our time, climate change threatens both natural ecosystems and human systems, with the potential to significantly disrupt patterns of resource consumption, industrial production, and economic growth. As a result, worldwide, regional, national, and local measures to control Greenhouse Gas concentrations in the Earth's atmosphere are being devised and implemented. To combat the escalating impacts of climate change, comprehensive climate action and responsible environmental stewardship are essential. Countries and communities must develop robust adaptation strategies and implement actions that address both current and future climate risks. This research paper explores the importance of climate action in mitigating greenhouse gas emissions, the role of environmental stewardship in safeguarding ecosystems, and the strategies required to establish sustainable development pathways. However, greater ambition and coordinated efforts are needed to cost-effectively manage environmental risks in the long term. Achieving these goals requires active participation from businesses, governments, and individuals, all of whom must take accountability for their carbon footprints and adopt sustainable practices to minimize environmental degradation. In the United Arab Emirates, national

P. Gunjan (✉)
Jayam SCM Consultants, TUV SUD, Chennai, India
e-mail: priya.gunja@gmail.com

J. Dsilva (ed.), *Sustainability Education, Youth and Innovation*,
https://doi.org/10.1007/978-3-032-21948-0_10

frameworks such as the UAE Green Agenda 2030 and the National Climate Change Plan 2017–2050 promote sustainability-driven economic growth, while initiatives like the Abu Dhabi Sustainable Finance Declaration encourage green investments and climate-resilient financial systems (MOCCAE, 2017; ADSFD, 2019).

Keywords Climate change · Climate action · Environment stewardship · Net zero · Case studies on climate action initiatives · Challenges and barriers · Conclusion and future outlook

Introduction

Key Climate Actions in the UAE

UAE'S Net Zero 2050 Initiative and Global Climate Leadership: The United Arab Emirates became the first country in the Middle East to announce a Net zero by 2050 strategic initiative demonstrating its commitment to reduce Greenhouse gas emissions and transitioning towards clean energy. Further strenghtening its position in in global climate action, the UAE hosted COP28 in the year 2023, highlighting its active role in international climate diplomacy, sustainable leadership, circular economy advancement and carbon reduction initiatives.

In January 2024, just weeks after COP28, the UAE submitted its first *long-term strategy* (LTS) to the UNFCCC, including its 2050 net zero target. The LTS officially communicates longer-term targets beyond 2030. In 2021 the UAE had already announced its intention to reach net zero by 2050, as part of the *UAE Net Zero 2050 Strategic Initiative*. The government has enshrined the target into law, provides a transparent assumption on the use of carbon dioxide removal in the target year, and establishes a legally binding review process of the target in five-year intervals.

Climate change poses severe risks to global ecosystems, economies, and communities. Human-induced climate change, including more frequent and intense extreme events, has caused widespread adverse impacts and related losses and damages to nature and people beyond natural climate variability. Greenhouse gas (GHG) emissions refer to the release of gases into Earth's atmosphere that trap heat and contribute to global warming and climate change. The primary greenhouse gases include Carbon dioxide, Methane, Nitrous oxide, CFCs etc. With the rising greenhouse gas (GHG) emissions and increasing environmental degradation, the need for proactive climate action and environmental stewardship has become crucial. Some development and adaptation efforts have reduced vulnerability, but the rise in weather and climate extremes has led to some irreversible impacts as natural and human systems are pushed beyond their ability to adapt. The objective of this updated technical paper is to compile information on key mitigation policies, practices and technologies to promote the principles, strategies, and policies essential for achieving a sustainable future.

Understanding Climate Action: Across the world, companies are implementing a wide range of measures to address climate change and fulfill their commitments under international frameworks such as United Nations Framework Convention on Climate Change (UNFCCC), the Kyoto Protocol and the Paris Agreement. Climate action today extends beyond environmental responsibility, it has become a global priority for achieving sustainable economic growth, energy security and long-term climate resilience. Climate action involves reducing Greenhouse gas emissions, lowering carbon footprint, promoting renewable and clean energy solutions. International agreements particularly the Paris Agreement have emphasized the importance of limiting the rise in global temepartes to well below 2 °C while pursuing efforts to restrict warming to 1.5 degrees centigrade above pre-industrial levels. The role of governments, corporations, and individuals in mitigating climate change through carbon neutrality, sustainable practices, and policy advocacy is analyzed in this section. The transition to clean and green technology is not just about energy—it's about transforming economies, improving health, and fostering sustainable development. Emerging solutions such as carbon capture & storage (CCS) hydrogen energy and renewable energy integration are gaining momentum as key pathways for reducing emissions and and accelerating the global energy transition. The UAE Green Agenda 2030 and National Climate Change Plan reflect the nation's commitment to sustainability-driven economic growth, climate resilience and environment stewardship.

There is a direct relation between global average temperatures and the concentration of greenhouse gases in the atmosphere, the key solution to the climate change problem rests in decreasing the amount of emissions released into the atmosphere and in reducing the current concentration of carbon dioxide (CO_2) by enhancing sinks and the effort to reduce the emissions and sinks can be done by adopting various measures. Taking everything into account, controlling GHG emissions at the micro level is an effective way to mitigate climate change. With regard to the impact of global/climate change, nations throughout the world are becoming more conscious of the risks involved, and governments and other organizations are implementing preventative steps through proper policy interventions as well as established aims and objectives. Companies began to monitor, quantify, and report their GHG emissions.

UAE's Climate Mitigation Measures and Net Zero Commitments: As part of its climate mitigation efforts, United Arab Emirates UAE has identified several opportunities to further improve the scope, target architecture and transparency of its net zero target. The UAE government has expanded the coverage of its climate commitments by including emissions from fluorinated gases (F-gases), international aviation, and international shipping under within its broader emissions reduction framework. The UAE's net zero ambitions are intended to contribute toward the global effort to limit temperature rise to 1.5 °C above pre-industrial levels as outlined under the Paris Agreement. Enhancing transparency and progress monitoring will further support the effectiveness of the country's long-term climate goals while helping address any existing gaps between current emissions pathways and science-based targets. At the same time, global energy transition pathways continue to emphasize the reducing dependence on fossil-fuel. According to the International Energy Agency {IEA (2023)} to achieve net zero initiative by 2050, there should be no new investments in oil and gas projects under a net zero scenario.

Through an increased use of renewable energy, the application of new technologies such as electric cars, or changes in practices or behaviours, such as driving less or changing one's diet. Further, including expanding of forests and other sinks to remove greater amounts of CO2 from the atmosphere or simply making improvements to cookstoves for cooking.

- Establishing a baseline for Greenhouse gas emission is the first critical step for achieving Climate action.
- The baseline serves as a starting point for tracking progress and setting achievable targets for reducing the impact.
- A clear and defined baseline will demonstrate transparency and credibility to the stakeholders, showing dedication towards addressing the environmental concerns. It also reinforces credibility in the eyes of regulators, investors, and the public. It reflects a tangible commitment to environmental responsibility and creates a foundation for long-term accountability.
- To achieve meaningful climate goals, it is essential to define specific, measurable objectives to determine the scope of the measurements and the accuracy of the data.
- Larger companies should prioritize their scope 3 emissions (CDP, 2022)
- Low-carbon transport continues to be a crucial area of focus in supporting scaled up mitigation action.
- By preventing deforestation, promoting reforestation, or using sustainable agricultural practices, these projects can help reduce emissions from land use changes.
- Nature based projects that restore, or conserve ecosystems (like forests, peatlands, mangroves, and wetlands) can capture and store large amounts of carbon dioxide (CO_2) from the atmosphere. This helps reduce the concentration of greenhouse gases and combat global warming.

Environmental Stewardship and Its Importance: For the Environmental stewardship UAE government has done mangrove plantation which absorbs CO_2. Marine and desert conservation projects are implemented to conserve biodiversity. And it refers to the responsible use and protection of natural resources through conservation, restoration, and sustainable management. Water conservation and sustainable agricultural practices are adopted in the UAE region. This section explores how organizations and individuals can contribute to biodiversity conservation, pollution reduction, and eco-friendly consumption patterns. Initiatives like carbon capture & storage (CCS) and hydrogen energy development aim to cut emissions. By adopting carbon management principles such as carbon offsetting, renewable energy practices and net zero strategies. Canada is leading and supporting low-carbon development action at the national and subnational levels. However, it is supporting diverse approaches to enable carbon pricing based on unique economic, policy and climate objectives and targets in various provinces.

Key Strategies for Climate Action and Stewardship: Countries often have their own regulations to control GHG emissions, such as carbon taxes, cap-and-trade systems, and emission standards for vehicles and industries.

One of the most prominent features of the Act is the creation of a carbon market, a mechanism enabling businesses to trade carbon credit. These credits represent quantifiable reductions in greenhouse gas emissions, thus incentivizing emission reductions and offsetting environmental impacts.

Various strategies defined in reducing GHG emissions and achieving net zero are defined below.

Transitioning to renewable energy sources: Masdar (Abu Dhabi's clean energy company) leads global renewable energy projects. Renewable Energy policies are fundamental in supporting both mitigation and adaptation goals. As a notable example, distributed RE can support adaptive capacity by increasing access to energy and strengthening the resilience of the overall electricity system to potential climate impacts and natural hazards. Renewable energy offers a sustainable solution to combat climate change by reducing carbon emissions.

- Countries are increasingly undertaking EE measures with the intention of speeding up technology deployment.
- Technology transfer is particularly vital in decarbonizing the industry sector, though it poses challenges in engaging all stakeholders effectively.
- The UAE is adopting a circular economy model for reducing waste and promoting recycling.
- Sustainable urban planning and green infrastructure
- Raising awareness on climate risks, green practices, and responsible consumption.
- Strengthening climate policies and regulations
- Enhancing corporate sustainability through ESG (Environmental, Social, Governance) frameworks.
- Science-based targets allow companies to work towards an overall goal by aligning corporate greenhouse gas emission reductions with global emission budgets generated by climate models. The initiative, a partnership of the Carbon Disclosure Project, the United Nations Global Compact, the World Resources Institute and the World Wildlife Fund, has worked to develop methodologies and tools to support companies in setting and working towards such targets.
- Implementing, monitoring and improving low-carbon transport implementation plans, including stakeholder engagement processes, roles and responsibilities for various agencies, funding sources for policy and technology actions, timelines, outreach and communication strategies, and monitoring, evaluating and reporting framework.
- Assessing the market and developing an electric vehicle (EV) deployment strategy including near- and long-term target markets.

- Carbon Capture and Storage technology is a promising technology for achieving net-zero emissions but requires further investment and innovation to become widely adopted.
- Closed loop systems, repairs and refurbishments can reduce waste and promote circular economy.
- Improving energy efficiency, implementing smart manufacturing practices, and encouraging partners to adopt sustainable practices

Case Studies on Successful Climate Initiatives: This section presents examples of countries and organizations that have successfully implemented climate action and environmental stewardship practices. Examples include:

- The Mohammed bin Rashid Al Maktoum Solar Park in Dubai aims to generate 5,000 MW by 2030.
- The Barakah Nuclear Plant provides clean electricity, reducing dependence on fossil fuels.
- Denmark's transition to renewable energy
- The European Union's Green Deal
- The Amazon Rainforest restoration projects
- Corporate carbon neutrality goals from leading global companies
- Bhutan—The World's First Carbon-Negative Country having balance economic growth with environmental sustainability.
- Walmart has made ambitious commitments to environmental sustainability, including reaching zero operational emissions (Scopes 1 and 2) by 2040 and achieving 100% renewable energy usage across its global operations. The company also focuses on sustainable agriculture, waste reduction, and energy efficiency within its operations. Walmart's sustainability strategy is grounded in reducing environmental impact while driving positive social outcomes, emphasizing both environmental and community resilience.
- Microsoft company will be carbon negative by 2030 which is shifting to 100% renewable energy by 2025.
- Xiaomi's Zero-carbon Philosophy has led it to make its products and services more efficient, economical, and environmentally clean. It is also reducing household greenhouse gas emissions and energy consumption by using smart-home solutions.
- Indigenous climate communication is a matter of justice and human rights. To close the communication gaps between people from different cultural backgrounds, language must be leveraged to provide access to innovative ideas for indigenous populations. The case studies from Bolivia and Nigeria presents some central themes for climate change stakeholders seeking to co-create climate-related solutions.
- Cultural Knowledge and Traditional Wisdom: Indigenous communities have developed intricate relationships with their environments over centuries. Their traditional knowledge and wisdom encompass a deep understanding of the local

ecosystems, weather patterns, and natural resources. Indigenous climate communication can help preserve and promote this invaluable wisdom, ensuring it is integrated into climate change mitigation and adaptation strategies.
- By amplifying indigenous voices, global climate communication will contain the perspectives and needs of marginalized communities. Climate change often disproportionately affects indigenous peoples due to their limited access to resources and opportunities. Indigenous-led climate communication can foster community engagement and empower local communities to participate in climate action actively.

Challenges and Barriers

Despite growing awareness and initiatives, several challenges hinder climate action and environmental stewardship. Especially in the UAE due to high dependence on fossils fuels for the economic growth, they make transitioning to clean energy complex. The UAE produces high amounts of waste, with low recycling rates and landfill dependency leads to methane emissions and environmental pollution. All these include financial constraints, political resistance, technological limitations, and societal behaviours. High initial investment costs for renewable energy, carbon reduction projects, and sustainable infrastructure limit large-scale adoption, especially in developing countries. Difficulty in tracking emissions, verifying carbon credits, and ensuring transparency weakens the credibility of climate action efforts. Resistance to adopting eco-friendly habits, misinformation, and lack of education slow down individual and community engagement. Some businesses prioritize short-term profits over sustainability, hence reducing real impact. Insufficient clean energy grids, limited access to sustainable materials, and reliance on fossil fuels make transitions difficult. Addressing these barriers requires global cooperation, innovation, and strong policy enforcement. The market price for carbon credits can fluctuate, affecting the project's profitability, future changes in carbon credit regulations or policies might impact the project's viability or returns.

Conclusion and Future Outlook

Climate action and environmental stewardship are critical for ensuring a habitable planet for future generations. Despite these barriers, the UAE is actively working on climate action through initiatives like the UAE Net Zero by 2050 strategy, Masdar (clean energy projects), and COP28 leadership. However, overcoming these challenges requires stronger policies, technological innovation, and public participation. Various initiatives and voluntary programs are implemented all over the world Climate action will become more effective by integrating indigenous knowledge, adapting to local contexts, promoting sustainable practices, empowering communities, and fostering intergenerational learning. Indigenous climate communication is vital in shaping our collective response to the climate crisis using equitable climate solutions. Stronger policies, increased climate finance, innovation in green technology, stricter accountability measures can only overcome the barrier in climate action stewardship. Governments, businesses, and individuals must collaborate to implement sustainable practices and policies. Public awareness and lifestyle shift toward eco-friendly practices further will strengthen these efforts. Climate action is a necessity towards mitigating climate change and ensuring a sustainable future. Access to clean energy can improve education, healthcare, and economic opportunities in these communities. We can protect our ecosystem by using it wisely and enhance economic resilience. The paper concludes with recommendations for integrating climate resilience into economic and social development plan. Align your goals with UN SDGs and Paris commitment, commit to net zero emissions, circularity and biodiversity by the targeted year. Through knowledge on sustainable practices and carbon reduction strategies only we can achieve our climate goals. Revamping our approach to create a positive impact using climate technology and best practices.

References

CDP. (2022). *Engaging the chain: Driving speed and scale.* https://www.cdp.net

Government of the United Arab Emirates. (2024). *UAE Net Zero by 2050 Strategic Initiative.* UAE Ministry of Climate Change and Environment. https://www.moccae.gov.ae/

Government of the United Arab Emirates. (2024). *Long-Term Strategy to the UNFCCC.* https://unfccc.int/

https://unfccc.int/process-and-meetings/conferences/un-climate-change-conference-united-arab-emirates-nov/dec-2023/about-cop-28?utm_source

Intergovernmental Panel on Climate Change. (2018). *Global Warming of 1.5°C – An IPCC Special Report.* https://www.ipcc.ch/sr15/

International Energy Agency. (2023). *Net Zero by 2050: A Roadmap for the Global Energy Sector.* https://www.iea.org/reports/net-zero-by-2050

International Energy Agency. (2023). *Net Zero by 2050: A Roadmap for the Global Energy Sector.* https://www.iea.org/reports/net-zero-by-2050

International Organization for Standardization. (2019). *ISO 14064-2:2019—Greenhouse gases—Part 2: Specification with guidance at the project level for quantification, monitoring and reporting of GHG emission reductions or removal enhancements.*

International Organization for Standardization. (2019). *ISO 14064-1:2018—Greenhouse gases—Part 1: Specification with guidance at the organization level for quantification and reporting of greenhouse gas emissions and removals.*

Microsoft Word—20230623—UAE Third Version of Second NDC_v15.docx.

UAE Energy Diversification. UAE Embassy in Washington, DC.

United Nations Development Programme. (2021). *Indigenous Peoples and Climate Change: Vulnerabilities, Adaptation, and Responses.* https://www.undp.org/

United Nations Framework Convention on Climate Change. (2015). *The Paris Agreement.* https://unfccc.int/process-and-meetings/the-paris-agreement/the-paris-agreement

Chapter 11
Green Agenda for Africa's Built Environment

Chinwe Ajene-Sagna, Michael Ivenso, and Emmanuel Falude

Abstract Africa's built environment is at a critical turning point as rapid population growth and urban expansion drive an unprecedented increase in construction activity. With the built environment sector being one of the largest contributors to greenhouse gas emissions and resource consumption, urgent action is needed to integrate sustainability into Africa's infrastructure development. This study highlights a "knowledge disconnect" between the scientific and economic communities, who recognize the urgency of the climate crisis and on-the-ground practitioners, who often prioritize immediate socio-economic concerns over long-term environmental resilience. To address this gap, the research advocates for radical leadership that can act as a bridge between policy frameworks and practical implementation. It proposes the establishment of a Green Policy Handbook to guide Africa's transition to sustainable infrastructure, incorporating green building certification standards, carbon markets, and tax incentives to drive innovation. Additionally, the study emphasizes the importance of capacity-building initiatives to educate stakeholders, promote sustainable construction practices, and mobilize grassroots engagement. A key component of the proposed approach is the strategic partnership with development finance institutions to secure funding for large-scale green projects. Through policy-driven, multi- stakeholder collaboration, Africa has the potential to develop ahead of traditional development models, and develop climate-smart urban landscapes that align with global climate goals while nurturing economic growth and social equity. This research provides a clear roadmap for sustainable urbanization, ensuring Africa's built environment is both financially and environmentally viable for future generations.

C. Ajene-Sagna (✉)
Adjunct-Faculty, Strategy Department Lagos Business School, Pan-Atlantic University, Lagos, Nigeria
e-mail: cajene-sagna@lbs.edu.ng

M. Ivenso
Director of Energy, Transportation and Infrastructure, National Council on Climate Change, Abuja, Nigeria

E. Falude
Director, Green Building Council of Nigeria, Abuja, Nigeria

J. Dsilva (ed.), *Sustainability Education, Youth and Innovation*,
https://doi.org/10.1007/978-3-032-21948-0_11

Keywords Climate change · Green policy · Sustainable development · Africa · Green building

Introduction

The world is currently experiencing an unprecedented era of poly-crisis, characterized by interconnected global calamities that require multifaceted and integrated solutions. At the heart of this interconnectedness sits the built environment, which is the supporting frame core to human existence, encompassing urban and rural communities, transportation networks, and centres of trade, commerce, education, religious, and recreational establishments.

In Africa, the built environment is amid an industrial and construction boom. The continent faces an infrastructure deficit of $130–170 billion annually, with a financing gap of $68–$108 billion (African Development Bank, 2022).

Driven by its demographic profile, Africa, the home to the youngest and fastest-growing population globally, is projected to host over 25% of the world's population by 2050.

At its current trajectory, this construction boom threatens to precipitate an environmental catastrophe. The global construction industry, the largest contributor of harmful greenhouse gases, consumes 40% of natural materials and is set to double by 2050.

Its environmental and health implications are extensive: respiratory diseases, air pollution, extreme weather, food supply disruptions, and increased wildfires. The African Development Bank reports that climate conditions already account for 5–15% of annual GDP losses, a trend expected to escalate without fundamental interventions.

A critical challenge is bridging the knowledge gap between scientific and economic communities and Africa's built environment ecosystem. While scientists and economists understand the urgency of the climate crisis, the built environment sector often prioritizes immediate socio- economic concerns, which this study addresses extensively.

Methodology

This study employs a multi-dimensional methodological approach, integrating both qualitative and quantitative research strategies to examine climate change challenges, policy responses, and opportunities for green innovation in Africa's built environment.

Philosophical and Theoretical Framework

The study is rooted in a transformative research paradigm, acknowledging the interconnectedness of social, economic, and environmental systems. It draws from:

- Critical Systems Theory, emphasizing interdisciplinary solutions to environmental issues.
- Sustainable Development Theory, promoting a balance between economic growth and environmental responsibility.
- Climate Resilience Frameworks, focusing on green building standards in Africa's urbanization.

Research Design

A mixed-methods research design was used to ensure a comprehensive, data-driven analysis of sustainability challenges and solutions. This includes:

1. Systematic Literature Review
   a. Academic research on climate change, sustainability, and construction.
   b. Grey literature from institutions like AfDB, IEA, UNPD, and Global Energy & CO_2 reports.
   c. Government policy analysis, including Nigeria's Climate Change Act (2021).
   d. Sectoral reports on climate resilience and green building innovations.
2. Qualitative Comparative Analysis (QCA)
   a. Case Studies of National Climate Change Policies, comparing Nigeria's Climate Change Act, Kenya's NCCAP, and South Africa's Green Building Policy.
   b. Stakeholder Mapping, analyzing perspectives from scientists, policymakers, financiers, and industry professionals.
3. Quantitative Data Analysis
   a. Greenhouse Gas (GHG) Emissions Modeling, focusing on Africa's construction sector and CO2 impact.
   b. Economic Impact Assessment, examining climate-induced GDP losses (5–15% annually) and infrastructure costs.
   c. Investment Projections, modeling green finance opportunities (e.g., carbon markets, green bonds, tax incentives).

Data Collection Methods

A combination of primary and secondary data sources ensures research depth:

- Semi-Structured Interviews with government officials, DFIs, and urban planning bodies on policy implementation and financing challenges.
- Policy Document Analysis, benchmarking Africa's green building policies against international standards(LEED, EDGE, IFC Greening Real Estate).
- Economic and Environmental Assessments, evaluating financial losses from climate change and cost-benefit comparisons of green vs. traditional buildings.

Analytical Approaches

The study employs:

- Thematic Content Analysis, identifying climate governance themes in interviews and policy documents.
- Critical Discourse Analysis, assessing how climate change is framed in African policies.
- Systems Thinking, mapping interdependencies between economic, environmental, and policy factors.
- Scenario Planning, forecasting sustainability trajectories under different policy and investment scenarios.

Findings

This research uncovers multi-dimensional insights that challenge conventional infrastructure and highlight the urgent need to integrate sustainability into Africa's built environment. These findings reveal both critical vulnerabilities and transformative opportunities, setting a path toward climate resilience, economic growth, and sustainable urbanization.

Demographic and Infrastructure Dynamics

Africa's population is expected to exceed 25% of the global total by 2050, requiring a massive infrastructure expansion to meet housing, transportation, and energy demands. However, current development models are failing to keep pace, making sustainability a crucial factor in future urban planning (United Nations Population Division, 2022).

Inadequacy of Traditional Infrastructure Models

Africa faces a $130–170 billion annual infrastructure deficit, with an additional financing gap of.

$68–108 billion. Traditional urban expansion models prioritize speed and cost-efficiency over sustainability, making cities vulnerable to climate-related disruptions.

Opportunity for Leapfrogging Sustainable Development

Unlike industrialized nations that must retrofit existing infrastructure, Africa can leapfrog high- carbon urbanization models by adopting climate-smart alternatives from the outset. Green construction technologies, energy-efficient materials, and sustainable urban planning can minimize environmental impact while driving economic growth. Investments in renewable energy, climate-resilient housing, and green public infrastructure can transform African cities into models of sustainability and resilience, reducing long-term costs and environmental risks (International Energy Agency, 2021; National Council on Climate Change Nigeria, 2022).

Climate Vulnerability Landscape

Despite contributing only 3.6% of global greenhouse gas (GHG) emissions, Africa bears a disproportionate share of climate-related consequences. The climate vulnerability index ranks nine out of the ten most at-risk countries globally within Africa. These vulnerabilities threaten economic stability, food security, and human settlements, amplifying existing socio-economic challenges.

Decreased Agricultural Productivity

Africa's reliance on rain-fed agriculture makes its food systems highly vulnerable to climate shifts. Rising temperatures, erratic rainfall, and soil degradation are already reducing crop yields, worsening food insecurity and increasing dependency on imports. Desertification and extreme weather events are shrinking arable land, driving rural-to-urban migration, which intensifies pressure on city infrastructure and social services.

Increased Frequency of Extreme Weather Events

Extreme weather events such as hurricanes, floods, and droughts are becoming more frequent and intense. Coastal cities like Lagos, Mombasa, and Accra face rising sea levels and storm surges, placing millions of lives and economic hubs at risk. The absence of climate-resilient infrastructure means that roads, buildings, and power grids are vulnerable to destruction, leading to escalating reconstruction costs. Lack of proactive adaptation strategies will continually undermine growth and longevity.

Disrupted Ecological Systems

Africa's ecosystems are under strain from rising temperatures, deforestation, and unsustainable urban expansion. Biodiversity loss, water scarcity, and land degradation threaten natural resilience, reducing the continent's ability to absorb carbon emissions. Rapid urbanization, often unregulated and poorly planned, has led to air pollution, inefficient waste management, and declining public health.

Escalating Economic Instability

Climate change already accounts for 5–15% of Africa's annual GDP losses, according to the African Development Bank (AfDB). Infrastructure damage, declining agricultural productivity, and energy inefficiencies strain national budgets and discourage foreign investment. Governments are forced to reallocate resources, often at the expense of climate-conscious economic planning, making these financial burdens escalate, deepening structural inequalities and socio-political instability (Ramanujam, 2024).

Knowledge Ecosystem Analysis

A significant knowledge gap exists between scientific climate research, economic policymaking, and real-world implementation in Africa's environment. This disconnect has slowed the adoption of green innovations, limited investment in sustainable solutions, and weakened public engagement with climate action.

Prioritization of Immediate Socio-Economic Challenges

Many governments and businesses prioritize short-term economic needs, such as infrastructure expansion, food security, and poverty alleviation over climate resilience. Climate action is frequently viewed as a luxury rather than a necessity, resulting in delayed policy implementation. However, this short-term focus overlooks the economic benefits of sustainability, such as job creation, long-term cost savings, and improved public health (Federal Republic of Nigeria 2021 and Federal Government Press Federal Republic of Nigeria, 2021).

Limited Localized Climate Data

Effective climate adaptation planning relies on accurate, region-specific data. However, Africa lacks localized climate models that can inform resilient urban planning, infrastructure investment, and disaster risk management. Many policymakers depend on outdated or global datasets that do not capture Africa's unique climate vulnerabilities (Climate Change Act, 2021).

Misconceptions About Climate Action Costs

One of the most significant barriers to sustainable development in Africa is the perception that green solutions are too expensive. Many governments and businesses focus on upfront costs, overlooking long-term financial benefits such as energy efficiency, operational savings, and increased property value. Green financing mechanisms like carbon credits, green bonds, and tax incentives are still underutilized (Global Energy and CO2 Status Report, 2022).

Perception of Climate Change as a Secondary Concern

In many African countries, climate change is viewed as a distant issue, often overshadowed by pressing economic and political challenges. However, failure to integrate climate resilience into national strategies will only deepen existing socio-economic crises.

Building a Climate-Resilient Future

The findings challenge traditional approaches to infrastructure development, making a strong case for integrating climate-smart solutions into Africa's built environment.

By closing the knowledge gap, mobilizing green financing, and mainstreaming climate adaptation strategies, African cities can unlock economic growth while ensuring environmental and social stability.

Conclusion

A key principle inferred from this study is that Africa can leapfrog outdated infrastructure models by adopting a strategic, policy-driven approach that integrates green building certification, tax incentives, and carbon markets. By doing so, the continent can drive economic transformation and mitigate environmental risks. However, the research also acknowledges several limitations such as; financial constraints, inconsistent policy enforcement, and the lack of localized climate data.

Theoretical implications lie in the contribution to sustainable urban planning discourse, emphasizing how Africa's development model can serve as a global benchmark for emerging economies. Practically, the research provides a roadmap for policymakers, urban planners, and investors, outlining how sustainable infrastructure can reduce climate risks while unlocking economic potential. To illustrate these frameworks, Exhibit. 11.1 presents visual documentation from the National Council on Climate Change (NCCC) handbook.

Exhibit. 11.1 NCOCC-Handbook

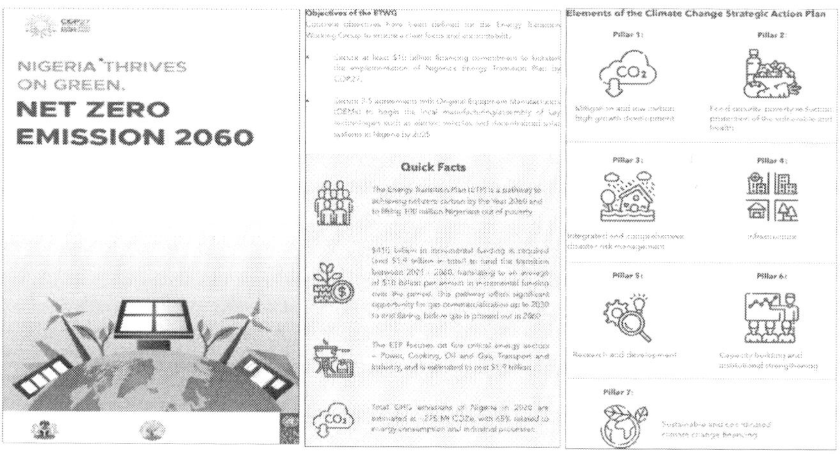

To ensure effective implementation of the proposed sustainability framework, this research recommends:

1. Governments must ensure compliance with green building codes, integrating sustainability into national development strategies.
2. Financial institutions should offer incentives for sustainable infrastructure projects, making climate-friendly investments more accessible.
3. Localized climate research should be prioritized to inform urban planning, resilience strategies, and infrastructure investments.
4. Businesses should be incentivized to adopt energy-efficient construction and invest in renewable technologies.
5. Educational initiatives should be expanded to train industry professionals, government officials, and local communities on sustainable urban planning practices.

References

African Development Bank. (2022). *Climate vulnerability index report 2022*. African Development Bank Group.

Federal Republic of Nigeria. (2021). *Climate change act, 2021*. Federal Government Press.

Global Energy and CO2 Status Report. (2022). *Buildings and energy consumption analysis*. International Energy Agency.

International Energy Agency. (2021). *Global status report for buildings and construction*. IEA Publications.

National Council on Climate Change Nigeria. (2022). *TNCOCC handbook*. Abuja. https://natccc.gov.ng/publications/TNCOCC-Handbook-Version-29-11-22.pdf

Ramanujam, M. (2024, November 14). Bringing cities back begins with buildings. *Forbes Business Council*. https://www.forbes.com/councils/forbesbusinesscouncil/2024/11/14/bringing-cities-. back-begins-with-buildings/

United Nations Population Division. (2022). *World population prospects 2022*. United Nations.

Chapter 12
Decarbonizing Urban Coastal Development with Floating and Adaptive Infrastructure

Rachel Zlatar

Abstract Coastal cities, which concentrate critical infrastructure and nearly 11% of the global population, face mounting threats from climate change. Conventional coastal development approaches have historically prioritized unsustainable methods and reactive defenses, such as land reclamation, seawalls, and dredging that are carbon-intensive and environmentally disruptive. These traditional methods often lead to permanent ecological footprints, including habitat loss, altered sediment dynamics, and high greenhouse gas emissions. In contrast, this extended abstract explores floating and adaptive infrastructure as a forward-thinking approach for decarbonizing urban coastal development. This study evaluates coastal infrastructure development by focusing on five key areas: land use and site selection, foundation and anchoring systems, construction materials and methods, carbon emissions during installation, and long-term maintenance and lifecycle impacts. By synthesizing findings from academic literature, policy analyses, technical reports, and diverse case studies, the study identifies innovative systems, low-carbon construction methods, and offshore applications that can help bridge the "coastal adaptation gap." This research demonstrates that floating and adaptive infrastructure with their innovative design, offer a promising pathway toward sustainable, low-carbon coastal cities that effectively balance environmental, economic, and socio-cultural needs.

Keywords Floating infrastructure · Smart coastal cities · Decarbonization · Blue economy · Climate resilience

Introduction

Coastal urbanization has concentrated populations, capital, and critical infrastructure in low-lying areas. The International Panel on Climate Change (IPCC) projects that global assets in 100 year floodplains could be valued between US$7.9 and US$12.7

R. Zlatar (✉)
The Ocean Edition, Dubai, UAE
e-mail: rachel@theoceanedition.com

J. Dsilva (ed.), *Sustainability Education, Youth and Innovation*,
https://doi.org/10.1007/978-3-032-21948-0_12

trillion under moderate scenarios, with even higher values under extreme conditions. Despite extensive adaptation planning since the IPCC's Sixth Assessment Report, many responses remain reactive—relying on static defenses such as seawalls and dredging (IPCC, 2021). These measures provide immediate flood protection but often disrupt sediment transport, cause coastal squeeze, and depend on energy-intensive materials (Aikins, 2024).

Massive investments in hard infrastructure are evident worldwide. Since 2000, cities have reclaimed 253,000 hectares of land (roughly the size of Luxembourg) driven by urban growth, rising sea levels, and the lure of high-profile projects (Barnard, 2024). Developments like Palm Jumeirah, The World Islands, Dubai Marina, and Bluewaters Island in the UAE and GCC exemplify this trend.

In contrast, floating and adaptive infrastructure offers a promising alternative. These flexible, modular solutions can integrate renewable energy, efficient water management, and sustainable construction practices. By reducing reliance on mass, high-impact construction methods, such approaches aim to decarbonize urban coastal development, better adapt to rising sea levels and extreme weather, and ultimately transform vulnerable urban areas into sustainable, low-carbon communities.

Methodology

The central research question is: How can floating and adaptive infrastructure contribute to decarbonizing urban development in coastal cities? This question is highly relevant both in theory and in practice, as coastal urbanization continues to accelerate while climate change intensifies environmental risks. Traditional coastal development methods, such as land reclamation, often involve substantial environmental disruption and high carbon emissions (Brady & Wallace, 2006). In contrast, floating infrastructure has emerged as a promising alternative that could mitigate these impacts while offering flexibility and adaptability. Addressing this research question contributes to both academic discourse on sustainable urban planning and practical decision-making for coastal development. To answer the research question, this study employs a mixed-methods approach that integrates a comprehensive review of academic papers, technical reports, policy documents, and case studies. This multifaceted framework is designed to capture both quantitative and qualitative dimensions of coastal infrastructure development. The evaluation focuses on five key areas: land use and site selection, foundation and anchoring, construction materials and methods, carbon emissions during installation, and long-term maintenance and lifecycle impacts. By integrating diverse sources of data and analyzing multiple dimensions of infrastructure development, the study provides a comprehensive assessment of both the economic and environmental viability of floating solutions. This, in turn, informs future planning and policy-making aimed at achieving sustainable, low-carbon coastal cities.

Findings

The analysis reveals that conventional coastal development methods, such as land reclamation, seawalls, and dredging incur significant environmental and carbon costs. To comprehensively assess these impacts, the evaluation examines five key areas: land use and site selection, foundation and anchoring systems, construction materials and methods, carbon emissions during installation, and long-term maintenance and lifecycle impacts.

Coastal zones are increasingly subject to erosion, which is primarily caused by human intervention. Instead of protecting coastlines, infrastructure like seawalls, breakwaters and roads disrupt natural sediment flows and accelerate beach loss (Kotsoni et al., 2017). Long-term monitoring has shown that in many regions, the rate at which coastal erosion strips away land far exceeds any natural or human-driven efforts to replenish it (Aikins, 2024). For those choosing to build along coastal sites with traditional solutions, they must commit to repeated and resource-heavy maintenance to counteract erosion as the persistent loss not only diminishes valuable coastal land and imperils infrastructure and properties, but also means that maintaining oceanfront developments will require ongoing, carbon-intensive restoration efforts. Additionally, the energy-intensive materials used in their construction (typically reinforced concrete and steel) lead to substantial carbon emissions, challenging the long-term sustainability of coastal developments (Aikins, 2024). Floating and adaptive solutions provide a sustainable alternative to static seawalls that disrupt natural sediment flow. They work alongside natural defenses, such as mangroves and oyster breakwaters which help blunt wave energy, filter water, and enhance biodiversity. As Danielle Bissett of the Billion Oyster Project states, "Sea level rise is inevitable. We have to allow the shoreline to adapt and be resilient rather than put up seawalls, which are obstructive," (Uteuova, 2022).

In response to these challenges, many regions have turned to land reclamation projects to expand usable coastal land. The same factors that make hard coastal defenses unsustainable like resource depletion, habitat loss, and escalating carbon emissions, also drive the need to rethink land reclamation. This has led to a growing interest in alternative approaches, such as floating solutions, which offer a more resource-efficient and environmentally friendly way to expand and protect coastal areas. Looking at material selection and use, conventional reclamation, offshore and construction projects traditionally rely on materials, such as cement and steel and involve intensive onsite activities, typically generating around 450 kg CO_2e/m^2, considering that cement production alone emits about 0.9 kg CO_2 per kg (Andrew, 2018). In contrast, floating systems use lower-carbon materials and offsite prefabrication, cutting emissions to approximately 350 kg CO_2e/m^2—a 20–25% reduction (Blue21, 2021). Notably, Japan's Environmental Island project, "Green Float," demonstrates an innovative solution by utilizing refined magnesium alloy. This alloy is produced by extracting magnesium—an abundantly available metal in seawater—and processing it into a lightweight, steel-like material that is robust and environmentally friendly for structural applications (Shimizu Corporation, n.d.).

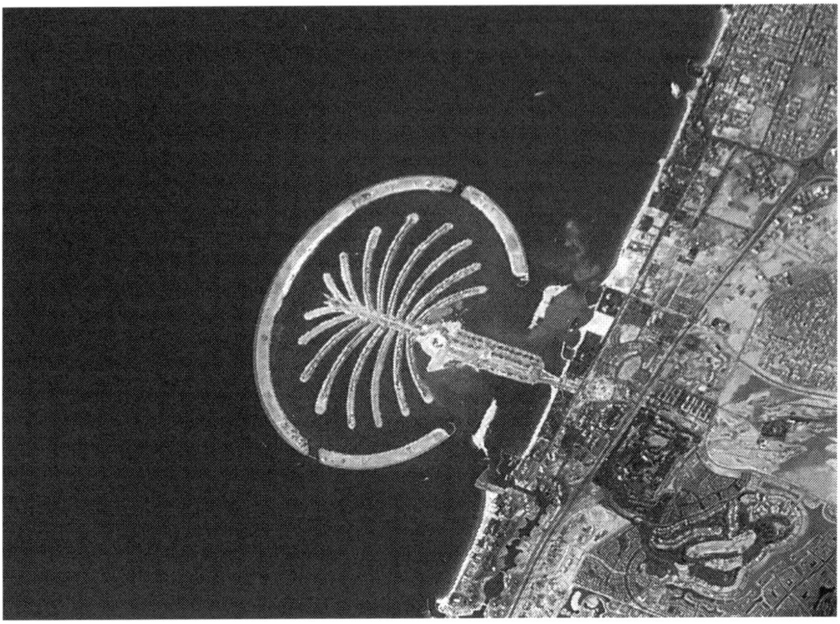

Fig. 12.1 Aerial view of palm Jumeirah (International Space Station, 2005)

A recent study showed the hallmark Palm Jumeirah Island's construction resulted in a striking 7.5 °C increase in water temperature and 0.5–0.8 μm increase in in water reflectance resulting from increased suspended sediments and chlorophyll (Mansourmoghaddam et al., 2022). A comparison conducted by HSB Marine demonstrates a comparison between conventional reclamation and emerging floating reclamation solution of the Palm Jumeirah in Dubai (referring to Fig. 12.1). Results comparing economic and environmental impacts as traditional land reclamation versus estimations as a floating project can be found in Table 12.1 (HSB Marine, 2023).

Conclusions

The research indicates that conventional coastal development methods, such as land reclamation, seawalls, and dredging, not only produce high carbon emissions and incur significant long-term maintenance costs but also exacerbate coastal erosion and disrupt natural sediment dynamics, as evidenced by projects like Palm Jumeirah. In contrast, floating and adaptive infrastructure offers a promising low-carbon alternative by reducing both installation emissions and lifecycle impacts while enhancing coastal resilience. These findings suggest that transitioning to floating solutions could

Table 12.1 Comparison of conventional reclamation and floating reclamation in palm Jumeirah project

	Conventional reclamation	Floating reclamation
Expanded land area	5.72 km^2	5.72 km^2
Amount of material used	120 million m^3 of sand	35 m^3 of concrete
Project cost	$12 billion USD	$8.4 billion USD
Time to completion	6 years	2.5 years
Environmental impact	–Increase of water temperature by 7.5 °C during the 19 years –Endangered marine life	–Because it is floating, it does not affect the marine life negatively

Source HSB Marine, 2023

broadly contribute to the decarbonization and sustainability of urban coastal development. However, exceptions exist in regions with unique geological or economic constraints, and the current body of empirical data on floating infrastructure remains limited. Consequently, further research is necessary to validate these approaches across diverse environmental conditions and to refine economic assessments under varying regional circumstances.

References

Aikins, E. (2024). Impact of sea erosion on sustainable coastal tourism development in Ghana. The case of Saltpond Beach resorts. *African Journal of Hospitality and Tourism Management, 4*, 52–73. https://doi.org/10.47963/ajhtm.v4i2.1563

Andrew, R. M. (2018). Global CO_2 emissions from cement production. *Earth System Science Data, 10*(1), 195–217. https://essd.copernicus.org/articles/10/195/2018/

Barnard, L. (2024, September 4). *It's reshaping the world*. Construction Briefing. https://www.constructionbriefing.com/news/out-of-the-blu/8038559.article

Blue21. (2021 January 6). *Floating islands development and deployment roadmap*. Blue21. https://www.blue21.nl/wp-content/uploads/2024/01/20210106-floating-islands-development-and-deployment-roadmap-1.pdf

Brady, K., & Wallace, A. (2006). Dredging and its environmental impacts. In *Environmental aspects of dredging and dredged material disposal* (pp. 1–28). CRC Press.

Intergovernmental Panel on Climate Change. (2021). *Sixth assessment report (AR6)*. IPCC. https://www.ipcc.ch/assessment-report/ar6/

Kotsoni, A., Dimelli, D., & Ragia, L. (2017). *Land use planning for sustainable development of coastal regions* (pp. 290–294). https://doi.org/10.5220/0006370802900294

Ma, K. -T., Noorizadegan, A., Chang, H. -C., Cobb, D., Liu, Y. -H., Chiang, M. -H.. (2024). *Review of anchor solutions for floating offshore wind turbines*. https://www.researchgate.net/publication/384845666_Review_of_Anchor_Solutions_for_Floating_Offshore_Wind_Turbines

Mansourmoghaddam, M., Malamiri, G., Reza, H., Rousta, I., Olafsson, H., & Zhang, H. (2022). Assessment of palm Jumeirah Island's construction effects on the surrounding water quality and surface temperatures during 2001–2020. *Water, 14*, 634. https://doi.org/10.3390/w14040634

Rougerie, et al. (2004). *Seawater air conditioning: A sustainable cooling approach*. [Journal/Report details].

Shimizu Corporation. (n.d.). *Environmental Island: Green float*. Retrieved February 3 2025. https://www.shimz.co.jp/en/topics/dream/content03/

South Pacific Applied Geoscience Commission. (2015). *Vulnerability index for South Pacific Island nations*. https://www.spac.org/vulnerability-index

Uteuova, A. (2022, August 12). *Seawalls 'do more harm than good' as scientists call for nature-based alternatives.* The Guardian. https://www.theguardian.com/us-news/2022/aug/12/seawalls-nature-based-alternatives-us

Part III
Decarbonizing Urban Development and Smart Cities

Chapter 13
Role of Built Environment in Promoting Quality of Higher Education in University Campuses—A Review

Zainab Rao and Mohd. Khalid Hassan

Abstract This review examines the macro-level aspects of campus planning and their impact on higher education quality, focusing on how comprehensive campus design strategies influence academic excellence, student experience, and institutional success. The analysis synthesizes current research on campus master planning principles, spatial organization, and the relationship between physical infrastructure and educational outcomes. This review underscores the important elements that need to addressed in campus planning to enhance the quality of built environment. Campus has been assessed on three latent variables such as urbanism, greenness and on-campus living with 10 indicators (Hajrasouliha, Landscape and Urban Planning, 158, 166–176, 2017). It has been observed that campus planning has enhanced the quality of built environment which ultimately impacted the student performance thus improving the quality of education in the universities. The paper discusses the integration of built environment assessment as one of the criteria in the higher education ranking systems.

Keywords Built environment · Higher education quality · University campus planning · Campus sustainability

Introduction

Built environment plays a crucial role in addressing the quality of life of inhabitants. This article aims to address the issue of built environment factors being overlooked in various higher education ranking systems globally. Various qualitative and quantitative factors of built environment quality are extracted from the available literature and suggested to score the quality of built environment. Campus planning

Z. Rao (✉) · Mohd. K. Hassan
Department of Architecture, Aligarh Muslim University, Aligarh, India
e-mail: zainabrao788@gmail.com

Mohd. K. Hassan
e-mail: arch5.khalidhasan@gmail.com

J. Dsilva (ed.), *Sustainability Education, Youth and Innovation*,
https://doi.org/10.1007/978-3-032-21948-0_13

at the macro level encompasses several critical dimensions that shape the educational environment. Research indicates that effective campus organization follows key planning principles including functional zoning, hierarchy of spaces, accessibility, and integration with the surrounding context (Campus Design of Universities An Overview, 2021). Studies have identified essential campus qualities that contribute to educational excellence, including walkability, compactness, greenness, and spatial configuration. Hajrasouliha (2017) developed a comprehensive framework for measuring campus qualities, demonstrating strong correlations between these physical attributes and student satisfaction, retention rates, and academic performance. Various qualitative and quantitative factors of built environment quality are extracted from the available literature and suggested to score the quality of built environment. Powered by Thomson Reuters, the Times Higher Education World University ranking is one of the most generally recognized organizations for evaluating colleges and universities (Anowar et al., 2015; https://www.timeshighereducation.com/, n.d.; Soh, 2013). Peer review accounts for 40% of the ranking of universities worldwide, major (primarily international) graduate recruiters for 10%, citations (per capita) of published academic papers for 20%, teaching staff: student ratios for 20%, and "international orientation" for 10% (Anowar et al., 2015; Ioannidis et al., 2007; Soh, 2013). The following indicators and their weightage of the QS World University Rankings: Academic peer review (40%) is given the highest (40%) weightage in this ranking method (Anowar et al., 2015; https://www.qs.com/, 2025); however, in regional ranking such as QS University Ranking: Asia, some of the weightages of the indicators were revised and gave 30% weightage to Academic peer review (Anowar et al., 2015; Ioannidis et al., 2007).

QS has been issuing world university rankings in cooperation with Times Higher Education (THE) since 2004, though it severed its collaboration with THE in 2010. Citation per faculty was divided into two categories - 'Paper per Faculty' and 'Citations per Paper' and category was given 15% weightage individually (Anowar et al., 2015). National Institutional Ranking Framework (NIRF) is an Indian HE institutional ranking system which initiated its first meeting in 2014. Rankings for universities and three academic areas—pharmacy, management, and engineering—were made public in 2016. Both the overall ranking and the general degree college ranking were implemented in 2017. Domain-specific rankings for architecture, medicine, and law were introduced in 2018. By 2019, there were six domain-specific rankings and three category-specific rankings. In 2020 and 2021, respectively, rankings for dental and research institutions were added (Rankings, n.d.). Three new categories—Open Universities, State Public Universities, and Skill Universities—will be added in 2024, bringing the Indian Rankings' portfolio up from 13 in 2023 to 16 (Rankings, n.d.). Teaching, Learning, and Resources (TLR), Research and Professional Practice (RP), Graduation Outcome (GO), Outreach and Inclusion (O&I), and Perception (PR) are the five general categories of characteristics that the NIRF uses to rate universities.

Methodology

Initially 36 research papers and review papers were selected from various databases such as Scopus, web of science and google scholar. The selection criteria include the focus of study at quality of built environment, campus planning and sustainability, higher education ranking systems and performance of student in a university campus. The paper attempts to find the nexus of quality of built environment, Higher education ranking, quality of education and campus planning. Global ranking system such as The Times Higher and QS ranking are taken into consideration at global level while NIRF (National Institutional Ranking Framework) -India has been considered in this review.

Findings

It has been observed from the literature that the quality of built environment and campus planning has been assessed in isolation. On the other hand, Higher education ranking systems overlooked built environment assessment criteria. Various papers discuss campus sustainability assessment tool as well but none of them address the integration of built environment assessment in current higher education ranking systems, neither at global nor at national level.

Conclusions

Various research papers discuss Quality of Built Environment, Higher education ranking systems, campus sustainability assessment tool as well but none of them addresses the integration of built environment assessment in current higher education ranking systems, neither at global nor at national level. Integration of Built assessment criteria as an evaluation indicator in Higher education ranking systems at least at national level will make the ranking intensive and dynamic in nature. İt will not only enhance the quality of infrastructure in government institutions built also will lead to sustainable campus planning.

References

Anowar, F., Helal, M. A., Afroj, S., Sultana, S., Sarker, F., & Mamun, K. A. (2015). A critical review on world university ranking in terms of top four ranking systems. *Lecture Notes in Electrical Engineering, 312*, 559–566. https://doi.org/10.1007/978-3-319-06764-3_72

Campus Design of Universities An Overview. (2021). *Journal of Design and Built Environment Vol 21*(31), 37–51.

Hajrasouliha, A. (2017). Campus score: Measuring university campus qualities. *Landscape and Urban Planning, 158*, 166–176. https://doi.org/10.1016/j.landurbplan.2016.10.007
(2025). https://www.qs.com/
(n.d.). https://www.timeshighereducation.com/
Ioannidis, J. P. A., Patsopoulos, N. A., Kavvoura, F. K., Tatsioni, A., Evangelou, E., Kouri, I., Contopoulos-Ioannidis, D. G., & Liberopoulos, G. (2007). International ranking systems for universities and institutions: A critical appraisal. *BMC Medicine, 5*. https://doi.org/10.1186/1741-7015-5-30
Rankings, I. (n.d.). *National institutional ranking framework*.
Soh, K. (2013). Times higher education 100 under 50 ranking: Old wine in a new bottle? *Quality in Higher Education, 19*(1), 111–121. https://doi.org/10.1080/13538322.2013.774795

Chapter 14
Investigating the Impact of Two-Phase Flow-Induced Vibration on Fouling in Water Desalination Membranes Using FSI and Deep Learning

Abdalellah O. Mohmmed and Fiseha M. Guangul

Abstract Desalination of seawater or highly saline wastewater requires an effective approach for addressing the freshwater crisis, which has significantly intensified due to the rapid growth of the global population and economic development. Membrane distillation is emerging as a promising technology and a vital contributor to achieving Sustainable Development Goal 6, which focuses on clean water and sanitation. The significance of this method stems from its ability to operate at low temperatures while maintaining a high removal rate for non-volatile components. However, it is subject to the polarization effect and membrane fouling. This study aims to develop a comprehensive computational framework based on Fluid Structure Interaction and AI deep learning techniques to optimize and predict membrane fouling while mitigating temperature polarization. The effect of slug two-phase flow on heat and mass transfer, along with membrane fouling, was thoroughly addressed. Multiple numerical simulations were conducted to collect diverse data on the coupling between slug flow and membrane structure. The effect of induced vibration on the fouling rate was evaluated. The results demonstrated that the generated bubbles within the slug flow enhanced distillation performance by 26%. A Recurrent Neural Networks (RNNs) algorithm based on three features—vibration amplitude, bubble length, and bubble frequency—was employed to predict membrane fouling. However, the prediction results indicate that additional validated data is required for greater accuracy and improved optimization.

Keywords Deep learning · Desalination · Direct contact membrane · Fluid-structure interaction · Optimization

---

A. O. Mohmmed (✉) · F. M. Guangul
School of Engineering, Infrastructure and Sustainability, De Montfort University, Dubai Internet City, Dubai, UAE
e-mail: abdalellah.mohammed@dmu.ac.uk

F. M. Guangul
e-mail: fiseha@dmu.ac.uk

J. Dsilva (ed.), *Sustainability Education, Youth and Innovation*,
https://doi.org/10.1007/978-3-032-21948-0_14

Introduction

The scarcity of clean water and energy represents two major obstacles to achieving sustainable development worldwide (Wang et al., 2019). From 2010 to 2035, it is anticipated that worldwide water usage will rise by about 1% annually, influenced by factors such as population growth, economic progress, and changing consumption habits (United Nations, 2023). The UAE authority forecasted that the yearly water demand would rise from 4.4 billion m3 in 2008 to 8.8 billion m3 by 2030 (Mohsen et al., 2016). As a result, the process of desalinating seawater has become essential in addressing freshwater shortages. However, traditional desalination methods are generally energy-demanding and significantly contribute to the worldwide emission of greenhouse gases (GHGs). In the Middle East, the share of total energy consumption dedicated to desalination, which was only 5% in 2016, is expected to increase to 15% by 2040 (Ayaz et al., 2022). The swiftly growing energy requirements of the desalination industry are poised to soon intensify climate-related issues (IEA, 2020). The IEA's World Energy Outlook Report indicates that the energy used by the desalination industry in the UAE is comparable to over 13 million tons of oil (IEA, 2020). The Dubai Electricity and Water Authority (DEWA) has granted a contract for building a Reverse Osmosis (RO) plant with a production capacity of 0.18 million m^3/d, which commenced operations in 2022. This strategic move aligns with DEWA's goal to boost its SWRO production capacity to 1.3 million m^3/d by 2030, thereby increasing its share from the current 13-41% (Ayaz et al., 2022). In 2017, DEWA's water production via Multi-Stage Flash (MSF) distillation was approximately 1.7 million m^3/d. The organization intends to sustain this production level without further dependence on this technology. Additionally, DEWA has outlined a strategy to ensure that by 2030, 100% of desalinated water will be produced using clean energy, achieved through the integration of renewable energy sources and waste heat (Ayaz et al., 2022). Improving operational efficiency and decoupling desalination from electricity production will not only yield cost savings of 13 billion AED but also result in a reduction of 44 million tons of greenhouse gases (GHGs).

Transitioning desalination processes from fossil fuel dependency to renewable energy sources is inadequate, as it merely shifts the cost burden from fuel to increased initial capital expenditure, while most environmental issues persist, albeit in different forms. Therefore, to effectively address the global challenges of water scarcity and pollution, common technologies utilized in industrial desalination applications include Multi-Effect Distillation (MED) (Harandi & Asadi, 2023), MSF (Shenvi et al., 2015), and RO (Al-Weshahi et al., 2013). Recent advancements in membrane distillation technologies have attracted significant scholarly interest, demonstrating considerable potential in applications such as desalination (Chen et al., 2023), wastewater treatment, metal extraction, and disinfection (Wang et al., 2019). Contemporary conventional technologies are predominantly characterized by their need for extensive pre-treatment and substantial energy consumption. Membrane distillation (MD), which employs hydrophobic microporous membranes, offers an effective method for concentrating and purifying seawater, highly concentrated brine, and industrial wastewater. Shaikh and Ismail (2023) using low-grade thermal energy, which is transported across the membrane in vapor form and subsequently condenses on the membrane to yield purified water. Membrane-based technologies are esteemed for their superior energy efficiency when compared to thermal methods, alongside their robustness and dependability. However, prevailing practices indicate that these energy-efficient systems are nearing the threshold of further efficiency gains, especially with respect to operational expenses (Ahmed et al., 2020), unless there is the adoption of substantial technological innovations. While membrane distillation (MD) technology is effective in treating high-concentration solutions, it is susceptible to membrane fouling when processing complex feed-liquid compositions. This fouling results in a decline in membrane permeance flux and a consequent reduction in water production capacity. During membrane distillation (MD) operations, multiple types of fouling may occur concurrently. The accumulation of foulants on membrane surfaces impairs pore functionality, diminishes the effective membrane area, and increases resistance to mass and heat transfer, thereby reducing membrane permeance flux. Furthermore, the presence of fouling elevates the likelihood of pore obstruction and membrane wetting, ultimately leading to a decline in MD performance (Gryta, 2008). Consequently, it is crucial to identify an effective strategy to mitigate inorganic fouling.

Researchers have explored various techniques to control fouling, including feed pretreatment (Shon et al., 2009), modification of membrane surfaces (Zhang & Angelidaki, 2013), cleaning with chemicals (Gryta, 2008), employing vibration and crystallization techniques (Zhong et al., 2018), and applying ultrasonic treatments (Hou et al., 2015). Recent studies have demonstrated that the introduction of gas bubbles represents one of the most effective techniques for managing fouling. This method involves the creation of a gas-liquid two-phase flow, which enhances the shear rate at the surface and optimizes the fluid dynamics in proximity to the membrane. Consequently, this approach not only mitigates the effects of temperature polarization but also reduces scale buildup on the membrane surface (Chen et al., 2014). Recognizing the practical importance of two-phase flow in the membrane distillation (MD) process and the current limited understanding of its mechanisms, this study examine the performance of heat and mass transfer, as well as the antifouling effects, in slug two-phase flow. Furthermore, it aims to establish a correlation between slug flow-induced vibration and the enhancement of transmembrane permeate flux performance, alongside the mitigation of membrane fouling.

Methodology

A comprehensive investigation was conducted to examine the impact of flow-induced vibration on the fouling of the membrane surface. As shown in Fig. 14.1, a numerical Fluid Structure Interaction (FSI) model was developed to elucidate the effect of slug two-phase flow on membrane fouling. This model incorporates both the fluid and structural domains. The Volume of Fluid (VoF) model was employed to simulate

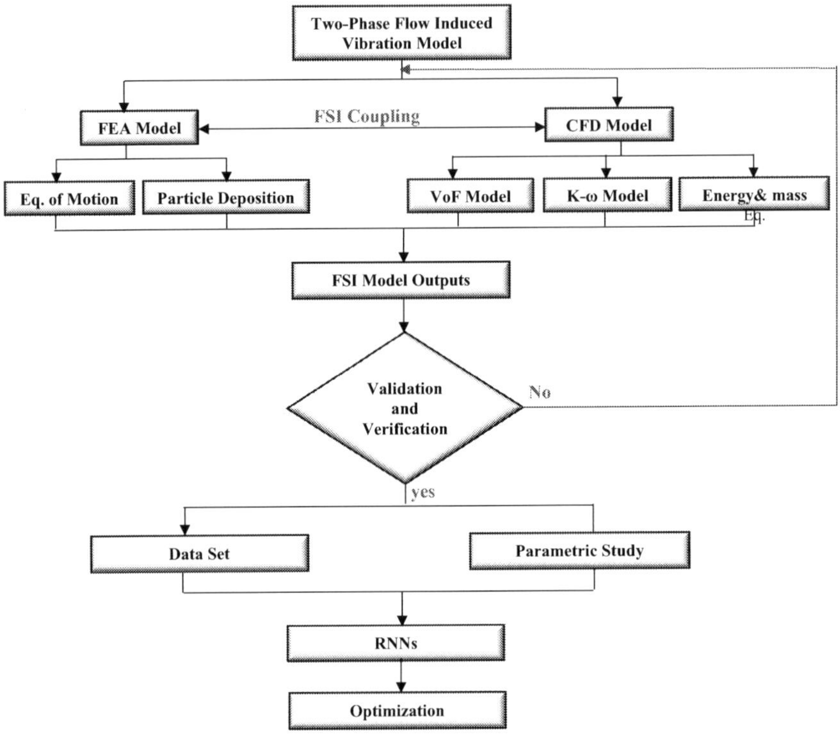

Fig. 14.1 Multiphase flow-induced vibration model applied for MD membrane system

the slug two-phase flow, while the K-ω model was utilized for turbulence modeling, and the energy equation was implemented to account for the temperature of the saline water. Pressure, bubble distribution, and velocity induced by the multiphase flow were mapped to the structural domain, and the effects of displacement, vibration frequency, and amplitude on membrane fouling were evaluated. Furthermore, Recurrent Neural Networks (RNNs) was employed to analyze the temporal distribution of vibration data. Various operational conditions and scenarios were examined to evaluate the optimal operational performance. Additionally, the collected data were processed using RNNs models to optimize the performance and predict the non-tested operational conditions.

Results and Discussion

The preliminary results of the FSI analysis, which integrated fluid and structural domains, are depicted in Fig. 14.2. Air bubbles injected into the system displayed random lengths and frequencies, while a steady temperature decline was observed with increasing distance from the feed inlet. A linear increase in fouling thickness over time was noted; this observation could be attributed to the use of a constant fouling factor. Implementing a particle deposition model could potentially enhance the accuracy of predictions. Vibration amplitude fluctuations were recorded within a range of ± 0.001 m. Notably, the influence of vibration on fouling was found to be negligible, a finding that stands in contrast to the conclusions of earlier studies by (Wan et al., 2024; Wang et al., 2022).

In this study, the permeate flux exceeds that observed in single-phase flow, attributable to an enhanced driving force and a reduced thermal boundary layer on the feed side. This enhancement is associated with improved flow mixing and elevated surface shear stress induced by the two-phase flow (Kim et al., 2023). Furthermore, a delay in the reduction of the permeate flux was observed following the injection of elongated slug bubbles. This phenomenon can be attributed to the anti-fouling effect exerted by these bubbles surrounding the membrane. The presence of these bubbles enhances shear stress and influences the deposition of crystal salt around the membrane (Dong et al., 2022). Vice versa, Table 14.1 illustrates the sample data utilized to train the RNNs; three features were selected to develop the AI prediction model. These features are vibration amplitude, bubble size, and bubble frequency. To enhance training stability, the input features and target data were normalized. A dropout layer with a dropout rate of 0.2 was implemented to mitigate overfitting, and a learning rate of 0.001 was employed to ensure improved convergence.

Figure 14.3 illustrates the prediction of fouling using RNNs along with the Mean Square Error (MSE) over time. The prediction exhibited a divergent trend from the actual membrane fouling, which could be attributed to the utilization of a constant fouling parameter, whereas the RNNs incorporated a random varying input variable.

Conclusions

Considering the practical relevance of membrane distillation and its vulnerability to polarization effects and membrane fouling, this investigation presented a computational framework integrating FSI and AI deep learning algorithms. In the study the impact of slug two-phase flow on the DCMD fouling and polarization were investigated. In addition, the study examines the induced vibration due to the presence of elongated slug bubbles and addresses the impact of vibration amplitude on membrane fouling. The permeate flux generated based on slug two-phase flow was improved by 26% compared to single-phase water flow. Nevertheless, upon examination of the predicted error (MSE), the RNNs algorithm's forecasting ability was found to be limited. Consequently, it is recommended to employ a combination of RNNs and CNNs to enhance fouling prediction and minimize error. Furthermore, supplementary data from diverse scenarios, such as considering parallel and counter flow, are necessary to expand the training set. Additional refinement of the learning rate and high-fidelity simulation are essential to ensure the accuracy of the collected data.

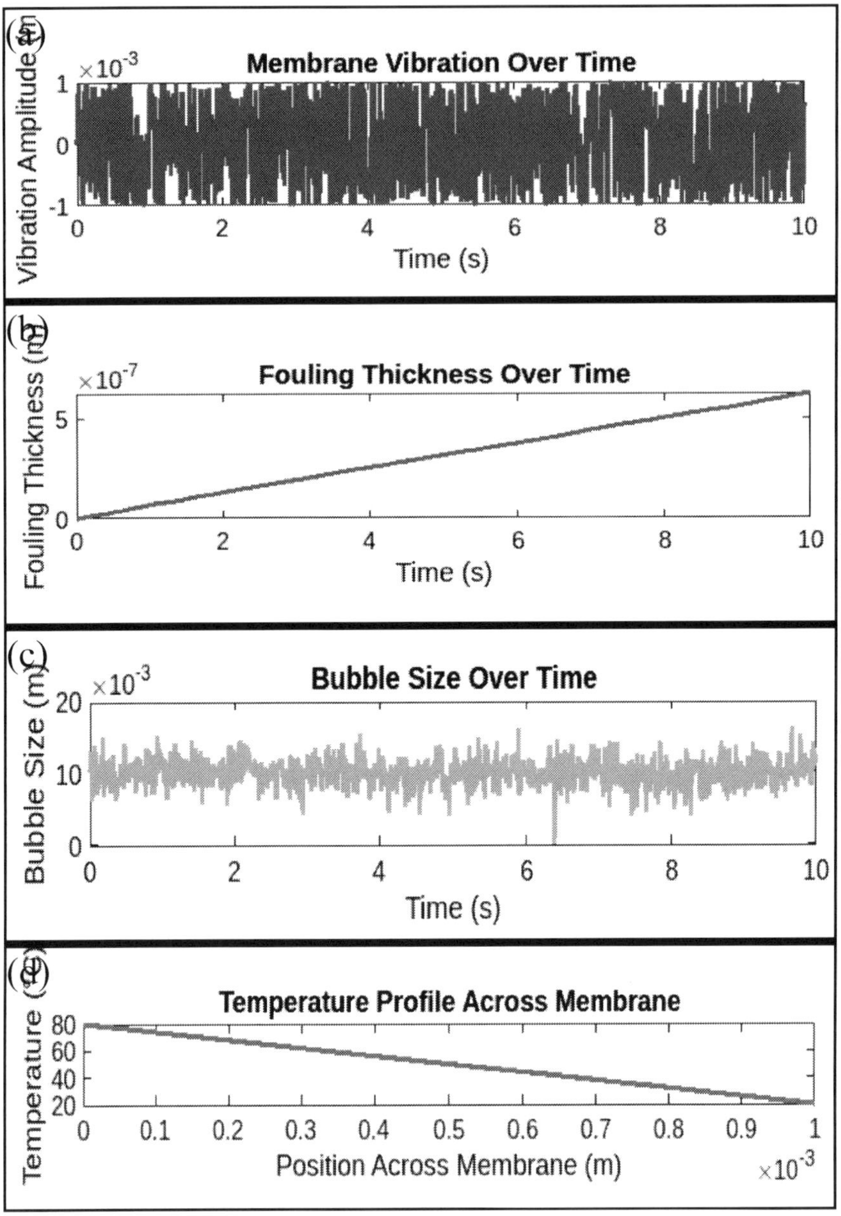

Fig. 14.2 Time series visualizations of **a** vibration amplitude **b** fouling thickness **c** bubble size **d** temperature profile

Table 14.1 Sample data used for RNNs

Time	Vibration	Fouling thickness	Bubble size	Bubble frequency
0	0	0	0.0126	6.0743
0.01001	0.00066453	0	0.013047	5.6969
0.02002	0.00099547	8.67e-10	0.01098	6.0743
0.03003	0.00095337	1.9601e-09	0.0060873	4.6526
0.04004	0.00097423	2.5404e-09	0.0097902	3.8031
0.05005	0.00054218	3.4942e-09	0.0068209	5.7249
0.06006	−0.0003662	3.864e-09	0.0090917	6.0516
0.07007	−0.0009868	4.197e-09	0.014004	7.3729

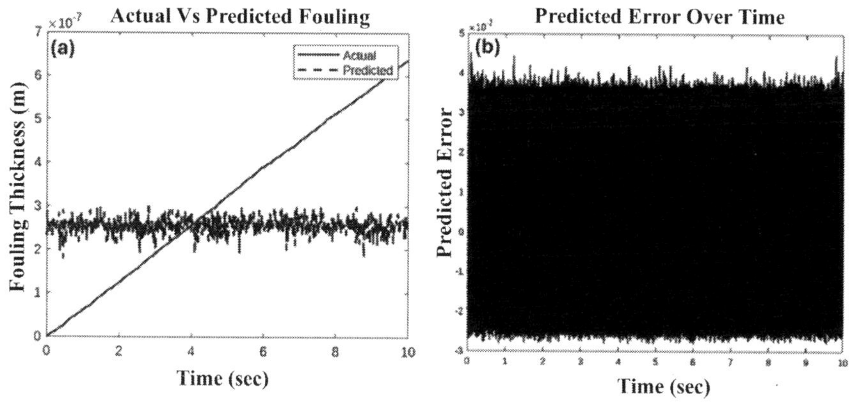

Fig. 14.3 Visualizations of **a** fouling prediction, **b** prediction error

References

Ahmed, F. E., et al. (2020). Hybrid technologies: The future of energy efficient desalination–A review. *Desalination, 495*, 114659.

Al-Weshahi, M. A., et al. (2013). Exergy efficiency enhancement of MSF desalination by heat recovery from hot distillate water stages. *Applied Thermal Engineering, 53*(2), 226–233.

Ayaz, M., et al. (2022). Sustainable seawater desalination: Current status, environmental implications and future expectations. *Desalination, 540*, 116022.

Chen, G., et al. (2014). Optimization of operating conditions for a continuous membrane distillation crystallization process with zero salty water discharge. *Journal of Membrane Science, 450*, 1–11.

Chen, H., et al. (2023). Ultrahigh-water-flux desalination on graphdiyne membranes. *Nature Water, 1*(9), 800–807.

Dong, C., et al. (2022). Performance intensification and anti-fouling of the two-phase flow enhanced direct contact membrane distillation for seawater desalination. *Desalination, 541*, 116059.

Gryta, M. (2008). Alkaline scaling in the membrane distillation process. *Desalination, 228*(1–3), 128–134.

Harandi, H. B., & Asadi, A. (2023). *Transport mechanisms in membranes used for desalination applications*.

Hou, D., et al. (2015). An ultrasonic assisted direct contact membrane distillation hybrid process for desalination. *Journal of Membrane Science, 476*, 59–67.

IEA, U. (2020). *Global energy review 2020.* Retrieved September, 10, 2020. https://www.iea.org/countries/ukraine

Kim, Y.-B., et al. (2023). Comprehensive experimental and theoretical investigations on the effect of microbubble two-phase flow on the performance of direct-contact membrane distillation. *Water Research, 229*, 119407.

Mohsen, M. S., et al. (2016). Energy options for water desalination in UAE. *Procedia Computer Science, 83*, 894–901.

Shaikh, J. S., & Ismail, S. (2023). Performance evaluation of a solar humidification dehumidification desalination system employing a multistage bubble column dehumidifier. *Solar Energy, 263*, 111933.

Shenvi, S. S., et al. (2015). A review on RO membrane technology: Developments and challenges. *Desalination, 368*, 10–26.

Shon, H., et al. (2009). Physico-chemical pretreatment to seawater reverse osmosis (SWRO): Organic characterization and membrane autopsy. *Desalination, 236*(1–3), 282–290.

United Nations. (2023). *The United Nations world water development report 2023: Partnerships and cooperation for water*. UNESCO.

Wan, H., et al. (2024). Crossflow-induced membrane vibration and its effects on fouling and scaling in direct contact membrane distillation. *Journal of Membrane Science, 694*, 122410.

Wang, C., et al. (2022). Insights on fouling development and characteristics during different fouling stages between a novel vibrating MBR and an air-sparging MBR for domestic wastewater treatment. *Water Research, 212*, 118098.

Wang, W., et al. (2019). Simultaneous production of fresh water and electricity via multistage solar photovoltaic membrane distillation. *Nature Communications, 10*(1), 3012.

Zhang, Y., & Angelidaki, I. (2013). A new method for in situ nitrate removal from groundwater using submerged microbial desalination–denitrification cell (SMDDC). *Water Research, 47*(5), 1827–1836.

Zhong, W., et al. (2018). Fouling mitigation in submerged VMD for the treatment of brackish groundwater concentrates with transverse vibration and crystallizer. *Desalination, 426*, 32–41.

Chapter 15
Carbon Accounting for Fabrication Factory—A Case Study of Global Energy Industries in the United Arab Emirates

Anto Joseph and Asifa Mahajabeen

Abstract Carbon emissions from factories significantly affect climate change. This study investigates the application of carbon accounting and offsetting mechanisms in the United Arab Emirates (UAE) factory sector. The study emphasizes the importance of precise measurements and sets up a framework that includes emissions from Scopes 1, 2, and 3. This framework is based on the Greenhouse Gas Protocol Corporate Accounting and Reporting Standard developed by the World Resources Institute (WRI) and the World Business Council for Sustainable Development (WBCSD). The framework builds on existing carbon accounting practices while considering the unique challenges of the UAE market. UAE factories can contribute to national sustainability goals and reduce emissions to save money. A Global Energy Industries case study details carbon accounting from April 2023 to April 2024 for Scope 1, 2, and 3 emissions. This case study is a practical illustration of implementing carbon accounting practices within the UAE context, highlighting opportunities for emission reduction and sustainability enhancement within individual factories.

Keywords Carbon accounting · Greenhouse gas protocol · Energy · UAE

Introduction

Compelling evidence demonstrates the acceleration of global temperatures due to greenhouse gases (GHGs) such as carbon dioxide (CO_2) accumulating in the Earth's atmosphere [1]. Countries, including the United Arab Emirates (UAE), are implementing policies, resource allocation, and legal frameworks to address this challenge. The Paris Agreement outlines Nationally Determined Contributions (NDCs) to reduce GHG emissions by at least 34% by 2030 and achieve net-zero emissions by 2050 [2, 3]. The metric used is the carbon dioxide equivalent (CO_2e), covering CO_2, methane (CH_4), nitrous oxide (N_2O), hydrofluorocarbons (HFCs), perfluorocarbons (PFCs), nitrogen trifluoride (NF_3), and sulfur hexafluoride (SF_6) [4].

A. Joseph (✉) · A. Mahajabeen
Department of Design and Architecture, Manipal Academy of Higher Education, Dubai, UAE

J. Dsilva (ed.), *Sustainability Education, Youth and Innovation*,
https://doi.org/10.1007/978-3-032-21948-0_15

The UAE has firmly committed to climate action by implementing initiatives to transition towards a low-carbon economy. Policies such as the UAE Net Zero 2050 Strategic Initiative highlight efforts to integrate renewable energy, enhance energy efficiency, and support carbon capture technologies. This paper examines the UAE's role in global climate action and the importance of adopting standardized frameworks to measure, manage, and mitigate GHG emissions [5].

Ghg Protocol and Emissions: Framework

The GHG Protocol provides guidelines for organizations to measure and report their emissions based on relevance, completeness, consistency, transparency, and accuracy [6]. Organizations establish a base year for emissions and monitor changes over time. Calculation methods involve identifying sources, selecting an approach, gathering data, and reporting emissions at the corporate level, with quality management systems ensuring accuracy.

Organizations must align their GHG reduction goals with national and international regulations, integrating voluntary and mandatory frameworks. Companies must also ensure data integrity in emissions inventories, especially when participating in carbon trading mechanisms or corporate sustainability reporting. This highlights businesses need to invest in reliable emissions-tracking tools and third-party verifications.

Organizational Boundaries: Organizations define GHG emissions by categorizing them into Scope 1 (direct emissions from owned or controlled sources), Scope 2 (indirect emissions from purchased energy), and Scope 3 (other indirect emissions such as business travel and supply chain) [8, 10]. The types of organizational boundaries about GHG emissions can include equity share and control approaches (Fig. 15.1).

Two main approaches define boundaries:

- Equity Share Approach: Emissions are allocated based on ownership share [9].
- Control Approach: Emissions are accounted for entirely when the company has financial or operational control [9].

Setting organizational boundaries helps ensure emissions are tracked accurately, promoting transparency in climate action reporting. Many multinational organizations use hybrid approaches better to capture emissions across complex supply chains and multi-stakeholder operations.

Steps In Identifying and Calculating Ghg Emissions

The following are the steps in identifying and calculating GHG Emissions (Fig. 15.2) [13]:

- Identifying emission sources (Scope 1, Scope 2, and Scope 3) [11, 12].
- Selecting a calculation method (emission factors, direct measurement, or stoichiometric calculations) [9].
- Data collection and emission factor selection for each scope [13].
- Applying calculation tools such as cross-sector and sector-specific tools [14, 15].

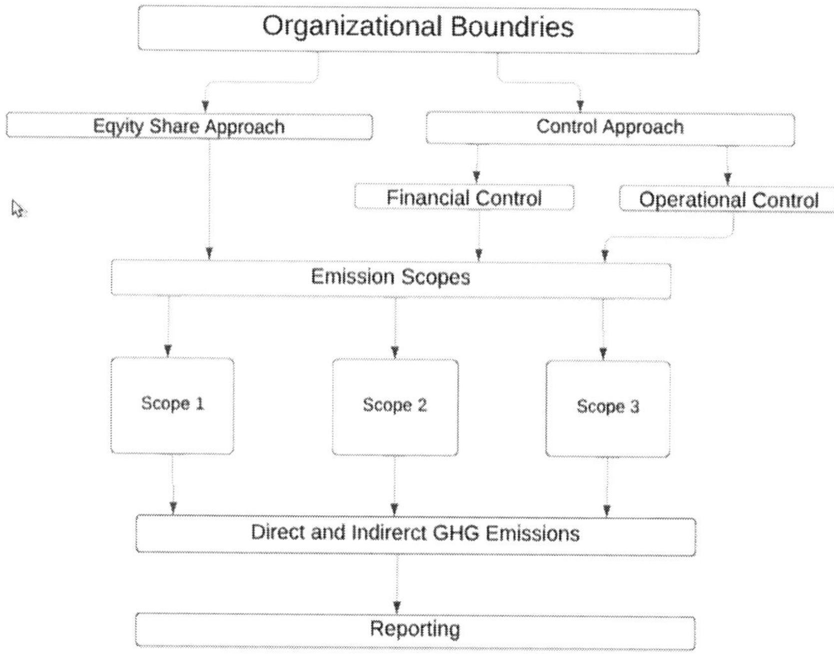

Fig. 15.1 Setting organizational boundaries

Fig. 15.2 Steps for GHG emission identifications

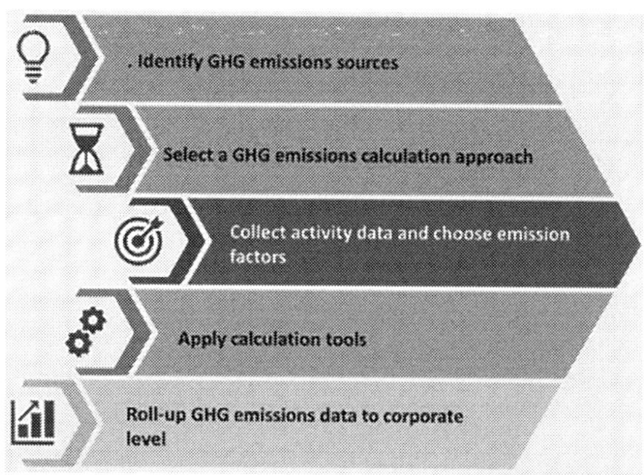

- Rolling up data to the corporate level using centralized or decentralized approaches.

Calculation methods must be adaptable to industry-specific needs. Sectors like oil and gas, transportation, and manufacturing require more robust direct monitoring systems than service-based industries where estimation methods are widely accepted. Investment in digital tools such as IoT-based monitoring, blockchain verification for carbon offsets, and AI-driven forecasting models can enhance accuracy and reliability in emissions calculations.

Importance of Setting GHG Targets

GHG targets help organizations manage risks, drive cost savings, and prepare for regulations [15–20]. Key reasons include:

- Minimizing Risks: Reducing emissions mitigates exposure to carbon pricing policies, regulatory penalties, and climate-related financial risks.
- Fostering Innovation: Emission reduction strategies can drive technological advancements in energy efficiency and low-carbon solutions.
- Ensuring Regulatory Compliance: Establishing internal targets prepares businesses for evolving environmental regulations, reducing potential liabilities.
- Enhancing Corporate Responsibility: Companies demonstrating leadership in sustainability gain competitive advantages, improve stakeholder trust, and attract environmentally conscious investors.
- Participating in Voluntary Programs: Organizations benefit from carbon credit programs, sustainability indexes, and international climate initiatives.

Framework for Setting, Achieving, and Tracking GHG Targets

Key steps include:

- Senior Management Commitment: Leadership support is essential.
- Defining target types: Absolute reduction vs. intensity-based targets.
- Setting boundaries: Determining which emissions and geographic areas to include.
- Choosing a base year and completion date: Fixed or rolling baseline approaches.
- Considering GHG offsets or credits: Evaluating carbon offset strategies.
- Establishing a double counting policy: Ensuring transparency in emission reductions.
- Tracking and reporting progress: Monitoring performance against targets.

Companies integrating sustainability into financial decision-making often adopt internal carbon pricing mechanisms, linking emissions reductions to profitability metrics. These measures help prioritize investments in low-carbon technologies.

Greenhouse Gas Accounting—Case Study

A case study was conducted at Globus Energy Industry LLC (GEI), a medium-scale steel fabrication factory in the UAE. The study assessed GHG emissions from June 1, 2023, to May 31, 2024 (FY2024). GEI aimed to quantify emissions, identify hotspots, and establish a baseline for reduction strategies. Table 15.1 outlines the essential metrics for the GEI, while Table 15.2 delineates the annual emissions for the UAE from FY2023 to FY2024.

Table 15.1 Summary of key performance indicators

Number of staff and faculty	35	$KgCO_2e$/Staff and faculty	121391.09
Premises area (m^2)	10,220	$KgCO_2e$	4,248,688.23

Table 15.2 Annual greenhouse gas emissions by source

Scope	Emissions ($KgCO_2e$) FY 2023–2024	Type
Scope 1: Direct GHG emissions	12212.30	Direct GHG emissions
Scope 2: Indirect GHG emissions from purchased electricity	7423.42	Indirect GHG emissions from purchased electricity
Scope 3: Other indirect GHG emissions	4229052.50	Other indirect emissions
Total GHG emissions	4,248,688.23	

Methodology

The study followed the GHG Protocol, incorporating:

- Relevance: Inventory reflecting company emissions.
- Completeness: Inclusion of all emission sources.
- Consistency: Data comparability over time.
- Transparency: Clear data reporting.
- Accuracy: Minimizing estimation errors.

System Boundaries

- Organizational boundaries: The control approach was used.
- Operational boundaries: Direct (Scope 1) and indirect (Scopes 2 and 3) emissions were assessed.

Table 15.3 shows the key figures for GEI. Table 15.4 summarizes the emission sources categorized within scope 1, derived from the data provided by the GEI. Table 15.5 summarizes the emission sources categorized in Scope 2. Table 15.6 provides an overview of the emissions sources considered in Scope 3.

Table 15.3 Key figures

Region	No of facilities	Status	Area(m^2)	Headcount
UAE	1	Owned	10,220	35
Total			10,220	35

Table 15.4 Overview of Scope 1 emission sources for 2023–2024

Category	Emission sources	Boundary
Stationary combustion	Generation of heat	Included
Mobile combustion	Company owned or leased vehicles	Included

Table 15.5 Overview of scope two emission sources for 2023–2024

Category	Emission sources	Boundary
Electricity	Purchased electricity	Included

Table 15.6 Overview of scope three emission sources for FY2024

Category	Emission sources	Boundary
Purchased goods and services	Purchased goods (raw materials) and services	Included
Capital goods	Production of capital goods (e.g., machinery and IT equipment)	Included
Waste generated in operations	Waste management of operational waste (landfilling, recycling, etc.)	Included
Business travel	Travel and accommodation of staff and faculty	Not applicable
Staff, faculty, and student commuting	Travel between home and work	Included

Measure Categories for GEI

Measure categories were defined through direct communication with professionals and desk research. Table 15.7 presents the findings.

Results

Total emissions for GEI in FY2024 were estimated at 4,248,688.23 kg CO_2e. The largest contributor was Scope 3 (4,229,052.50 kg CO_2e). Figures 15.3 and Fig. 15.4 illustrate emissions distribution, with purchased electricity contributing 7423.42 kg CO_2e.

Table 15.8 illustrates the key figures for GHG emissions (in Kg CO_2e) by the GHG Protocol. Emission factors are referred from IPCC [7].

Key Interpretations: Three emissions, mainly from construction and commuting, were dominant. Sustainable construction and alternative transportation are priority areas. While purchased electricity's impact is minor, renewable energy adoption remains crucial.

Limitations

Table 15.7 Data collection methods and the specific data needed for the measure categories in the context of the Steel factory

Sl	Measure category	Scope	Data collection Method	Data collected
1	Stationary combustion	Scope-1	Personal communication via call, email and meeting	The volume of gas consumed from invoices
2	Mobile combustion	Scope-1	Personal communication via email	Total liters from fuel bills
3	Electricity	Scope-2	Personal communication via email	Total KWH from electricity bills
4	Purchased goods and services	Scope-3	Personal communication via email	Invoice of purchased goods
5	Capital goods	Scope-3	Personal communication via email	Count
6	Waste generated	Scope-3	Personal communication via email	Weight of scrap
7	Staff, faculty, and student commuting	Scope-3	Personal communication via email	Total liters from fuel bills

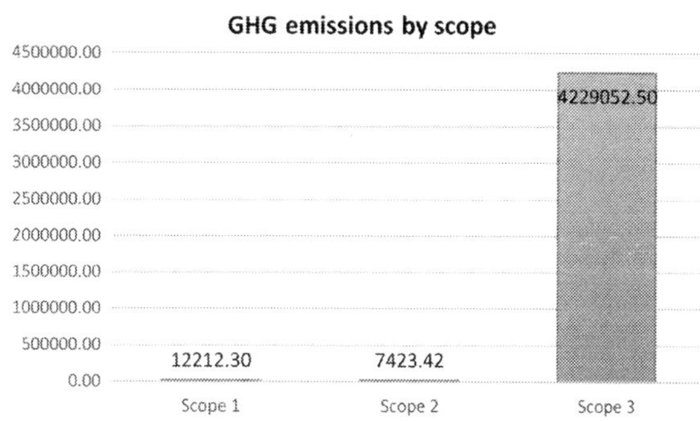

Fig. 15.3 GHG emissions by scope

- Limited Scope 3 Analysis: A comprehensive Life Cycle Assessment (LCA) is needed.
- Data Accessibility: Improved access to government databases would enhance accuracy.

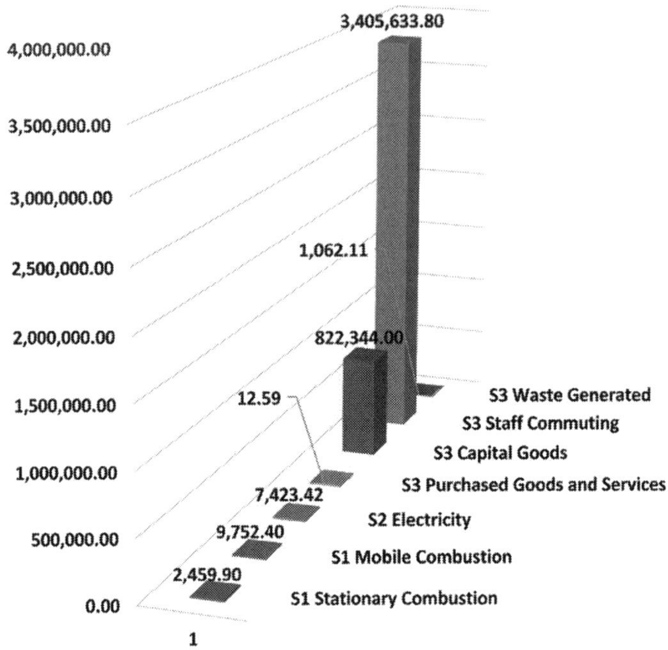

Fig. 15.4 Categories in the Context of GEI

Table 15.8 Greenhouse gas emissions by scope and activity for FY2024

Activity	Consumption	Unit	Emission Factor	Emissions(Kg CO_2e)
Scope 1: Direct GHG emissions				
Stationary combustion				
LPG	1.58	m^3	1556.9 kg CO_2e per m^3	2459.90
Mobile combustion				
Petrol	4515	Litters	2.16 kg CO_2e per litter	9752.40
Scope 2: Indirect GHG emissions from purchased electricity, heating, and cooling				
Electricity				
Grid	12,799	KWH	0.58 kg CO_2e per kWh	7423.42
Scope 3: Other indirect emissions				
Purchased goods and services				
Water	22,220	Gallon	5.6672E-07	12.59
Capital goods				
Construction	4,000,000	AED	0.000205031	820,123.40
Laptop	2	No. of devices	0.258890469	517.78
Computers	5	No.of devices	0.34056338	1702.82
Staff commuting				
Car(petrol)	50,000	km	0.068112676	3,405,633.80
Generated general waste	50,000	kg	2.12423E-05	1062.11
Total GHG emissions				4,248,688.23

Conclusion

GEI's total emissions were 4,248,688.23 kg CO_2e, with capital goods and commuting as primary contributors (82% of footprint). Renewable energy integration and transitioning to electric vehicles are recommended. Scope 1 and 2 reduction targets were modeled following the Science-Based Targets Initiative (SBTi). Setting a Scope 3 target is not mandatory but encouraged.

Future Research

Future research could explore:

1. Carbon Taxation: Evaluating its impact on emissions and reduction strategies.
2. Carbon Offsetting: Assessing feasibility and effectiveness within GEI's framework.

GEI can strengthen its emission reduction strategies and contribute to broader sustainability goals by addressing these limitations and future research avenues.

References

1. Alromaizan, M., Afy-Shararah, M., Jagtap, S., Litos, L., & Salonitis, K. (2023). Developing a carbon accounting tool for SMEs in the Agri-food sector. *Procedia CIRP, 116*, 492–497.
2. Baratta, A., Cimino, A., Longo, F., Solina, V., & Verteramo, S. (2023). The impact of ESG practices in industry with a focus on carbon emissions: Insights and future perspectives. *Sustainability, 15*(8), 6685.
3. Blumberg, G., & Sibilla, M. (2023). A carbon accounting and trading platform for the UK construction industry. *Energies, 16*(4), 1566.
4. Braglia, M., Di Paco, F., Gabbrielli, R., Grassi, C., & Marrazzini, L. (2024). Greenhouse gas emissions deployment (GED): A novel lean method for mitigating greenhouse gas emissions in industrial environment. *Sustainable Production and Consumption.*
5. Doś, A., Błach, J., Lipowicz, M., Pattarin, F., & Flori, E. (2023). Institutional drivers of voluntary carbon reduction target setting—Evidence from Poland and Hungary. *Sustainability, 15*(14), 11155.
6. Greenhouse gas accounting report American University of Sharjah (2023). *A greenhouse gas emissions inventory for a university campus*. https://www.aus.edu/sites/default/files/final_american_university_of_sharjah_-_ghg_report_2022.pdf
7. Intergovernmental Panel on Climate Change (IPCC). (2006). *2006 IPCC guidelines for national greenhouse gas inventories*. Institute for Global Environmental Strategies.
8. Intergovernmental Panel on Climate Change (IPCC).(n.d.). *Sixth assessment report*. https://www.ipcc.ch/-The Paris Agreement https://unfccc.int/process-and-meetings/the-paris-agreement (specifically Article 6).
9. Kaplan, R. S., & Ramanna, K. R. (2016). *Implementing carbon pricing: A guide for effective corporate action*. Greenleaf Publishing.
10. Kasperzak, R., Kureljusic, M., Reisch, L., & Thies, S. (2023). Accounting for carbon emissions—Current state of sustainability reporting practice under the GHG protocol. *Sustainability, 15*(2), 994.
11. Mazhar, M. (2023). Understanding greenhouse gas management: Scopes and boundaries for carbon footprinting. In *The Handbook of Carbon Management* (pp. 265–287). Routledge.
12. Segovia-Hernández, J. G., Hernández, S., Cossío-Vargas, E., & Sánchez-Ramírez, E. (2023). Challenges and opportunities in process intensification to achieve the UN'S 2030 agenda: Goals 6, 7, 9, 12 and 13. *Chemical Engineering and Processing-Process Intensification*, 109507.
13. The UAE's Nationally Determined Contribution (NDC). (n.d.). https://unfccc.int/sites/default/files/NDC/2023-07/Third%20Update%20of%20Second%20NDC%20for%20the%20UAE_v15.pdf.
14. Tomar, S. (2023). Greenhouse gas disclosure and emissions benchmarking. *Journal of Accounting Research, 61*(2), 451–492.
15. United States Environmental Protection Agency (EPA): Understanding Global Warming Potentials (n.d.). https://www.epa.gov/ghgemissions/inventory-us-greenhouse-gas-emissions-and-sinks
16. van Benthem, A. A., Crooks, E., Giglio, S., Schwob, E., & Stroebel, J. (2022). The effect of climate risks on the interactions between financial markets and energy companies. *Nature Energy, 7*(8), 690–697.
17. Wong, C. W., Wong, C. Y., Boon-Itt, S., & Tang, A. K. (2021). Strategies for building environmental transparency and accountability. *Sustainability, 13*(16), 9116.
18. World Resources Institute & World Business Council for Sustainable Development. (2011). *Greenhouse gas protocol corporate accounting and reporting standard*. https://www.wri.org/initiatives/greenhouse-gas-protocol

19. Yaman, C. (2022). Greenhouse Gas Management. In *The Palgrave Handbook of Global Sustainability* (pp. 1–18). Springer International Publishing.
20. Yavuzdeğer, A., Esenboğa, B., Tümay Ateş, K., Demirdelen, Ö., & Yüksel, M. (2024). Assessing the carbon footprint of plastic bottle blow Mold based on product life cycle for managing the Mold industry's carbon emission. *Sustainability, 16*(5), 2149.

Chapter 16
Hybrid Renewable Energy System (HRES) for Wadi Haqeel: Integration and Impact Analysis

Talha Ubaid and Muhammad Wajid Saleem

Abstract The Wadi Haqeel community, in Ras Al Khaimah heavily depends on electricity supplied by the Federal Electricity and Water Authority (FEWA) leading to heightened issues and rising expenses linked to fossil fuels usage. This study explores the viability and advantages of implementing a Hybrid Renewable Energy System (HRES) that includes panels, wind turbines and rainwater harvesting systems to decrease the community's dependency, on fuels reduce carbon emissions and contribute to UAEs sustainability objectives. This study seeks to harness the energy in the area (receiving between 1900 and 2300 kWh/m^2 annually) along, with the moderate wind resources available to design and evaluate the energy production capabilities and cost effectiveness of a Hybrid Renewable Energy System customized for Wadi Haqeel. The approach involves utilizing simulation models to forecast energy generation from panels and wind turbines and evaluate their combination, with rainwater collection for long term sustainability purposes. Primarily findings indicate that HRES could significantly diminish reliance, on the electricity grid enforce energy security and support the UAEs Vision 2021 for energy efforts.

Keywords Hybrid renewable energy system · Sustainability · Solar energy · Wind energy · Rainwater harvesting · Life cycle assessment

Introduction

The increasing energy needs in the UAE are causing worries about the long-term sustainability of using fuels for electricity generation in regions such, as Wadi Haqeel (Sameera Perera, n.d.). The Federal Electricity and Water Authority (FEWA) has

T. Ubaid (✉)
School of Engineering and Sustainable Development, De Montfort University Dubai, Dubai, UAE
e-mail: p2835808@alumni365.dmu.ac.uk

M. W. Saleem
College of Engineering and Energy, Abdullah Al Salem University, Khaldiya, Kuwait
e-mail: muhammad.wajid@aasu.edu.kw

J. Dsilva (ed.), *Sustainability Education, Youth and Innovation*,
https://doi.org/10.1007/978-3-032-21948-0_16

been the electricity supplier, in the UAE as the population continues to increase; nevertheless, relying on fuels leads to increased carbon emissions and escalating energy prices. Near the Jebel Jais Mountain range lies wadi Haqeel boasting energy prospects, with annual solar radiation levels ranging from 1900 to 2300 kWh/m^2 and moderate wind speeds of about 5 m/s (PVMARS, 2024; Tezer et al., 2017).

This initiative seeks to tackle the economic obstacles by suggesting a Hybrid Renewable Energy System (HRES). This system combines energy with wind power and rainwater collection to lessen dependence on power sources. In doing so it enhances energy reliability. Supports the sustainability objectives of the UAE (Hassan et al., 2023; Iqbal et al., 2014).

Methodology

Researching includes gathering data on the amount of sunlight received from the sun (irradiance) wind velocities shown in Figs. 16.1 and 16.2, in the region and the availability of rainwater, for Wadi Haqeel (Global Solar Atlas, 2024). We created a simulation model, with software tools to enhance the setup and dimensions of panels and wind turbines (Tezer et al., 2017). The study estimates the energy needs of 30 households in Wadi Haqeel and evaluates how well the system performs in settings. Furthermore it examines the impact of integrating rainwater harvesting on water conservation and its role, in promoting community sustainability (Arabia Weather Inc, 2024).

The Life Cycle Assessment (LCA)s purpose was to evaluate the effects of the HRES throughout its lifecycle—from manufacturing and set up to usage and disposal (Kar et al., 2021). The system's effectiveness was also evaluated against the energy solutions offered by FEWA with a focus, on saving energy costs and reducing carbon emissions while ensuring cost effectiveness (Hassan et al., 2023).

Results and Discussion

The test findings show that the suggested HRES setup could produce 64 thousand kilowatt hours of electricity each year from solar energy and 52.56 thousand kilowatt hours, from wind turbines. This would meet a part of Wadi Haqeel's energy requirements (PVMARS, 2024). The rainwater collection system should provide water for drinking purposes and help promote sustainability, in general (Swain, 2024). After reviewing the cost benefit analysis results, for the HRES system it seems that it is financially feasible. The estimated time for the investment to be recouped is 10 years when considering both costs and potential energy savings (Iqbal et al., 2014; Alshehhi, 2024).

The environmental advantages are significant well; it is expected that CO_2 emissions will decrease by 30% when compared to the existing FEWA energy provision

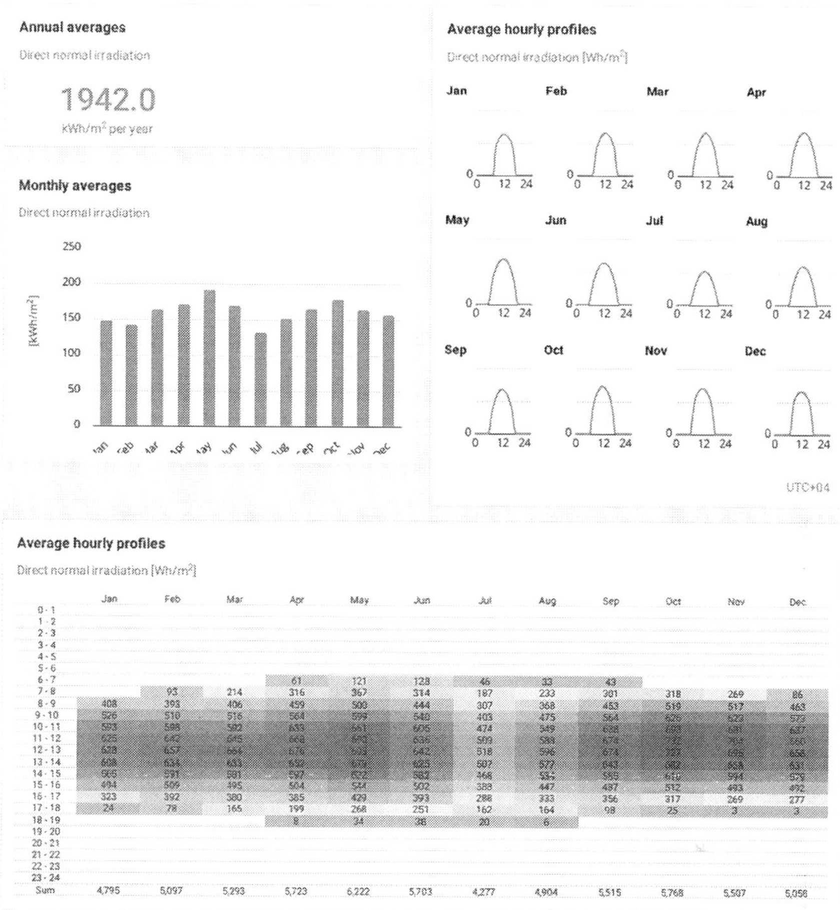

Fig. 16.1 Data visualization of annual and monthly solar irradiation, with detailed hourly distribution across different months (GSA_Report_Ras al-Khaimah-2)

of 180,000 kWh annually (Global Solar Atlas, 2024; Tezer et al., 2017). The LCA results highlight that while the initial manufacturing and installation of solar panels and wind turbines contribute to emissions, the operational phase greatly offsets these through renewable energy generation (Kar et al., 2021).

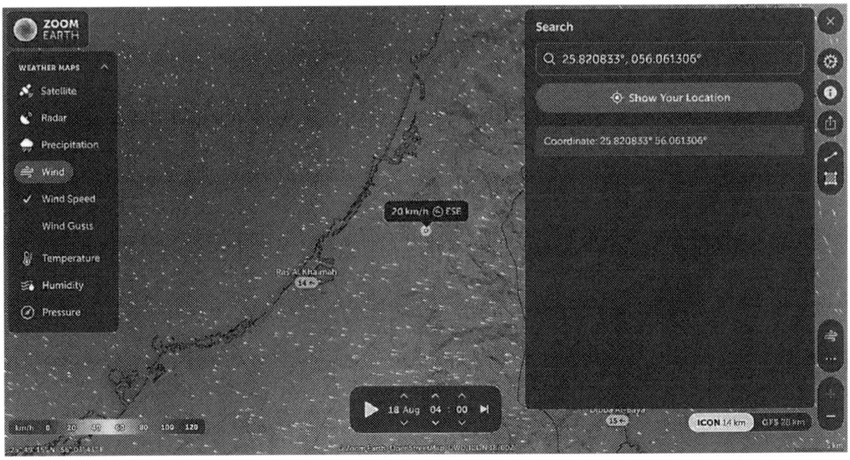

Fig. 16.2 Wind speed data (Wind analysis)

Conclusion

The implementation of a Hybrid Renewable Energy System (HRES), in Wadi Haqeel is an advancement in the journey towards energy in isolated and rural regions of the UAE. The system's efficient utilization of local renewable sources—alongside the opportunity for rainwater collection—proves to be a practical and enduring answer to the energy and water requirements of the community. This endeavour also sets a precedent for projects throughout the UAE by contributing to the nations lasting energy objectives and mitigating effects.

References

Arabia Weather Inc. (2024).

Global Solar Atlas. (2024). https://globalsolaratlas.info/map

GSA_Report_Ras al-Khaimah-2.

Hassan, Q., Algburi, S., Sameen, A. Z., Salman, H. M., & Jaszczur, M. (2023). A review of hybrid renewable energy systems: Solar and wind-powered solutions: Challenges, opportunities, and policy implications. In *Results in engineering* (Vol. 20). https://doi.org/10.1016/j.rineng.2023.101621

Hossin, K., & Alshehhi, H. (2024). Energy consumption behavior analysis in the UAE educational buildings for sustainable economy: A case study of Ras Al Khaimah Schools. *International Journal of Energy Economics and Policy*, *14*(2). https://doi.org/10.32479/ijeep.15360

Iqbal, Z., Soyumer, G., C Kazim, W. (2014). Design and implementation of 59 kWp solar hybrid mini-grid in SOLAB, Ras Al Khaimah. *2014 IEEE 40th Photovoltaic Specialist Conference, PVSC 2014*. https://doi.org/10.1109/PVSC.2014.6925496

Kar, S., Chanda, C. K., C Banerjee, S. (2021). Modelling simulation and characteristics of a mono PERC solar cell under environmental condition. In *IEEE Green Technologies Conference*, April 2021. https://doi.org/10.1109/GreenTech48523.2021.00017

PVMARS. (2024). *PVMARS | PVMars Solar Energy Storage Manufacturer*. https://www.pvmars.com/product/2kva-2kw-wind-turbine-cost/

Sameera Perera, W. C. (n.d.). *District cooling for Al Hamra Village in Ras Al Khaimah-United Arab Emirates (UAE)*. https://aurak.ac.ae/files/rakric/Phd%20C%20Master%20Thesis/2011/CSEM-MR%2011-2.KTH_Perrera_District%20cooling%20Al%20Hamra%20Village.pdf

Swain, S. (2024, June 14). *Water conservation strategies to increase water efficiency*. EcoMENA Echoing Sustainability in MENA. https://www.arabiaweather.com/en

Tezer, T., Yaman, R., & Yaman, G. (2017). Evaluation of approaches used for optimization of stand-alone hybrid renewable energy systems. In *Renewable and sustainable energy reviews* (Vol. 73). https://doi.org/10.1016/j.rser.2017.01.118

Wind analysis. https://zoom.earth/

Chapter 17
Investigating the Impacts of Energy-Efficient Design on Sustainable Construction Projects in the Kingdom of Saudi Arabia (KSA)

Walid Nassereddine

Abstract The current study explores the influence of energy-efficient designs on some aspects of construction in the Kingdom of Saudi Arabia (KSA) regarding life-cycle costs and sustainability. The country is going through a transformative period of economic and infrastructural evolution. This evolution is heavily driven by Vision 2030, whose aim is to diversify the economy and position of the nation towards commerce and sustainability. Using a qualitative case-study approach, interviews are employed as a data collection method, with a sample size of 16 construction management stakeholders, to investigate the challenges of adoption, benefits, and challenges of energy-efficient principles. The study employed thematic analysis for data interpretation. Findings revealed that while energy-efficient designs enhance initial investment costs, they significantly reduce lifecycle costs through energy savings, lower maintenance costs, and increased project value. These designs also improve sustainability indices and contribute to the national economy and environmental resilience by integrating advanced technologies and renewable resources. Additionally, the study highlights implications for labour practices, emphasising the need for workforce adaptation to energy-efficient construction.

Keywords Energy-efficient design · Construction sustainability · Energy conservation · Construction management

Introduction

The construction sector is a significant contributor to environmental degradation, which is released to add to global warming and greenhouse effects. In regions like KSA, where there is ongoing heavy construction to develop tourist spots, the carbon

W. Nassereddine (✉)
Beirut, Lebanon
e-mail: walid.nassereddine@hotmail.com

J. Dsilva (ed.), *Sustainability Education, Youth and Innovation*,
https://doi.org/10.1007/978-3-032-21948-0_17

and water footprint is immensely increasing, contributing to rising global warming, reduction in water and non-renewable resources, and greenhouse gas emissions.

As a response, KSA construction projects are rapidly changing their strategies. They opted for energy-efficient design and its principles to reduce the footprints and enhance and contribute to sustainability. However, there remains a gap between a clear understanding of energy-efficient building designs and practices and the benefits of these designs on sustainability. This study addresses this critical gap of understanding impacts and its real-time and practical benefits in the construction projects of the KSA, and it also addresses;

- To assess the cost-effectiveness of energy-efficient designs on the lifecycle costs of sustainable construction projects in the KSA
- To determine the environmental benefits of employing energy-efficient design principles in reducing the carbon footprint of construction projects in the KSA
- To enumerate the long-term sustainability benefits achieved from energy-efficient design principles and strategies in the KSA
- To identify the key challenges faced by key stakeholders for implementing energy-efficient design in sustainable construction projects in the KSA
- To explore the potential of energy-efficient design in improving the performance of buildings in extreme weather conditions of the KSA

Research revealed that in the KSA, there is progress while constructing concepts to understand the role of energy-efficient design in sustainability; however, some reviewed evidence highlighted the increasing emphasis on sustainability globally (Gan et al., 2020), general sustainable strategies gained momentum without typical addressing of regional differences like local resource constraints (Ahmed, 2022), extreme climate conditions (Khan et al., 2024) and socio-economic and cultural factors influencing construction practices (Gopi et al., 2022). Moreover, there is limited empirical research to evaluate the long-term economic (Gopi et al., 2022) and environmental impacts of energy-efficient designs in KSA's construction projects (Li et al., 2021). There are recent underexplored elements which can give perspective to energy-efficient designs in KSA and align them with Vision 2030. Within these underexplored areas, Life Cycle Cost Analysis (LCCA) were found to be key drivers of sustainable economic growth and determine future costs an liabilities (Altaf et al., 2023). New methods for life cycle costs (LCC) analysis of energy-efficient designs have been presented (Liu & Mi, 2017). There are levels of features, which are probabilistically predicted in terms of seismic damage in energy-efficiency. Some other features are also included in lifecycle costs, which are repair, replacement costs. These underexplored areas are policy frameworks that lack comprehensive studies to examine the perspectives of key stakeholders like architects, engineers, and project managers in adopting sustainable practices (Häkkinen et al., 2015). Therefore, these gaps can lead to the motivation of decoding actionable strategies and case studies to foster the design tailored to KSA's needs.

Growth and expansion are important aspects of any industry, and similarly, the construction sector of KSA is proceeding towards it. However, the industry is famous for its hazardous outcomes in terms of emissions, environmental impacts and economy. According to Gan et al. (2020), buildings are responsible for up to 40% of anthropogenic greenhouse gas (GHC) emissions, which are more closely associated with human activities during construction, the primary driver of global warming. Research on lowering building energy use, carbon emissions, and lifetime costs began to offer solutions and suggestions for this problem. As a result, climate change and the increasing emphasis on sustainable building practices—of which energy efficiency is becoming a key component—have been the subject of an intriguing early evaluation of building environmental assessment in recent years.

The interconnectedness of building and environmental responsibility has revolutionized into a pivotal dimension of construction design and practices. Their innovative design and construction techniques within the realm of green building design focus on strategies to enhance energy efficiency (Umoh et al., 2024). Green building design has become imperative in construction and it emphasizes the significance of reducing ecological footprint like carbon footprint of buildings (Ochieng et al., 2014). Various strategies like passive design, green building, and efficient insulation systems are creating structures to harness natural resources intelligently. They minimize energy consumption and bring healthier indoor environment and innovative construction. Research also highlighted the implementation of recycled matter that are locally procured. It scrutinizes emerging technologies like bio-concrete and healing materials for aiding these improvements (Ojha et al, 2025). Through these identified measures, up to 25% environmental impact can be reduced as a response to minimizing the greenhouse gases emissions with sustainable energy (Zhang et al., 2022). However, research gaps remain as what the strategies were and what the benefits were.

Methodology

The research employs a qualitative research design to investigate the influence of energy-efficient design and strategies on construction projects in the KSA. It adopts a pragmatic research philosophy, which combines subjective and objective data to assess practicality like Triple Bottom Line (TBL) and LEED certification. An inductive approach aided in generating data directly, which enables a comprehensive understanding of patterns and themes related to energy-efficient designs. The research used a case study as a strategy for comprehending real-world construction projects, which provided a detailed analysis of energy-efficient designs in the KSA.

Data collection was another crucial step, which was performed using semi-structured interviews with 16 purposively selected individuals. For selecting participants, their location/residence of KSA and work experience of green construction projects were taken as inclusion criterion. These individuals were architects, engineers, policymakers and sustainability consultants working in the KSA. These participants are sampled based on their experience, regional presence, and involvement in energy-efficient construction projects. The interviews also focus on their perspectives on experience and address or lived challenges while adapting to these systems.

After data collection, the analysis was performed using thematic analysis via Nvivo12 software (Braun & Clarke, 2006). The analyses determined the recurring themes and sub-themes from the responses and organised that qualitative data into interpretable categories. Thus, this method allowed us to delve deeper into the impact, influence and challenges.

Findings

Braun and Clarke (2006) presented a subtle framework for qualitative research design as it helps in identifying, analysing and reporting specific codes and patterns within the datasets. It employs coding and categorising collected data systematically (shown in Table 17.1).

Thematic Analysis

Stage 1—Familiarisation

According to Braun and Clarke (2006), thematic analysis involves transcribing the interview responses, which provides written data for reading and examinations. This step is crucial in ensuring that the researcher gains a richer perspective of the collected data, establishing a robust base for further analysis. Transcription has a significant role in the process of familiarisation. Further, familiarisation with the responses allowed to develop an intimate understanding of the material, which can pave the way for systematic coding and themes.

Stage 2—Coding

For phase 2, 'coding', the software 'Nvivo 12' generates codes, which are based on and formed from research questions and objectives for formulating themes and sub-themes (shown in Fig. 17.1). Each category provides information against the research questions and objectives.

Stage 3—Themes

Phase 3 of Braun and Clarke (2006) is searching for potential themes, and it involves considering the recurring ideas, words and responses from the codes generated. During this initial coding and interview processing, 18 interviews were repeatedly gone through thoroughly again, and from this phase, three (3) main themes and twelve (12) sub-themes were identified, each representing the multifaceted version of data

Table 17.1 Braun & Clarke (2006)'s thematic analysis

S#	Phases	Definition
1	Familiarization	Going through data repeatedly, and gain deeper understanding
2	Coding	Identifying patterns and assigning descriptive codes
3	Themes	Group coding into potential themes according to research objectives/aims/questions
4	Reviewing themes	Checking themes against the data extracted and determine accuracy (Braun & Clarke, 2006)
5	Defining and naming themes	Refinement of themes and sub-themes and name them (Braun & Clarke, 2006)
6	Write-up	Use direct quotations and examples to present thematic analysis

Name	Files	Reference
(RQ1) Influence of energy-efficient designs on lifecycle costs of construction projects	16	35
Added Value of Construction Projects	4	4
Cost-Benefit Analysis	13	21
Reduction In Maintenance Needs	10	10
(RQ2) Influence of energy-efficient designs on sustainability of construction projects	16	60
Economic Sustainability	7	7
Minimization of Environmental Impacts	11	14
Reduction in Energy Consumption	8	8
Smart Buildings	13	20
Standard of Sustainability	8	11
(RQ3) Challenges	16	40
Extreme Weather Conditions	10	10
High Initial Costs	11	13
Lack of Availaibility of Materials and Technologies	8	11
Resistance from Stakeholders To Adopt Change	6	6

Fig. 17.1 Coding & themes

collected. These themes differ in their notion but have the same context, reflecting the diverse perspectives and benefits of the designs.

Stage 4—Reviewing themes and sub-themes

Braun and Clarke (2006) focus on analysing and developing initial themes. During this phase, the reviewed themes and subthemes initially coded against the dataset are found reliable and deemed consistent with the attainment of research questions. This iterative process enables the elimination of overlapping themes, but this study did not need it.

Stage 5—Naming Themes

Naming and describing themes in qualitative study requires presenting themes with the researchers' definitions of what the researcher understood by these themes. Each topic is carefully examined to ensure that it covers the main content of the knowledge it represents.

Table 17.2 has been developed to showcase the main themes and their descriptions.

Stage 6—Report Write-Up

Braun and Clarke's (2006) sixth stage of thematic analysis, "Writing," aims to present the findings coherently and consistently by connecting the themes identified in the research questions and information.

Table 17.2 Themes & its description

Main themes	Descriptions
RQ1—Influence of energy-efficient designs of lifecycle costs on construction projects	
Added value of construction projects	Improves project functionality, appeal and marketability
Cost–benefit analysis	Monitoring and balancing the short-term and long-term financial savings achieved
Reduction in maintenance needs	Reduce the demands, assures durability and effective operations
RQ2—Influence of energy-efficient designs on sustainability of construction projects	
Smart buildings	Incorporate technological advancements for inprovements
Minimization of environmental impacts	Reduces carbon footprints and promotes eco-friendly practices
Economic sustainability	Make sure of financial feasibility with environmental considerations
Reduction in energy consumption	Optimizes use of non-renewable energies to reduce operational costs
Standards of sustainability	Regulations and features must align with international sustainability guidelines
RQ3—Challenges	
High initial costs	Needs substantial financial investment in materials and technologies
Extreme weather condition'	Complicates effectiveness of designs and its implementation
Lack of availability of materials and technologies	Lowers feasibility in certain markets
Resistance from stakeholders to adopt change	Halts innovation and solution due to traditional reluctance

Discussion

The findings from the study indicated that construction requires sustainable techniques to play a significant role in reducing project risks, which particularly involve financial, environmental and operational domains. Further analysis narrowed down to the point that integration of sustainable techniques like energy-efficient design, material optimisation, and management of waste significantly manage budget overruns, delays, and regulatory compliance risks.

The study revealed three (3) main themes and 12 sub-themes. Theme one specified the value-added benefits (sub-theme 1), cost–benefit analysis (sub-theme 2), and reduced costs for maintenance needs (sub-theme 3). Findings showed that energy-efficient designs can significantly increase the added value of the buildings due to its resilience, appeal and cost-effectiveness. The cost-effectiveness may not be assessed initially as initial costs are higher, the total cost of ownership is lower due to reduced energy consumption, maintenance costs and operational costs. For instance, it lowers maintenance needs and its costs by having durable materials and systems designed for increasing energy efficiency, referring to fewer repairs and maintenance. For example, good insulation, improved heating and cooling, and electric windows not only reduce the building's overall energy consumption.

Second theme specified transformative aspects of energy-efficient designs like constructing smart buildings (sub-theme 4), minimizing the environmental impacts using passive design, energy-intensive mechanical system and light and ventilation system (sub-theme 5), economic sustainability (sub-theme 6), reducing energy consumption using HVAC system, natural lighting and high-performance insulation (sub-theme 7), and maintaining the standards of sustainability (sub-theme 8). Standards of sustainability play a vital role while guiding the adoption and energy implementation in construction projects. According to technology, electrical energy's development and production has become more efficient and easier to use, making it more widespread and more stable.

Third theme is specified challenges and barriers that hinder the progress of sustainable construction projects in KSA. There are several challenges identified while reviewing interviews like extreme weather conditions of KSA (sub-theme 9), lack of availability of materials and technologies (sub-theme 10), high initial costs (sub-theme 11), and resistance from stakeholders to adopt change (sub-theme 12). As we know, KSA has harsh climatic conditions due to it being a desert, has difficult construction. It also poses a significant challenge for implementing designs. Further, it lacks producing locally sourced materials and technologies, and they rely on imports from international markets. This increases the cost margins. Thus, it seems less cost-effective initially due to higher initial costs, which also creates resistance to change.

The findings from the study indicated that construction requires sustainable techniques for playing a significant role to reduce project risks, which particularly involve financial, environmental and operational domains. Further analyzing narrowed down to the point that integration of sustainable techniques like energy-efficient design, material optimization and management of wastes significantly manage budget overruns, delays and regulatory compliance risks.

Conclusions

The research's energy-efficient design estimates rely on prospective descriptive and interventional studies. There may be restrictions on this study. As a result, they are increasingly susceptible to biases and confounding, which could have affected the results. The generalisation and applicability of data and findings may not be appropriate in other situations due to its exclusive focus on construction projects and the Kingdom of Saudi Arabia. Due to its qualitative nature, it only offers a glimpse of construction management. Additionally, despite the qualitative interview-based methodology's apparent merits, the output is likely to be skewed because of the excessive sharing of feelings, opinions, views, and uncertainty.

The study was supported by the triple bottom line (TBL) and LEED certification model. However, because energy-efficient systems are readily available, future research could use the most recent case studies on these designs. It is proposed that stakeholders in these projects could look up to these studies, which use the most recent climate forecasts, their impact on building designs, and their efforts to promote more climate-sustainable projects. On a deeper level, significant elements of the cost, the effects of environmental factors, and the potential benefits in extreme weather conditions are only provided in verbatim qualitative and descriptive/narrative form; these could be enhanced if more accurate data becomes available in the form of case studies.

References

Altaf, M., Alaloul, W. S., Musarat, M. A., & Qureshi, A. H. (2023). Life cycle cost analysis (LCCA) of construction projects: Sustainability perspective. *Environment, Development and Sustainability, 25*(11), 12071–12118.

Ahmed, K. J. (2022). LEED certificate as a green building rating system: A study of middle east and Asia countries.

Braun, V., & Clarke, V. (2006). Using thematic analysis in psychology. *Qualitative Research in Psychology, 3*(2), 77–101.

Gan, V. J., Lo, I. M., Ma, J., Tse, K., Cheng, J. C., & Chan, C. (2020). Simulation optimisation towards energy-efficient green buildings: Current status and future trends. *Journal of Cleaner Production, 254*, Article 120012. https://doi.org/10.1016/j.jclepro.2020.120012

Gopi, B., Ramesh, G., & Logeshwaran, J. (2022). The fuzzy logical controller based energy storage and conservation model to achieve maximum energy efficiency in modern 5g communication. *ICTACT Journal on Communication Technology, 13*(3), 2774–2779.

Häkkinen, T., Kuittinen, M., Ruuska, A., & Jung, N. (2015). Reducing embodied carbon during the design process of buildings. *Journal of Building Engineering, 4*, 1–13.

Khan, S. I., Al Dodaev, Z., & Haque, A. (2024). The integration of renewable energy sources into mechanical systems, focusing on efficiency and reliability—A Case Study. *Control Systems and Optimization Letters, 2*(2), 172–177.

Li, L., Shi, D., Hou, R., Li, H., Pan, M., & Han, Z. (2021). To talk or to work: Flexible communication compression for energy efficient federated learning over heterogeneous mobile edge devices. In *IEEE INFOCOM 2021—IEEE Conference on Computer Communications* (pp. 1–10). IEEE.

Liu, M., & Mi, B. (2017). Life cycle cost analysis of energy-efficient buildings subjected to earthquakes. *Energy and Buildings, 154*, 581–589. https://doi.org/10.1016/j.enbuild.2017.08.056

Ochieng, E. G., Wynn, T. S., Zoufa, T., Ruan, X., Price, A., & Okafor, C. (2014). Integration of sustainability principles into construction project delivery. *Architectural Engineering Technology, 3*, 1.

Ojha, A., Kumar, A., Das, D., & Bandyopadhyay, T. K. (2025). Bio-concrete and its self-healing capacity to next-generation sustainable architecture: A comprehensive review. *Exon, 2*(1), 1–15.

Umoh, A. A., Adefemi, A., Ibewe, K. I., Etukudoh, E. A., Ilojianya, V. I., & Nwokediegwu, Z. Q. S. (2024). Green architecture and energy efficiency: A review of innovative design and construction techniques. *Engineering Science & Technology Journal, 5*(1), 185–200.

Zhang, M., Yan, T., Wang, W., Jia, X., Wang, J., & Klemeš, J. J. (2022). Energy-saving design and control strategy towards modern sustainable greenhouse: A review. *Renewable and Sustainable Energy Reviews, 164*, Article 112602. https://doi.org/10.1016/j.rser.2022.112602

Chapter 18
The Future of Urban Mobility: Challenges and Opportunities in Autonomous Public Transport

Dalia Hafiz

Abstract The advent of autonomous public transport (APT) signifies a transformative evolution in urban mobility, driven by advancements in automation, artificial intelligence, and sustainability efforts. APT offers significant benefits, including enhanced safety, reduced congestion, and lower emissions, while simultaneously addressing last-mile connectivity and accessibility challenges. However, its widespread adoption is contingent upon overcoming regulatory, technological, and societal barriers. This paper examines the historical development, advantages, challenges, and prospects of APT, with a focus on environmental impacts, economic feasibility, and public perception. It also discusses policy considerations and regulatory frameworks necessary for seamless integration of AV. The findings suggest that APT can be pivotal in shaping sustainable urban transport systems, provided that technological advancements and societal acceptance progress with regulatory measures.

Keywords Autonomous public transport · Urban mobility · Autonomous vehicles · Sustainability · Smart cities

Introduction

The introduction of autonomous public transportation (APT) represents a significant shift in urban mobility, driven primarily by technological advancements and the need for efficient and sustainable transportation systems. Public transport plays a vital role in shaping metropolitan areas and influences both quality of life and urban dynamics. Integrating autonomous vehicles (AVs) into public transport systems is expected to enhance safety by reducing human error, a significant factor contributing to traffic congestion and accidents (Iclodean et al., 2020). These vehicles are designed to

D. Hafiz (✉)
College of Architecture, Art and Design, Ajman University, Ajman, UAE
e-mail: d.hafiz@ajman.ac.ae

J. Dsilva (ed.), *Sustainability Education, Youth and Innovation*,
https://doi.org/10.1007/978-3-032-21948-0_18

maintain optimal distances between one another and strictly follow traffic regulations, thereby improving both road safety and efficiency.

Urban environments particularly benefit from autonomous shuttle buses, as they can navigate congested areas where traditional transport methods struggle to meet increasing demands. Pilot programs on university campuses and in industrial zones demonstrate the potential of these vehicles in controlled settings. However, widespread adoption depends on demonstrating their reliability, efficiency, and societal acceptance. Furthermore, legal frameworks surrounding AVs vary across regions, influencing their implementation and operation (Iclodean et al., 2020). Social implications, including job displacement and public perception, must also be considered as AVs become integral to public transport networks.

Methodology

This study employs a historical and comparative analysis to examine the development, benefits, and challenges of autonomous public transport. The research synthesizes data from academic sources, industry reports, and case studies to trace the evolution of AV technology in public transport. A chronological review highlights key milestones, from early automation experiments in the 1920s to modern-day applications. Comparative assessments of pilot programs and real-world implementations also provide insights into operational efficiencies, public acceptance, and regulatory frameworks. The methodology also incorporates an evaluative approach to assess the economic, environmental, and social implications of AVs, drawing on recent studies and empirical data.

Historical Developments in Autonomous Public Transport

The development of autonomous public transport has evolved over several decades. Early automation experiments date back to the 1920s, with Francis P. Houldina demonstrating remote vehicle control. By 1939, at the New York World's Fair, Norman Bel Geddes, backed by General Motors, showcased a vehicle operating via magnetic fields, envisioning the future of automated transportation (Katona & Juhasz, 2020). RCA Labs further advanced automation in the 1950s with vehicles interpreting road signals for autonomous steering and braking.

The 1980s marked a turning point, with Mercedes-Benz introducing vision-based automation for non-driver vehicles that successfully operated on public roads. The United States Department of Transportation established the first legislative frameworks in 1997, providing a legal basis for AV integration. Schiphol Airport's "Park Shuttle," introduced in December 1997, was one of the earliest examples of AVs in public use. Since the 2000s, rapid advancements in automation technology have led to the widespread testing and deployment of AVs in public transport systems (Katona & Juhasz, 2020).

Benefits of Autonomous Public Transport

Autonomous public transport presents multiple benefits, including enhanced safety, environmental sustainability, and economic efficiency. Automated buses offer more consistent and predictable driving patterns, reducing accident risks. Research indicates that passengers perceive AVs as safer and more reliable than traditional buses (Seredynski, Nielsen, Ekman, & Johansson, 2023). Additionally, AVs, particularly electric autonomous buses, contribute to sustainability by eliminating tailpipe emissions and aligning with urban policies aimed at reducing greenhouse gas emissions and energy consumption.

Economic benefits include potential cost reductions in public transport operations. Eliminating the need for drivers may lower operational costs and, in turn, reduce passenger fares. This economic advantage and improved efficiency position AVs as a viable solution for modern urban mobility challenges (Seredynski et al., 2023).

Challenges and Lessons Learned

Despite the advantages, implementing autonomous buses presents challenges. Public perception of security remains a significant concern, particularly regarding safety at night or in isolated areas where digital assistance may not be sufficient. Additionally, the displacement of bus drivers raises socioeconomic concerns, though new job opportunities in AV maintenance and supervision may emerge (Seredynski et al., 2023).

Passenger acceptance also depends on how AVs integrate with existing transport systems. Studies indicate that users appreciate the benefits of automation but do not foresee significant changes in their travel behavior due to the existing sufficiency of public transport. Addressing these concerns requires incorporating information and communication technology (ICT) solutions to enhance security and improve passenger confidence.

Environmental Impact of Autonomous Public Transport

Autonomous public transport can potentially reduce emissions and significantly improve urban mobility. Studies indicate that a single AV can replace 7–11 conventional vehicles, leading to an emission reduction of up to 94% compared to traditional vehicles. This impact is further amplified when AVs operate on electric power sourced from renewable energy.

Moreover, AV integration optimizes fleet management, reducing traffic congestion and emissions. In Austin, Texas, autonomous taxi fleets are projected to reduce emissions by 60%. However, the effectiveness of these reductions depends on factors such as fleet size, charging infrastructure, and operational strategies. Urban land use could also be transformed as AVs reduce the need for parking spaces, allowing for more green areas and pedestrian-friendly zones.

Social and Economic Implications

The deployment of autonomous shuttles as a service (ASaaS) addresses last-mile transportation challenges by enhancing urban mobility. AVs provide accessibility to underserved communities, improving connections to essential services such as hospitals and retail centers. Their integration can also reduce road accidents, as human error remains a leading cause of traffic incidents (Bucchiarone, Battisti, Marconi, Maldacea, & Ponce, 2021).

Economically, AVs promise cost savings for municipalities and transport operators by reducing labour costs and optimizing fleet management. Public–private partnerships can further drive innovation and investment in autonomous transport solutions. However, concerns over job displacement necessitate workforce retraining and policy interventions to support affected workers (Bucchiarone et al., 2021).

Public Perception and Acceptance

Public acceptance is crucial for the success of autonomous public transport. A study in Singapore revealed a high level of user acceptance, with participants expressing willingness to adopt AVs due to their reliability and efficiency (Chng & Cheah, 2020). However, concerns persist regarding technical reliability, accident liability, and unexpected operational scenarios (Hilgarter & Granig, 2020). Effective communication, transparency, and user education are key to improving trust in AVs.

Limitations and Regulatory Issues

The technological efficiency of AVs remains a challenge, requiring continuous advancements in sensor technologies and AI-driven decision-making. Data security and privacy concerns also arise, as AVs collect vast amounts of information susceptible to cyber threats (Acharya, 2014). Additionally, inconsistent regulations across regions hinder the seamless deployment of AVs. Policymakers must develop comprehensive legal frameworks to address liability, safety, and ethical concerns (Kotliarenko, 2020).

Findings

Autonomous public transport (APT) has evolved from early automation experiments in the 1920s to modern AI-driven systems, offering significant benefits in terms of safety, sustainability, and economic efficiency. By eliminating human error, AVs enhance road safety, reduce greenhouse gas emissions, and optimize operational costs (Seredynski et al., 2023). However, security concerns, job displacement, and public skepticism remain barriers to widespread adoption (Bucchiarone et al., 2021). Regulatory inconsistencies and data security risks further complicate implementation, requiring continuous technological advancements and comprehensive legal frameworks (Acharya, 2014; Kotliarenko, 2020).

These findings underscore the potential of APT in reshaping urban mobility while highlighting the need for continued technological advancements, policy development, and societal engagement to ensure successful implementation.

Conclusion and Future Directions

Autonomous public transport represents a paradigm shift in urban mobility, offering safety, environmental, and economic benefits. However, successful implementation depends on addressing technological, legal, and social challenges. Future research should focus on refining AV technology, developing inclusive policies, and ensuring public trust to facilitate the widespread adoption of AV technology.

References

Acharya, A. (2014). *Are we ready for driverless vehicles? Security vs. Privacy—A social perspective*. Retrieved from http://arxiv.org/abs/1412.5207

Bucchiarone, A., Battisti, S., Marconi, A., Maldacea, R., & Ponce, D. C. (2021). Autonomous Shuttle-as-a-Service (ASaaS): Challenges, opportunities, and social implications. *IEEE Transactions on Intelligent Transportation Systems, 22*(6), 3790–3799. https://doi.org/10.1109/TITS.2020.3025670

Chng, S., & Cheah, L. (2020). Understanding autonomous road public transport acceptance: A study of Singapore. *Sustainability, 12*(12), 4974. https://doi.org/10.3390/su12124974

Hilgarter, K., & Granig, P. (2020). Public perception of autonomous vehicles: A qualitative study based on interviews after riding an autonomous shuttle. *Transportation Research Part F: Traffic Psychology and Behaviour, 72*, 226–243. https://doi.org/10.1016/j.trf.2020.05.012

Iclodean, C., Cordos, N., & Varga, B. O. (2020). Autonomous shuttle bus for public transportation: A review. *Energies, 13*(11), 2917.

Katona, G., & Juhasz, J. (2020). The history of the transport system development and future with sharing and autonomous systems. *Communications-Scientific Letters of the University of Zilina, 22*(1), 25–34.

Kotliarenko, V. I. (2020, April). Legal regulation in the field of automated automobile transport and road tests of autonomous vehicles. In *IOP Conference Series: Materials Science and Engineering* (Vol. 819, No. 1, p. 012033). IOP Publishing.

Part IV
Health, Wellbeing, and Sustainable Communities

Chapter 19
Communicating Health and Sustainable Development Research to Policymakers: A Policy Roadmap for Evidence-Informed Decision Making

Immanuel Azaad Moonesar

Abstract This perspective examines the critical intersection between health, sustainable development, and policy communication, building upon insights from a policy workshop titled "Communicating Sustainable Development Research/Data to Policymakers." As the MENA region undergoes significant economic transformation with health emerging as a key growth engine, effective communication between researchers and policymakers becomes paramount. This paper presents a comprehensive framework for translating complex health and sustainability research into actionable policy insights, addressing communication barriers, and establishing clear pathways for evidence-informed decision making. The analysis incorporates best practices for policy brief development, stakeholder engagement strategies, and knowledge translation mechanisms specifically relevant to the evolving health economy landscape in the Middle East.

Keywords Health policy communication · Sustainable development · Evidence-based policy · Knowledge translation · MENA health economy

Introduction

The convergence of health and sustainable development has emerged as a defining characteristic of 21st-century policy making, particularly in the Middle East and North Africa (MENA) region, where countries are actively diversifying their economies through health sector investment (Moonesar & Hickman, 2017). As governments across the region invest billions in healthcare infrastructure, medical tourism, and health technology, the need for effective communication between researchers and policymakers has never been more critical.

I. A. Moonesar (✉)
Mohammed Bin Rashid School of Government, Dubai, UAE
e-mail: immanuel.moonesar@mbrsg.ac.ae

J. Dsilva (ed.), *Sustainability Education, Youth and Innovation*,
https://doi.org/10.1007/978-3-032-21948-0_19

The policy workshop "Communicating Sustainable Development Research/Data to Policymakers," conducted at the SEE Global Sustainability Summit in Dubai, highlighted a persistent challenge: the significant gap between research production and policy implementation in sustainability and health domains. This challenge is particularly acute in rapidly transforming economies where policy decisions must be made quickly to capitalize on emerging opportunities while ensuring sustainable development outcomes.

Research indicates that policymakers face substantial barriers when attempting to incorporate scientific evidence into decision-making processes (Young & Quinn, 2007). These barriers become more pronounced in the health and sustainability nexus, where complex interdependencies between environmental, social, and economic factors require sophisticated understanding and communication strategies.

The Policy Communication Challenge in Health and Sustainability

Understanding Policy Triggers and Contexts

Public policy emerges from various triggers, ranging from strategic leadership priorities to external crises (Simon, 2018). In the MENA context, several factors have accelerated the integration of health and sustainability considerations into policy frameworks.

Strategic Economic Diversification: Countries like the UAE, Saudi Arabia, and Qatar have identified health as a cornerstone of their economic diversification strategies, necessitating evidence-based policy development to guide substantial investments in healthcare infrastructure and medical tourism initiatives.

Climate and Health Nexus: Rising temperatures, water scarcity, and environmental degradation across the region have created urgent public health challenges that require integrated policy responses based on robust research evidence.

Post-pandemic Recovery: The COVID-19 pandemic highlighted the critical importance of health system resilience and the need for policies that integrate health security with sustainable development goals.

Research Communication Barriers

Analysis of research communication challenges reveals several systematic barriers that impede effective knowledge translation in the health and sustainability domain.

Technical Complexity: Health and sustainability research often involves sophisticated methodologies, complex data sets, and nuanced findings that are difficult to translate into policy-relevant insights (Morris, 2019).

Temporal Misalignment: Research timelines rarely align with policy decision-making cycles, creating challenges for timely evidence integration into policy processes.

Institutional Disconnects: Limited formal mechanisms exist for facilitating regular dialogue between research institutions and policy-making bodies, particularly in emerging health economy sectors.

Cultural and Language Barriers: In the MENA region, effective communication must navigate multiple languages, cultural contexts, and institutional frameworks, adding complexity to knowledge translation efforts.

A Framework for Effective Health and Sustainability Policy Communication

The Knowledge Translation Imperative

Knowledge translation, defined as "a dynamic and iterative process that includes synthesis, dissemination, exchange and ethically sound application of knowledge to improve health outcomes, provide more effective health services and strengthen health care systems" (Canadian Institutes of Health Research, 2021), provides a foundational framework for addressing communication challenges.

In the context of health and sustainable development, knowledge translation must extend beyond traditional health system boundaries to encompass economic, environmental, and social policy domains. This requires sophisticated communication strategies that can effectively convey complex interdependencies and trade-offs inherent in sustainable development policy choices.

The Four Pillars of Effective Policy Communication

Based on workshop insights and international best practices, four key pillars emerge for effective policy communication in health and sustainability domains:

1. Establishing Clear Communication Channels

Governments and research institutions must develop systematic platforms for regular dialogue between researchers and policymakers. In the UAE's context, this includes leveraging existing initiatives like the Mohammed Bin Rashid School of Government's policy forums and the UAE's Future Government initiatives to create structured opportunities for evidence-sharing and dialogue.

2. Developing Standardized Knowledge Translation Products

Policy briefs, executive summaries, and policy notes must be designed specifically for busy policymakers who require concise, actionable information. These products should follow established formatting guidelines that prioritize clarity, visual appeal, and actionable recommendations (Moonesar, 2024).

3. Strategic Media Engagement and Communication

Effective policy communication extends beyond direct researcher-policymaker interactions to include strategic engagement with media outlets, social platforms, and public forums that shape policy discourse and public opinion.

4. Capacity Building and Training Programs

Both researchers and policymakers require enhanced skills for effective communication across disciplinary and institutional boundaries. This includes training researchers in policy communication and helping policymakers develop greater research literacy.

Best Practices for Policy Brief Development

Structure and Content Guidelines

Effective policy briefs in the health and sustainability domain must adhere to specific structural and content guidelines that maximize impact and usability.

Executive Statement: The opening section should clearly articulate the policy challenge, proposed solution, and expected outcomes using accessible language free from technical jargon.

Evidence Base: Research findings must be presented using clear visualizations, compelling statistics, and concrete examples that demonstrate real-world implications for health and sustainability outcomes.

Policy Recommendations: Recommendations should be specific, actionable, and clearly linked to stated policy objectives. Each recommendation should include implementation timelines, resource requirements, and success metrics.

Implementation Considerations: Policy briefs must address practical implementation challenges, including stakeholder coordination requirements, potential resistance factors, and risk mitigation strategies.

Visual Communication and Design

Modern policy briefs must incorporate sophisticated visual design elements that enhance comprehension and retention.

Infographics and Data Visualization: Complex health and sustainability data should be presented through clear charts, graphs, and infographics that highlight key trends and relationships.

Case Studies and Examples: Real-world examples from the MENA region and other comparable contexts help policymakers understand practical applications and potential outcomes.

Clear Typography and Layout: Professional design elements, consistent formatting, and logical information hierarchy improve readability and professional credibility.

Sectoral Applications: Health Economy Integration

Medical Tourism and International Competitiveness

As MENA countries develop medical tourism capabilities, policy communication must effectively convey complex relationships between healthcare quality, infrastructure investment, regulatory frameworks, and economic outcomes. Research on medical tourism economics, patient safety protocols, and international benchmarking must be translated into actionable policy insights that guide strategic investments and regulatory decisions.

Health Technology and Innovation Policy

The rapid advancement of health technologies, including telemedicine, artificial intelligence applications, and digital health platforms, requires sophisticated policy frameworks that balance innovation promotion with patient safety and data protection concerns. Effective communication strategies must help policymakers understand technology capabilities, regulatory requirements, and economic implications.

Sustainable Healthcare Infrastructure

Integration of sustainability principles into healthcare infrastructure development requires complex policy coordination across multiple sectors. Policy communication must effectively convey research findings on energy-efficient healthcare facilities, waste management systems, and sustainable supply chains while addressing economic feasibility and implementation challenges.

Regional Considerations and Cultural Adaptation

MENA-Specific Communication Strategies

Effective policy communication in the MENA region must account for unique cultural, institutional, and linguistic contexts.

Multilingual Communication: Policy briefs and communication materials should be available in Arabic and English, with cultural adaptation of content and examples to reflect regional contexts and priorities.

Islamic Finance and Health Policy: Communication strategies must address the intersection of Islamic finance principles with health sector investment and policy development, particularly relevant for sukuk financing of healthcare infrastructure.

Cross-Border Collaboration: Regional health initiatives and pan-GCC cooperation mechanisms require communication strategies that facilitate knowledge sharing and policy coordination across national boundaries.

Stakeholder Engagement and Multi-level Governance

The complex governance structures characteristic of federal systems like the UAE require sophisticated stakeholder engagement strategies that account for federal, emirate, and local-level policy responsibilities. Effective communication must address multiple decision-making levels and facilitate coordination across jurisdictional boundaries.

Measuring Impact and Effectiveness

- Communication Success Metrics

Effective policy communication requires systematic evaluation of impact and effectiveness:

- Policy Uptake Indicators: Tracking the integration of research findings into policy documents, legislative proposals, and strategic planning processes provides concrete evidence of communication effectiveness.

- Stakeholder Engagement Metrics: Measuring participation in policy forums, workshop attendance, and ongoing dialogue between researchers and policymakers helps assess relationship-building success.
- Media Coverage and Public Discourse: Monitoring media coverage of health and sustainability policy issues provides insights into broader communication reach and influence.
- Long-term Policy Outcomes: Evaluating the implementation success of evidence-informed policies provides the ultimate measure of communication effectiveness.

Future Directions and Recommendations

Technological Innovation in Policy Communication

Emerging technologies offer new opportunities for enhancing policy communication effectiveness.

Digital Platforms and Interactive Tools: Virtual reality applications, interactive data dashboards, and digital simulation tools can help policymakers better understand complex health and sustainability relationships.

Artificial Intelligence and Personalization: AI-powered systems can customize policy communication content for specific audiences, ensuring relevant and targeted information delivery.

Social media and Digital Engagement: Strategic use of social media platforms and digital communication channels can expand reach and facilitate broader stakeholder engagement in policy development processes.

Institutional Development and Capacity Building

Sustained improvement in policy communication requires institutional innovation and capacity building.

Policy Communication Centers: Establishing specialized units within research institutions and government agencies dedicated to knowledge translation and policy communication.

Professional Development Programs: Creating formal training programs for researchers and policymakers that build cross-disciplinary communication skills and mutual understanding.

International Collaboration Networks: Developing partnerships with international institutions and organizations to share best practices and learn from global experiences in policy communication.

Conclusion

The integration of health considerations into sustainable development policy frameworks represents both an unprecedented opportunity and a significant challenge for the MENA region. Success in capturing the economic potential of the health economy while achieving sustainable development goals depends critically on effective communication between researchers and policymakers.

The framework presented in this analysis—emphasizing clear communication channels, standardized knowledge translation products, strategic media engagement, and capacity building—provides a roadmap for enhancing policy communication effectiveness. However, implementation success requires sustained commitment from both research institutions and government agencies to invest in relationship building, skill development, and institutional innovation.

As the MENA region continues its economic transformation journey, the quality of communication between researchers and policymakers will significantly influence the success of health sector investments and the achievement of sustainable development objectives. The investment in effective policy communication represents not just an academic exercise, but a critical component of successful economic diversification and sustainable development strategy.

The path forward requires recognition that policy communication is both an art and a science, requiring a sophisticated understanding of research evidence, policy processes, and stakeholder dynamics. By embracing this complexity and investing in communication excellence, the MENA region can maximize the potential of its health economy transformation while contributing to global knowledge on effective policy communication in rapidly evolving economic contexts.

References

Canadian Institutes of Health Research. *Guide to knowledge translation planning at CIHR: Integrated and end-of-grant approaches*. Government of Canada Publications. https://cihr-irsc.gc.ca/e/45321.html

Moonesar, I. A. (2024). *Public policy as a catalyst for health equity*. MedEdge MEA, July–August 2024 edn. https://issuu.com/mededgemea/docs/fourth_edition_read_only

Moonesar, I. A., & Hickman, C. L. (2017). How the UAE can reduce the prevalence of obesity among the youth: A health policy perspective. *Advances in Obesity, Weight Management & Control, 6*(4), 64–68. https://medcraveonline.com/AOWMC/how-the-uae-can-reduce-the-prevalence-of-obesity-among-the-youth---a-health-policy-perspective.html

Morris, Z. S. (2019). Research communication strategies for policy impact: A systematic review. *Policy Sciences*, 52(3), 287–305. https://link.springer.com/article/10.1007/s11077-019-09352-0
Researchers. Overseas Development Institute.
Simon, C. A. (2018). *Public policy: Preferences and outcomes* (4th ed.). Routledge. https://www.routledge.com/Public-Policy-Preferences-and-Outcomes/Simon/p/book/9781138499439
Young, E., & Quinn, L. (2007). *Making research evidence matter: A guide to policy engagement for researchers*. Overseas Development Institute. https://odi.org/en/publications/making-research-evidence-matter-a-guide-to-policy-engagement-for-researchers/

Chapter 20
Measuring What Matters: A Framework for Social Sustainability in Buildings and Cities

Gayathri Unnikrishnan

Abstract Social sustainability remains one of the least defined and least measured aspects of sustainability in the built environment, often overshadowed by carbon reduction and environmental metrics. This study addresses that gap by introducing the Liveable Future Framework, a structured approach for embedding social sustainability into material selection, urban planning, and ESG strategies. Using a mixed-methods approach, the research draws on global best practices, industry standards, and case studies from cities including Stockholm, Singapore, San Francisco, and Dubai. The framework is built on five pillars, People, Product, Places, Philanthropy, and Processes, offering a holistic model that aligns human well-being, equity, and resilience with environmental performance. Findings highlight persistent challenges, including the lack of measurable social impact indicators in ESG reporting, weak integration of public health metrics into urban policy, and the disconnect between sustainable material guidelines and social responsibility. By positioning social sustainability as a measurable and strategic imperative, the framework provides policymakers, designers, and businesses with practical tools to align net-zero ambitions with community well-being. The study demonstrates how inclusive design, equitable access to green space, and ethical sourcing can be embedded into the development process, strengthening community resilience and enhancing quality of life. This work positions social sustainability not as an optional add-on, but as an essential driver of truly sustainable cities.

Keywords Social sustainability · ESG · ESG frameworks for Human Well-Being · Public health · Well-Being

---

G. Unnikrishnan (✉)
Liveable, CA, USA
e-mail: gayathri@makeitliveable.com

J. Dsilva (ed.), *Sustainability Education, Youth and Innovation*,
https://doi.org/10.1007/978-3-032-21948-0_20

Introduction

The UAE's commitment to Net Zero 2050, Masdar City, and the Green Economy for Sustainable Development places it at the forefront of global sustainability efforts. Currently, sustainability has focused predominantly on environmental metrics such as energy efficiency and emissions reduction, often overlooking social sustainability (Deloitte, 2023). This paper argues that net-zero strategies must integrate social equity, inclusivity, and community well-being to create truly sustainable urban environments.

Historical Context of Social Sustainability in the Built Environment

The concept of social sustainability within environmental design has evolved significantly since the 1970s when community participation in urban development was first recognized as a critical factor in shaping resilient and inclusive spaces (Burra Charter, 1979). Initially, sustainability frameworks prioritized environmental efficiency over social concerns. However, the introduction of the Burra Charter (1979) and later the United Nations Sustainable Development Goals (2015) emphasized the need to integrate social equity, health, and governance into sustainability strategies.

This shift has led to the rise of ESG frameworks that mandate social impact assessments alongside environmental metrics, driving organizations to redefine sustainability as a multi-dimensional effort (Deloitte, 2023; UN, 2015).

Integration of Social Sustainability into ESG Strategies

In recent years, ESG (Environmental, Social, and Governance) frameworks have required organizations to assess and report their social impacts alongside environmental performance, fostering transparency and accountability (World Green Building Council, 2024).

This shift reflects a broader understanding that sustainable design must not only address the physical and environmental aspects of buildings but also their long-term social implications. Key ESG rating agencies now include social sustainability metrics such as equity in housing, access to green spaces, and community well-being indicators (Deloitte, 2023). However, a lack of standardized frameworks remains a challenge, necessitating industry-driven approaches to measure social impact effectively.

Industry-Driven Approaches to Social Sustainability

To ensure meaningful integration of social sustainability into urban development, an industry-driven approach is necessary. This approach prioritizes collaboration between policymakers, corporate entities, and local communities to align sustainable design principles with socio-economic needs.

For example, stakeholder engagement workshops and multi-sector partnerships have been instrumental in creating equitable and resilient urban environments (Shirazi & Keivani, 2020). Moreover, private-sector initiatives like the WELL Building Standard and LEED v5 have further reinforced the need for structured evaluation of social impact in the built environment.

Methodology

This research follows a mixed-method approach, combining:

1. Literature Review: Analysis of existing ESG frameworks, sustainability charters (Burra Charter, UN SDGs), and industry standards (LEED v5, WELL Building Standard, Living Building Challenge) to establish best practices for integrating social sustainability into environmental design.
2. Case Study Analysis: Examination of global cities that successfully integrate social sustainability into urban planning, including:
    - Stockholm, Sweden: Equitable access to green spaces and Stockholm Wood City timber development.
    - Singapore: Smart city policies that enhance social connectivity and livability.
    - San Francisco, USA: Office-to-housing conversions and equity-driven urban planning.
    - Dubai, UAE: Social sustainability in the Dubai 2040 Urban Master Plan and UAE Green Agenda 2030.
3. Industry Benchmarking: Review of certifications and ESG reporting standards (GRI, GRESB, WELL, and LEED v5) to assess how companies and policymakers currently measure social impact within built environments.

Findings

Despite advancements in net-zero strategies, ESG reporting, and green building certifications, social sustainability remains critically underdeveloped in the built environment. The three major gaps identified are:

- Lack of Social Impact Metrics in ESG Reporting: 90% of ESG reports primarily focus on carbon emissions and energy efficiency, while less than 20% include measurable social impact indicators (Deloitte, 2023).
- The Disconnect Between Urban Development and Public Health: Studies show that urban design significantly affects mental and physical health, yet <30% of global urban sustainability policies explicitly integrate public health indicators (Shirazi & Keivani, 2020).
- Failure to Link Sustainable Materials with Social Responsibility: Most sustainable material selection guidelines prioritize carbon impact and circularity but fail to account for social sustainability factors like labor practices, indoor air quality, and supply chain ethics.

The findings from this research contribute to:

1. Policy & Urban Planning: Establishing a framework for integrating social sustainability metrics into national sustainability policies,
2. Public Health & Well-Being: Demonstrating how urban design, green space access, and sustainable healthcare policies directly impact mental and physical well-being.
3. Corporate ESG Strategies: Providing a roadmap for businesses to enhance ESG disclosures by including measurable social sustainability indicators
4. Community Resilience: Offering practical strategies for designing inclusive, accessible, and climate-resilient communities, particularly relevant for cities facing extreme climate challenges.

The Liveable Future Framework

This research introduces the Liveable Future Framework shown in Fig. 20.1, a structured approach to integrating social sustainability into material selection, urban planning, and environmental design. The framework bridges the gap between where organizations start and where the market is moving, ensuring that sustainability strategies align with evolving ESG expectations, human well-being, and urban development goals.

The 5Ps of Social Sustainability form the foundation of this framework, ensuring sustainability strategies comprehensively address both environmental and human impact:

- People: Prioritizing well-being, inclusivity, and workforce development, ensuring human-centered sustainability in organizations and urban design.
- Product: Emphasizing sustainable and ethical sourcing of materials, ensuring that built environments promote responsible consumption and production while minimizing environmental and social harm.
- Places: Designing resilient, accessible, and culturally aligned urban environments, fostering inclusive and equitable communities where social sustainability is embedded into urban planning and infrastructure development.

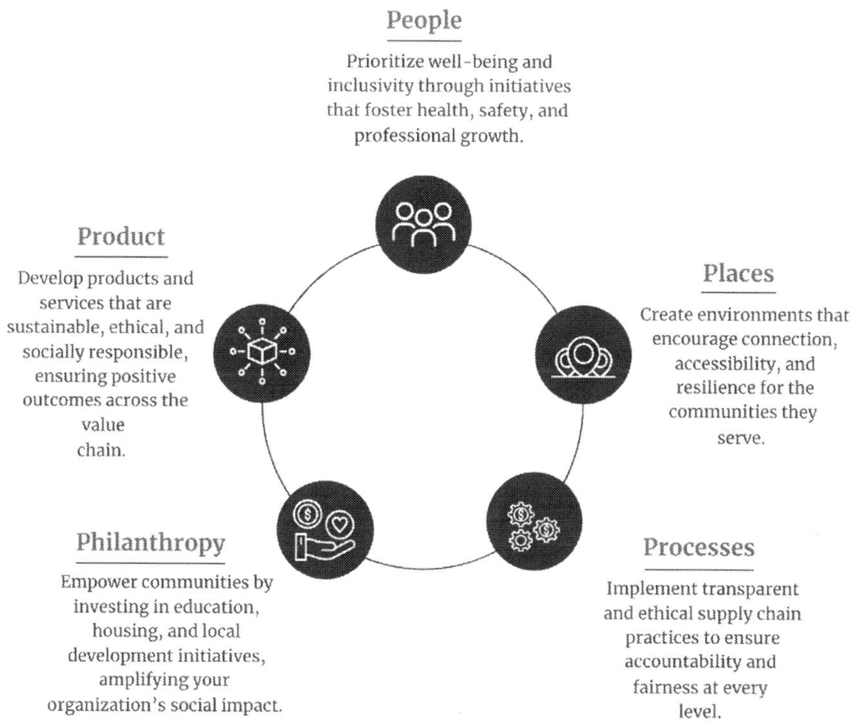

Fig. 20.1 Liveable future framework

- Philanthropy: Encouraging businesses and urban developers to invest in community development, education, and social equity initiatives, ensuring that corporate responsibility efforts contribute meaningfully to sustainable cities and communities.
- Processes: Implementing transparent, ethical, and data-driven governance models, ensuring long-term accountability in sustainability efforts and enabling organizations to track, report, and continuously improve their social impact strategies.

Stakeholder Engagement and Governance in Social Sustainability

A crucial element of integrating social sustainability into ESG strategies is effective stakeholder engagement, ensuring that diverse perspectives are included in decision-making processes.

Industry-driven approaches emphasize collaboration with local communities, policymakers, and businesses to establish shared sustainability goals as shown in Fig. 20.2 (World Green Building Council, 2024).

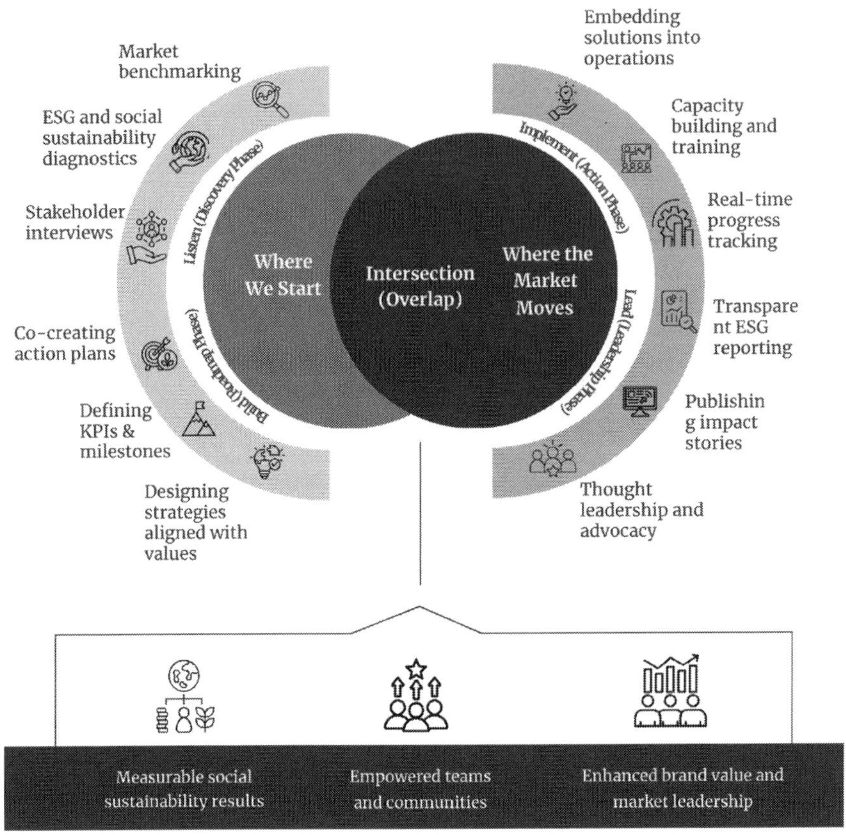

Fig. 20.2 Application of Liveable future framework in sustainability strategies

However, organizations often struggle to balance environmental and social priorities, particularly when ESG metrics prioritize carbon reduction over human-centered design (Deloitte, 2023). Effective governance structures must support multi-stakeholder collaboration to ensure sustainability policies reflect both environmental imperatives and social equity considerations.

Conclusions

This paper sets out to address a critical gap in the built environment: the underrepresentation of social sustainability in urban planning and ESG strategies. Through a review of global frameworks, case studies, and benchmarking, this research introduced the Liveable Future Framework to support cities and organizations in embedding measurable, people-centered sustainability into practice.

By embedding social sustainability into urban planning, green spaces, and material selection, this paper directly contributes to:

- Sustainable wellbeing and health: Ensuring indoor air quality and built environment design promote mental and physical well-being.
- Green space access: Advocating for equitable distribution of parks, urban greenery, and community spaces.
- Community resilience to environmental crises: Strengthening disaster preparedness and social cohesion through inclusive design.

Case study analysis reinforces the framework's practical relevance. Stockholm's integration of equitable green space access through its Wood City development exemplifies the "Places" pillar, showing how inclusive design boosts livability. Singapore's smart city infrastructure, aimed at strengthening social connectivity, reflects the "Processes" and "People" components. Closer to the UAE, Dubai's 2040 Urban Master Plan, with its emphasis on community resilience and livability, aligns with all five Ps—offering a compelling regional model for operationalizing the framework.

In sum, this research demonstrates that social sustainability is not a soft metric but a measurable, strategic imperative, and the Liveable Future Framework offers the tools to act on it. As the UAE and other nations advance toward net-zero targets, integrating social sustainability will be essential for creating cities that are not only carbon–neutral but truly livable.

References

Burra Charter. (1979). *The Burra Charter: The Australia ICOMOS charter for places of cultural significance*. ICOMOS Australia

Deloitte. (2023). *ESG trends shaping the future of sustainable finance*. Deloitte Insights

Shirazi, M. R., & Keivani, R. (2020). *Promoting social sustainability of urban neighborhoods*. Springer.

United Nations. (2015). *Transforming our world: The 2030 agenda for sustainable development*. United Nations.

World Green Building Council. (2024). *Creating better places for people: The role of the built environment in social sustainability*. WorldGBC.

Chapter 21
Landscape of Care: Intensification of the Quantity and Quality of Urban Green Spaces for Health and Well-Being

Fang Yun Koh, Christine Yogiaman, Jason Lim, and Kai Zheng

Abstract Tapping into the transformative potential of people-plant interactions to achieve optimal population health, this thesis looks to transform urban green spaces, frequently aestheticised and limited in functionality, into intensive rehabilitative assemblies in the context of high-density urban living. Straddling both culture and environment, urban green spaces are prime to be redesigned as the upkeep of health and well-being becomes ever more pertinent with an ageing population. Through intensification in the quantity and quality of urban green spaces close to the home, not only are traditional notions of environmental aesthetics disrupted, the agency to better health and well-being is returned to the individual and community.

Keywords Urban green spaces · People plant interactions · Self-sustaining infrastructure · Health and well-being

Introduction

It is commonly espoused that nature, in it multitude of forms, play a pivotal role in promoting health and well-being. The innate affinity, emotional connection and even spiritual bonds that humans hold with nature are theorised to be based on the phylogenetic development of the modern human, and the Biophilia Hypothesis asserts that the propensity to affiliate with other living organisms is a fundamental and genetically based human need.

F. Yun Koh · J. Lim · K. Zheng
Singapore University of Design and Technology, Singapore, Singapore
e-mail: jason_lim@sutd.edu.sg

K. Zheng
e-mail: kai_zheng@sutd.edu.sg

C. Yogiaman (✉)
Canadian University of Dubai, Dubai, UAE
e-mail: christine.yogiaman@cud.ac.ae

J. Dsilva (ed.), *Sustainability Education, Youth and Innovation*,
https://doi.org/10.1007/978-3-032-21948-0_21

A rapidly urbanised world threatens the natural landscape and one's genetic disposition to connect with nature, thereby compromising the maintenance, treatment and recovery of the mind and body. Therapeutic landscapes in environmental psychology and health geography points to interrelations of landscape, culture and health (Gesler, 1992). Gesler posits that "physical and built environments, social conditions and human perceptions combine to produce an atmosphere which is conducive to healing". This paper aims to re-evaluate existing green spaces, in terms of availability and functionality, beyond the notion of health as disease-free and more widely encompassing individual agency and in physical, mental, and social aspects, and postulates a new processual landscape of health integrated into the existing environment to augment health and well-being.

Methodology

With Singapore as a case study, how can the environment be designed to optimise population health, given its challenges with land scarcity and high population density? This paper examines this question, with the following objectives: (1) evaluating the quantity and quality of green urban spaces available, (2) determine accompanied human factors contributing to well-being and (3) to suggest key strategies in enhancing health amongst its population.

Quantity and Distribution of Green Space

The metrics of evaluation are formulated firstly with the basis that nature must be available and accessible. Those with intrinsic motivations will actively seek out green spaces and are less so hindered by distance. However, with the graduate decline of physical mobility and control with age, the opportunities available to access green spaces is likely to reduce. Therefore, it is crucial to ensure that green spaces are widely available and well distributed to maintain connectedness even in changing circumstances.

From literature review, the quantity of green space is measured against Park Provision Ratio (PPR), which refers to the area of park space available per 1,000 population. Green spaces that are included in the calculation refers to publicly accessible green spaces, such as gardens, nature parks, parks, park connectors. It does not include privately owned spaces accessible by a few, such as golf courses.

Despite Singapore's strive to achieve its liability aim, Singapore's 2020 PPR of 0.78 ha/1,000 population and 2030 goal of 0.8 ha/1,000 population remains below the recommended value of 0.9 ha/1,000 population recommended by Wold Health Organisation (WHO). Globally, a study in 2012 has found that the PPR of the average of 13 high-density north-American cities to be 2.44 more than triple that of Singapore's PPR of 0.75, and at the top, Stockholm with PPR of 7.38 (Tan et al., 2013).

Singapore grapples with the downward pressures of high population density and space constrains that makes it difficult to increase its PPR.

Quality of Green Space

Beyond numbers and spatial distribution of green spaces, one of the gaps identified in existing literature on the enhancement of urban green is concerning the quality of green spaces and the level of interaction permitted to an individual with surrounding green.

One study looks into People-Plant Interactions and break down the categories of interaction with green (Fig. 21.1).

1. Peaceful Abiding

Peaceful abiding refers to "interact[ing] with plants in an attentive and restful state" (Delsesto, 2019). The constant overdrive and engagement of the urbanised lifestyle deplete the limited psychological resources that a person holds, and it, therefore, becomes important that attention is replenished for the mind to be restored to a state of equilibrium by means of mental and physical wandering (Kaplan, 1995).

2. Tactile Stimulation

Tactile Stimulation refers primarily to the physical sensations of touch when working with the earth and plant material. Aligning the body with the slow and subtle movements of the plant, these daily encounters ground the human body and mind, becoming almost a therapeutic modality to self-introspection. Studies have also demonstrated that the physicality of touch can improve mental health and immune functioning. Contact with plant material results in a significant reduction of oxygenated hemoglobin in the brain, triggering an unconscious calming effect (Koga & Iwasaki, 2013). Exposure to the soil bacterium Mycobacterium Vaccae helps regulate the neurotransmitters that affect an individual's emotional state via the release of serotonin, a natural antidepressant that strengthens immunity (Lowry et al., 2016).

3. Physical Exertion

Gardening involves a variety of compound movements that health practitioners recognise as beneficial to the upkeep of the body. Fundamentally, permutation of the six basic motions of pushing and pulling, walking, lifting, stretching gripping, and bending (Park & Shoemaker, 2009), linked together to form a rhythm and routine that exist within the body and with space and time. Repetitive ritualistic behaviour is a built-in hardwired system to induce calm (Eilam et al., 2011), and often, the individual might lose their internal sense of time, or in some cases, experience a different rhythm of time. Repetitive tasks also have the potential to empower, by providing "some modicum of control back into the hands of the individual—a manoeuvre that

Fig. 21.1 People-plant interaction: peaceful abiding, tactile stimulation and physical exersion

helps to improve confidence and self-assurance in situations that would be otherwise out of our control" (Eilam et al., 2011).

This paper looks to utilise People-Plant Interactions as determinants of "quality" urban green spaces. A "good quality" green space will support all three types of interactions.

Human Factors

According to the Self Determination Theory (SDT), humans have three predominant psychological needs, namely autonomy, competence, and relatedness, that govern intrinsic motivation (Deci & Ryan, 2000). Even though there are many positive physiological and psychological health effects associated with working with green and being in nature, fulfilment is not a guaranteed by-product of participation (Quested et al., 2018). When offerings and needs become misaligned, well-being may be undermined.

Findings

By means of systematic pictorial study of scraped images from keyword associations of urban green spaces in Singapore, available urban green spaces are evaluated. Interviews with participants of community gardens yield understanding on community roles and systems.

1. Web-Based Pictorial Analysis:

Parks in Singapore are found to be largely be categorised as aesthetic and manicured, with the general population interacts passively with the environment and interactions largely falling under Peaceful Abiding.

While increasing in numbers, smaller plots of urban spaces serving as community gardens and guerrilla gardens see a more functional role to a select group of people, with the interactions largely falling under Tactile Stimulation and Physical Exertion.

The search 'Horticulture Therapy' yielded images that display a wider variety of interactions compared to other keyword searches. However, this might be due to intentionally organised activities to promote wellness and health. Without professional assistance and input in a structured session, the average individual might not be able to access the benefits of People-Plant Interactions.

Gardening for wellness in Singapore lacks the elements of spontaneity and agency. There is need for a new type of green urban space that fully integrates all elements of People-Plant Interactions to improve "quality" of green spaces. To address quantity of green spaces available. The intention is to intensify small plots of empty spaces between residential apartments by "multiplying" the ground.

2. Community Interviews:

Interviews with members of Ground-Up Initiative (GUI) Singapore and community garden at Bedok residential estate in Singapore, regarding how community gardens improve well-being, has highlighted that well-being is also directly influenced by feelings of social connection, autonomy and confidence in their abilities to nurture beautiful and productive gardens. These responses strongly corroborate with the aforementioned study, which finds that a community garden that may successfully promote well-being must embody these elements (Quested et al., 2018): purposeful activities, sense of community, working together, knowledge and skills, inclusivity, connection, ownership (Fig. 21.2).

From the findings collected, besides improving "quality" and "quantity" of green urban spaces, urban spaces must support agency and connectedness to be able to promote well-being within communities (Fig. 21.3).

In an environment characterised by rapid, dense, and constantly evolving cityscapes, it is crucial that the intervention is designed to be adaptable, able to fit into irregularly shaped sites, and easily assembled or disassembled as needed. The proposed ecosystem comprises of 8 m × 8 m × 8 m modular prefabricated steel frame that may be agglomerated and connected along any of its 6 faces, forming the skeletal frame that will be populated with components from a kit of parts supporting rainwater harvesting, energy storage, soil renewal, soil beds for planting and logistical transport of heavy items. These systems are large and intentioned to be powered by hand, with collective efforts of the community, despite the backdrop of automation. These "inefficiencies", with human as key machine mechanism, maximises well-rounded interactions with nature and creates marked awareness on community events around the daily operation of the garden thereby strengthening social cohesion (Kaplan & Kaplan, 1989).

Recognising the natural decline of the human body, a collaborative system which matches strengths, interests and abilities of an individual to sequenced movements necessary to support the plant's growth cycle, a young individual has opportunities to become deeply involved in environmental stewardship and the transformation of urban green spaces whilst supporting a less mobile individual in sustained participation across different life stages and circumstances.

Fig. 21.2 Key Word Association Study to understand the characteristics of common types of urban green spaces in Singapore

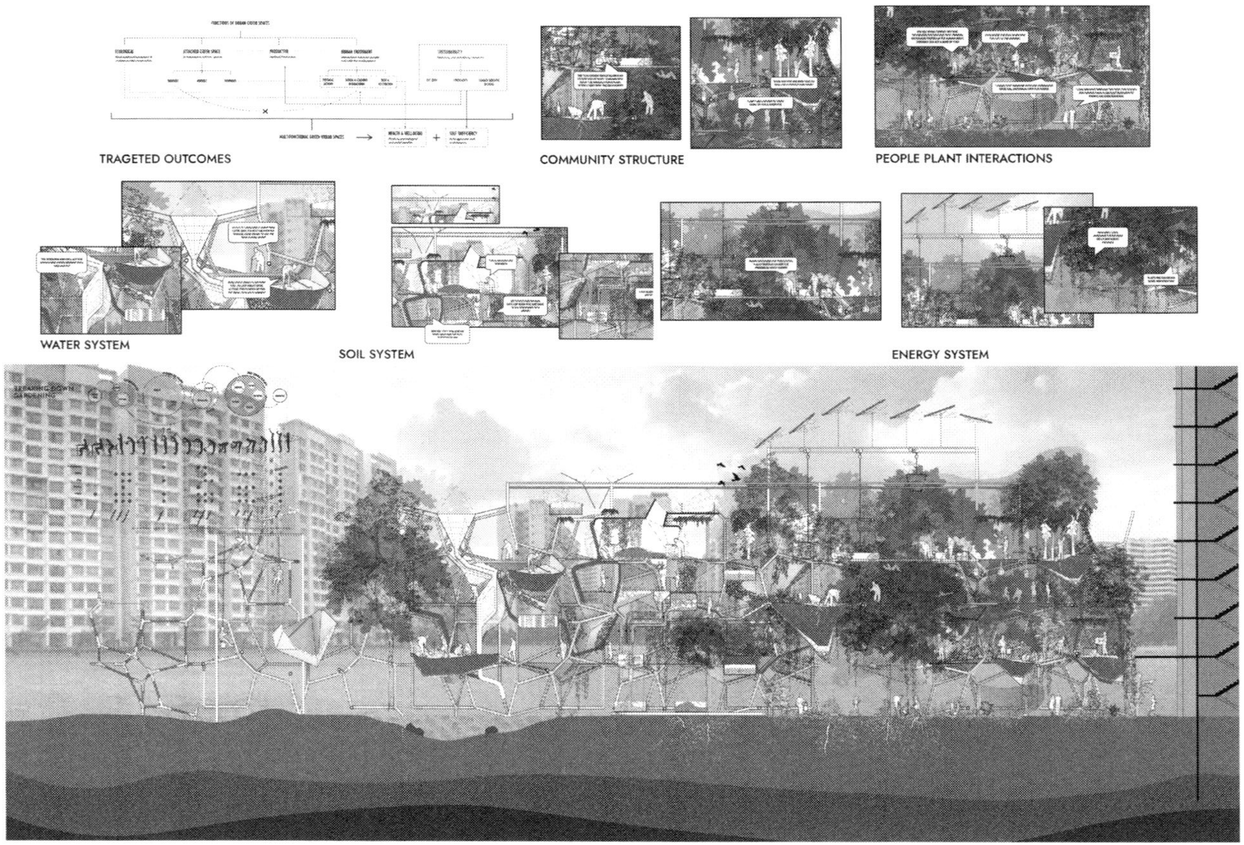

Fig. 21.3 Proposed new landscape of green that targets people-plant interaction for optimized health and well-being

Limitations

The proposed framework postulates strong intensification of the quantity and quality green which hinges on the adoption by the community. Since individual motivations and goals are varied, the resulting outcomes on spatial development will be challenging to implement and govern. Perhaps, however, the intention is not to restrict the trajectories of development. The natural evolution of this new landscape from use, to disuse and reuse should be celebrated.

Conclusion

Despite the above mentioned challenges in execution, underlying concepts that form the basis of the explored proposal remain applicable in Singapore's pursuit of livability with its urban-greening programme. Beyond the aesthetic visuals of green, it is necessary to consider that green can be fundamental and pivotal in enhancing preventive health and rehabilitation of a city and its population.

References

Deci, E. L., & Ryan, R. M. (2000). The '"what"' and '"why"' of goal pursuits: Human needs and the self-determination of behavior. *Psychological Inquiry, 11*, 227–268. https://doi.org/10.1207/S15327965PLI1104-01

Delsesto, M. (2019). People–plant interactions and the ecological self. *New Phytologist Foundation, 2*(3), 201–211. https://doi.org/10.1002/ppp3.10087

Eilam, D., Izhar, R., & Mort, J. (2011). Threat detection: Behavioral practices in animals and humans. *Neuroscience and Biobehavioral Reviews, 35*(4), 999. https://doi.org/10.1016/j.neubiorev.2010.08.002

Gesler, W. M. (1992). Therapeutic landscapes: Medical issues in light of the new cultural geography. *Social Science and Medicine, 34*, 735–746. https://doi.org/10.1016/0277-9536(92)90360-3

Kaplan, S. (1995). The restorative benefits of nature: Toward an integrative framework. *Journal of Environmental Psychology, 15*, 169–182. https://doi.org/10.1016/0272-4944(95)90001-2

Kaplan, R., & Kaplan, S. (1989). *The experience of nature: A psychological perspective*. Cambridge University Press.

Koga, K., & Iwasaki, Y. (2013). Psychological and physiological effect in humans of touching plant foliage—using the semantic differential method and cerebral activity as indicators. *Journal of Physiological Anthropology, 32*(7). https://doi.org/10.1186/1880-6805-32-7

Lowry, C. A., Hollis, J. H., de Vries, A., Pan, B., Brunet, L. R., Hunt, J. R. F., Paton, J. R. F., van Kampen, E., Knight, D. M., Evans, A. K., Rook, G. A. W., & Lightman, S. L. (2016). Identification of an immune-responsive mesolimbocortical serotonergic system: Potential role in regulation of emotional behavior. *Neuroscience, 146*(2), 756–772. https://doi.org/10.1016/j.neuroscience.2007.01.067

Park, S.-A., & Shoemaker, C. (2009). Observing body position of older adults while gardening for health benefits and risks. *Activities, Adaptation & Aging, 33*(1), 31–38. https://doi.org/10.1080/01924780902718582

Quested, E., Høgersen-Ntoumani, C., Uren, H., Hardcastle, S., & Ryan, R. M. (2018). Community gardening: Basic psychological needs as mechanisms to enhance individual and community well-being. *Ecopsychology, 10*(3), 173–180. https://doi.org/10.1089/eco.2018.0002

Tan, P. Y., Wang, J., & Sia, A. (2013). Perspectives on five decades of the urban greening of Singapore. *Cities, 32*, 24–32. https://doi.org/10.1016/j.cities.2013.02.001

Chapter 22
Urban Shape and Public Spaces to Promote Active Mobilities and Better Breathing: The Cases of Barcelona and Paris

Patrizia Ingallina and Marco Carlotti

Abstract To improve air quality and achieve greater decarbonization, cities are increasingly aiming to reduce car dependency and promote active mobility modes—such as walking and cycling. This article explores how urban transformations aimed at fostering active mobility intersect with the historical urban form, through a comparative analysis of two European cities: Barcelona (Spain) and Paris (France). It presents preliminary findings from an ongoing mixed-methods study, combining qualitative observations, stakeholder interviews, and an extensive review of scientific and institutional documentation. Initial results indicate that Barcelona has implemented a more context-sensitive, incremental pedestrianization process that aligns with its orthogonal historic urban fabric. In contrast, Paris faces unintended negative externalities—such as traffic congestion and increased pollution in adjacent areas—due to top-down, sectoral pedestrianization measures that often overlook the complexity of its historical morphology. The article argues that successful transitions toward active mobility must be grounded in three key principles: (1) respecting the primary urban texture, composed of public spaces (streets, squares, and plazas) and their structural, social, and functional roles; (2) ensuring prior enhancement of public transportation systems to facilitate modal shifts; and (3) introducing active mobility infrastructures that are contextually adapted, accessible, and designed to prevent user conflicts.

Keywords Urban shape · Public space · Active mobility · Decarbonization

P. Ingallina (✉) · M. Carlotti
Sorbonne Université, Médiations - Sciences des Lieux, Sciences des Liens, Paris, France
e-mail: patrizia.ingallina@sorbonne-universite.fr

M. Carlotti
e-mail: marco.carlotti@etu.sorbonne-universite.fr

J. Dsilva (ed.), *Sustainability Education, Youth and Innovation*,
https://doi.org/10.1007/978-3-032-21948-0_22

Introduction

First and foremost, the city must be understood not merely as a physical entity but as a complex web of interrelated relationships. At its foundation lies sociability, which constitutes the very essence of urban life and finds its most tangible expression in public spaces—often described as the "most democratic spaces of the city" (Devillers, 1994). As such, it is imperative that these spaces remain accessible and enjoyable for all users (Gehl, 2010), with particular attention paid to ensuring high air quality.

The importance of air quality in public space design became especially evident during the COVID-19 pandemic, when lockdown measures led to a notable decline in pollutant emissions and corresponding improvements in air quality (Saha et al., 2022; Venter et al., 2020). These improvements were largely attributed to the drastic reduction in transportation activity—a sector responsible for approximately 20% of global carbon emissions (Ritchie, 2020)—as confirmed by numerous cross-regional studies (Arter et al., 2024; Vichova et al., 2021). This context has reignited debates about the imperative to reduce private car usage and instead promote active modes of mobility—walking, cycling, and the use of micro-mobility devices such as scooters. These modes of transport offer dual benefits: they have positive effects on individual health (Kong et al., 2024) and significantly reduce CO_2 emissions (Brand et al., 2021).

In light of the critical role active mobility plays in decarbonizing cities and improving urban air quality, this article explores how cities are attempting to foster these modes through transformations with regard to urban shape. Specifically, it analyzes two European cities—Barcelona (Spain) and Paris (France)—that have each implemented ambitious policies aimed at reducing air pollution by limiting private automobile use and promoting lower-emission transport options such as walking and cycling. Both cities have witnessed a growing shift from private vehicle usage toward active mobility, albeit through divergent strategies shaped by their respective urban morphologies (Àrea de Barcelona, 2024; Ville de Paris, 2025).

This paper argues that Barcelona represents a more successful and context-sensitive approach, having implemented a gradual and well-integrated pedestrianization strategy that aligns with its historic orthogonal street grid. In contrast, Paris has pursued a more forceful pedestrianization approach, often altering its historically layered and complex urban fabric in ways that generate friction and contestation. Drawing from these two cases, the article concludes that reducing the dominance of automobiles in urban space requires a phased and spatially intelligent strategy—one that considers (1) the underlying urban morphology, (2) the prior enhancement of public transportation infrastructure, and (3) the tailored integration of active mobility systems compatible with the city's spatial structure.

Methodology

This paper aims to examine the relationship between urban form, public space, active mobility, and air quality, with a particular focus on how historical urban morphology—an essential dimension for envisioning the city of the future (Pinon, 1992)—is considered in policy initiatives designed to promote active mobility and improve air quality. It builds upon two previous presentations by the authors (Ingallina, 2022; Ingallina & Carlotti, 2023).

This study adopts a mixed-methods methodology, with research tools developed progressively. It began with the authors' experiential observations of urban transformations in Paris and their perceived impacts on certain neighborhoods. This initial phase involved in-situ observations and photographic documentation, revealing mobility-related conflicts and congestion around newly transformed public spaces designed to reduce car traffic. To deepen this analysis, ongoing interviews have been conducted with local shopkeepers and residents' associations formed in response to the city's interventions—namely the Beaumarchais-Bastille Asphyxiés (BBA) collective, addressing issues related to Place de la Bastille, and the Association des riverains du Quartier Daumesnil Reuilly, which has voiced concerns over the limited pedestrianization of Place Félix Éboué (12th arrondissement), following a logic similar to that of Bastille.

Finally, the research draws on a review of scientific literature, policy reports, and technical publications concerning urban form, public space, active mobility, and air quality. Many of the analyzed reports originate from municipal agencies in Paris and Barcelona, providing empirical data on the effectiveness of implemented measures.

This study remains ongoing and reports preliminary findings.

Findings

Barcelona: A Progressive Pedestrianization in a Regular Urban Framework

One of the most polluted districts in Barcelona is the Eixample, a 7.48 km^2 neighborhood designed by Ildefons Cerdà in the nineteenth century. Characterized by a strict orthogonal grid (hippodamian layout), the Eixample is composed of uniformly square residential blocks, each measuring 113.3 m per side. Its wide roadways and geometric layout have historically facilitated the flow of vehicular traffic, making it the epicenter of car circulation in the city—up to 350,000 vehicles daily (Mas & Kohan, 2022), positioning it as the most traffic-congested and polluted district in Barcelona. This level of exposure to vehicular pollution is linked to roughly 225 premature deaths annually, not including traffic-related fatalities. The Example also contributes significantly to Barcelona's distinction as the European city with the highest vehicle density, with 5,844 vehicles per km^2 (Mas & Kohan, 2022).

At the city scale, air pollution in Barcelona is responsible for over 1,000 premature deaths each year, more than 2,000 serious health conditions, and over 1,000 childhood asthma cases, according to the Barcelona Public Health Agency (Rico et al., 2019). The same studies highlight transportation as the dominant source of CO_2 emissions, underscoring its role not only in environmental degradation but also as one of the leading public health threats in the region.

In response to the adverse impacts of traffic congestion and air pollution, Barcelona began implementing the "superblock" (superilles) model as early as 1993, starting in the historic El Born district. This form of tactical urbanism seeks to curb air and noise pollution, mitigate climate change effects, and reclaim public space for pedestrians and active mobility (Ajuntament de Barcelona, 2016). The Example, due to its orthogonal street layout, has become the primary site for scaling up this model. A standard superblock consists of a 400 m by 400 m unit comprising nine residential blocks. Within these, through-traffic is redirected to perimeter roads, while interior streets prioritize pedestrians, cyclists, and local access, transforming the urban fabric into one centered on social and ecological functions.

Although two pilot superblocks were created in the Gràcia district in 2005, the model gained momentum during the administration of Ada Colau, Mayor of Barcelona from 2015 to 2023. Under her leadership, the superblock initiative became a core component of the city's urban transformation agenda. The Poblenou superblock, launched in September 2016, was followed by those in Sant Antoni and Horta within the same year. These interventions were part of the broader "Let's Fill Streets with Life" (Omplim de vida els carrers) program, launched in 2016, which aimed to extend the superblock model citywide. By 2021, six superblocks had been completed, three were under construction, and nine more were in various stages of planning.

The impact of superblocks have been generally positive, although spatially uneven. A study conducted by the Agència de Salut Pública de Barcelona (2021), as part of the "Health in the Streets" (Salut als Carrers) program, evaluated the superblocks of Poblenou, Sant Antoni, and Horta. The results, while mixed, pointed to a perceived reduction in both air and noise pollution. A significant drop in nitrogen dioxide (NO_2) concentrations—an average decrease of 14.57 $\mu g/m^3$—was recorded only in the Sant Antoni superblock, while no measurable changes were observed in Horta, indicating that the effectiveness of superblocks may vary depending on local conditions and implementation strategies.

According to Kostandinovic (2018) for C40, this plan aligns with Barcelona's ambition to reduce its per capita carbon emissions by 40% by 2030 compared to 2005 levels.

Paris: A Forced Pedestrianization in a Complex Urban Texture

With over 2,500 premature deaths annually attributed to nitrogen dioxide exposure, Paris ranks as the fourth most affected European city, according to a study by

Khomenko et al. (2021). Despite a marked increase in bicycle use—accounting for 11.2% of trips in 2024 compared to just 4.3% by private car (Ville de Paris, 2024)—the city's intensive pedestrianization efforts have also generated visible congestion, as observed in key areas such as Place de la Bastille (11th arrondissement) and Place Félix Éboué (12th arrondissement). In some locations, this shift has coincided with a rise in local air pollution levels, exemplified by the controversial pedestrianization of the Seine riverbanks.

Public space conflicts are particularly pronounced in Paris—more so than in peer cities such as Amsterdam, Barcelona, or Berlin—complicating the coexistence of diverse users, especially vulnerable populations including the elderly and persons with disabilities (Conseil National des Transports, 2004). These tensions stem largely from a politically driven push to pedestrianize as many zones as possible, including high-traffic areas, without sufficient consideration for three fundamental principles of urban morphology:

1. Urban texture, as informed by typo-morphological theory (Rossi, 1984), which emphasizes the structural hierarchies of roadways (primary, secondary, tertiary) as the skeleton of the city. Roads, in this view, are enduring and "permanent" elements of urban form and should be preserved accordingly.
2. The regularity of the urban framework where street patterns are shaped by topographical and geographical conditions (e.g., winding streets on slopes, straight avenues on flat terrain).
3. The threefold value of public space—historical, functional, and social—where squares and plazas are not only places of collective identity and appropriation but also function as traffic distribution nodes.

The failure to respect the first principle was particularly evident in the 2019 removal of the expressway along the Seine. This intervention disrupted a key east–west urban thoroughfare, displacing traffic onto smaller central streets and triggering increases in congestion and pollution as shown in Fig. 22.1. AirParif (2017) recorded up to a 15% rise in pollution levels in the eastern part of Paris. The case of the Seine riverbanks illustrates the pitfalls of sectoral, isolated planning decisions that treat vital infrastructure as detached from the broader urban network.

The second and third principles were similarly neglected in the redesign of historic public squares. These spaces—originally conceived to manage traffic flows and serve as places of neighborhood identity—have been altered in ways that undermine both function and social meaning. A notable example is the forced pedestrianization of one side of Place de la Bastille, where car traffic is now prohibited. Our fieldwork indicates not only a sharp rise in congestion in the surrounding area but also challenges in reclaiming the space for public life (Fig. 22.2). The square's form has been compromised: the newly created space lacks coherence, suffers from excessive noise, and is marked by conflict between users (car users, scooter riders, cyclists, pedestrians). Noise pollution has so significantly affected local residents that they have formed an advocacy group (Association Bastille Beaumarché Asphyxiés) to protest the changes. According to interviews with members of this association conducted

Fig. 22.1 Pedestrianized Seine riverbanks. Photo by P. Ingallina

in 2025, local merchants have also experienced reduced foot traffic and diminished economic activity due to the area's decreased attractiveness.

Although air quality data specific to Place de la Bastille remain unavailable, evidence from on-site observations, testimonies from local associations, and an interview with the head of another community group—engaged in efforts to preserve the Place Félix Éboué (Fig. 22.3), currently undergoing a similar transformation—suggest a worsening of congestion. The results of an online public survey are currently pending and are expected to provide further insights into the socio-spatial impacts of these urban interventions.

Conclusions

Ultimately, while the reduction of car space in cities is both necessary and urgent, it must be embedded in a gradual, nuanced, and context-sensitive planning approach. Pedestrianization policies (and, similarly, greening initiatives), though widely appreciated and politically valorized, can prove ambivalent when not grounded in a deep understanding of urban uses and the existing spatial structure. Poorly designed

Fig. 22.2 Traffic jam on Place Félix Eboué (12th arrondissement of Paris, France). Photo by P. Ingallina

or hastily implemented, such interventions risk disrupting the balance of public spaces, undermining their appropriation by users, or generating unintended negative effects—such as displaced pollution, increased traffic congestion, and more.

Accordingly, three key actions must be undertaken: (1) respecting the primary urban texture—composed of public spaces such as roads, squares, and plazas—and acknowledging their multiple structural, social, and functional roles within the city; (2) improving the public transportation system as a prerequisite for modal shifts; and (3) introducing active mobility options that are well-adapted to the existing urban form, easily accessible, and designed to minimize user conflicts. Urban interventions

Fig. 22.3 Community meeting to promote a new redevelopment project for the Place Félix Éboué. Photo by M. Carlotti

should not be driven solely by the desire to demonstrate political action through scattered initiatives.

Rather, emphasis must be placed on understanding how public spaces and urban form—reflecting embedded urban practices—can structurally and politically influence the success or failure of mobility transitions aimed at advancing both ecological sustainability and human well-being.

References

Agència De Salut Pública de Barcelona. (2021). *Report of results of the environmental and health effects of the Superblocks model in Barcelona. Salut Als Carrers (SAC) project*. Retrieved June 20, 2025, from https://www.aspb.cat/wp-content/uploads/2021/10/English-ASPB_salut-carrers-resultsreport-Superblocks.pdf

AirParif. (2017). *Voies sur berge: La qualité de l'air, un an après*. Retrieved June 20, 2025, from https://www.airparif.fr/actualite/2017/voies-sur-berge-la-qualite-de-lair-un-apres

Ajuntament de Barcelona. (2016, May). *Let's fill streets with life*. Retrieved June 20, 2025, from https://ajuntament.barcelona.cat/ecologiaurbana/sites/default/files/en_gb_MESURA%20GOVERN%20SUPERILLES.pdf

Àrea de Barcelona. (2024, July 8). *Mobility in the Barcelona area exceeds 20 million journeys on a weekday and two out of three are already made using sustainable transportation*. Retrieved

June 20, 2025, from https://ce.atm.cat/en/web/portal-atm/w/np_la-mobilitat-a-l-%C3%A0rea-de-barcelona-supera-els-20-milions-de-despla%C3%A7aments-en-dia-feiner-i-dos-de-cada-tres-ja-es-fan-amb-modes-sostenibles-2

Arter, C. A., Buonocore, J. J., Isakov, V., Pandey, G., & Arunachalam, S. (2024). Air pollution benefits from reduced on-road activity due to COVID-19 in the United States. *PNAS Nexus, 3*(1), pgae017. https://doi.org/10.1093/pnasnexus/pgae017

Brand, C., Dons, E., Anaya-Boig, E., Avila-Palencia, I., Clark, A., de Nazelle, A., Gascon, M., Gaupp-Berghausen, M., Gerike, R., Götschi, T., Iacorossi, F., Kahlmeier, S., Laeremans, M., Nieuwenhuijsen, M. J., Pablo Orjuela, J., Racioppi, F., Raser, E., Rojas-Rueda, D., Standaert, A., … Int Panis, L. (2021). The climate change mitigation effects of daily active travel in cities. *Transportation Research Part D: Transport and Environment, 93*, Article 102764. https://doi.org/10.1016/j.trd.2021.102764

Conseil National des Transports. (2004). *Une voirie pour tous. Sécurité et cohabitation sur la voie publique au-delà des conflits d'usage.*

Devillers, C. (1994). *Le projet urbain.* Pavillon de l'Arsenal.

Gehl, J. (2010). *Cities for people.* Island Press.

Ingallina, P. (2022, November 7). *Transports alternatifs dans la ville de demain: quelles mobilités pour la réappropriation des espaces publics ?* [Conference presentation]. Demain, les transports, Collège de France, Paris, France. https://www.college-de-france.fr/fr/agenda/grand-evenement/demain-les-transports/transports-alternatifs-dans-la-ville-de-demain-quelles-mobilites-pour-la-reappropriation-des-espaces

Ingallina, P., & Carlotti, M. (2023, October 23–24). *Breathe better in the city: Public spaces and active mobility. The cases of Paris and Barcelona* [Conference presentation]. Mitigating the effects of global warming and unlocking new opportunities to achieve sustainability. Sorbonne University Abu Dhabi.

Khomenko, S., Cirach, M., Pereira-Barboza, E., Mueller, N., Barrera-Gómez, J., Rojas-Rueda, D., de Hoogh, K., Hoek, G., & Nieuwenhuijsen, M. (2021). Premature mortality due to air pollution in European cities: A health impact assessment. *Lancet Planet Health, 5*, e-121–e-134. https://doi.org/10.1016/S2542-5196(20)30272-2

Kong, H., Wu, J., & Li, P. (2024). Impact of active mobility on individual health mediated by physical activities. *Social Science & Medicine, 348*, 116834. https://doi.org/10.1016/j.socscimed.2024.116834

Kostandinovic, N. (2018, March). *The implementation of the Superblocks programme in Barcelona: Filling our streets with life.* C40 Cities. https://www.c40.org/case-studies/barcelona-superblocks/

Mas, J. G., & Kohan, C. J. (2022, July). *Air pollution in the city.* Barcelona Metròpolis. https://www.barcelona.cat/metropolis/en/contents/air-pollution-the-city

Pinon, P. (1992). *Composition urbaine.* Éditions du Service technique de l'urbanisme—Villes et Territoires.

Rico, M., Font, L., Arimon, J., Marí, M., Gómez, A., & Realp, E. (2019). *Informe qualitat de l'aire de Barcelona.* Agència de Salut Pública de Barcelona. https://www.aspb.cat/wp-content/uploads/2020/10/Informe_qualitat-aire-2019.pdf

Ritchie, H. (2020, October 6). *Cars, planes, trains: Where do CO_2 emissions from transport come from?* OurWorldinData.org. Retrieved June 20, 2025, from https://ourworldindata.org/co2-emissions-from-transport#article-citation

Rossi, A. (1984). *The architecture of the city.* The MIT Press.

Saha, L., Kumar, A., Kumar, S., Korstad, J., Srivastava, S., & Bauddh, K. (2022). The impact of the COVID-19 lockdown on global air quality: A review. *Environmental Sustainability (Singapore), 5*(1), 5–23. https://doi.org/10.1007/s42398-021-00213-6

Venter, Z. S., Aunan, K., Chowdhury, S., & Lelieveld, J. (2020). COVID-19 lockdowns cause global air pollution declines. *Proceedings of the National Academy of Sciences, 117*(32), 18984–18990. https://doi.org/10.1073/pnas.2006853117

Vichova, K., Veselik, P., Heinzova, R., & Dvoracek, R. (2021). Road transport and its impact on air pollution during the COVID-19 pandemic. *Sustainability, 13*(21), 11803. https://doi.org/10.3390/su132111803

Ville de Paris. (2024, June 3). *Le vélo surpasse la voiture à Paris et en petite couronne (et de loin).* Retrieved June 20, 2025, from https://www.paris.fr/pages/inedit-le-velo-surpasse-la-voiture-a-paris-et-en-petite-couronne-et-de-loin-26832

Ville de Paris. (2025, February 2). *Comment se sont déplacés les Parisiens en 2023.* Retrieved June 20, 2025, from https://www.paris.fr/pages/le-bilan-des-deplacements-a-paris-en-2023-27604

Chapter 23
Building Health System Resilience: The Launch of Health Services Management Centre (HSMC) Dubai

Ayat Abu-Agla and Ross Millar

Abstract Global health systems are experiencing major financial and operational stress as they respond to post pandemic environments and the challenge of climate change. Inadequate investment in health systems now coexists alongside the pressing need to review the nature of health systems in terms of their role and mission. This is particularly the case within the Eastern Mediterranean region (EMR). Countries are being tasked with restructuring and transforming health systems and their health workforce to meet the needs of universal health coverage (UHC) and sustainable development goals (SDG). Realising the need to decolonise global health and reimagine education, it is high time to re-think how we teach, train, and build capacities that are fit-for-purpose and fit-for-practice to lead resilient health systems. How we achieve such an objective requires a different approach to research, education, and development. A desk-based review of policy documents and a series of stakeholder consultations regarding the needs and priorities for health system resilience in the EMR. The HSMC Dubai aims to support health policy development and translate policy into practice for the EMR. A decolonising approach is required that engages with different communities and provides more culturally sensitive approaches to health and well-being. HSMC Dubai aims to advance sustainable development and build health system resilience through the launch of an MSc Global Health System Leadership and a research agenda that provides a regional and global hub for challenges such as climate sustainability, digital innovation, and population health.

Keywords Global health · Health system · Health system leadership · Sustainable development goals · Universal health coverage · Decolonization · Eastern Mediterranean region

---

A. Abu-Agla (✉)
University of Birmingham, Dubai, UAE
e-mail: a.s.y.abu-agla@bham.ac.uk

R. Millar
University of Birmingham, Birmingham, UK

J. Dsilva (ed.), *Sustainability Education, Youth and Innovation*,
https://doi.org/10.1007/978-3-032-21948-0_23

Introduction

In an era of polycrises, the COVID-19 pandemic has drawn attention to how chronic underinvestment in our health systems compromises human health and leads to serious economic and social setbacks. The World Health Organisation (WHO) estimates a projected shortfall of 10 million health workers by 2030, mostly in low- and lower-middle income countries. Without a robust, resilient, and well-equipped health workforce, there can be no progress on the health-related Sustainable Development Goals (SDGs). Regional disparities, gender inequalities, recruitment and retention shortages, migration, and mental health are illustrative of the challenges facing a sustainable global health workforce.

We have witnessed the Eastern Mediterranean region's (EMR) most vulnerable countries falling further behind, as they lack the fiscal space to make the necessary investments in the human resources for health to build more resilient health systems. Furthermore, countries are being tasked with restructuring and transforming health systems to meet the needs of universal healthcare coverage (UHC) and sustainable development goals (SDG). In the publication 'Building resilient health systems to advance universal health coverage and ensure health security in the Eastern Mediterranean Region' the following priorities are identified [1]:

- Strengthening health emergency and disaster risk management in line with the plan of action for ending the COVID-19 pandemic and preventing and controlling future health emergencies in the EMR;
- Optimizing ministries of health and building institutions for public health;
- Establishing primary health care-oriented models of care;
- Enhancing and scaling up a fit-for-purpose, fit-to-practice health workforce;
- Promoting equity and enhancing financial protection;
- Improving access to medicines, vaccines, and health products; and
- Fostering an integrated approach in policy, planning and investments for long-term health system resilience.

In our view, "building back" fit-for-purpose health systems will require greater consideration of the social, economic, and environmental factors to support health system resilience. There is pressing need to review the nature of health systems in terms of their role and mission. Furthermore, it is high time to re-think how we teach, train, and build capacities in health systems.

The decolonisation of global health education, for example, has emerged as a critical perspective to challenge the dominance of Eurocentric health system paradigms by advocating for inclusive, equitable, and contextually grounded approaches to knowledge production. Decolonising education involves rebalancing power dynamics by centring voices from the Global South, valuing indigenous knowledge systems, and fostering reciprocal partnerships in research and training. Scholars argue that decolonisation must be embedded in curricula, pedagogy, and institutional structures to dismantle systemic inequities and promote social justice in global health [2, 3]. The WHO has also underscored the importance of culturally sensitive and context-specific education to improve health outcomes and policy effectiveness [4, 5].

The integration of the United Nations Sustainable Development Goals (SDGs) into health education is vital for equipping future leaders with the competencies needed to address complex global challenges. SDG 3 (Good Health and Well-being), SDG 4 (Quality Education), SDG 5 (Gender Equality), and SDG 8 (Decent Work and Economic Growth) are particularly relevant to the development of resilient and equitable health systems. The WHO's Global Strategy on Human Resources for Health and the Working for Health 2022–2030 Action Plan emphasize the need for education systems to align with the SDGs to build a competent, motivated, and equitably distributed health workforce [5, 6].

The COVID-19 pandemic exposed deep vulnerabilities in global health systems, highlighting the urgent need for resilience, adaptability, and equity. Building back better requires a paradigm shift towards health systems that are both fit-for-purpose designed to meet current and emerging health needs and fit-for-practice capable of being implemented effectively in diverse contexts. A whole-of-government and whole-of-society approach is essential to achieve this transformation, involving coordinated action across sectors, inclusive governance, and community engagement.

Bridging the gap between education and practice is central to any agenda. By aligning health education with the principles of decolonisation and the SDGs, and by fostering interdisciplinary and cross-sectoral collaboration, it is possible to cultivate a health workforce capable of leading transformative change and ensuring health systems are prepared for future challenges.

Methodology

The University of Birmingham, founded in 1900, was England's first civic or 'redbrick' university. The University of Birmingham's Health Service Management Centre (HSMC) was founded in 1972. The Centre is one of the foremost centres of research, education, and development for health and social care organisations.

Building on this heritage, the Health Services Management Centre (HSMC) Dubai has been established in recognition of the need to develop contextually relevant education, research and development for the EMR and beyond.

Various engagement activities have informed the development of the HSMC Dubai. Its development has been informed by a range of desk reviews, stakeholder mapping and global, regional, and national consultations. In the following section, we summarise these results as follows.

Findings

HSMC Dubai has been developed with stakeholder consultation sessions that have included at a national level the Dubai Health Authority (DHA), National Institute for Health Specialities (UAE) and Emirati Public Health Association; at a regional level the Arab Board of Health Specialisations, Eastern Mediterranean Public Health Network (EMPHNET), International Academy of Public Health (IAPH), MENA Health Policy Forum and Arab Public Health Association; and at a global level the Health System Global and World Health Organisation (HQ, EMRO and WCO) Health workforce and Health system departments.

To complement these endeavors, HSMC Dubai is informed by and aligns with global health policy and university strategy agendas in the following areas.

HSMC Dubai supports the World Health Organisation's Global Strategy on Human Resources for Health: Workforce 2030, Working for Health 2022–2030 Action Plan, the identified global and eastern Mediterranean region essential public health functions, the Global Competency framework for universal health coverage, the Salah declaration on universal health coverage, and the Seventy-fourth World Health Assembly resolution WHA74.14 in 2021: Protecting, safeguarding and investing in the health and care workforce and EMRO Regional Flagship report [6–12].

HSMC Dubai has proactively engaged with the University of Birmingham Dubai Academic Strategy 2021–26 to connect, create, and grow. This is summarised as follows.

Pillar One Connect

The scale of growth that we aim to achieve can only be realised by connecting HSMC Dubai with the University of Birmingham, the UAE, and the wider region. We are situating the Dubai campus within a series of University of Birmingham global networks. HSMC Dubai has been leading the university health and well-being subject group engagement with Dubai Health Authority (DHA) to facilitate a coordinated university wide response and demonstrate implementation of the partnership. In recognition of that, UoBD HSMC was awarded the DHA Partnership trophy in November 2023 and to commemorate the partnership, an MoU was signed by the Director General DHA with the UoBD. At regional level, UoBD HSMC Dubai signed an MoU with our regional stakeholder the Eastern Mediterranean Public Health Network's International Academy of Public Health (IAPH) to strategically collaborate in areas of joint interest based on the eastern Mediterranean regional health systems priorities on education, research, knowledge transfer.

Pillar Two Create

- HSMC Dubai are creating platforms for research and development with the following aims:
- To provide a regional and global hub for research excellence in health systems.

- To provide thought leadership in health system challenges such as workforce development, climate sustainability, digital innovation, and population health.
- To build cross University collaboration between Birmingham and Dubai.
- To promote an interdisciplinary and collaborative approach to UoBD research and development.
- To demonstrate the role of UoBD as a leading research-intensive campus in the region, we are creating a platform for how evidence contributes to providing solutions for global challenges. We have conducted a priority setting exercise to better understand the health system needs of the EMR. The exercise will be translated into a series of implementation workshops over the next year.

Pillar Three: Grow

The MSc Global Health System Leadership launched in January 2025 will facilitate the development of knowledge, skills and professional behaviours for current and future health leaders. The modules embedded in the programme structure focus on Global Health Systems and Policy; Leadership and Management in Global Health; Implementation Science; Strategic Planning; Global Health Economics and Emergency Preparedness Response and Resilience [13]. To complement this development, HSMC Dubai successfully hosted the first UoBD campus Continuous Professional Development (CPD) course in January 2024 piloting its Global Health Systems and Policy module. High level candidates including the Director of Strategy and Governance, Director of Medical Education and Research, Director of Health Economics and Insurance, and Public Health Departments and their teams formed part of the first cohort.

Conclusion

The Health Services Management Centre (HSMC) Dubai aims to continue the proud tradition of civic contribution to university, health, and social care organization and workforce. In line with our mission to 'support health policy research and development, deliver excellent education and put policy into practice', we are passionate about facilitating a transformational agenda that can respond to EMR regional health system priorities and population needs with a culturally sensitive approach.

By fostering interdisciplinary collaboration, engaging with diverse stakeholders, and promoting evidence-informed decision-making, HSMC Dubai aims to bridge the gap between policy and practice. Its work is rooted in the belief that resilient, equitable, and sustainable health systems can only be achieved through inclusive leadership, capacity building, and a commitment to social justice. As the region continues to navigate complex health challenges from workforce shortages to climate-related health threats, the Centre stands ready to contribute thought leadership, innovative solutions, and a platform for meaningful dialogue and action.

References

1. WHO. (2023). *Building resilient health systems to advance universal health coverage and ensure health security in the Eastern Mediterranean region.* https://applications.emro.who.int/docs/Build-resilient-health-systems-UHC-EMR-eng.pdf
2. Abimbola, S., et al. (2021). The foreign gaze: Authorship in academic global health. *BMJ Global Health, 6*(5), e006620. https://doi.org/10.1136/bmjgh-2021-006620
3. Khan, M., Abimbola, S., Aloudat, T., & Capobianco, E. (2022). Decolonising global health in 2022: A roadmap to move from rhetoric to reform. *BMJ Global Health, 7*(3), e008731. https://doi.org/10.1136/bmjgh-2022-008731
4. WHO. (2008). *Closing the gap in a generation: Health equity through action on the social determinants of health.* World Health Organization. https://apps.who.int/iris/handle/10665/43943
5. WHO. (2025). *Health system strengthening interventions to improve the health of displaced and migrant populations in the context of climate change.* World Health Organization. https://iris.who.int/handle/10665/382023
6. WHO. (2022). *Working for health 2022–2030 action plan.* World Health Organization.
7. WHO. (2016). *Global strategy on human resources for health: Workforce 2030.* 9789241511131-eng.pdf.
8. WHO. (2021). *21st century health challenges: Can the essential public health functions make a difference?* 9789240038929-eng.pdf.
9. WHO. (2022). *Global competency framework for universal health coverage.* 21256_Global Competency Framework for UHC_For Web.
10. WHO. (2018). *Salalah declaration on universal health coverage 2018 the road to universal health coverage in the Eastern Mediterranean region.* salalah_uhc_declaration_-_final.pdf.
11. WHO. (2021). *Protecting, safeguarding and investing in the health and care workforce.*
12. WHO. (2024). *Regional flagship initiative 2: Investing in a resilient health workforce.* EMRC71B-eng.pdf.
13. Millar, R., Abu-Agla, A., & Badr, E. (2025). Building capacity and capability for science diplomacy: Challenges in decolonizing the curriculum for global health system leadership. *Frontiers in Public Health, 13*, 1441351. https://doi.org/10.3389/fpubh.2025.1441351

Chapter 24
Effective Citizen Science: Playing a Role in Marine Life Restoration

Ahmed Nabil

Abstract Great barrier reef of Australia lost around 65% of its corals, there are immense efforts exerted by scientists and marine biologists to protect and maintain the remaining 35%. Red Sea and Arabian Gulf are highly threatened as well. The philosophy of "protection is always preferred over restoration" is a luxury that no longer applies to most regions. Now Restoration is necessary to preserve the biodiversity and functionality of reefs and ensure the sustainability of their resources. This paper presents an effective citizen science program for a Sustainable Future with two directions: (a) Raising awareness (b) Practical efforts. The Program is designed to give participants an introduction to Coral Reef Ecology and Restoration. Participants should have full knowledge about climate change and its impact on marine life. Conservation strategies and how to contribute in restoration efforts. Participants should leave a positive impact by creating a new marine life (Artificial and/or natural habitat). To present effective Citizen science, the program structure consists of raising awareness over workshops and practical efforts over field study projects. As this aim of this study, the participants shall be capable of Identifying and Monitoring corals for health condition and status Conducting advanced research and/or restoration projects Identifying rare or endangered species of coral. In addition, create new habitat for corals and marine animals—artificial and/or natural. The programme introduces participants to Coral Reef Ecology and Restoration, ensuring they: understand climate change and its impact on marine life, Learn conservation strategies and how to engage in restoration, Contribute to the creation of new marine habitat (artificial and/or natural) and The objective is to facilitate effective citizen science. It is a holistic approach regarding ecosystem, focusing on the long-term sustainability and adaptability of the ecosystem rather than just trying to add more corals into areas where they will not survive.

Keywords Citizen · Science · Marine · Conservation · Sustainability

---

A. Nabil (✉)
Alpha Consultancy FZC, Sharjah, UAE
e-mail: alphadivinguae@gmail.com

J. Dsilva (ed.), *Sustainability Education, Youth and Innovation*,
https://doi.org/10.1007/978-3-032-21948-0_24

Introduction

Current Coral Reef Situation

Coral reefs around the world are in critical decline due to climate change, human activities, and other environmental pressures. The Florida and Caribbean barrier reefs have been completely lost, while the Great Barrier Reef of Australia has lost about 65% of its corals, despite extensive conservation efforts. Other regions, including the Red Sea, Eastern Africa, and the Arabian Gulf, are also facing severe threats.

As protection alone is no longer enough, coral restoration has become essential to preserve the biodiversity and functionality of these ecosystems, ensuring the sustainability of marine resources. While All efforts are diverted toward setting strategies and resources for marine conservation, there is a big gap in terms of human resources and in-site capacity. Our abstract is about filling this gap, building capacity and generate "early career ocean professionals" and effective citizen science.

Environmental Education for a Sustainable Future

In response to this urgent situation, rises the urgent need of skilled professionals. This discovery reminds us of how much we still need to learn about the ocean's inhabitants and the importance of collecting data directly from the field. Coral Reefs, in addition to being fascinating, are key indicators of the health of our seas, yet many of their habits and characteristics remain unknown.

The Effective Citizen Science Program was developed to educate the public and junior consultants on coral reef ecology, increase awareness of the challenges reefs face, and promote practical restoration. By equipping students, researchers, and volunteers with both knowledge and hands-on experience, the program fosters active engagement in reef surveys, conservation initiatives, and restoration projects, ultimately contributing to the creation of a sustainable future for coral reefs.

Methodology

The program is structured into two main components: raising awareness and practical efforts. Together, these components provide participants with comprehensive knowledge about coral reefs, their ecological significance, and the scientific skills required to assess reef health and contribute to restoration activities.

The awareness component consists of three theory-based workshops. The first introduces participants to coral reef ecosystems, including fish, invertebrates, and reef survey methods used to assess reef health. The second workshop focuses on human impacts, such as pollution, overfishing, and habitat destruction, while also examining reef substrates such as coral, algae, reef rock, and sand. The third workshop explores the marine biodiversity of the Arabian Gulf, highlighting key species such as sharks, dolphins, and dugongs, and offering insight into the region's unique ecosystem.

The practical efforts component emphasizes field-based activities that enable participants to apply their knowledge in real-world settings. Volunteers are trained in coral identification at the taxonomic level, biodiversity monitoring through reef surveys, and threat assessments including bleaching, disease, and predation. They also engage in coral health monitoring to detect early signs of stress, while more advanced participants conduct research and restoration projects such as coral transplantation and reef rehabilitation. Efforts also extend to identifying rare and endangered coral species, ensuring targeted protection. Habitat creation forms a central pillar of the program, where participants help design and restore both artificial and natural habitats, including coral nurseries and transplantation projects. Coral farming further complements these initiatives by cultivating corals in controlled environments before introducing them to degraded reef systems, thereby supporting sustainable marine habitats.

The program adopts a holistic approach to restoration, focusing not merely on adding corals to degraded areas but on enhancing overall ecosystem health. Attention is given to factors such as water quality, biodiversity, and resilience, ensuring long-term adaptability in the face of environmental change.

Ethical guidelines are integrated throughout, ensuring that all restoration activities respect ecosystems and communities while aligning with sustainable conservation principles.

Impact and Outcomes

By the conclusion of the program, participants will have gained a thorough understanding of coral reef ecology, the effects of climate change on marine life, and the strategies necessary for reef conservation. They will also acquire hands-on skills in reef health assessment, coral species monitoring, and restoration practices, enabling them to contribute directly to the protection and rehabilitation of coral ecosystems. Importantly, participants will have taken part in the creation of new marine habitats, thereby leaving a tangible and positive impact on local ecosystems.

The program places particular emphasis on inclusivity and empowerment, focusing on engaging girls and women to help close the gender gap in marine conservation, with mothers serving as role models within their communities. Similarly, it prioritizes youth participation, preparing the next generation of leaders in marine sustainability. By instilling a sense of empowerment and responsibility, the program ensures that participants leave not only with skills and knowledge but also with the motivation to champion marine protection and restoration in their communities.

In addition, the program supports broader environmental efforts by advancing marine conservation projects, strengthening environmental protection initiatives, and promoting gender equity within the field of marine science.

Conclusion

In response to the ongoing decline of coral reefs, the proposed Effective Citizen Science Program provides a comprehensive approach to Marine Life and Coral Reef conservation, blending education, hands-on activities, and ethical restoration efforts. By raising awareness about the threats facing coral reefs and involving participants in practical conservation activities, the program offers a sustainable solution to the challenges faced by coral ecosystems. Through this initiative, participants not only gain valuable knowledge and experience but also contribute to the restoration and preservation of one of the planet's most important ecosystems, ensuring a healthier marine environment for future generations.

Bibliography

Jameson, S. C., Ammar, M. S. A., Saadalla, E., Mostafa, H., & Riegl, B. (2008). A quantitative damage assessment of diving sites in the Egyptian Red Sea during a period of severe anchor damage: A baseline for restoration and sustainable tourism management. *Journal of Sustainable Tourism, 16*(4), 395–413. https://doi.org/10.2167/jost758.0

Medio, D., Ormond, R. F. G., & Pearson, M. (1997). Effects of briefings on rates of damage to coral by scuba divers. *Biological Conservation, 79*(1), 91–95. https://doi.org/10.1016/S0006-3207(96)00108-4

Orams, M. B. (1999). *Marine tourism: Development, impacts and management*. Routledge.

Chapter 25
The Role of Green Spaces in Promoting Sustainable Communities' Challenges and Opportunities in the UAE

Elena Babii, Bushra Al Derbi, Tommaso Buffa, and Meera Al Blooshi

Abstract As cities in the United Arab Emirates (UAE) face rapid urbanization and environmental challenges, integrating green spaces in urban planning and communities has become crucial. This paper explores the benefits of well-designed green spaces in the UAE's arid climate, focusing on their roles in improving wellbeing, reducing urban stress, promoting physical activity, fostering social interaction, and providing equitable access to recreational areas. It also investigates the best sustainability practices for integrating green spaces, the challenges and opportunities in maximizing their potential, and their contribution to carbon dioxide (CO_2) sequestration in alignment with Dubai's sustainability initiatives. A qualitative approach was employed, utilizing case studies, site observations, and interviews with key stakeholders to analyze green spaces in Dubai designed over the past decade. Findings indicate that well-planned green spaces improve air quality, increase outdoor thermal comfort, support biodiversity, and promote physical and mental well-being. However, challenges such as high-water consumption, inappropriate plant species, and lack of shade persist due to ineffective design and planning. This study highlights the need for tailored sustainability strategies, including xeriscaping, smart irrigation technologies, and the use of drought-resistant native plants. Recommendations emphasize the importance of incorporating green spaces into urban design, prioritizing sustainable initiatives, and promoting green space equity. These strategies align with the UAE's vision for sustainable development, contributing to resilient and livable cities.

Keywords Green spaces · Sustainable community · Well-being · CO_2e · Urban planning · Carbon sequestration

---

E. Babii (✉) · B. Al Derbi · T. Buffa · M. Al Blooshi
Parsons, Gulf Towers, Dubai, UAE
e-mail: Elena.babii@parsons.com

B. Al Derbi
e-mail: Bushra.alderbi@parsons.com

T. Buffa
e-mail: Tommaso.buffa@parsons.com

M. Al Blooshi
e-mail: Meera.alblooshi@parsons.com

J. Dsilva (ed.), *Sustainability Education, Youth and Innovation*,
https://doi.org/10.1007/978-3-032-21948-0_25

Introduction

The UAE is experiencing rapid urbanization and significant environmental sustainability challenges, making the integration of green spaces increasingly important. The swift population and economic growth in the UAE have driven the need for strategic planning initiatives to enhance spatial structural quality. Among the governmental initiatives aiming to optimize the utilization of public spaces and infrastructure, Dubai Municipality launched in 2021 the 'Dubai 2040 Urban Master Plan' which aims to optimize public spaces and infrastructure, aligning with the emirate's economic priorities and sustainability goals. Multiple projects have been announcing that focus on greening the city. Projects like the 64 km-long sustainable urban corridor 'Dubai Green Spine,' aim to create a continuous network of green corridors and parks, fostering biodiversity, improving air quality, and providing accessible recreational areas (Zawya, 2024). These projects will address numerous benefits of the green spaces on the social well-being of its residents.

The International Association of Horticultural Producers (2023) (AIPH) mentions: "Dubai is at the threshold of becoming the world's largest and most sustainable oasis." (AIPH, 2023).

All these announcements and initiatives are commendable, but potentially more important will be the correct landscape design and execution of these projects.

Industry experts are starting to notice that there is a gap in guidance tailored to green spaces in the UAE. Existing sustainability certification rating systems like Al Safat in Dubai and Estidama in Abu Dhabi focus on buildings and communities but lack best landscape practices. Presenting a challenge for aligning green initiatives with long-term sustainability goals. A tailor-made sustainability framework would provide clarity and ensure consistent implementation, supporting the UAE's commitment to environmental stewardship.

Scope and Objectives

This paper explores the theme of "Health, Well-being, and Sustainable Communities", particularly in the UAE's climate. The primary focus areas include:

- Investigating the advantages of green spaces.
- Identifying challenges and opportunities in maximizing potential green spaces.
- Providing recommendations on challenges and opportunities while designing green spaces.
- Analyzing how green spaces can contribute to CO_2 sequestration.

The scope of this research entails an analysis of Dubai's green spaces, including parks, urban streets, cultural districts, and residential areas. This study employs a combination of qualitative methods, involving examining secondary data, interviews with stakeholders, and site observations. This study intends to enrich the understanding of the connection between well-designed green spaces and their advantages. It highlights the numerous benefits of green spaces concerning health, well-being, and community.

Methodology

This paper employs a qualitative approach to examine the role of urban green spaces and sustainable communities. This methodology includes case studies from previous projects, site observations, and interviews with stakeholders such as urban planners, sustainability experts, and landscape architects.

The case studies explore existing UAE urban planning strategies and green space projects. Site observations were conducted to assess the green spaces, landscape, and planting features, record physical attributes, visitor behavior, and interactions with green spaces. Additionally, interviews with stakeholders provided firsthand perspectives on challenges and opportunities.

This comprehensive approach enables the authors to derive conclusions and formulate critical sustainable solutions.

Findings

Benefits of Urban Green Spaces

Well-designed and well-maintained green spaces offer environmental, social, and economic advantages. In Dubai's climate, these areas improve air quality by diminishing airborne pollutants, enhance outdoor thermal comfort and assist in carbon sequestration by absorbing CO_2 emissions.

In a study by Masdar, The National and the World Future Energy Summit, a survey was performed where UAE residents stated: "public green space in the cities where they live make a major contribution to their personal happiness". The findings from the survey found that 65% of respondents said "existing green public space in their city added to their personal happiness. More than a third of respondents cited mental health benefits.

87% regularly visit public green spaces, while those who don't say this is because there are not enough of them (28.3%), current facilities are poor (27%), or they are too far away (22%).

More than 500 people, almost all living in the UAE, took the online poll, with 72% living in Abu Dhabi and 21% in Dubai". This survey forms part of the report, People and Spaces: The role of urban green spaces in cultivating innovation, sustainability and health and well-being in Arab cities (The National, 2020).

Carbon Sequestration, Air Quality and Outdoor Thermal Comfort Improvement

Urban green spaces absorb carbon dioxide through trees, shrubs, and other types of vegetation. Urban forests and parks serve as crucial carbon sinks. Studies show that they can capture between 1.2 to 1.6 tons of carbon per hectare each year. Green spaces enhance air quality by filtering out pollutants like particulate matter, sulfur dioxide, and nitrogen dioxide (Research Floor, 2024).

Case Study 1

In 2024, Parsons' landscape architects conducted a microclimate study for a confidential project in Dubai to encourage walkability while enhancing green spaces and outdoor thermal comfort. This microclimate study was used and analyzed as a case study for this research paper. The architects used methods such as: Site visits to collected site-specific climatological data and ground surface temperatures to benchmark simulations; Analytic Simulations which provided comparisons between baseline and proposed conditions for exposed sites and Computational Fluid Dynamics (CFD) Simulation to model wind flows and urban heat island effects. An observation across all the studied locations highlights the critical role ground material plays in outdoor comfort.

1. Unshaded asphalt roads can exceed 40 °C (Fig. 25.1).
2. Concrete has the potential to get hot when unshaded (Fig. 25.1).
3. Sand gets hot when unshaded (Fig. 25.1).
4. In terms of thermal comfort, high Standard Predicted Mean Vote Modified (SPMV) values were predicted at the majority of existing sites.

Another finding from the site surveys was the importance of shading:

1. Shaded concrete was 5–10 °C cooler (Fig. 25.2).
2. Fully shaded sand was up to 10 °C cooler (Fig. 25.2).
3. Tree shading in parks improved thermal comfort by ~30% (Fig. 25.2).
4. Partial shade was found to be beneficial.

Parks and open spaces are arguably the typology where enhancing thermal comfort is most important. They may also be used by those who are sensitive to heat (e.g., children and elderly).

The study's findings underscore the importance of design and material selection in urban green spaces to enhance outdoor thermal comfort. Shading plays a crucial role in reducing ground surface temperatures and mitigating urban heat island effect. Cooler outdoor environments encourage physical activity, social interaction, and community engagement. Sequestering CO_2 through trees, shrubs, and other vegetation significantly reduce greenhouse gas emissions, contributing to cleaner air and healthier environment.

The connection between air quality, outdoor thermal comfort, and the theme of "Health, Well-being, and Sustainable Communities" is evident.

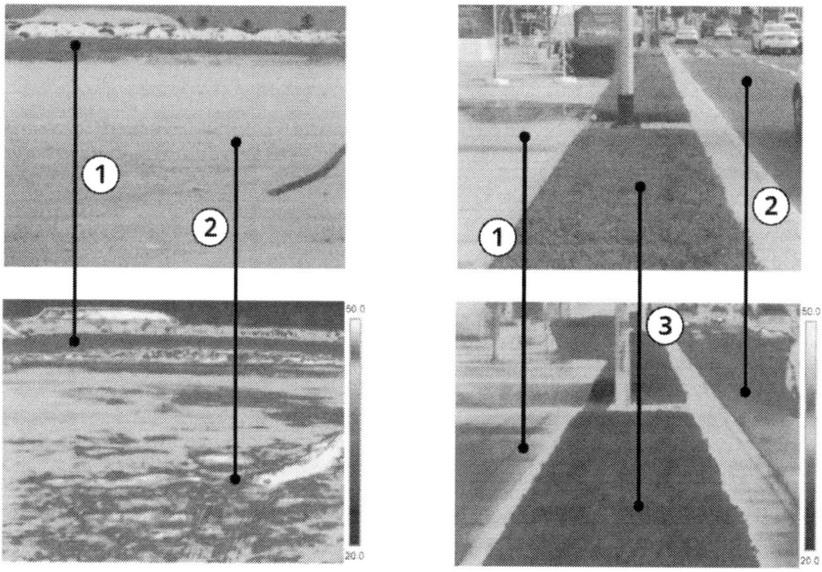

1. Grass (~30oC)
2. Sand (~40oC)

1. Unshaded concrete pavement (~35oC)
2. Unshaded road (~45oC)
3. Unshaded shrub (~25oC)

Fig. 25.1 Microclimate simulation by Parsons, Dubai open spaces 2024. *Note that the thermographic images were captured in April 2024. Summer temperatures would be much higher*

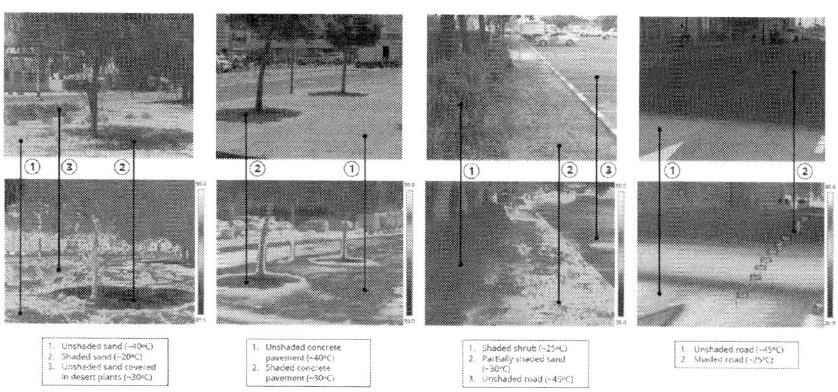

Fig. 25.2 Microclimate simulation by Parsons, Dubai open spaces 2024

Case Study 2

An urban development in the UAE designed to be a fully integrated residential community that emphasizes green living, and green infrastructure.

The purpose is to assess its key feature of integrating intelligent controllers and cutting-edge irrigation technologies. This irrigation system delivers water directly to the roots of plants at low flow rates and regular intervals. Pressure-reducing valves and flow sensors were included to identify and address any issues before leading to water loss.

Water Conservation Measures

Over-watering is more than just wasting a resource; it can seriously harm the environment and the health of landscapes. Excess water washes away fertilizers and pesticides, pollutes waterways, and cuts off oxygen to plant roots, promoting issues like fungus and rot. To address these challenges, the project implemented the following measures:

1. Seasonal Adjustment: The irrigation system was designed to adjust run times based on seasonal needs.
2. Smart Controllers: The project incorporated smart controllers that gather weather data to automatically adjust watering times.
3. Operating Pressure: The project set optimal pressure for each irrigation zone to minimize misting and water loss.
4. Low Volume: Emitters and drip systems were employed to deliver water directly to the roots of plants.
5. Matched Precipitation Rates: The irrigation system was designed to provide adequate water to each zone while avoiding over-watering.

Plant Selection

Alongside the water management strategy, "Planting Palette and Strategy" focuses on using native and climate-adapted species to make the most of resources. An example of this is the Ghaf tree, which thrives in the UAE environment and also offers shade. The plant palette is designed to be diverse, featuring a variety of species that are well-suited to the site, while carefully avoiding any invasive plants that could harm the ecosystem. Additionally, the strategy incorporates swales in areas prone to flooding.

Diverse plant life supports local wildlife, promoting biodiversity and creating a balanced ecosystem within the community. The central park serves as a communal space where residents can gather, socialize, and participate in community events. The selection of native and adaptive trees will have a positive impact on the overall sustainable development, contributing to ecological balance. The success of such projects hinges on the selection and analysis by landscape and sustainability experts.

Case Study 3

As part of Parsons' project portfolio, Case Study 3 is a project located in Dubai, and has been selected as a prime example of sustainable design, microclimate strategies and its overall positive impact on wellbeing. This project has also been awarded with LEED for Cities and Communities and WELL Communities—both Gold Rating.

This project leverages passive strategies to mitigate the region's harsh climate, reduce the urban heat island effect, and enhance outdoor comfort. This project promotes physical activity, biophilia, and overall well-being by integrating outdoor spaces designed for year-round use.

The project's microclimate strategy incorporates:

1. The landscape design incorporates tall trees with wide canopies that provide ample shade.
2. Hardscape materials were selected to regulate urban temperatures, particularly in hot climates where reflectivity, absorptivity, and emissivity determine their thermal performance.
3. Planting and softscape were selected to allow cool winds to penetrate while blocking the more violent wind gusts that can raise temperatures and carry dust.

Shading plays a crucial role in enhancing outdoor comfort in urban spaces and promoting an active, community-focused lifestyle. Trees serve as the primary shading solution, offering a cost-effective, ecological, and aesthetically pleasing approach to mitigating urban heat. Play areas are 90% shaded, walkways feature continuous shade for 80% of their length, while secondary walkways maintain 60% shade coverage.

Challenges of Previously Designed Green Spaces in Dubai

Green spaces in Dubai suffer from poor design and inadequate planning, resulting in high water consumption, inappropriate plant species, and lack of shade. The climate exacerbates these issues. The absence of a standardized sustainability framework has led to inconsistencies in the design and maintenance of urban green spaces across the city.

According to ARN News Dubai Eye 103.8 (2025) shown in Fig. 25.3, and the Dubai Municipality (2025) Dubai achieved a significant green milestone by planting 216,500 new trees in 2024, representing a 17% increase from the previous year. "Key greening projects included major intersections and bridges along Sheikh Zayed Road, Al Jamayel Street, Al Khawaneej Street, Al Khail Road, Sheikh Mohammed Bin Zayed Road and Latifa Bint Hamdan Street". While these initiatives aim to enhance intersections and bridges, they raise questions about their sustainability.

Fig. 25.3 ARN News, @DMunicipality/X ARN News (2025)

Why UAE Requires Tailored Strategies for Green Open Spaces

Common international sustainability strategies cannot be applied without considering the local context. There is an immediate requirement for a customized approach to sustainable initiatives that reflect the existing and anticipated climate conditions. Methods like xeriscaping, intelligent irrigation techniques, and the selection of drought-resistant native plant species are vital for boosting the sustainability of urban green spaces. Solutions like shaded pathways and green/vertical gardens enhance outdoor thermal comfort.

Climate Challenges and Future Considerations

Challenges such as rising temperatures, increased heat island effects, and water scarcity remain significant obstacles. Research and policy development are needed to create adaptive, climate-resilient urban landscapes. A sustainability framework will align these initiatives with Dubai's larger environmental objectives, ensuring that urban green spaces remain sustainable and advantageous.

Recommendations and Opportunities

Policy Makers, Consultants and Contractors

- Incorporate green infrastructure into new developments and urban renewal projects.
- Introduce a locally tailored tool to calculate carbon sequestration.
- Ensure that green spaces are equitably distributed in areas needed to encourage walkability and connectivity.
- Create approaches in urban areas to tackle issues like heat islands, air pollution, and resilience.

Softscape

- Plant a significant number of trees.
- Design parks and green areas.
- Choose native and adaptive plant species.
- Incorporate shading in design.
- Balance canopy density with wind penetration.
- Combine artificial shade structures with vegetation.
- Incorporate wind-responsive design.

Hardscape

- Avoid unshaded asphalt.
- Use materials with high albedo.
- Vegetated/naturalized surfaces will tend to remain the coolest.
- Select concrete as an alternative to asphalt.

Irrigation

- Incorporate green infrastructure through passive features.
- Utilize advanced irrigation technologies and smart water management controls.
- Use low-volume irrigation systems to deliver water to roots.

Environmental

- Consider the impact of surroundings on thermal comfort.
- Prevent developments from negatively affecting green spaces.
- Consider the sun path when placing trees.
- Ensure running and cycling paths are shaded.
- Place seating areas near tree clusters/shading devices.
- Implement targeted active cooling measures.

UAE Initiatives in Increasing Green Spaces

The country has made progress in increasing its green spaces through government initiatives, urban planning projects, and sustainability goals. Initiatives like 'Our Green Emirates,' 'Ghaf: The Symbol of Our Environment' (Abu Dhabi City Municipality, 2024), and the 'Dubai Greenery' initiative demonstrate the UAE's commitment to expanding green spaces and a sustainable urban environment. Projects like the Dubai Green Spine, which aim to transform the Sheikh Mohammad Bin Zayed Road into a 64 km sustainable corridor, demonstrate an innovative green approach (Zawya, 2024).

Investing in carbon sequestration can bring economic benefits by reducing healthcare costs and lower energy costs. Green spaces can attract tourism and improve residents' quality of life.

Literature Gaps and Future Research

There is a lack of evidence on the green space design in the UAE and their effectiveness in sequestering carbon. Further research is needed to develop strategies.

Conclusion

Urban green spaces are essential for addressing the environmental, social, and health challenges posed by rapid urbanization and climate change in the UAE. Dubai's 2040 Urban Master Plan and the Dubai Green Spine projects highlight the steps being taken to integrate green infrastructure into urban planning.

Challenges such as high-water consumption, the use of inappropriate plant species, and insufficient shading persist due to suboptimal design and planning.

This paper provides specific recommendations on softscape, hardscape, environmental considerations, and strategies for policymakers, consultants, and contractors. By adopting these approaches, cities can create healthier, more vibrant communities that are better equipped to cope with the impacts of climate change. The recommendations emphasize the necessity of incorporating green spaces into urban design. These strategies align with the UAE's vision for sustainable development and contribute to the creation of resilient and livable cities.

References

Abu Dhabi City Municipality. (2024, September 24). *Abu Dhabi City Municipality enhances the aesthetics of Ghaf trees as part of the 'Ghaf: The symbol of our environment' initiative.* https://www.dmt.gov.ae/en/adm/Media-Centre/News/24sep2024

Dubai Eye 103.8. (2025, January 10). *Dubai sees major rise in green spaces with over 216,000 new trees.* https://www.dubaieye1038.com/news/local/dubai-sees-major-rise-in-green-spaces-with-over-216000-new-trees/

Dubai Municipality. (2025, January 27). *216,500 trees planted in 2024 to enhance sustainability.* https://www.dm.gov.ae/2025/01/27/216500-trees-planted-in-2024-to-enhance-sustainability

International Association of Horticultural Producers. (2023, October 23). *Dubai, UAE: Nature-based city.* https://aiph.org/green-city-case-studies/dubai-uae-nature-based-city

Research Floor. (2024). *The role of urban green spaces in mitigating climate change: An integrative review of ecological, social, and health benefits.* https://er.researchfloor.org/wp-content/uploads/2024/04/The-Role-of-Urban-Green-Spaces-in-Mitigating-Climate-Change-An-Integrative-Review-of-Ecological-Social-and-Health-Benefits.pdf

The National. (2020, January 19). *Public green space in our cities makes us happy but we need more, say UAE residents.* https://www.thenationalnews.com/uae/environment/public-green-space-in-our-cities-makes-us-happy-but-we-need-more-say-uae-residents-1.962040

Zawya. (2024, July 24). 64-km Dubai Green Spine project unveiled. https://www.zawya.com/en/business/energy/64-km-dubai-green-spine-project-unveiled-d303tll8

Chapter 26
Environmental, Social and Governance (ESG) Considerations in Health Technologies and Innovations with a Focus on the GCC and MENA Regions

Yacine Hadjiat and Carlos Montana

Abstract Digital health innovations are cost-effective and inclusive solutions for healthcare. Given the challenges in achieving the Sustainable Development Goals (SDGs), digital health technologies have the potential to accelerate progress in providing health solutions. However, while digital health research has advanced, there is limited focus on how these solutions interact with socio-economic factors, and the risk of generating more inequality. This paper discusses the trade-offs of Environmental, Social and Governance (ESG) risks in digital health, with a focus on the GCC and MENA regions. While examining digital health through the lens of Environmental, Social, and Governance (ESG) factors, we propose a theoretical framework to evaluate their potential risks and benefits. The cost of developing these technologies is high in terms of time, money, and opportunity, making it essential to consider their real-world impacts. This research will provide an overview of ESG factors in health technologies innovations, highlight possible trade-offs between them, and propose a framework for evaluating these factors in practice. It starts with a short literature review of cases in Dubai, the GCC, and the MENA region, showcasing how these factors influence the region's health technologies innovation. The paper

Y. Hadjiat
College of Medicine, Mohammed Bin Rashid University of Medicine and Health Sciences, Dubai, UAE

Department of Physiopathology and Clinical Pharmacology, French National Institute of Health and Medical Research (INSERM), Boulogne-Billancourt, France

Paris Cité University, Paris, France

Y. Hadjiat
e-mail: yacine.hadjiat@dubaihealth.ae

C. Montana (✉)
Art and Design Department, Rochester Institute of Technology, Dubai, UAE
e-mail: crmcad@rit.edu

Queensland University of Technology, Brisbane, Australia

J. Dsilva (ed.), *Sustainability Education, Youth and Innovation*,
https://doi.org/10.1007/978-3-032-21948-0_26

concludes by emphasizing the need for a more structured approach to incorporating ESG considerations into digital health innovation frameworks.

Keywords Digital health · Health technologies innovation · Environmental · Social and governance (ESG) · Sustainability · GCC · MENA regions

Introduction

Digital health technologies and innovations is a rapidly growing area that uses technology to help diagnose, monitor, and treat patients, as well as to collect and share their health information. It has evolved from simple online apps for medical advice and wellbeing into a wide range of new technologies, including digital treatments, artificial intelligence, wearable devices, digital health indicators, and telemedicine. These innovations aim to improve access to healthcare, especially for people who are often left out, and to create faster, more efficient connections between doctors and patients using data-driven tools.

Simultaneously, digital health technologies and innovations—especially those using AI—use a lot of energy. The shift to clean energy isn't happening fast enough, and since many countries still depend on fossil fuels for power, electricity continues to be a big contributor to climate change. So, digital technologies are not environmentally neutral and can have a high energy impact.

Thus, this paper explores the research question: How can we better evaluate environmental, social and governance (ESG) considerations in health technologies innovation, and especially with a focus on the GCC and MENA regions?

Methodology

The research includes mixed methods, summarized in four points below.

1. **Scoping Literature Review**

A targeted review of academic literature was conducted to understand the current state of digital health with a focus on ESG-related themes such as sustainability, equity, governance, and environmental impact. Key databases included PubMed, Scopus, and Google Scholar, along with policy reports from regional institutions and organizations like the GCC Board Directors Institute and WHO.

2. **Thematic Analysis**

Relevant findings were organized using thematic coding to identify key ESG challenges and opportunities within digital health. Themes such as carbon emissions, access disparities, data governance, and climate-health linkages were used to inform the development of the framework.

3. **Framework Adaptation and Synthesis**

The Digital Health Equity Framework (Crawford & Serhal, 2020) was selected as a foundational model and adapted to include environmental and governance dimensions.

4. **Contextualization to GCC and MENA Regions**

Regional case examples and policy initiatives—such as ESG indices, climate policy, and digital health adoption strategies—were analyzed to tailor the framework to the socio-economic and political realities of the GCC and MENA regions.

Literature Review

The integration of Environmental, Social, and Governance (ESG) considerations into digital health initiatives is gaining momentum globally, with particular emphasis in regions like Dubai, the United Arab Emirates (UAE), the Gulf Cooperation Council (GCC), and the broader Middle East and North Africa (MENA). This brief literature review explores the sustainability and ESG factors in digital health within these regions.

Digital health technologies have emerged as promising solutions to address healthcare challenges within the MENA region. For example, a scoping review by El-Jardali et al. (2023) indicates that these technologies can enhance the quality, accessibility, and availability of healthcare services. In parallel to digital health, the GCC region has witnessed a surge in ESG-related activities in general. As noted by GCC Board Directors Institute & APCO Worldwide (2022), the Dubai Financial Market, for instance, launched the UAE Index for ESG in collaboration with S&P Dow Jones Indices and the Hawkamah Institute for Corporate Governance to gauge UAE-listed companies' commitments to ESG principles. Similarly, the Abu Dhabi Securities Exchange introduced ESG disclosure guidelines for listed companies, reflecting a regional commitment to sustainability.

The UAE has demonstrated significant attention to global health and climate change over the past two decades and has recently hosted COP 28. Multiple policies, strategies, and bodies have been developed to align with the SDGs, showcasing the nation's dedication to integrating sustainability into various sectors, including healthcare. According to Alkhaldi et al (2023), "global health and climate change became the most two notable priorities on the government agenda and its strategic thinking is that both priorities can no longer be overlooked".

Digital health technologies could help fight climate change by cutting greenhouse gas emissions and making healthcare easier to access. Rahimi-Arsabili et al. (2022) found that using virtual care can lower the environmental impact of health systems. These tools can also help spot diseases and health risks early, reduce exposure to dangers, and bring care to places that need it most. At the same time, the energy needs of digital solutions, especially AI based, are high, and the transition to clean energy is slower than is required. Many nations around the world continue to rely on fossil fuels to generate energy and thus electricity remains a major cause of climate change (Asongu et al, 2020). Thus, digital technologies are not carbon neutral and can have a high energy cost. Technologies like blockchain are energy hungry and produce between 22–29 million metric tons of carbon dioxide.

The integration of sustainability and ESG considerations into digital health technologies and initiatives is gaining traction in Dubai, the UAE, the GCC, and the broader MENA region. While significant progress has been made, challenges remain, particularly concerning infrastructure, privacy, and the need for rigorous evaluation. Addressing these challenges through patient-centered design, effective monitoring, and adherence to data protection regulations can enhance the sustainability of digital health interventions. Furthermore, aligning digital transformation efforts with ESG principles can contribute to broader sustainability goals, fostering economic growth and improved healthcare outcomes in the region.

Theoretical Proposal

An ESG Framework for Considering Digital Health Innovations

For digital solutions in healthcare and the SDGs, this paper presents an equity framework as a means of evaluating or constructing designs and innovative solutions. This framework may be adapted to consider 'future thinking', in effect considering the health impacts climate change and adaptation will have on the health of countries and communities—this is the true ESG approach. Aspects of climate or environmental issues cannot be considered in isolation. Social and Governance concerns are both drivers of climate change effects, actions and mitigants. For instance, a simple example is where changing temperatures impact disease prevalence, rainfall levels impact water borne diseases, and lack of water may introduce additional diseases. This impacts the health system's ability to manage the scope and levels of diseases in communities. Foresight is essential—what is predicted, what will we mitigate and how will this affect people's health. The closure of a fossil fuel energy source may positively impact pollution, but could lead to a severe economic downturn, lack of access to alternative energy, increased unemployment, and an increased level of people being unable to access preventative healthcare and treatment options. Extreme weather events have health outcomes.

The Digital Health Equity Framework, adapted from Crawford and Serhal (2020) applied to ESG lists the following factors (Fig. 26.1) to consider when designing, implementing and maintaining digital health innovations. The framework's factors fall mainly in the social factor area but can be complemented by environmental and governance factors to provide an integrated view.

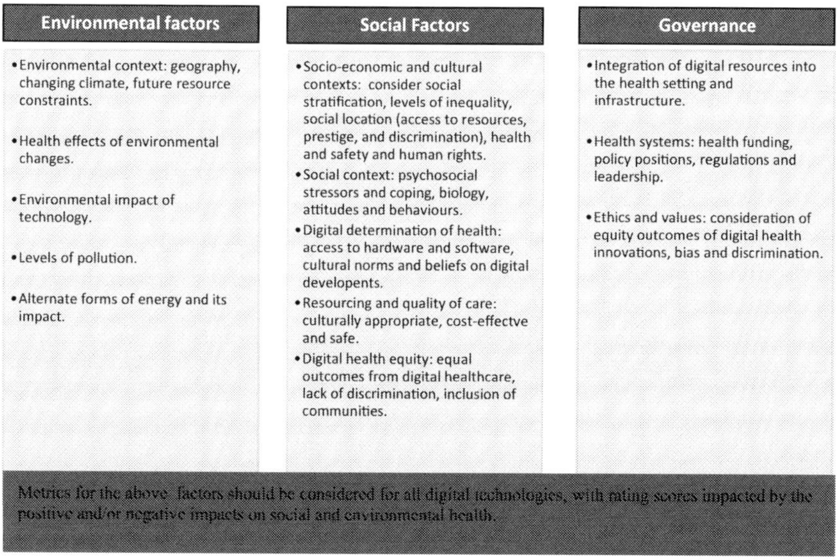

Fig. 26.1 The digital health equity framework. Adapted from Crawford and Serhal (2020)

Conclusions

At present a comprehensive framework of ESG in digital health does not exist with a comprehensive taxonomy, related measures or indicators. This is an evolving and developing area, and this research sets out the foundational parameters in which it may start to be developed. Digital health innovations continue to emerge, and health systems are rapidly implementing them for clinical care. The potential for digital health technologies to be used to improve the health of individuals and communities as well as to reduce health disparities, makes them an important consideration in Health and Social development. Having said this, the balance between these innovations and the energy they may consume, hardware and its recyclability for digital health should all be considered.

References

Alami, H., Rivard, L., Lehoux, P., Ag Ahmed, M. A., Fortin, J. P., & Fleet, R. (2023). Integrating environmental considerations in digital health technology assessment and procurement: Stakeholders' perspectives. *Digital Health, 9*, 20552076231219110. https://doi.org/10.1177/20552076231219113

Alkhaldi, M., Moonesar, I. A., Issa, S. T., Ghach, W., Okasha, A., Albada, M., Chelli, S., & Takshe, A. A. (2023). Analysis of the United Arab Emirates' contribution to the sustainable development goals with a focus on global health and climate change. *International Journal of Health Governance, 28*(4), 357–367. https://doi.org/10.1108/IJHG-04-2023-0040

Asongu, S. A., le Roux, S., & Biekpe, N. (2020). The criticality of growth, urbanization, electricity, and fossil fuel consumption to environmental sustainability in Africa. *Science of the Total Environment, 712*, Article 136376. https://doi.org/10.1016/j.scitotenv.2019.136376

Crawford, A., & Serhal, E. (2020). Digital health equity and COVID-19: The innovation curve cannot reinforce the social gradient of health. *Journal of Medical Internet Research, 22*(6), Article e19361. https://doi.org/10.2196/19361

El-Jardali, F., Bou-Karroum, L., Jabbour, M., & Sinha, C. (2023). Digital health in fragile states in the Middle East and North Africa (MENA) region: A scoping review of the literature. *PLoS ONE, 18*(4), Article e0285226. https://doi.org/10.1371/journal.pone.0285226

GCC Board Directors Institute, & APCO Worldwide. (2022). ESG in the GCC region: Accelerating integration. https://gccbdi.org/sites/default/files/2023-01/APCO-GCC-BDI-ESG_Report_2022.pdf

Rahimi-Ardabili, H., Magrabi, F., & Coiera, E. (2022). Digital health for climate change mitigation and response: A scoping review. *Journal of the American Medical Informatics Association, 29*(12), 2140–2152. https://doi.org/10.1093/jamia/ocac134

Part V
Social Innovation and Inclusive Sustainability

Chapter 27
Social Innovation as a Catalyst for Inclusive and Equitable Sustainability: A New Paradigm for Global Change

Sameera Fernandes, M. S. Sheeja, and Shubha Chandra

Abstract Social innovation has emerged as a transformative force in addressing systemic challenges related to sustainability, equity, and inclusive growth. This study explores how social entrepreneurship, grassroots movements, and inclusive business models contribute to sustainable development while bridging socio-economic disparities. By employing a mixed-methods approach, this research identifies key mechanisms through which social innovation fosters environmental justice, poverty alleviation, and gender equity. Findings highlight the role of community-led initiatives, digital transformation, and innovative financing models in creating inclusive sustainability frameworks. The study concludes with policy recommendations and strategic insights for institutional stakeholders to scale social innovation for a just transition towards a sustainable future (Mulgan in Social innovation: How societies find the power to change. Policy Press, 2019); Nicholls and Murdock in Social innovation: Blurring boundaries to reconfigure markets. Palgrave Macmillan, 2012).

Keywords Social innovation · Inclusive growth · Sustainability · Gender equity · Social justice

Introduction

Sustainability and equity have traditionally been approached as separate yet critical issues in the discourse of global development. Sustainability emphasizes environmental stewardship and long-term resource management, whereas equity focuses on social justice, economic inclusion, and the fair distribution of opportunities (Raworth, 2017; Sachs, 2020). However, contemporary challenges such as climate change, socio-economic disparities, and political instability necessitate a holistic approach that integrates both dimensions. The rise of social innovation offers a

S. Fernandes (✉) · M. S. Sheeja · S. Chandra
Garden City University, Bengaluru, India
e-mail: sameera.fernandes@gmail.com

J. Dsilva (ed.), *Sustainability Education, Youth and Innovation*,
https://doi.org/10.1007/978-3-032-21948-0_27

promising solution by addressing systemic barriers through collaborative, adaptive, and community-driven interventions (Westley et al., 2021).

Social innovation, as defined by Murray et al. (2010), refers to novel strategies, concepts, and solutions designed to meet social needs while fostering new social relationships. This concept aligns closely with the principles of sustainable development, which emphasize economic prosperity, environmental responsibility, and social inclusion (Yunus, 2017; Sachs, 2020). Social innovation serves as a catalyst for transforming grassroots movements, policy initiatives, and sustainable business models into inclusive, equitable frameworks that foster systemic change (Nicholls & Murdock, 2012).

Objectives

This study aims to explore the role of social innovation in fostering inclusive and equitable sustainability. The primary objective is to examine how social entrepreneurship, grassroots movements, and financial models contribute to sustainable development. Additionally, the research seeks to analyze policy interventions that enable social enterprises to scale their impact. Lastly, the study aims to provide actionable insights for policymakers, investors, and stakeholders to integrate social innovation within national sustainability strategies (Smith, 2013; Tencati & Zsolnai, 2012).

Methodology

This study employs a mixed-methods approach, integrating both qualitative and quantitative research techniques to ensure a comprehensive understanding of how social innovation fosters inclusive and sustainable development. The research follows an interdisciplinary framework incorporating sustainability science, social entrepreneurship theory, and policy analysis (Bocken et al., 2014; Geels, 2002).

The study is guided by three key research questions:

How does social innovation contribute to fostering inclusive and sustainable development?

What are the primary mechanisms through which social enterprises and grassroots initiatives drive systemic change in sustainability?

How can financial models and policy interventions support the scaling of social innovation?

These questions are relevant to both theory and practice, providing insights into the role of social entrepreneurship, economic policy, and sustainability frameworks (Elkington, 1997; Hart & Milstein, 2003).

Table 27.1 Social enterprises by region

Region	Social enterprises	Employment generated
Africa	320	12,000
Asia	540	23,000
Europe	610	34,000
North America	430	21,000
South America	290	13,000

Table 27.2 Impact score by type of innovation

Type of innovation	Impact score (out of 10)
Financial inclusion	8.5
Renewable energy	9.2
Sustainable agriculture	8.8
Waste management	7.9
Social services	9.0

Findings

Findings indicate that social enterprises significantly contribute to sustainable development by addressing poverty reduction, gender equity, and environmental conservation as per Table 27.1 (Nicholls & Murdock, 2012). Businesses such as Grameen Bank and Patagonia's B Corp initiatives demonstrate that financial sustainability can be achieved alongside social impact. Additionally, community-driven solutions such as solar energy cooperatives in sub-Saharan Africa and urban farming projects in low-income areas showcase the power of participatory models in sustainability as per Table 27.2. These initiatives not only reduce carbon footprints but also create job opportunities for vulnerable populations (Meadows, 2008; Piketty, 2014).

Discussion

The transformative potential of social innovation lies in its ability to integrate economic, environmental, and social objectives. Social enterprises often address gaps left by traditional market and governmental systems, particularly in underserved communities. Digital transformation has also amplified the reach and efficiency of social innovations, enabling scalable solutions through technology-driven platforms (Castells, 2010; Christensen et al., 2015).

Moreover, innovative financing models such as impact investing, green bonds, and blended finance have proven critical in supporting social enterprises. These models not only provide necessary capital but also align investment objectives with sustainable and inclusive growth goals (Bugg-Levine & Emerson, 2011; Jackson & Harji, 2013).

Conclusion and Suggestions

This research underscores the transformative potential of social innovation in fostering inclusive sustainability. By integrating social impact into economic and environmental initiatives, social enterprises can create systemic change, addressing issues of poverty, gender inequality, and climate resilience simultaneously (Sen, 1999; Stiglitz, 2010).

To further enhance the impact of social innovation, policymakers should establish targeted regulatory frameworks that incentivize social enterprises and grassroots movements. Governments should provide financial support through subsidies, tax incentives, and accessible microfinance options to enable social enterprises to expand their reach. Additionally, international cooperation between governments, non-profits, and corporations must be strengthened to create a globally recognized policy ecosystem that supports sustainable social innovation (United Nations, 2015; World Economic Forum, 2020).

The role of financial institutions is also critical in fostering inclusive sustainability. Investors should be encouraged to integrate environmental, social, and governance (ESG) criteria into their investment strategies to support social enterprises. Impact investing, green bonds, and blended finance models should be expanded to ensure that innovative business models focused on social and environmental impact receive the necessary funding to scale their solutions effectively (Porter & Kramer, 2011; Goyal & Joshi, 2011).

In conclusion, social innovation offers a dynamic and transformative approach to achieving sustainability and equity. By implementing robust policy frameworks, financial incentives, capacity-building programs, and technological advancements, stakeholders can maximize the impact of social enterprises, ensuring long-term, inclusive sustainability for future generations (Schwab, 2016; Ostrom, 1990).

References

Alvord, S. H., Brown, L. D., & Letts, C. W. (2004). Social entrepreneurship and societal transformation. *Journal of Applied Behavioral Science, 40*(3), 260–282.

Bocken, N. M. P., Short, S. W., Rana, P., & Evans, S. (2014). A literature and practice review to develop sustainable business model archetypes. *Journal of Cleaner Production, 65*, 42–56.

Bornstein, D. (2007). *How to change the world: Social entrepreneurs and the power of new ideas.* Oxford University Press.

Borzaga, C., & Defourny, J. (2001). *The emergence of social enterprise.* Routledge.

Chesbrough, H. (2006). *Open innovation: The new imperative for creating and profiting from technology.* Harvard Business School Press.

Defourny, J., & Nyssens, M. (2010). Social enterprise: The landscape of social entrepreneurship in Europe. *Social Enterprise Journal, 6*(2), 202–228.
Drucker, P. F. (1985). *Innovation and entrepreneurship: Practice and principles*. Harper & Row.
Elkington, J. (1997). *Cannibals with forks: The triple bottom line of 21st-century business*. Capstone Publishing.
Freeman, R. E. (1984). *Strategic management: A stakeholder approach*. Pitman.
Geels, F. W. (2002). Technological transitions as evolutionary reconfiguration processes: A multi-level perspective and a case-study. *Research Policy, 31*(8–9), 1257–1274.
Goyal, P., & Joshi, M. (2011). A study of social entrepreneurship and sustainable development in India. *International Journal of Business and Management, 6*(5), 97–103.
Haigh, N., & Hoffman, A. J. (2012). Hybrid organizations: The next chapter of sustainable business. *Organizational Dynamics, 41*(2), 126–134.
Hart, S. L., & Milstein, M. B. (2003). Creating sustainable value. *Academy of Management Executive, 17*(2), 56–67.
Holling, C. S. (2001). Understanding the complexity of economic, ecological, and social systems. *Ecosystems, 4*(5), 390–405.
Howaldt, J., & Schwarz, M. (2010). *Social innovation: Concepts, research fields, and international trends*. Sozialforschungsstelle Dortmund.
Kanter, R. M. (1999). From spare change to real change: The social sector as beta site for business innovation. *Harvard Business Review, 77*(3), 122–132.
Kuratko, D. F. (2016). *Entrepreneurship: Theory, process, practice*. Cengage Learning.
Leadbeater, C. (1997). *The rise of the social entrepreneur*. Demos.
Leisinger, K. M. (2007). Corporate philanthropy: The "top of the pyramid." *Business and Society Review, 112*(3), 315–342.
Mair, J., & Marti, I. (2006). Social entrepreneurship research: A source of explanation, prediction, and delight. *Journal of World Business, 41*(1), 36–44.
Martin, R. L., & Osberg, S. (2007). Social entrepreneurship: The case for definition. *Stanford Social Innovation Review, 5*(2), 28–39.
Meadows, D. H. (2008). *Thinking in systems: A primer*. Chelsea Green Publishing.
Mulgan, G. (2019). *Social innovation: How societies find the power to change*. Policy Press.
Murray, R., Caulier-Grice, J., & Mulgan, G. (2010). *The open book of social innovation*. The Young Foundation.
Nicholls, A., & Murdock, A. (2012). *Social innovation: Blurring boundaries to reconfigure markets*. Palgrave Macmillan.
Nicholls, A. (2006). *Social entrepreneurship: New models of sustainable social change*. Oxford University Press.
Ostrom, E. (1990). *Governing the commons: The evolution of institutions for collective action*. Cambridge University Press.
Piketty, T. (2014). *Capital in the twenty-first century*. Harvard University Press.
Porter, M. E., & Kramer, M. R. (2011). Creating shared value. *Harvard Business Review, 89*(1–2), 62–77.
Prahalad, C. K., & Hammond, A. (2002). Serving the world's poor, profitably. *Harvard Business Review, 80*(9), 48–57.
Raworth, K. (2017). *Doughnut economics: Seven ways to think like a 21st-century economist*. Chelsea Green Publishing.
Reynolds, P. D., & Curtin, R. (2011). *New business creation: An international overview*. Springer.
Sachs, J. D. (2020). *The ages of globalization: Geography, technology, and institutions*. Columbia University Press.
Schumpeter, J. A. (1934). *The theory of economic development: An inquiry into profits, capital, credit, interest, and the business cycle*. Harvard University Press.
Schwab, K. (2016). *The fourth industrial revolution*. Crown Business.
Seelos, C., & Mair, J. (2005). Social entrepreneurship: Creating new business models to serve the poor. *Business Horizons, 48*(3), 241–246.

Sen, A. (1999). *Development as freedom*. Knopf.

Shane, S. (2003). *A general theory of entrepreneurship: The individual-opportunity nexus*. Edward Elgar Publishing.

Smith, A. (2013). Social innovation: Blurring boundaries to reconfigure markets. *Innovation: Management, Policy & Practice, 15*(1), 3–16.

Spreckley, F. (2011). *Social audit toolkit*. Local Livelihoods.

Stiglitz, J. E. (2010). *Freefall: America, free markets, and the sinking of the world economy*. W. W. Norton & Company.

Tencati, A., & Zsolnai, L. (2012). The collaborative enterprise. *Journal of Business Ethics, 111*(2), 231–238.

Thompson, J. L. (2002). The world of the social entrepreneur. *International Journal of Public Sector Management, 15*(5), 412–431.

United Nations. (2015). *Transforming our world: The 2030 agenda for sustainable development*. United Nations Publications.

Weerawardena, J., & Mort, G. S. (2006). Investigating social entrepreneurship: A multidimensional model. *Journal of World Business, 41*(1), 21–35.

Westley, F., Antadze, N., & Riddell, D. (2021). The evolution of social innovation: Building resilient and equitable systems. *Stanford Social Innovation Review, 19*(3), 26–33.

World Economic Forum. (2020). *The global competitiveness report 2020*. World Economic Forum.

Yunus, M. (2017). *A world of three zeros: The new economics of zero poverty, zero unemployment, and zero net carbon emissions*. PublicAffairs.

Zahra, S. A., Gedajlovic, E., Neubaum, D. O., & Shulman, J. M. (2009). A typology of social entrepreneurs. *Journal of Business Venturing, 24*(5), 519–532.

Zott, C., & Amit, R. (2010). Business model design: An activity system perspective. *Long Range Planning, 43*(2–3), 216–226.

Chapter 28
Bridging the Gap to UAE's Vision 2030/2050: A Framework for Embedding Sustainability in Organizational Learning & Development

Joice Reena Pais

Abstract The UAE's Vision 2030 and 2050 set ambitious sustainability goals, including renewable energy and a net-zero future. These forward-thinking policies necessitate immediate action, requiring not just strategic planning but also a skilled workforce capable of driving sustainability and ESG (Economic, Social & Governance) initiatives. However, many organizations lack such initiatives or rely on outdated learning models that neglect these critical areas, creating a gap between National Vision and Workforce preparedness. By exploring how organizations can integrate sustainability and ESG principles into their corporate learning and development programs, this study aims to bridge above identified gap. It identifies key barriers and guides policymakers and corporate leaders in updating training programs to prepare employees for both short-term and long-term sustainability objectives. This study employs a qualitative approach, incorporating a small-scale survey and drawing on literature reviews and case study analysis. Thematic analysis is used to extract insights into training gaps, policy influence, and best practices. Findings of this focused study include a four-stage framework: Engagement, Education, Empowerment, and Evolution. While this framework contributes to sustainability training literature it also offers practical insights for organizations. It also enables alignment with national policies, workforce development, and sustainability innovation aligned with UAE Vision 2030 and 2050.

Keywords Sustainability training · UAE Vision 2030 · ESG integration · Organizational learning · Workforce development

---

J. R. Pais (✉)
Terra Numen, Dubai, UAE
e-mail: info@terranumen.net

J. Dsilva (ed.), *Sustainability Education, Youth and Innovation*,
https://doi.org/10.1007/978-3-032-21948-0_28

Introduction

The UAE has positioned itself as a global leader in sustainability through Vision 2030 and Vision 2050, emphasizing renewable energy, economic diversification, and a knowledge-based, net-zero future. While policy frameworks provide direction, their effectiveness is contingent on workforce readiness. Building on organizational learning theories (Argyris & Schön, 1978; Senge, 1990[1]) and Green HRM (Renwick et al., 2012), this study investigates how corporate learning programs can be redesigned to bridge this gap. By presenting a practical framework for integrating sustainability and ESG (Economic, Social & Governance) principles into corporate training programs, this study offers actionable insights for policymakers. While it expands academic discourse on sustainability-driven corporate learning and workforce transformation, it greatly supports economic resilience as well as environmental sustainability.

Problem Statement: Despite the UAE's ambitious sustainability policies, many organizations are failing to incorporate effective Sustainability and ESG principles into their workforce development strategies. This leads to absence of skills and knowledge necessary to meet nation's sustainability objectives.

Objectives: This study aims to:

1. Identify challenges in embedding sustainability within UAE corporate training
2. Analyze regional and global best practices
3. Develop a framework aligning corporate learning with UAE Vision 2030
4. Provide actionable strategies for bridging training gaps and fostering sustainable workforce development.

Scope: The study examines large and medium-sized UAE based organizations' training programs across sectors like finance and heavy industry, focusing on challenges and initiatives related to Vision 2030 and 2050.

Expected Research Results: The study is expected to highlight the gaps between policy ambitions and corporate execution. It is also expected to present a structured framework for sustainable corporate learning.

Methods: A qualitative research approach is applied here, integrating a comprehensive literature review, case study analysis of UAE corporations as well as review of policy documents. Wherever applicable and possible, a compact survey is also conducted.

Theoretical Foundations: From Learning Organizations to Green HRM

Organizational learning theories (Argyris & Schön, 1978; Senge, 1990), sustainability-focused learning (Molnar & Mulvihill, 2003; Siebenhüner & Arnold, 2007), and Green HRM (Renwick et al., 2012) emphasize the crucial role of corporate training in driving sustainability performance, fostering a culture of continuous learning with improved sustainability performance (Al-Ali, 2022).

---

[1] Senge, P. (as cited in the original paper, implying 1990 work)

The Rationale for Embedding Sustainability in Organizational Learning

The UAE's Vision 2030 and Vision 2050 emphasize a shift toward a knowledge-based, net-zero economy. Studies highlight the urgency of integrating sustainability into corporate training, as traditional Learning and Development (L&D) programs mainly remain compliance-driven and lack innovation. Without strategic updates, companies' risk being unprepared to meet sustainability targets.

Literature Review Based on Policy Documents

UAE policies mandate sustainability integration in corporate practices. Key frameworks include Vision 2030 (economic diversification, renewable energy), Vision 2050 (net-zero transition), UAE Energy Strategy 2050 (carbon reduction targets), LT-LEDS (Long-Term Low Emission Development Strategies), the Estidama Pearl Rating System (Green building rating system by the Abu Dhabi Urban Planning Council) and national environmental strategies such as those outlined by Environment Agency Abu Dhabi (2024). These policies highlight the need for corporate training programs that extend beyond compliance and embed ESG principles for long-term impact.

Empirical Insights and Case Studies from the UAE

Some of the large organizations like ADNOC, DP World, DEWA have implemented comprehensive and structured training programs on Sustainability and ESG principles. However, there are many mid and small-scale organizations those lack effective learning strategies, despite senior management commitments.

Frameworks and Best Practices for Green Training

Frameworks for green training, such as the four-dimensional approach by Sult et al. (2023), provide guidance for integrating ESG into L&D, emphasizing awareness, communication, embedding, and evaluation. Green HRM research highlights the importance of continuous learning and responsible leadership for improved sustainability performance (Alansaari et al., 2022; Usman et al., 2024).

Synthesis and Implications for UAE Organizations

UAE businesses must embed sustainability into L&D to achieve Vision 2030 & 2050 targets. Both policymakers as well as corporate leaders should prioritize strategic sustainability training to bridge the policy-execution gap through dynamic learning systems.

Methodology

This study investigates how sustainability and ESG principles are integrated into corporate training programs within UAE organizations, employing a primarily qualitative research approach. Given the time constraints, the research mainly derives from secondary data sources, including a comprehensive literature review, analysis of policy documents, and case studies of a few organizations based in UAE. (e.g., DP World, ADNOC, Masdar, DEWA, select ENOC divisions, Emirates Aviation Catering).

However, a small-scale survey ($n = 25$) was conducted to gain preliminary insights into current training practices and perceived challenges. The survey included multiple-choice questions on training topics, Likert scale questions on the importance of sustainability integration, and open-ended questions on perceived challenges. Quantitative survey data were analyzed descriptively, while qualitative data from open-ended questions were coded for thematic analysis.

Data Collection

Data collection involved a review of academic articles, industry reports, and government policy documents (e.g., UAE Vision 2030, Ministry of Energy and Infrastructure (2023) Vision 2050, LT-LEDS) including national climate strategies developed by Ministry of Climate Change and Environment (2023) to establish the current state of knowledge. Case studies were analyzed using publicly available information (like annual reports, press releases) to provide empirical insights.

Data Analysis

Thematic analysis of literature and case study data identified recurring themes related to barriers, best practices, and policy influence. These themes informed the development of the 4E Sustainability Training Framework comprising Engagement, Education, Empowerment, Evolution.

Justification

The qualitative approach is well-suited for exploring complex organizational practices in sustainability training. Reliance on secondary data, supplemented by the survey, provides a robust foundation for the proposed framework and aligns with the objectives of the study.

Ethical Considerations

As this study primarily uses publicly available data, ethical concerns are minimal. Future primary data collection will adhere to informed consent and confidentiality protocols.

Findings

The literature review and case study analysis reveal a persistent gap between the UAE's ambitious sustainability policies (e.g., Vision 2030, Vision 2050, LT-LEDS) and the on-the-ground reality of corporate training programs. Many organizations still rely on traditional, compliance-based approaches that lack a deep integration of sustainability and ESG principles, leaving the workforce underprepared to drive innovative, low-emission solutions.

Survey Findings

A small-scale survey (n = 25) explored current training practices and perceived challenges. As depicted in Graph 28.1, while 56% of organizations reported including sustainability or ESG topics in their training, a significant 32% were unsure, while 12% explicitly stated they do not include these topics. This highlights a significant gap in the understanding and implementation of sustainability training across organizations.

As depicted in Graph 28.2, this gap is further widened due to the fact that some of the reported inclusion of "sustainability" reflects a focus on Corporate Social Responsibility (CSR) activities rather than a comprehensive integration of ESG principles. For instance, while 56% of organizations reported including sustainability topics, the most frequently mentioned topics were "Corporate Social Responsibility" (44%), while crucial areas like "Renewable Energy" and "Climate Change and Net-Zero" were only included by 24% of the organizations. This indicates a compliance-driven approach rather than a genuine commitment to sustain.

Consistent with the broader literature, Tag Cloud 28.1 shows the qualitative data from the open-ended question "Primary Challenges in Integrating Sustainability and ESG Content". A common theme such as Lack of expertise was seen among most of the participants. Apart from that were Resources Constraints, Low employee

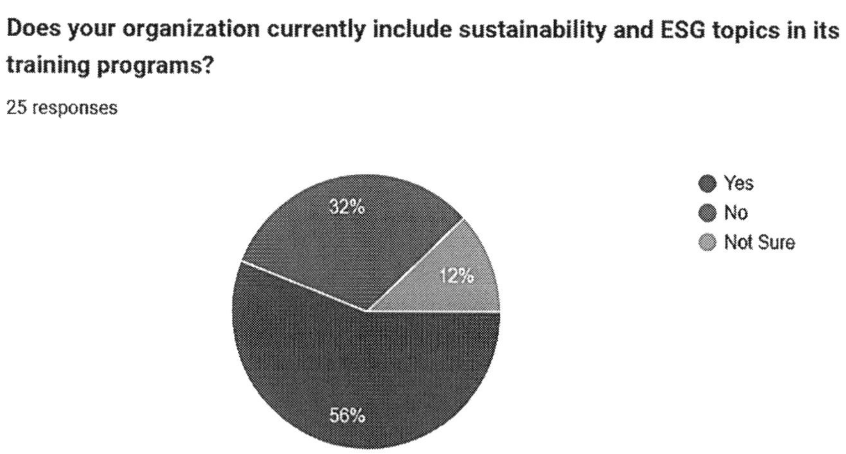

Graph 28.1 Sustainability training inclusion

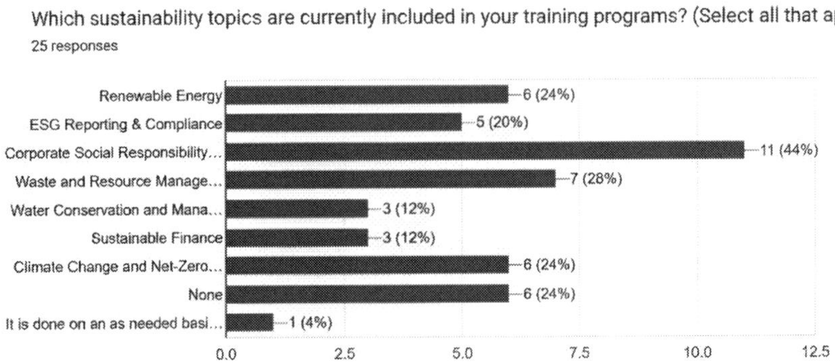

Graph 28.2 Sustainability topics covered

engagement, Lack of awareness among employees as well as Lack of awareness among the senior leadership team.

As illustrated in Graph 28.3, an overwhelming majority (over 80%) of survey respondents recognize the importance of sustainability training. This highlights a strong awareness of sustainability's relevance. Yet, the existence of implementation challenges suggests that recognizing the value of sustainability training does not automatically translate into its effective execution. Another open-ended survey question "Do you believe that enhanced sustainability training can contribute significantly to

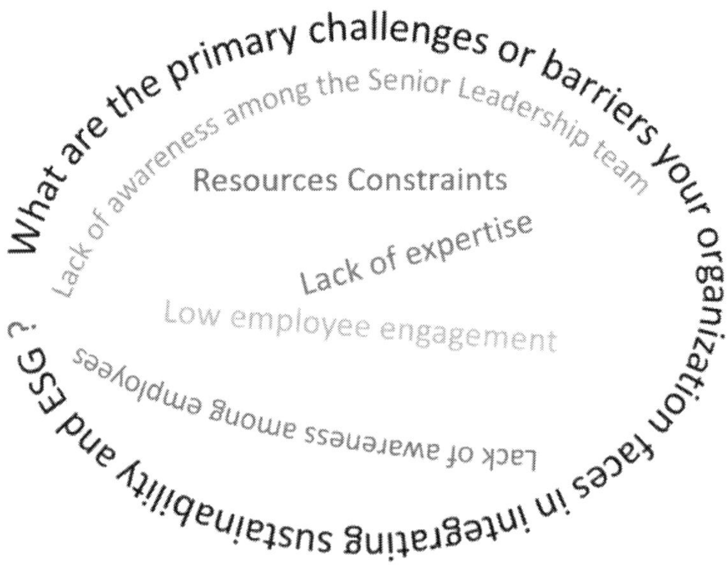

Tag Cloud 28.1 Challenges to integrating sustainability and ESG

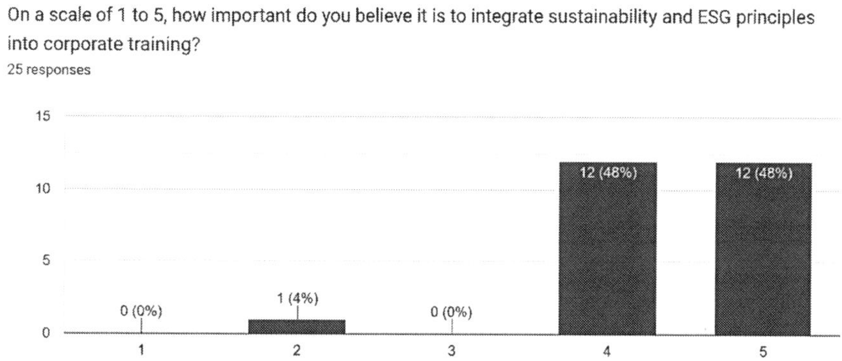

Graph 28.3 Importance of training in sustainability topics

your organization's Sustainable growth? Please explain briefly", where over 65% of the participants responded positively, compliments this importance of integration.

Tag Cloud 28.2 shows response to the open-ended question "What additional Support or Resources to Improve Integration?", implying perceived solution by the participants. While 'External expertise' was among the recurring theme, "specialized training modules", "in-person training" as well as "workshops" to enhance their training programs were suggested. Respondents also emphasized the need for "awareness classes" and integrating "sustainability targets/KPIs" at both organizational and individual levels. These findings highlight a desire for a more comprehensive and integrated approach to sustainability training.

Key Findings from Literature Review and Case Studies

Inadequate Integration: Analysis of few large corporations and mid-sized companies indicates that even among organizations with some sustainability initiatives, many still rely on legacy training models that do not fully integrate Sustainability and ESG content.

Barriers to Effective Training: Consistent with the survey findings, common challenges identified in the literature and case studies include limited resources, organizational inertia, and insufficient leadership commitment.

Positive Impacts When Done Right: Case studies demonstrate that organizations with well-designed sustainability training programs experience improved operational efficiency (Usman et al., 2024), enhanced environmental performance, increased employee engagement, and better overall ESG outcomes.

Proposed Framework

To bridge the gap between policy and practice, this study proposes the 4E Sustainability Training Framework, an adaptation of a cyclical model for embedding sustainability into organizational learning (refer Fig. 28.1).

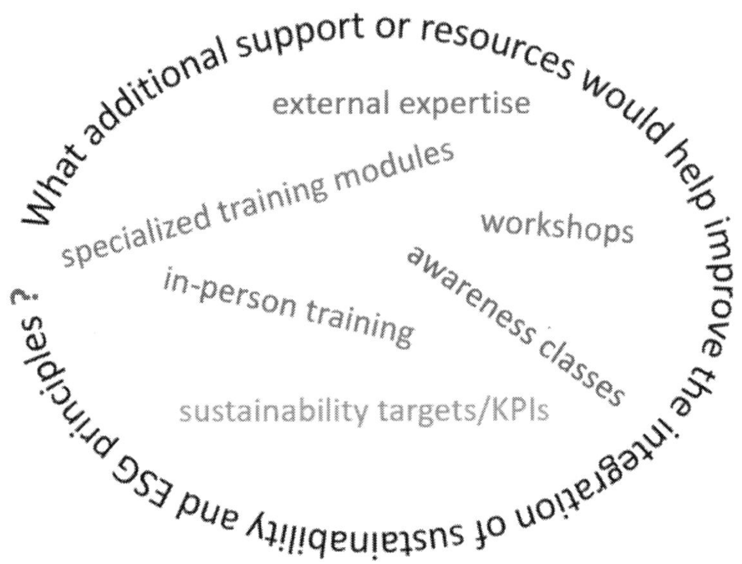

Tag Cloud 28.2 Support for integrating sustainability and ESG

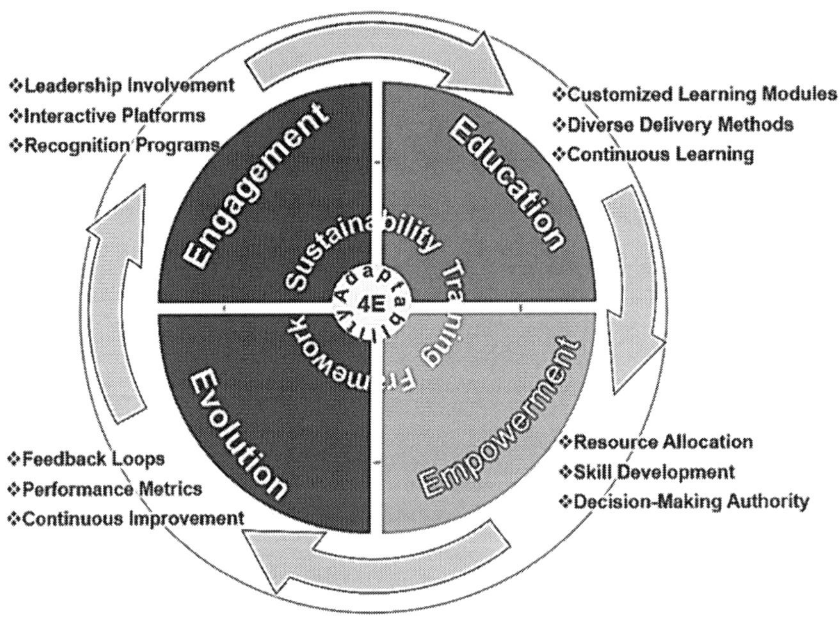

Fig. 28.1 4E sustainability training framework

Engagement: Building buy-in and fostering a culture of sustainability through visible leadership involvement, interactive platforms for communication and collaboration, and employee recognition programs are crucial first steps. An organization's culture, defined by its shared values, beliefs, and behaviors, significantly influences its social sustainability impact. A culture that prioritizes ethical practices, diversity, equity, inclusion, and employee well-being fosters positive social outcomes. Leaders play a pivotal role in championing sustainability by visibly integrating it into their decisions and actions, clearly communicating its importance, and setting high expectations. Practical examples include companies like Patagonia and Unilever, where deeply embedded sustainability values have boosted employee morale, customer loyalty, and overall social impact. This ultimately contributes to long-term value creation and a positive reputation.

Education: Equipping employees on Sustainability and ESG through workshops, online modules, and customized learning. For example, organizations can offer hands-on workshops on climate change, interactive modules on renewable energy or simulations of ESG risk assessment, tailored by industry and role, to stay updated on best practices.

Empowerment: Translating learning into action by allocating resources to support sustainability initiatives, developing relevant skills through hands-on training and simulations, and granting clear decision-making authority to enable implementation.

Evolution: Ensuring training program relevance through feedback loops to gather input from participants, performance metrics (KPIs) to track impact, and a commitment to continuous improvement and adaptation to emerging challenges and opportun.

The 4E framework provides a structured, adaptable roadmap for developing impactful sustainability training, cultivating a knowledgeable and motivated workforce aligned with both short-term and long-term sustainability visions.

Argument: By involving leadership team, customized continuous learning options, by empowering the employees and by way adapting to continuous improvement, the 4E framework bridges the gap between policy ambitions and implementation. This strengthens the UAE's commitment to a sustainable future, fostering economic resilience and global environmental stewardship.

Conclusions

This study highlights the misalignment between UAE sustainability ambitions and the current state of corporate training. With many organizations still relying on conventional, compliance-driven training that lacks substantive ESG integration, a workforce ill-prepared for the nation's sustainability objectives is being produced. The 4E Framework (Engagement, Education, Empowerment, Evolution) offers a transformative approach to bridge this divide, laying the groundwork for a culture that champions sustainability.

The role of organizational culture as a catalyst for social sustainability cannot be overstated. A well-defined culture that prioritizes ethical conduct, diversity, and employee well-being not only amplifies social impact but also enhances stakeholder trust and drives long-term value creation. Leaders are instrumental in this transformation, setting the tone and direction for a sustainable ethos.

Suggestions for Practice

Strategic Alignment: Organizations should align their training policies with national sustainability strategies (e.g., Vision 2030, Vision 2050) and integrate sustainability as a core element of workforce development.

Targeted Training Programs: Develop specialized training modules and utilize diverse, interactive delivery methods to address industry-specific sustainability challenges and cater to different learning styles and roles, from juniors to key management personnel (KMPs). These programs should go beyond theoretical knowledge and incorporate practical, hands-on experiences that enable employees to apply sustainability principles in their daily work.

Leadership Development: Invest in leadership development programs to equip KMPs with the knowledge and skills to champion sustainability initiatives and foster a culture of continuous improvement. Leaders should not only be knowledgeable about sustainability but also be able to model sustainable behaviors and inspire their teams to embrace them.

Organizational Culture: Organizations should strive to create a culture that values sustainability. This can be achieved through:

- Clear Communication: Regularly communicate the importance of sustainability and how it aligns with the organization's values and goals.
- Employee Engagement: Involve employees in sustainability initiatives and decision-making processes.
- Incentive Programs: Develop incentive programs that reward employees for sustainable behaviors and contributions.
- Culture Audits: Conduct regular culture audits to assess the organization's progress in embedding sustainability values and identify areas for improvement.

By implementing these strategies, organizations can create a workplace where sustainability is not just a policy but a deeply ingrained value.

Challenges for Implementation

Implementing the 4E framework can face hurdles too like awareness gaps, resource constraints, and change resistance across public and private sectors. While government and large entities can leverage regulations, SMEs may struggle. Partnerships with government, private institutes, independent subject matter experts and incentivized public–private collaborations can effectively bridge these gaps and ensure successful implementation.

Suggestions for Future Research

Empirical Validation: Future studies should aim to validate the 4E framework across diverse organizational settings within the UAE, using both quantitative and qualitative data.

Longitudinal Impact Assessment: Research should track the long-term impact of sustainability training on organizational performance and ESG outcomes, providing insights into the sustained effectiveness of such initiatives.

Comparative Analysis: Future research may focus on comparative cross-sectoral analysis to investigate sector-specific challenges, evaluate industry best practices across industries, and develop context specific frameworks for implementation.

In conclusion, the integration of sustainability into corporate training, underpinned by a robust organizational culture, is pivotal for realizing the UAE's strategic sustainability goals. This study serves as a stepping stone towards a more sustainable and prosperous future, emphasizing the need for continuous improvement and adaptation in corporate sustainability practices.

References

Al-Ali, T. Z. (2022). *Toward the micro-foundations of environmental sustainability: Fostering pro-environmental behavior among employees to boost corporate environmental performance in the United Arab Emirates*. Doctoral dissertation, United Arab Emirates University. https://scholarworks.uaeu.ac.ae/all_dissertations/192/

Alansaari, A., Hilmi, M. F., & Abudaqa, A. (2022, September). *Investigating the role of green training and leadership, organizational learning and organizational commitment towards green performance: A case of public firms in UAE*. https://www.researchgate.net/publication/363195462

Argyris, C., & Schon, D. (1978). *Organizational learning: A theory of action perspective*. https://www.jstor.org/stable/40183951

Molnar, E., & Mulvihill, P. R. (2003, March). *Sustainability-focused organizational learning: Recent experiences and new challenges*. https://www.researchgate.net/publication/227619561

Renwick, D., Redman, T., & Maguire, S. (2012, January). *Green human resource management: A review and research agenda*. https://www.researchgate.net/publication/230537826

Siebenhüner, B., & Arnold, M. (2007, July). Organizational learning to manage sustainable development. *Business Strategy and the Environment, 16*(5), 299–311. https://doi.org/10.1002/bse.579

Sult, A., Wobst, J., & Lueg, R. (2023). The role of training in implementing corporate sustainability: A systematic literature review. *Corporate Social Responsibility and Environmental Management*. https://doi.org/10.1002/csr.2560

Usman, M., Akhtar, M. W., Zahoor, N., Khan, M. A. S., & Adomako, S. (2024). Triggering employee green activism through green human resource management: The role of green organizational learning and responsible leadership. *British Journal of Social and Economic Review, 4*(2), 1–18. https://pure-oai.bham.ac.uk/ws/portalfiles/portal/248981682/bse.4028.pdf

Policy Documents

Ministry of Energy and Infrastructure. (2023, July). *Updated UAE Energy Strategy 2050*. https://u.ae/en/about-the-uae/strategies-initiatives-and-awards/strategies-plans-and-visions/environment-and-energy/uae-energy-strategy-2050

Environment Agency—Abu Dhabi. (2024, May). *ENVIRONMENT VISION 2030*. https://u.ae/en/about-the-uae/strategies-initiatives-and-awards/strategies-plans-and-visions/environment-and-energy/environment-vision-2030-abu-dhabi

Ministry of Climate Change & Environment. (2023). *The United Arab Emirates' First Long-Term Strategy (LTS)—Demonstrating Commitment to Net Zero by 2050*. https://unfccc.int/sites/default/files/resource/UAE_LTLEDS.pdf

Chapter 29
Eco Genius: Combining Ikigai and Design Thinking for Regenerative Sustainability

Tawheed Bashar

Abstract Addressing today's sustainability challenges requires holistic action integrating spirituality and ancient wisdom to mitigate the negative effects of modernisation (Luetz and Nunn in Sustainability Science 18:2035–2042, 2023). The longevity and well-being of Okinawans in Japan are attributed to Ikigai, an ancient Japanese concept centred on living with a purpose (https://www.bluezones.com/explorations/okinawa-japan/). In contrast, Design Thinking is a modern innovation and problem-solving methodology that prioritises human needs and values. By merging these two concepts, this paper proposes a purposeful action framework for sustainability. Existing literature highlights the successful integration of Ikigai and Design Thinking in addressing urgent social challenges, such as in Indonesia and Japan (Christianto and Smarandache in Sustainable Machine Intelligence Journal 7, 2024. https://doi.org/10.61356/smij.2024.77101). The Eco Genius framework introduced in this paper explores the combination by placing the environment—planet Earth—at the centre of solutions, using Ikigai and Design Thinking to foster deeper connections between people and the planet. The Framework enables individuals and organisations to take purposeful action toward sustainability. It also presents findings from an Eco Genius pilot program conducted in a primary school setting. While initial outcomes have shown increased engagement and sense of ownership, further applied research is needed to refine and simplify the framework for effective implementation.

Keywords Regenerative sustainability · Eco genius framework · Ikigai · Design thinking · Purpose-led innovation · Environmental empathy · Human-centred design · Sustainability education

T. Bashar (✉)
Eco Genius Multi Activites Co.Ltd, Khartoum, Sudan
e-mail: tawhed.bashar0@gmail.com

J. Dsilva (ed.), *Sustainability Education, Youth and Innovation*,
https://doi.org/10.1007/978-3-032-21948-0_29

Introduction

Exploring spiritual values and narratives can help cultivate a more balanced relationship between humans and nature. This requires a deeper understanding of the intersection between spirituality and sustainability. Integrating these two dimensions can create synergies that strengthen holistic development efforts, generating a collective impact greater than the sum of their individual contributions. Indigenous communities lived sustainably for thousands of years, making it crucial to engage with, examine, and incorporate their perspectives as foundational principles for place-based approaches to future resilience (Luetz & Nunn, 2023).

This paper introduces a framework that combines Ikigai, an indigenous Japanese philosophy centred on purpose, with Design Thinking, a structured approach to innovation and problem-solving. By synthesising insights from existing literature and practical applications, it explores both concepts, examines their interconnectedness, and ultimately proposes a framework for purpose-driven sustainability—one that prioritises both human well-being and planetary health.

This framework aims to inspire individuals and organisations to take purposeful action for sustainability, foster empathy, and deepen their understanding of both people and the planet. It presents a rationale based on two methodologies—Ikigai and Design Thinking—that have demonstrated positive outcomes across diverse contexts. Additionally, it outlines an actionable framework for adopting purposeful, innovative strategies for sustainability. The paper concludes with key findings and insights for future application.

Methodology

This extended abstract is a synthesis of the literature review of journals, books, and guides. Furthermore, since the topic is human-centred, the paper reflects opinions from online blogs of relevant communities and activists.

The proposed framework in Sect. 2.4 emerged through observations and years of practicing Design Thinking and Ikigai for regenerative, sustainable transformation through strategy development, coaching, mentoring, and stakeholder engagement interventions. Furthermore, a pilot case study of the Eco Genius framework was conducted to validate the effectiveness of the proposed framework. This section describes the concepts of Ikigai and design thinking, explores their synergies, and presents the suggested framework alongside the pilot study.

Ikigai

The remarkable longevity and good health of the Okinawan community are attributed to Ikigai, their deep sense of purpose and fulfilment in life (Blue Zones, 2008). Ikigai is a Japanese term commonly interpreted in English literature as "purpose in life" or "life worth living" (Kotera et al., 2021). At a deeper level, it encompasses a mental and spiritual state in which individuals perceive their lives as meaningful and valuable while striving toward their goals (What Does Ikigai Mean? 2025).

Basil and Suganya (2023) describe Ikigai as emerging at the intersection of four core elements: what one loves, what one is good at, what the world needs, and what one can be paid for. These components reflect a balance between internal fulfilment and external contribution, often guiding individuals toward lives of greater meaning and societal value. This conceptual foundation, inspired a quadrant-based model of Ikigai is illustrated (Fig. 29.1). This adapted version maps Ikigai across four interconnected zones—Passion, Mission, Vocation, and Profession (Basil and Suganya, 2023)—positioned along axes of personal fulfilment, financial sustainability, skill proficiency, and impact on people and planet.

Beyond its practical applications, Ikigai holds profound cultural and philosophical significance, reflecting the values of interconnectedness, harmony, and balance that are deeply ingrained in Japanese culture. It encourages aligning personal aspirations with societal needs and embracing a mindful, purposeful approach to life. Philosophically, Ikigai prompts reflection on fulfilment, happiness, and meaningful existence, guiding individuals to cultivate self-awareness, gratitude, and a deeper understanding of their role in society (Christianto & Smarandache, 2024).

Fig. 29.1 Ikigai Quadrant Framework: Personal and planetary purpose mapped across skill, impact, and sustainability dimensions. Created by the author, conceptually inspired by Basil and Suganya (2023)

Design Thinking

Design Thinking is a problem-solving methodology focused on empathy, creativity, and collaboration to address complex challenges. It's a human-centred approach aimed at understanding the needs, desires, and experiences of end-users. Unlike traditional methods, it encourages curiosity, experimentation, and learning from failure to foster innovation.

The Design Thinking process consists of five stages: Empathise, Define, Ideate, Prototype, and Test (Fig. 29.2). The Empathise stage involves an immersive understanding of the users' needs and challenges gained through interviews, observations, and immersion in their context. The Define stage synthesises insights, needs, and patterns into a clear, actionable problem or opportunity statement (Hasso Plattner Institute of Design at Stanford, 2010), ensuring a human-centred approach. In Ideate, a range of ideas and potential solutions is generated. Prototyping involves creating tangible representations of these solutions to explore their feasibility and functionality. Lastly, in the Testing stage, user feedback is gathered to refine and improve the prototype. The process is repeated iteratively to reach an optimal solution.

Design Thinking prioritises end-user experiences in the design process, focusing on understanding their behaviours, preferences, and challenges. Unlike traditional

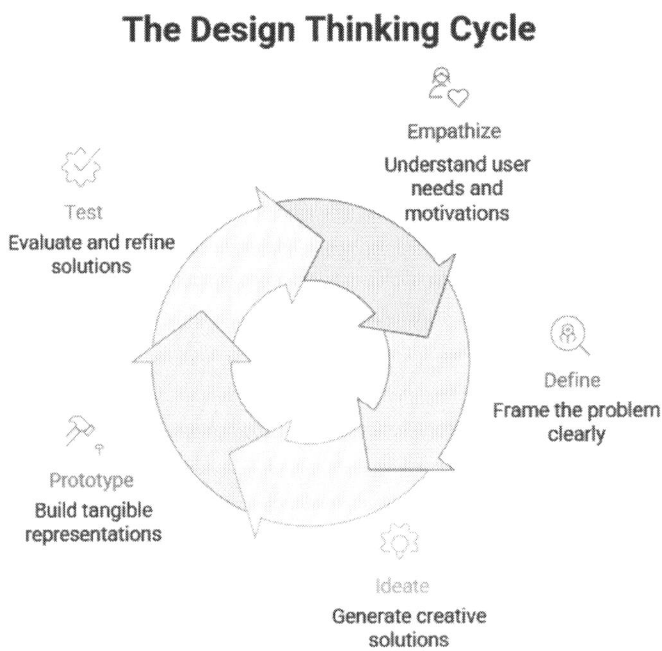

Fig. 29.2 Illustration of the Design Thinking process imagined based on the widely accepted stages of the methodology. Designed by the author using Generative AI illustration tools (Hasso Plattner Institute of Design at Stanford, 2010)

problem-solving methods that emphasise technical or business requirements, this approach begins with empathy, allowing designers to develop insights that redefine challenges and opportunities, and drive ideation and solution development. By centring on users, Design Thinking ensures that solutions are functional, meaningful, and impactful, ultimately creating innovations that resonate with their needs (Christianto & Smarandache, 2024).

Synergies of Ikigai and Design Thinking

Both Ikigai and Design Thinking adopt a human-centred approach, emphasising understanding and addressing individuals' needs, aspirations, and lived experiences (Christianto & Smarandache, 2024). They also explore how personal fulfilment can contribute to the greater good. For example, reflecting on what one loves and excels at aligns with Design Thinking's empathy principle and is refined through continuous iteration, prototyping and testing. Similarly, understanding what the world needs fosters a broader sense of empathy, helping individuals connect their passions with societal and environmental demands—ultimately shaping their Ikigai.

Design Thinking provides a structured methodology for individuals to explore and actualise their Ikigai through systematic problem-solving and creative exploration. Conversely, integrating Ikigai principles into Design Thinking enhances innovation by embedding a deeper sense of purpose and meaning—both personally and globally. This synergy ensures that solutions address challenges while promoting self-awareness and personal fulfilment, enabling individuals to be valuable members of society (Christianto & Smarandache, 2024).

Ikigai and Design Thinking for Sustainability: The Eco Genius Framework

This paper introduces Eco Genius (Fig. 29.3), a framework integrating Design Thinking and Ikigai to address sustainability challenges by considering human and planetary needs. Eco Genius recognises the interconnectedness between individuals and their environment, guiding practitioners and users through a structured five-stage process: Empathize, Define, Ideate, Implement, and Validate. By combining Design Thinking's problem-solving methodologies with Ikigai's purpose-driven philosophy and self-awareness, this framework provides a holistic approach to sustainability, connecting personal strengths and aspirations with the broader goal of ecological and social well-being. The following sections outline the stages of the Eco Genius framework and their role in fostering meaningful, impact-driven innovation.

Empathize: This stage explores both human and planetary perspectives to foster a deeper understanding of individual purpose in relation to environmental needs. At

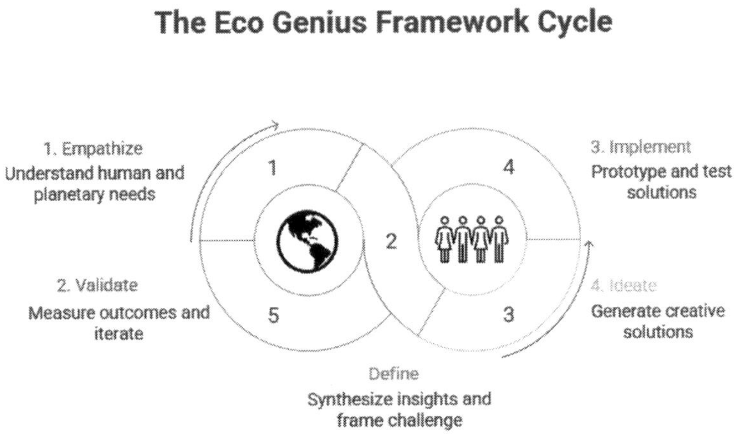

Fig. 29.3 The Eco Genius framework cycle imagined using AI generation tools

this stage, practitioners and users engage with stakeholders, communities, and the environment through immersive experiences, using Design Thinking tools such as journey mapping, stakeholder mapping, and interviews; exploring the four components of Ikigai—what individuals love and excel at (understanding humans) and what the planet needs and rewards (understanding the planet). Effective empathy requires immersion, documentation, and curiosity-driven exploration while avoiding biases and premature conclusions. Individual users may rely partially on self-reflection but are encouraged to reflect with other relevant stakeholders, while practitioners facilitate and guide the process for users through active listening, iterative questioning, and deep observation of both humans and the planet to generate insights for the next stage.

Define: In this stage, users and/or practitioners synthesise patterns, insights, and needs to form a clear perspective on what individuals love, what they excel at, and what the world requires (Hasso Plattner Institute of Design at Stanford, 2010). This stage helps outline preliminary ideas about users' passions and missions and their connection to the surrounding environment, pressing global challenges concerning the users. Framing a well-defined question at this stage facilitates the transition to the next step. For example: How might we utilise an individual's hip-hop dancing passion to address their concern for the declining hummingbird population? (An issue that the same person is concerned about)?

Ideate: At this stage, practitioners generate ideas using Design Thinking techniques like brainstorming and brainwriting to address the question formulated in the Define stage. This encourages creativity, divergent thinking, open-mindedness and exploration of unconventional ideas. However, prioritising ideas based on their importance and feasibility is recommended to ensure effective solutions (IBM Design Thinking Field Guide V3.4, 2017).

Implement: This stage involves prototyping and testing within the Design Thinking process (Brown & Wyatt, 2010). Here, practitioners immerse users in lived experiences to foster empathy for the planet, validate initial observations, and refine their understanding of what individuals excel at in real-world situations. It can be viewed as a simulation or a lived experience to explore and refine the Ikigai. This can be pilot programs, co-creation projects that simulate real-life scenarios, or planned actions that test how the proposed approach is perceived and applied.

Validate: In this final step, practitioners monitor and measure the outcomes of the implementation stage and integrate the findings into the Empathise, Define, or Implement stages. This iterative process helps refine an individual's Ikigai, deepens the understanding of the planet and global needs, and explores further areas of expertise. Additionally, this stage evaluates whether a proposed idea effectively addresses the identified problem and enhances comprehension of user experiences.

In summary, the Eco Genius framework offers a structured approach for aligning human talents with planetary needs. This model is applicable in educational, and professional settings, as well as personal development efforts, fostering a deeper connection between individual purpose and global sustainability.

Case Study: Eco Genius Pilot in a Primary School

Overview

Alfaitori School in Alhaj Yousif, Khartoum, Sudan, serves children aged 5–12 from low-income backgrounds, with historical examination pass rates below 10%. Student engagement is low, particularly in scientific subjects. The Eco Genius pilot aimed to foster environmental responsibility and active learning.

Implementation Approach

A comprehensive assessment of the school environment and students' collective Ikigai—motivations, needs, and aspirations—was conducted through interviews, discussions, observations, and immersive experiences. The Eco Genius framework was implemented in four phases:

1. Empathy Phase – A storytelling session on the success of an Indian community's waste management efforts (leveraging students' interest in Indian drama) was followed by a treasure hunt, encouraging mindful exploration of school facilities.
2. Reflection & Problem Identification – Guided discussions led students to recognise the importance of responsible waste practices.
3. Ideation & Action – Working in groups, students used available materials (paper, colours, play dough, and school resources) to ideate and prototype solutions.
4. Validation – Each group presented a concept, demonstrating their understanding and proposing improvements to the school's waste system.

Key Outcomes

- Based on observations and facilitator feedback, 95% of students remained engaged, showing sustained interest.
- Final presentations reflected a deeper understanding of school cleanliness challenges and improvement opportunities.
- Students exhibited increased confidence in their ability to take responsible action and a sense of ownership as observed post-experience.

Challenges and Limitations

- Unstable Conditions & Resource Constraints – COVID-19 disruptions, limited personnel, and school infrastructure, coupled with a one-month gap between preparation and implementation, may have affected engagement.
- Traditional Learning Mindset – Students and school staff accustomed to rote learning, rather than creative thinking, posed implementation challenges.
- Limited Iterations – The process ideally requires multiple iterations, but time and resource constraints allow only one.
- Data Documentation & Synthesis – Capturing all observations while facilitating was unfeasible, and data synthesis was not fully explored.

Findings

Engaging communities in action-oriented approaches is essential for fostering regenerative, restorative, and reparative efforts toward a sustainable planet. The effectiveness of such an approach depends on the target group and community context.

Ikigai and Design Thinking have demonstrated positive outcomes, including enhanced health and longevity, fostering motivation for meaningful contributions to the world, and driving innovative, human-centred solutions. While Ikigai is philosophy, and Design Thinking is widely used in business innovation, Christianto & Smarandache (2024) have explored their convergence to form a purposeful framework. In Japan in the 1950s, the convergence of design thinking and Ikigai successfully addressed childhood malnutrition challenges. Similarly, in Indonesia, this approach has been applied to develop sustainable food systems and strengthen local economies.

Encouraging environmental empathy can emerge in solutions that are both sustainable and human-centred. The first iteration of the Eco Genius framework demonstrated positive shifts in students' attitudes, awareness, and sense of responsibility. Observations suggest that this approach cultivates environmental empathy and serves as a sustainable model for community engagement, with lasting impact beyond the hands-on experience. This could play a role in accelerating sustainability efforts, particularly in regions where environmental attitudes and behaviours that pose significant challenges.

Conclusion

The complex challenges our planet faces today require a regenerative approach that aligns human purpose with environmental needs. The Eco Genius framework introduces a novel and holistic approach to sustainability by integrating the introspective philosophy of Ikigai with the structured, human-centred methodology of Design Thinking. This combination supports regenerative transformation by fostering purpose-driven action, environmental stewardship, and social innovation.

The pilot study demonstrated the framework's potential to cultivate empathy, ownership, and creativity in learners—linking personal motivation with ecological responsibility. Its strengths lie in promoting experiential learning and embedding sustainability into an individual's sense of purpose.

However, the model's reliance on skilled facilitation may hinder scalability in under-resourced settings. Its iterative nature can be difficult to sustain in unstable environments, and adapting it to rigid educational or cultural systems requires thoughtful contextualisation.

Future applied research should explore how Eco Genius can be mainstreamed and simplified for broader application across diverse communities. Scaling through digital tools—such as AI-enabled feedback loops—could enhance insight generation and synthesis. Additionally, integrating indigenous spiritual knowledge may deepen its cultural relevance and regenerative potential.

Ultimately, Eco Genius reframes sustainability not as a technical challenge but as a relational journey—rooted in empathy, cultural wisdom, and purpose—offering a compelling blueprint for transformative education and purposeful climate action.

Acknowledgements The Young Water Solutions organisation funded the pilot implementation presented in this paper as part of the Young Water Fellowship–East Africa 2019, a seed funding program supporting social innovations in water, sanitation, and hygiene (WASH) solutions.

Special thanks to Alfaitori School, located in a marginalised area of Khartoum, Sudan, for its cooperation and participation. Appreciation is extended to the school director, Mr Elawalid, for his support throughout the implementation. The author also gratefully acknowledges the contribution of Nadia Abdallah, co-facilitator and project partner, whose collaboration was instrumental to the success of the pilot.

References

Basil, F., & Suganya, M. (2023). Unveiling the secrets of Ikigai: A study on the Japanese philosophy of a meaningful life. *International Journal of Novel Research and Development (IJNRD), 8*, 417. https://www.ijnrd.org/papers/IJNRD2310352.pdf

Blue Zones. (2008). Okinawa, Japan. Blue Zones. https://www.bluezones.com/explorations/okinawa-japan/

Brown, T., & Wyatt, J. (2010). Design thinking for social innovation. *Development Outreach, 12*(1), 29–43. https://doi.org/10.1596/1020-797X_12_1_29

Christianto, V., & Smarandache, F. (2024). The convergence of Ikigai and design thinking: Crafting a purposeful framework. *Sustainable Machine Intelligence Journal, 7*. https://doi.org/10.61356/smij.2024.77101

Dam, R. F., & Siang, T. Y. (2024). What is design thinking and why is it so popular? *The Interaction Design Foundation*. https://www.interaction-design.org/literature/article/what-is-design-thinking-and-why-is-it-so-popular

Hasso Plattner Institute of Design at Stanford. (2010). An Introduction to Design Thinking Process Guide. Stanford University. https://web.stanford.edu/~mshanks/MichaelShanks/files/509554.pdf

IBM Design Thinking Field Guide V3.4. (2017). SlideShare. https://www.slideshare.net/slideshow/ibm-design-thinking-field-guide-v34-73702608/73702608

Kotera, Y., Kaluzeviciute, G., Garip, G., & Chamberlain, K. J. (2021, March 2). Health Benefits of Ikigai: A Review of Literature. https://www.researchgate.net/publication/349725406_Health_Benefits_of_Ikigai_A_Review_of_Literature

Luetz, J. M., & Nunn, P. D. (2023). Spirituality and sustainable development: An entangled and neglected relationship. *Sustainability Science, 18*(4), 2035–2042. https://doi.org/10.1007/s11625-023-01347-8

What does Ikigai mean? (2025). *Ikigai Tribe*. https://ikigaitribe.com/ikigai

Chapter 30
Redefining Sustainability Metrics: The Strategic Imperative for a UAE-Centric Inclusive Rating System

Sameera Fernandes, M. S. Sheeja, and Shubha Chandra

Abstract The evolving landscape of global sustainability highlights the urgent need for rating systems that reflect regional priorities and socio-economic contexts (United Nations Development Programme in Inclusive growth and sustainable development in the Middle East. UNDP Publications, 2023). This paper advocates for a UAE-Centric Inclusive Sustainability Rating System (UCISRS), addressing the strategic gaps in traditional global frameworks like Fitch and S&P (Fitch Ratings in Global sustainability report. Fitch Ratings, 2023; S&P Global in Corporate sustainability assessment methodology. S&P Global, 2024). Through a comparative analysis, this study underscores the UAE's unique socio-economic dynamics, geopolitical influence, and vision for sustainable growth, which current systems overlook (UAE Ministry of Climate Change and Environment in National sustainability strategy 2050. Government of the UAE, 2023). The findings reveal that a customized rating system is not just beneficial but critical to driving the UAE's leadership in global sustainability. This paper proposes a UAE-specific UCISRS aligned with SEE Rating principles, designed to foster holistic development while bridging the inclusivity gap prevalent in global rating systems.

Keywords UAE sustainability ratings · Strategic sustainability · Inclusive growth · Social innovation · Global rating gaps

Introduction

In an era where sustainability defines national competitiveness, traditional rating agencies like Fitch and S&P fall short in capturing the strategic nuances of rapidly transforming economies like the UAE. These global systems prioritize economic stability over social equity, neglecting critical dimensions such as inclusive growth, gender equity, and environmental justice. As the UAE accelerates its diversification

S. Fernandes (✉) · M. S. Sheeja · S. Chandra
Garden City University, Bengaluru, India
e-mail: sameera.fernandes@gmail.com

J. Dsilva (ed.), *Sustainability Education, Youth and Innovation*,
https://doi.org/10.1007/978-3-032-21948-0_30

agenda under Vision 2030, there is an urgent need for a customized rating system that reflects its unique socio-economic fabric, geopolitical aspirations, and leadership role in the global sustainability discourse. This paper explores why the UAE must take ownership of its sustainability narrative through a UAE-Centric Inclusive Sustainability Rating System (UCISRS).

Methodology

This research employs a mixed-methods approach, combining qualitative and quantitative analyses. A comparative framework evaluates existing global sustainability rating systems (Fitch, S&P) against the proposed UAE-Centric Inclusive Sustainability Rating System (UCISRS). Primary data were collected through expert interviews with sustainability leaders in the UAE, while secondary data analysis involved reviewing policy documents, sustainability reports, and academic literature. Thematic analysis identified key gaps in current rating methodologies, while statistical tools assessed the correlation between inclusive sustainability practices and socio-economic indicators.

Findings

The comparative analysis highlighted in Table 30.1 reveals significant gaps in global rating systems:

1. Strategic Blind Spots: Global ratings overlook the UAE's unique geopolitical role and strategic sustainability initiatives.
2. Limited Social Metrics: Fitch and S&P prioritize economic stability over social equity, neglecting factors such as gender parity, inclusive growth, and community resilience.
3. One-Size-Fits-All Approach: Global ratings apply uniform criteria, disregarding regional diversity and local socio-economic conditions.
4. Inadequate Representation of Grassroots Innovations: Community-led sustainability initiatives, pivotal in the UAE's context, are underrepresented.
5. Lack of Alignment with National Visions: Fitch and S&P do not account for region-specific sustainability goals like the UAE Vision 2030, which emphasizes economic diversification, innovation, and social development.
6. Static Assessment Models: Traditional rating systems rely heavily on historical data, failing to capture dynamic and rapidly evolving sustainability practices in emerging economies.
7. Insufficient Environmental Justice Considerations: Global rating agencies often overlook the distributional impacts of environmental policies, which are critical in assessing equitable sustainability.

Table 30.1 Comparative analysis of global rating systems versus UCISRS

Criteria	Fitch and S&P rating systems	UAE-centric inclusive sustainability rating system (UCISRS)
Focus	Economic stability, governance, fiscal health	Inclusive growth, social equity, environmental justice, strategic sustainability
Social metrics	Limited focus on gender, labor rights, and equity	Comprehensive social indicators including gender parity, labor rights, community well-being
Alignment with national visions	Generic global standards, lacks regional adaptation	Aligned with UAE Vision 2030, focusing on economic diversification and innovation
Dynamic adaptability	Static models based on historical data	Dynamic, real-time adaptability to evolving sustainability practices
Representation of innovations	Neglects grassroots and technological innovations	Strong emphasis on digital transformation, AI, and social entrepreneurship
Environmental justice	Limited consideration of distributional impacts	Holistic environmental justice framework assessing equitable policy impacts
Inclusivity	One-size-fits-all approach across regions	Tailored to UAE's diverse socio-economic and cultural context
Technological integration	Minimal integration of modern tech in assessments	Advanced metrics incorporating digital technologies and sustainable innovations

8. Neglect of Technological and Social Innovations: The transformative role of digital technologies, AI, and social entrepreneurship in advancing sustainability is not adequately reflected.

The proposed UCISRS incorporates:

- Strategic Sustainability Metrics: Aligning with UAE Vision 2030 and regional geopolitical dynamics.
- Social Equity Metrics: Gender equity, labor rights, and community well-being.
- Environmental Justice: Assessing the distributional impacts of environmental policies.
- Inclusive Growth Indicators: Evaluating policies for SMEs, blue-collar workers, and vulnerable populations.

Conclusion

The study concludes that a UAE-Centric Inclusive Sustainability Rating System (UCISRS) is not just desirable but a strategic imperative. The case for change is both urgent and necessary, driven by the UAE's rapid economic diversification, ambitious Vision 2030 goals, and its role as a global sustainability leader (UAE Ministry of Climate Change and Environment, 2023). The adoption of UCISRS will fill critical gaps in traditional rating systems, which often overlook the unique socio-economic, environmental, and cultural dynamics specific to the UAE (Fitch Ratings, 2023; S&P Global, 2024).

Principles and Generalizations Inferred from the Results:

The primary principle inferred from this research is that sustainability must be context-specific, reflecting local socio-economic, environmental, and cultural dynamics. The UCISRS embodies this principle by focusing on inclusive growth, environmental justice, and social equity. Generalizations from the study suggest that one-size-fits-all global sustainability rating systems are inadequate for capturing the complexity of rapidly evolving economies like the UAE.

Exceptions, Problems, or Limitations:

While UCISRS offers a comprehensive framework, there are limitations. First, data availability and standardization across industries may pose challenges, potentially affecting the accuracy and consistency of ratings. Second, the initial implementation phase may face resistance from stakeholders accustomed to traditional rating systems. Additionally, continuous updates will be required to adapt to evolving sustainability trends, necessitating robust governance mechanisms.

Theoretical and Practical Implications:

Theoretically, this research contributes to the growing body of knowledge on sustainability rating systems, emphasizing the need for localized approaches. Practically, UCISRS provides a roadmap for policymakers, businesses, educational institutions, and industries to integrate sustainability into their core strategies. It will serve as a strategic tool for decision-making, investment analysis, and policy formulation, fostering a culture of continuous improvement and sustainability leadership.

Conclusions Drawn and Recommendations:

The UAE stands at the forefront of sustainable development, and the adoption of UCISRS will reinforce its leadership position. Key recommendations include:

Institutionalize UCISRS: Establish a dedicated body to govern and update the rating system, ensuring alignment with national and global sustainability goals.

Capacity Building: Invest in training programs for stakeholders across sectors to enhance understanding and effective application of the rating system.

Data Infrastructure: Develop robust data collection and analysis mechanisms to support accurate and reliable sustainability assessments.

Stakeholder Engagement: Foster collaboration among government agencies, businesses, academia, and civil society to promote widespread adoption and continuous improvement of UCISRS.

Global Benchmarking: While rooted in the UAE's context, UCISRS should aim for international recognition, positioning the UAE as a global model for inclusive sustainability.

The Way Forward:

Policy Integration: The UAE government can leverage UCISRS to inform national sustainability policies, ensuring that economic growth is inclusive, equitable, and environmentally sustainable. It will align with global sustainability goals while reflecting UAE-specific priorities (United Nations Development Programme, 2023).

Business Sector Impact: For businesses, UCISRS offers a comprehensive framework that goes beyond traditional financial metrics. It will enhance corporate reputation, attract sustainable investments, and drive long-term value creation through improved ESG (Environmental, Social, and Governance) performance (OECD, 2023).

Educational Advancement: In the education sector, UCISRS can serve as a critical tool for curriculum development, fostering a generation of sustainability-conscious leaders. Academic institutions can incorporate these metrics to promote research and innovation in sustainability practices (World Economic Forum, 2024).

Industrial Transformation: Industries will benefit from clear sustainability benchmarks, enabling them to adopt best practices, improve operational efficiency, and reduce environmental impact. UCISRS will guide industries towards resource optimization and sustainable growth (International Energy Agency, 2023).

Community Engagement: The framework will encourage community-led sustainability initiatives, enhancing social cohesion and resilience. It ensures that sustainability efforts are inclusive, addressing the needs of vulnerable populations and promoting gender equity (UN Women, 2024).

The Case for Change:

The global sustainability landscape is evolving rapidly, and the UAE must proactively shape its narrative. Traditional rating systems like Fitch and S&P are not equipped to capture the dynamic socio-economic transformations within the UAE. The need for UCISRS is the need of the hour—it will serve as a catalyst for policy reforms, sustainable investments, and community-driven development. Its adoption will position the UAE as a global benchmark for inclusive sustainability, driving economic prosperity, social equity, and environmental stewardship.

UAE-Centric Inclusive Sustainability Rating System (UCISRS) Framework

The proposed UCISRS framework shown in Fig. 30.1 is designed to address the gaps in global rating systems, focusing on inclusivity, socio-economic relevance, and sustainability leadership:

Core Dimensions:

Environmental Sustainability: Climate action, resource efficiency, and biodiversity conservation.

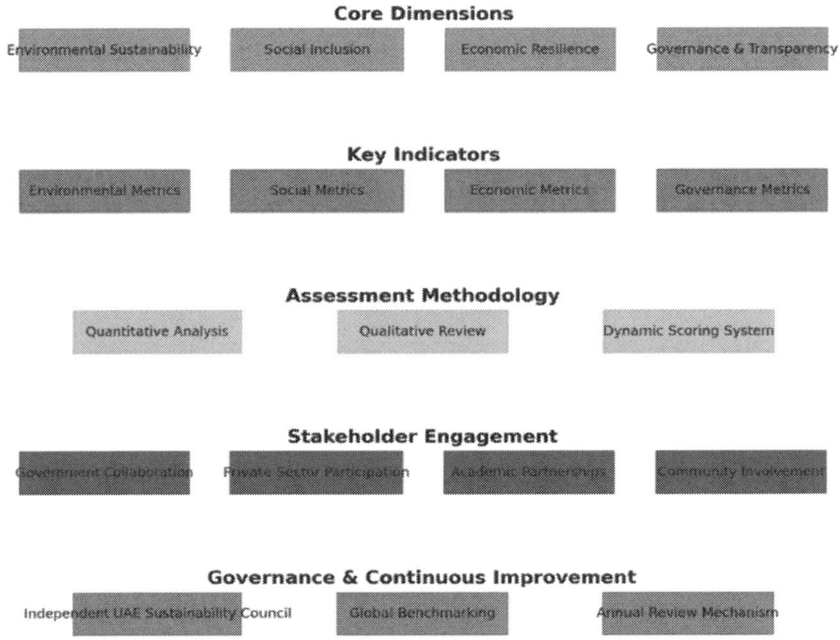

Fig. 30.1 Framework for sustainability assessment, core dimensions, indicators, methodology, stakeholder engagement, and governance

- Social Inclusion: Gender equity, labor rights, community engagement, and social justice.
- Economic Resilience: Inclusive economic growth, innovation, SME development, and diversification.
- Governance & Transparency: Ethical leadership, policy effectiveness, data transparency, and regulatory compliance.

Key Indicators:

- Environmental Metrics: Carbon footprint, renewable energy adoption, water and waste management.
- Social Metrics: Gender diversity in leadership, labor practices, education access, and public health indicators.
- Economic Metrics: GDP diversification ratio, SME contribution, employment rates, innovation index.
- Governance Metrics: Anti-corruption measures, policy transparency, regulatory quality.

Assessment Methodology:

- Quantitative Analysis: Data-driven metrics based on verified national statistics and international benchmarks.

- Qualitative Review: Stakeholder interviews, policy reviews, and case studies for holistic insights.
- Dynamic Scoring System: Real-time adaptability to reflect emerging sustainability trends and practices.

Stakeholder Engagement:

- Government Collaboration: Integration with national development plans and regulatory frameworks.
- Private Sector Participation: Incentives for businesses to adopt sustainable practices.
- Academic Partnerships: Research-driven insights for continuous improvement.
- Community Involvement: Grassroots feedback mechanisms to ensure inclusivity.

Governance and Continuous Improvement:

- Establish an Independent UAE Sustainability Council to oversee implementation, compliance, and updates.
- Regular benchmarking against international best practices to ensure global relevance.
- Annual Review Mechanism to update indicators and methodologies in response to new challenges and opportunities.

References

Agyeman, J. (2013). *Introducing just sustainabilities: Policy, planning, and practice*. Zed Books.

Bebbington, J., Larrinaga, C., & Moneva, J. M. (2014). Corporate social reporting and reputation risk management. *Accounting, Auditing & Accountability Journal, 21*(3), 337–361. https://doi.org/10.1108/09513570810863932

Brundtland, G. H. (1987). *Our common future: Report of the World Commission on Environment and Development*. Oxford University Press.

Carroll, A. B. (1999). Corporate social responsibility: Evolution of a definitional construct. *Business & Society, 38*(3), 268–295. https://doi.org/10.1177/000765039903800303

Delmas, M. A., & Burbano, V. C. (2011). The drivers of greenwashing. *California Management Review, 54*(1), 64–87. https://doi.org/10.1525/cmr.2011.54.1.64

Dyllick, T., & Muff, K. (2016). Clarifying the meaning of sustainable business: Introducing a typology from business-as-usual to true business sustainability. *Organization & Environment, 29*(2), 156–174. https://doi.org/10.1177/1086026615575176

Elkington, J. (1997). *Cannibals with forks: The triple bottom line of 21st-century business*. Capstone.

Epstein, M. J. (2008). *Making sustainability work: Best practices in managing and measuring corporate social, environmental, and economic impacts*. Berrett-Koehler Publishers.

Fitch Ratings. (2023). *Global sustainability report*. Fitch Ratings.

Geels, F. W. (2002). Technological transitions as evolutionary reconfiguration processes: A multi-level perspective and a case-study. *Research Policy, 31*(8–9), 1257–1274. https://doi.org/10.1016/S0048-7333(02)00062-8

Global Reporting Initiative. (2020). *GRI standards: Global sustainability reporting guidelines*. https://www.globalreporting.org

Hart, S. L., & Milstein, M. B. (2003). Creating sustainable value. *Academy of Management Executive, 17*(2), 56–67. https://doi.org/10.5465/ame.2003.10025194
International Energy Agency. (2023). *Sustainable industrial transformation report*. IEA Publications.
Meadows, D. H., Meadows, D. L., & Randers, J. (2004). *Limits to growth: The 30-year update*. Chelsea Green Publishing.
Moser, S. C., & Dilling, L. (2007). *Creating a climate for change: Communicating climate change and facilitating social change*. Cambridge University Press.
OECD. (2023). *Business and sustainability in emerging markets*. Organisation for Economic Co-operation and Development.
Porter, M. E., & Kramer, M. R. (2011). Creating shared value. *Harvard Business Review, 89*(1/2), 62–77.
Raworth, K. (2017). *Doughnut economics: Seven ways to think like a 21st-century economist*. Chelsea Green Publishing.
Sachs, J. D. (2015). *The age of sustainable development*. Columbia University Press.
S&P Global. (2024). *Corporate sustainability assessment methodology*. S&P Global.
Schaltegger, S., & Wagner, M. (2011). Sustainable entrepreneurship and sustainability innovation: Categories and interactions. *Business Strategy and the Environment, 20*(4), 222–237. https://doi.org/10.1002/bse.682
Sen, A. (1999). *Development as freedom*. Oxford University Press.
Smith, W. K., Gonin, M., & Besharov, M. L. (2013). Managing social-business tensions: A review and research agenda for social enterprise. *Business Ethics Quarterly, 23*(3), 407–442. https://doi.org/10.5840/beq201323327
Spangenberg, J. H. (2011). Sustainability science: A review, an analysis, and some empirical lessons. *Environmental Conservation, 38*(3), 275–287. https://doi.org/10.1017/S0376892911000270
Stiglitz, J. E., Sen, A., & Fitoussi, J. P. (2009). *Report by the Commission on the Measurement of Economic Performance and Social Progress*. OECD.
UAE Ministry of Climate Change and Environment. (2023). *National sustainability strategy 2050*. Government of the UAE.
UN Global Compact. (2021). *Principles for responsible investment: Impact investing and ESG frameworks*. United Nations.
UN Women. (2024). *Gender equity and sustainability: A global perspective*. United Nations Entity for Gender Equality and the Empowerment of Women.
United Nations Development Programme. (2023). *Inclusive growth and sustainable development in the Middle East*. UNDP Publications.
World Economic Forum. (2024). *The future of education for sustainable development*. WEF Publications.

Chapter 31
A Conceptual Framework for Enhancing Inclusivity for People with Autism Spectrum Disorder (ASD) in Sustainable Urban Planning in the United Arab Emirates (UAE)

Agam Podi Kalindu Dhaneesha Mendis

Abstract Urban spaces have proven to be significant for the well-being of the community by enhancing both physical and mental health. Unfortunately, these spaces have mostly been designed and built with neurotypical individuals in mind. This exclusion creates a challenge regarding accessibility for people with Autism Spectrum Disorder (ASD). Due to the difficulty in processing everyday sensory information and cognitive functions, people with ASD face significant challenges in urban spaces. While the United Arab Emirates (UAE) has taken steps toward inclusivity, there is a lack of a systematic approach to ASD-inclusive urban planning. This study proposes a conceptual framework through literature synthesis that will guide how to enhance the inclusivity of people with ASD in sustainable urban planning in the UAE. The conceptual framework consists of four main components contributing to exploring the gaps to achieve inclusivity: 'the challenges faced by people with ASD,' 'stakeholder involvement in urban planning,' 'technological and innovation assessment,' and 'the development and implementation of policies relating to ASD and sustainable urban planning.' The study further recommends analysing these components to identify existing gaps associated with inclusivity and to develop suitable, actionable recommendations to enhance ASD-friendly, sustainable urban planning in the UAE. This study provides insights for policymakers and practitioners to ensure that inclusive urban planning is not confined to specific facilities but is instead integrated into the core of urban design, promoting independence and social inclusion for people with ASD.

Keywords Autism spectrum disorder · Inclusivity · Sustainable urban planning · United Arab Emirates

---

A. P. K. D. Mendis (✉)
Westford University College, Tijara 3 Building, Al Taawun Street, Sharjah P.O. Box 61110, United Arab Emirates
e-mail: dhaneeshamendis@gmail.com; agam.k@westford.org.uk

J. Dsilva (ed.), *Sustainability Education, Youth and Innovation*,
https://doi.org/10.1007/978-3-032-21948-0_31

Introduction

Studies on people with Autism Spectrum Disorder (ASD) have been progressing across various disciplines since its early appearance during the 1910s to 1940s (Sarraf, 2024). With the recent rise of the neurodiversity paradigm, studies on ASD have been exploring new frameworks that extend beyond traditional medical descriptions (Pellicano & Houting, 2022). According to the American Psychiatric Association (2013), ASD is identified as a prevalent neurodevelopmental disorder that is marked by difficulties in social communication, as well as controlled and repetitive behaviours, interests, or activities (AlQahtani, 2023). Moving away from the conventional medical standpoint that views ASD as a 'disorder' needing treatment or a 'deficit' that should be addressed, the neurodiversity paradigm understands ASD as not just a normal variant of human existence but also as a cultural phenomenon (Jaarsma & Welin, 2012; Sarraf, 2024). Within this paradigm, people with ASD, as part of a marginalised community, aim to achieve recognition for a shared identity and the rights that differentiate their community (Sarraf, 2024).

People with ASD often experience obstacles in their interactions with the environment around them (Tola et al., 2021). Numerous studies have highlighted that an altered understanding of sensory stimuli and the processing of environmental information is a significant issue in ASD. The significance of varying understandings of sensory stimuli from the environment is also highlighted in the fifth edition of the Diagnostic and Statistical Manual of Mental Disorders (DSM-5) updates, which acknowledged sensory difficulties among the behavioural patterns that were overlooked in previous editions: hyper- or hypo-reactivity to sensory input or unusual interest in sensory aspects of the environment (e.g., adverse response to specific sounds or textures, excessive smelling or touching of objects, visual fascination with lights or movement) (American Psychiatric Association, 2013). Additionally, personal testimonies from individuals with ASD stress the importance of recognising the atypical perception of sensory stimuli from the environment as one of the main challenges associated with autism (Beale-Ellis, 2017; Grandin, 1995; Tola et al., 2021). To date, the need to reconsider living spaces based on the particular requirements of individuals who operate in an atypical way has become increasingly significant in the broader discussion, as well as in the fields of urban planning and design (Sarraf, 2024; Tola et al., 2021).

Various scholars have explored the relationship between autism and the built environment, highlighting the significant role environmental factors play in the daily experiences of people with ASD (Black et al., 2022; Bölte et al., 2019; Kenna, 2022; McAllister et al., 2022; Tola et al., 2021; Toronyi, 2021). Nevertheless, while interdisciplinary research acknowledges these challenges, a critical gap remains in integrating urban planning theory to systematically address the autistic community's right to access and participate fully in urban life. According to Sarraf (2024), for neurodiverse communities, spatial justice means creating environments that are accessible, participatory, and empowering. Despite the rapid urbanisation of the United Arab Emirates (UAE) and its commitment to sustainable and smart city development, urban planning policies lack explicit guidelines for accommodating people with ASD. This gap is particularly concerning given the rising prevalence of ASD in the UAE (AlQahtani, 2023). Therefore, this study aims to develop a conceptual framework to enhance inclusivity for people with ASD in sustainable urban planning in the UAE.

Methodology

This study is primarily based on a thorough literature review. It is necessary to review the literature to collect current information in the field of research and pave the way for the discovery of new information (Saunders et al., 2016). Hence, a narrative literature review was conducted to achieve the aim of this study, which is to develop a conceptual framework to enhance inclusivity for people with ASD in sustainable urban planning in the UAE. Key terms such as 'neurodiversity, inclusive urban design, sustainable urban development, and smart city initiatives' were utilised in Google searches to collect the relevant data from the articles. Manual content analysis was conducted to review and analyse the selected articles, following the opportunity for the author to review them extensively.

Literature Findings

Challenges Faced by People with ASD in Urban Spaces.

Urban spaces, commonly known as public spaces within cities, consist of parks, roads, shopping malls and various open areas that foster community engagement. These spaces have proven to be significant for the well-being of the community by enhancing both physical and mental health (Bowler et al., 2010). Unfortunately, as McAllister et al. (2022) claim, urban spaces have been designed and built with neurotypical individuals in mind. This exclusion creates a challenge regarding accessibility for the neurodiverse communities. Due to the difficulty in processing everyday sensory information and cognitive functions, people with ASD encounter significant challenges in urban spaces (Black et al., 2022; McAllister et al., 2022; Sarraf, 2024; Tola et al., 2021).

Many urban spaces are created without considering the sensory understandings of people with ASD, subjecting them to overwhelming noise, bright lights, and unpredictable movement patterns, which can lead to feelings of anxiety and discomfort (Kenna, 2022). According to McAllister et al. (2022), public transportation, a crucial component of urban mobility, often lacks sensory-friendly features such as quiet areas, structured waiting spaces, and clear visual directions. Another significant challenge is spatial navigation. People with ASD find it difficult to navigate complex and unstructured settings since conventional urban designs are built on unspoken social signals and do not include clearly defined pathways or ASD-friendly navigation systems (Sarraf, 2024; Smith, 2015; Tola et al., 2021). This exclusion restricts their ability to move independently, thereby heightening their dependence on caregivers and diminishing their engagement in public spaces (McAllister et al., 2022). Moreover, social exclusion continues to be a significant concern for people with ASD. The lack of adaptive play areas for children with ASD, tranquil recreational spaces, and ASD-friendly community programmes leads to fewer opportunities for social engagement for these people, further isolating them from urban life (Black et al., 2022). These challenges highlight the pressing requirement for inclusive urban planning for people with ASD.

Several countries have integrated autism-friendly urban spaces to enhance accessibility for these communities. Studies on urban accessibility in Derry/Londonderry, Northern Ireland, emphasise the importance of transition zones and gradual shifts in sensory inputs that help people with ASD adjust between high-stimulation and quiet environments (McAllister et al., 2022). In addition, cities such as Copenhagen and Singapore have introduced sensory-friendly green spaces, predictable street layouts, and autism-inclusive public transportation to enhance urban accessibility (Black et al., 2022). In Dubai, several public places such as Dubai Frame, Children's City, Dubai International Airport, some police stations, and libraries have received Autism-Friendly Certification (AFC), incorporating autism-friendly services and accommodating the needs of people with ASD (Dubai Autism Center, 2025). However, these efforts remain fragmented, as they do not yet form a city-wide standard for urban planning, resulting in inconsistent accessibility across various urban areas. Additionally, global best practices provide valuable insights into how cities in the UAE can expand beyond localised efforts and adopt ASD-inclusive policies on a larger urban scale.

The Need for Inclusion of People with ASD In Sustainable Urban Planning - Road to Develop a Conceptual Framework for UAE.

Due to the increasing prevalence of ASD in the UAE and other countries in the Gulf Cooperation Council (GCC), it is significant to design urban areas that cater for the needs of neurodiverse individuals (AlQahtani, 2023). The biopsychosocial framework introduced by the World Health Organisation (WHO, 2001) pinpoints that environmental barriers, rather than disabilities themselves, limit the participation of autistic individuals (Sarraf, 2024). This factor corresponds with spatial justice theories, highlighting the rights of people with ASD to equitable access to urban spaces (Black et al., 2022; Bölte et al., 2019; Lefebvre, 1996). Nevertheless, urban planning frameworks in many countries, including the UAE, have not yet completely integrated ASD-friendly design principles, highlighting the need for a structured approach.

According to Tola et al. (2021), a stakeholder-driven approach is key to ensuring inclusivity in sustainable urban planning. Effective solutions involve collaboration between policymakers, urban planners, ASD specialists, autistic individuals themselves, and their families. Participatory urban design has demonstrated the benefits of including neurodiverse perspectives in city planning in Stockholm, Singapore, Copenhagen, and Northern Ireland (Kenna, 2022; McAllister et al., 2022;

Sarraf, 2024; Tola et al., 2021). Likewise, the UAE can build on such approaches by integrating multi-stakeholder engagement into its urban development strategies.

Existing urban planning policies in the UAE demonstrate a robust commitment to sustainability and smart city development; however, explicit guidelines for accessibility and sustainability (ASD) are still lacking. While the UAE has taken steps toward autism inclusion through the AFC, these attempts remain isolated and limited to specific places. To build a holistic and inclusive city, ASD-friendly universal design principles and technology-driven approaches need to be embedded into the building regulations, transportation infrastructure, and urban master plans in the UAE.

To bridge these gaps, this study proposes a conceptual framework that aligns spatial justice, sustainability, stakeholder collaboration, and smart technology to create ASD-inclusive urban spaces in the UAE.

Conceptual Framework

The proposed conceptual framework, as shown in Fig. 31.1, is developed based on the literature synthesis.

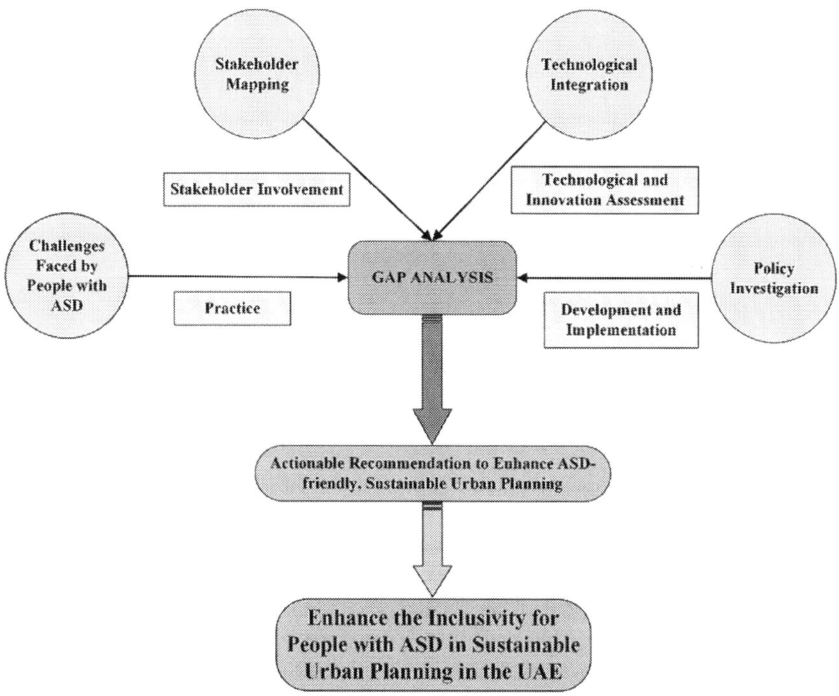

Fig. 31.1 Conceptual framework

According to Fig. 31.1, the four main components contributing to exploring the gaps to achieve inclusivity are 'the challenges faced by people with ASD,' 'stakeholder involvement in urban planning,' 'technological and innovation assessment,' and 'the development and implementation of policies relating to ASD and sustainable urban planning. An analysis of these components of the conceptual framework will reveal the gaps associated with inclusivity, which, in turn, will inform the development of suitable, actionable recommendations to enhance ASD-friendly, sustainable urban planning in the UAE.

Discussion

This study has explored the gap between ASD inclusivity and sustainable urban planning in the UAE. The literature review has identified several critical challenges, such as sensory overload, lack of wayfinding, and social exclusion for people with ASD. While some initiatives, such as the AFC programme, mark progress, inclusivity remains limited to specific facilities rather than being integrated into broader policies. Global examples from Stockholm, Singapore, Copenhagen, and Northern Ireland show that ASD-inclusive planning is achievable when embedded into national policies and smart city strategies. However, a structured policy approach is lacking in the UAE, emphasising the need for a comprehensive framework. This study aimed to develop a conceptual framework to guide policy, design, and technological interventions for creating more accessible and inclusive urban spaces in the UAE.

Conclusion and Way Forward

This study, as part of an ongoing research project, aims to bridge the gap between ASD inclusivity and sustainable urban planning in the UAE. According to the literature review, while the UAE has made progress in ASD-inclusive urban planning, the efforts remain isolated rather than systemic. Global best practices highlight the need for city-wide ASD-inclusive policies. However, a structured policy approach for ASD inclusivity in UAE urban planning is currently lacking, highlighting a critical gap. Moving forward, the UAE must embed Universal Design Principles and technological advancements into urban planning, integrate ASD-friendly features into sustainable smart city initiatives, and enhance stakeholder collaboration. As a way forward, the proposed conceptual framework will guide how to enhance the inclusivity of people with ASD in sustainable urban planning. By implementing this conceptual framework, this study will provide a roadmap to develop ASD-friendly, inclusive recommendations for sustainable urban planning in the UAE and beyond. Future work from this ongoing research project will also explore primary research methods, such as key informant interviews, case studies, stakeholder network analysis, and technology and innovation assessment, to validate and refine the proposed framework with actionable recommendations. This research aligns with the United Nations Sustainable Development Goals (SDGs), particularly SDG 11: Sustainable Cities and Communities, by promoting social inclusion, equity, and innovation. By ensuring that ASD-inclusive urban planning becomes an integral part of sustainability and smart city strategies, this study contributes to creating more accessible, equitable, and thriving urban environments in the rapidly evolving landscape of the UAE.

References

AlQahtani, O. (2023). Autism spectrum disorder: Where does the Gulf Region stand? An overview of ASD in the Arab Gulf Region: The UAE as a regional model. *Research in Autism Spectrum Disorders, 108*, Article 102247. https://doi.org/10.1016/j.rasd.2023.102247

American Psychiatric Association. (2013). *Diagnostic and Statistical Manual of Mental Disorders* (5th ed.). American Psychiatric Association.

Beale-Ellis, S. (2017). *Sensing the city*. Jessica Kingsley Publisher.

Black, M. H., McGarry, S., Churchill, L., D'Arcy, E., Dalgleish, J., Nash, I., Jones, A., Tse, T. Y., Gibson, J., Bölte, S., & Girdler, S. (2022). Considerations of the built environment for autistic individuals: A review of the literature. *Autism, 26*(8), 1904–1915. https://doi.org/10.1177/13623613221102753

Bölte, S., Girdler, S., & Marschik, P. B. (2019). The contribution of environmental exposure to the etiology of autism spectrum disorder. *Cellular and Molecular Life Sciences, 76*(7), 1275–1297. https://doi.org/10.1007/s00018-018-2988-4

Bowler, D. E., Buyung-Ali, L. M., Knight, T. M., & Pullin, A. S. (2010). A systematic review of evidence for the added benefits to health of exposure to natural environments. *BMC Public Health, 10*(1), 456. https://doi.org/10.1186/1471-2458-10-456

Dubai Autism Center. (2025, February 7). *Certified Organisations*. https://www.dubaiautismcenter.ae/afp2/

Grandin, T. (1995). *Thinking in pictures and other reports from my life with autism*. Doubleday.

Jaarsma, P., & Welin, S. (2012). Autism as a natural human variation: Reflections on the claims of the neurodiversity movement. *Health Care Analysis, 20*(1), 20–30. https://doi.org/10.1007/s10728-011-0169-9

Kenna, T. (2022). Cities of neurodiversity: New directions for an urban geography of neurodiversity. *Area, 54*(4), 646–654. https://doi.org/10.1111/area.12803

Lefebvre, H. (1996). *Writings on cities* (E. Kofman & E. Lebas Eds.). Wiley-Blackwell.

McAllister, K., McBeth, A., & Galway, N. (2022). Autism spectrum condition and the built environment. *Cities & Health, 6*(6), 1164–1178. https://doi.org/10.1080/23748834.2022.2139210

Pellicano, E., & den Houting, J. (2022). Annual research review: Shifting from 'normal science' to neurodiversity in autism science. *Journal of Child Psychology and Psychiatry, 63*(4), 381–396. https://doi.org/10.1111/jcpp.13534

Sarraf, M. (2024). Autism in urban planning: In search of a theoretical framework. *Cities & Health*, 1–16. https://doi.org/10.1080/23748834.2024.2404802

Saunders, M., Lewis, P., & Thornhill, A. (2016). *Research methods for business students* (7th ed.). Pearson Education Limited.

Smith, A. D. (2015). Spatial navigation in autism spectrum disorders: A critical review. *Frontiers in Psychology, 6*. https://doi.org/10.3389/fpsyg.2015.00031

Tola, G., Talu, V., Congiu, T., Bain, P., & Lindert, J. (2021). Built environment design and people with Autism Spectrum Disorder (ASD): A scoping review. *International Journal of Environmental Research and Public Health, 18*(6), 3203. https://doi.org/10.3390/ijerph18063203

Toronyi, D. (2021). Hidden geographies: Design for neurodivergent ways of hearing and sensing. *Cities & Health, 5*(1–2), 133–137. https://doi.org/10.1080/23748834.2019.1627059

World Health Organisation. (2001). *International classification of functioning, disability and health.* World Health Organisation.

Chapter 32
Assessing the Impact of Climate Change Adaptation Measures on Health Equity Across Vulnerable Populations

Ibrahim Shah, Vijaya Kittu Manda, Chitra Saruparia, and Muhammad Younus

Abstract Climate change poses significant challenges to global health, disproportionately affecting vulnerable populations and worsening existing health inequities. This review explores the complex link between health equity and climate change adaptation, emphasizing the need for inclusive and equitable strategies to mitigate climate-related health impacts. The study highlights the disproportionate burden of climate change on marginalized communities. This study examines the direct and indirect impacts of climate phenomena on health while examining the inequitable allocation of vulnerabilities and risks among various socio-cultural and economic factions. Current adaptation measures, such as disaster preparedness, climate-resilient healthcare systems, and community-based initiatives, are assessed for effectiveness and equity considerations. Key barriers, including socioeconomic disparities, limited access to resources, and insufficient policy integration, are discussed, along with opportunities for advancing equity-focused adaptation. By synthesizing evidence from global and local case studies, this review highlights the importance of integrating health equity into climate adaptation efforts to ensure sustainable and fair outcomes for all. Addressing these challenges requires interdisciplinary collaboration, participatory approaches, and targeted investments to build resilience among the most at-risk populations. The findings provide actionable insights for policymakers

I. Shah
Lahore School of Nursing, The University of Lahore, Sargodha, Pakistan
e-mail: Ibrahim.shah@lsn.uol.edu.pk

V. K. Manda (✉)
PBMEIT, Visakhapatnam, India
e-mail: vijaykittu@hotmail.com

C. Saruparia
Faculty of Humanities and Social Sciences, National Law University Jodhpur, Rajasthan, India
e-mail: Chitra.saruparia@nlujodhpur.ac.in

M. Younus
Department of Government Affairs and Administration, Universitas Muhammadiyah Yogyakarta, Yogyakarta, Indonesia
e-mail: mohammedyounusghazni@gmail.com

J. Dsilva (ed.), *Sustainability Education, Youth and Innovation*,
https://doi.org/10.1007/978-3-032-21948-0_32

integrating equity into global climate adaptation frameworks. This review contributes by synthesizing equity-centered adaptation strategies across LMIC contexts.

Keywords Climate change adaptation · Health equity · Vulnerable populations · Climate resilient healthcare · Environmental justice

Introduction

Climate change represents one of the most significant global health challenges of our time, with its impacts falling disproportionately on vulnerable and marginalized populations. This systematic review examines the critical intersection between climate change adaptation and health equity, particularly in low- and middle-income countries with the most acute challenges. While climate change affects all populations, its health impacts—including increased exposure to extreme weather events, shifting patterns of infectious diseases, and compromised food and water security—are not distributed equally across society, also shown in Fig. 32.1. Socioeconomic disparities, limited access to healthcare resources, and varying adaptive capacities create a complex web of vulnerabilities that exacerbate existing health inequities. This study synthesizes current evidence on climate adaptation strategies incorporating health equity considerations, evaluating their effectiveness in reducing health disparities among marginalized communities. By examining global case studies and adaptation frameworks, we aim to provide actionable insights for policymakers and healthcare practitioners working to build climate-resilient health systems.

Figure 32.1 illustrates of the adverse effects of climate change on health.

Review of Literature

Climate change poses enormous challenges to global health (WHO, 2023). It disproportionately affects vulnerable populations and exacerbates health inequities (Khanal et al., 2023; Kosanic et al., 2022). The effect of climate vulnerability on different individuals and communities depends on structural and socioeconomic advantages within society. Privileged racial and ethnic groups often have greater access to resources and institutional support and are considered trustworthy and provided safety when vocal about their problems (Wilkerson, 2020). Vulnerability arises from differential exposure, sensitivity, and limited adaptive capacity, with economically disadvantaged groups often facing barriers to mitigation and advocacy. Using Health Impact Assessments (HIA) and a climate vulnerability framework, strategies and indicators can be developed to effectively monitor and address these inequities (Haigh et al., 2020). The disproportionate effects of climate change on vulnerable populations

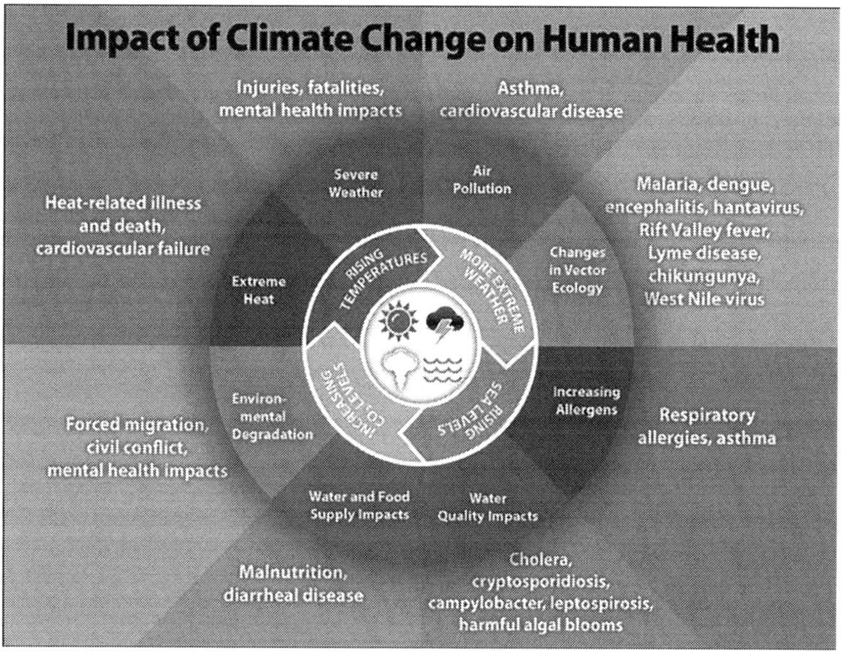

Fig. 32.1 Impact of climate change on human health (CDC). *Source* https://www.cdc.gov/climateandhealth/effects/default.htm

are highlighted in case studies from Bangladesh, Puerto Rico, and the Sahel, advocating for implementing integrated climate justice frameworks and community-based adaptation strategies to mitigate these challenges (Yadav et al., 2025).

With the intensification of climate change, studies have explored the intersection of climate change and human rights, focusing on the challenges faced by Indigenous peoples, women, children, and people with low incomes. These groups are more susceptible to climate-related risks, and the research emphasizes the crucial role of international instruments, like the UNFCCC and human rights treaties, in establishing strong legal safeguards and fostering greater public participation to ensure a just and equitable future (Sini & Singh, 2025). Climate change significantly impacts vulnerable groups, including the elderly in East China, through rising temperatures, extreme weather events, and deteriorating air quality. Direct effects include heat-related illnesses, floods, and respiratory diseases, while indirect effects exacerbate cardiovascular and mental health challenges. Heat-related illness encompasses various health results varying from dehydration to heat stroke (Epstein et al., 2019).

Research has shown that climate change is responsible for cardiovascular health risks (Peters et al., 2021), kidney disease (Borg et al., 2021), mental health (APA, 2019), allergies and respiratory health (D'Amato et al., 2016), infectious diseases

(BMJ, 2020), pregnancy risks (Bekkar et al., 2020), and global food and water insecurity (Romanello et al., 2021).

Adaptation measures, however, carry their risks. Measures like household insurance and property protection may inadvertently increase inequalities, while community engagement and planning have been found to reduce health disparities. Researchers emphasize that adaptation strategies must be tailored to local contexts to prevent exacerbating inequalities among vulnerable populations (Grace, Turner et al., 2024). The literature also underlines the vulnerability of certain groups, including children and the elderly, particularly susceptible to climate-induced harm. Women who face preexisting physical, economic, and political vulnerabilities encounter more significant hardships during and after climate-induced disasters, complicating their recovery (Malak et al., 2020; Das et al., 2026).

Abbass et al. (2022) reviewed the sectoral assessment of climate change mitigation and adoption approaches globally in different sectors and the related economic costs. They highlighted the role of government in crafting climate policy. Improved research on regional disparities and vulnerable populations is critical to developing targeted health adaptation strategies (Kinay et al., 2019).

Researchers have examined the intersection of social determinants, climate risks, and health disparities in Germany, focusing on adaptation strategies for vulnerable populations. It emphasizes the need for inclusive, equity-centered climate adaptation policies, advocating for participatory approaches to reduce climate-related health inequities. It calls for systemic changes to bolster the resilience and well-being of at-risk communities (Bolte et al., 2023). Building on these findings, cross-disciplinary frameworks exploring psychological and structural processes have illuminated the fundamental drivers of climate-related health inequities. There is an urgent need to understand the intricate role of climate change, particularly its effects on natural and human systems, as a key factor contributing to health disparities and is significant in devising effective strategies for addressing health inequities in the face of ongoing climate changes (Pearson et al., 2023).

A case study of an equity-oriented adaptation program evaluated the implications of the Environmental Justice Index in identifying socially vulnerable communities with climate change adaptation needs. The research found that relying on a single index to select "disadvantaged" communities is restrictive and fails to fully capture the complexities of environmental and climate justice. This underscores the need for more nuanced, comprehensive approaches in selecting vulnerable populations for adaptation strategies (Turek-Hankins et al., 2020).

Research Gaps Identified

There are three research gaps identified:

1. **Empirical Knowledge Deficit**: While climate adaptation strategies are widely studied, there is a lack of research specifically analyzing their impact on health equity,

especially in low- and middle-income countries (LMICs). The study fills this gap by synthesizing global case studies and assessing equity-focused adaptation measures.

2. **Policy-Implementation Gap**: There is a disconnect between climate change adaptation policies and their real-world implementation in vulnerable communities. The study highlights how existing policies often fail to incorporate social determinants of health and offers insights for improving policy frameworks to ensure equitable health outcomes.

3. **Theoretical Integration Gap**: Climate adaptation research often fails to incorporate interdisciplinary perspectives, such as environmental justice, social determinants of health, and vulnerability frameworks. The study bridges this gap by integrating multiple theoretical perspectives to provide a holistic understanding of climate-induced health disparities.

Methods

Design

A systematic review (SR) is an extensive and organized way to gather research data on a particular topic. It uses precise and reproducible methods to find, choose, and evaluate relevant studies to reduce bias and increase the reliability of the findings (Lame, 2019; Pollock & Berge, 2018). After the data has been synthesized, the results are analyzed, considering the research question. This phase considers the evidence's application, advantages, and disadvantages. PRISMA principles are used in this review to ensure transparency and present the techniques, findings, and conclusions (Rao, 2023).

Aim and Research Question

This systematic review aims to explore the intersection of health equity and climate change adaptation, focusing on the disproportionate impacts of climate change on marginalized communities.

Developing a precise research topic is the first stage in a study, often organized using the PICO (Population, Intervention, Comparison, Outcome) framework. This stage is crucial for ensuring clarity and focus in the review process (Martinez et al., 2023). The study question was formulated using a structured approach known as the PICO framework.

RQ: How do climate change adaptation strategies incorporating health equity influence health disparities and health outcomes in marginalized communities in low- and middle-income countries?

P Marginalized communities affected by climate change (Low -Middle-income countries).

I Climate change adaptation strategies incorporating health equity.

C No treatment.

O Reduction in health disparities and improved health outcomes.

Outcomes Measured

The primary outcome of the present review was:

- Reduction in health disparities

The secondary outcomes of this systematic review were:

- Improved health outcomes.

Search Strategy

The present review included studies that met the following criteria, and the inclusion criteria were:

- Marginalized communities in low- and middle-income countries (LMICs).
- Climate change adaptation strategies incorporating health equity include policy interventions, community-based initiatives, healthcare infrastructure improvements, and climate-resilient public health programs.
- Studies report on health outcomes, including reduced health disparities and improvements in health due to climate change adaptation.
- Peer-reviewed research articles, reports, or case studies provide empirical data. Both quantitative, qualitative, and mixed-methods studies will be included.
- Studies published in English.

The exclusion criteria of the study are as follows:

- Studies focusing on high-income countries or populations not significantly affected by climate change.
- Studies focusing on climate change mitigation strategies without addressing adaptation or health equity.
- Studies focus solely on non-health outcomes (e.g., economic or environmental impacts) without addressing health impacts.
- Opinion pieces, editorials, commentaries, or narrative reviews lack original empirical data.
- Studies published in languages other than English (unless translation is available).

The following electronic databases were searched:

- PubMed (1957 to Jan 2025);
- Cochrane Library (1996 to Jan 2025);
- Embase (2002 to week 18, Jan 2025);
- Scopus (2016 to Jan 2025);
- CINAHL (1982 to Jan 2025);

Search Methods

A preliminary search was conducted using the Medline database via the EBSCOhost platform to identify appropriate search terms that would yield studies relevant to the systematic review question. After determining suitable search terms, these terms were subsequently used to perform searches across all three selected databases.

Once the search terms were finalized, they were applied collectively across all five selected databases.

- "climate adaptation" OR "climate resilience"
- "health disparities" OR "social determinants of health" OR "environmental justice"
- low-income populations" OR "socially vulnerable populations"

("climate change adaptation" OR "climate resilience") AND ("health equity" OR "health disparities") AND ("marginalized communities" OR "vulnerable populations") AND ("health outcomes" OR "disease burden").

Data Extraction

Data extraction systematically collects relevant information from the selected studies to answer the research question (Schmidt et al., 2021). Data was extracted using a data extraction table (Appendix 2). Extracted data is presented in Table 1: Characteristics of Included Studies and Table 2: Outcomes Measured. An extraction tool was utilized to ensure that the relevant data was accurate and applicable to the research question, as adapted from PRISMA (2020) (Page et al., 2021). To ensure accuracy and reliability, the extracted data was independently reviewed by two reviewers, following best practices as outlined by (Stoll et al., 2019). This dual-review process helps minimize errors and biases, ensuring the integrity of the data.

Data Analysis and Synthesis

In this review, the data analysis and synthesis will be conducted systematically by integrating findings from the relevant studies to identify common themes and trends related to the impact of climate change adaptation strategies on health equity in marginalized communities. The following steps will guide the analysis:

Study Selection and Extraction

Studies focusing on climate change adaptation strategies incorporating health equity in low- and middle-income countries will be selected. Key data will be extracted, including the adaptation strategies implemented, the socioeconomic and cultural characteristics of the communities studied, and the reported health outcomes.

Data Organization

The extracted data will be organized into categories based on common themes, such as:

- Types of climate change adaptation strategies (e.g., policy interventions, community-based initiatives, infrastructure improvements).
- Health disparities addressed (e.g., chronic diseases, mental health, access to healthcare).
- Social determinants of health (e.g., income, education, housing, environmental conditions).
- Geographic location and population characteristics (e.g., low-income, rural, indigenous communities).

Synthesis of Findings

A qualitative synthesis will be conducted to identify patterns and trends across the studies. This will include:

- Analyzing how health equity considerations are integrated into climate change adaptation strategies.
- Identifying the types of health outcomes that are most affected by these strategies.
- Assessing the effectiveness of these strategies in reducing health disparities in marginalized communities.

Evaluation of Adaptation Strategies

The review will evaluate the success of different adaptation strategies in promoting health equity by comparing the health outcomes achieved in communities where equity-focused strategies were implemented. Emphasis will be placed on improvements in overall health, reduction in health disparities, and addressing social determinants of health.

Identification of Gaps and Recommendations

The analysis will also identify gaps in the literature, such as areas where evidence is lacking or where current adaptation strategies fall short in addressing health equity. Based on these gaps, future research and policy development recommendations will be made to integrate health equity into climate change adaptation.

Quality Appraisal

Critical appraisal is watchfully and systematically examining research to assess its reliability, value, and relevance to direct professionals' vital clinical decision-making (Al-Jundi & Sakka, 2017). The Evidence-Based Librarianship (EBL) critical appraisal checklist was utilized to appraise the quality of the included studies.

Discussion

In this scoping review, relevant articles from the existing literature have been synthesized that focus on the relationship between adaptation measures of climate change and health equity. Additionally, it discusses how various climate adaptation strategies influenced health outcomes across diverse populations, emphasizing the differential impacts on vulnerable communities. The environmental health determinants and their linkage with non-communicable diseases (NCDs) were frequently investigated. Moreover, additional health outcomes impacted by climate change beyond NCDs have also been explored; our sample identified a couple of integrated frameworks, community-based adaptation strategies, and equity-centered climate adaptation policies emphasizing participatory approaches to promote climate adaptation. The current literature supports the need for a transnational approach to prioritizing human rights and social justice and ecosystem-based adaptation strategies coupled with integrating climate resilience into disaster planning to promote a more resilient and agile system that promotes climate adaptation and health inquiry. Additionally, we identified critical gaps in the literature that include the limited focus on morbidity outcomes, inadequate stratification by sub-populations, and a noticeable lack of research from low- and middle-income countries (LMICs) and low-income countries (LICs). These findings call for future research priorities and underline the necessity of equity-focused policies and strategies to mitigate climate-related health inequities effectively.

Environmental Health Determinants and Non-Communicable Diseases (NCDs)

Environmental health determinants and their connection with non-communicable diseases (NCDs) were reported commonly in the literature. Most of the articles emphasized addressing environmental health determinants, particularly concerning non-communicable diseases (NCDs). Research indicates climate-induced environmental changes significantly exacerbate health disparities, disproportionately impacting vulnerable populations already facing social inequities (Ziegler & Muchira, 2023). Environmental determinants such as deteriorating air quality, food insecurity, and limited access to safe drinking water substantially increase the prevalence of NCDs like cardiovascular diseases, respiratory disorders, and chronic kidney disease (Borg et al., 2021; Peters & Schneider, 2021; Smirnova et al., 2023). For example, increased exposure to pollutants due to climate change disproportionately burdens socioeconomically disadvantaged communities, heightening health inequities by exacerbating underlying chronic health conditions (Boeckmann & Zeeb, 2016). This intersection highlights a critical public health challenge posed by climate change. Exacerbating existing health disparities, particularly among vulnerable and socioeconomically disadvantaged communities, necessitates a comprehensive, equity-centered approach to environmental health policy and healthcare delivery. Understanding environmental health risks is crucial for making informed decisions to prevent NCDs.

Other Health Outcomes Beyond NCDs

Climate change impacts health through various mechanisms beyond non-communicable diseases (NCDs) through various pathways; the heightened frequency of extreme weather events and environmental degradation have direct implications for infectious diseases, mental health, maternal and child health, and nutritional status (Bekkar et al., 2020; Cianconi et al., 2020; Romanello et al., 2021). Specifically, the elderly, children, Indigenous communities, and women experience compounded vulnerabilities during climate-induced disasters (Khine & Langkulsen, 2023; Ngcamu, 2023). The mental health consequences, including anxiety, depression, and trauma-related disorders, further illustrate the broad scope of climate-induced health impacts, demanding comprehensive health system preparedness and interventions (Raker et al., 2020). Understanding these relationships is vital for developing comprehensive strategies that address health and educational outcomes in climate change. The current literature evidence reveals that climate change poses a multifaceted threat to health, influencing various outcomes from respiratory diseases to maternal and child health.

Implementing Integrated Frameworks and Community-Based Adaptation Strategies

Implementing integrated frameworks and community-based adaptation strategies has emerged as a critical approach to mitigating climate-related health disparities. The literature strongly advocates for adaptive governance mechanisms that include diverse stakeholders, integrating traditional ecological knowledge and community-driven responses (Akamani, 2023). Community-based participatory action research fosters resilience by empowering local populations, particularly in rural areas, enhancing their capacity to respond to and recover from climate stressors (Bailie, 2023). Evidence from adaptation programs emphasizes the effectiveness of participatory approaches in achieving sustainable and equitable health outcomes, highlighting the essential role of local context in guiding adaptation measures (Turner et al., 2024). Implementing integrated frameworks and community-based adaptation strategies is crucial for addressing climate change effectively. It empowers local communities to manage climate-related risks based on their priorities and capacities.

Equity-Centered Climate Adaptation Policies and Participatory Approaches

Equity-centered climate adaptation policies represent another significant area of focus. Such policies advocate for systemic changes incorporating social determinants of health, emphasizing participatory methods and equitable access to resources (Bolte et al., 2023). However, there remains a notable gap between policy formulation and practical implementation. This gap underscores the need for enhanced advocacy and inclusive policy frameworks that directly involve affected communities in decision-making, ensuring that climate actions do not inadvertently reinforce existing inequities (Turek-Hankins et al., 2020).

Ecosystem-Based Adaptation and Climate Resilience in Disaster Planning

Advocating for ecosystem-based adaptation and integrating climate resilience into disaster planning is essential for sustainable climate responses. Ecosystem-based strategies, such as maintaining healthy wetlands and forest ecosystems, have shown promise in reducing climate vulnerabilities while promoting biodiversity and ecosystem services. (Akamani, 2023). Integrating resilience-building into disaster response and healthcare infrastructure planning further mitigates the long-term health impacts of climate-induced disasters, especially in communities with limited adaptive capacities (Raker et al., 2020). In conclusion, addressing climate change adaptation through a health equity lens requires multifaceted approaches, incorporating comprehensive policies, integrated frameworks, community engagement, and international collaboration rooted in human rights and environmental justice. Future research and policy efforts should prioritize closing gaps in empirical knowledge, policy implementation, and theoretical integration to combat climate-induced health disparities and build resilient communities globally effectively.

Policy Implications

An increasing amount of research shows that climate change adaptation should transcend technocratic solutions and integrate context-specific, community-driven strategies. Conventional and indigenous adaptation strategies have shown significant potential in improving resilience, especially in resource-limited environments. These methodologies frequently depend on indigenous ecological knowledge and sustainable resource management, rendering them both economically efficient and culturally suitable for at-risk groups (Al-Falahi et al., 2023). Incorporating these practices with contemporary adaption measures can enhance the inclusivity and efficacy of climate responses.

The susceptibility to climate change is unevenly distributed and is profoundly influenced by demographic and societal factors. Senior populations, especially in highrisk zones like flood-prone coastal communities, encounter exacerbated risks stemming from physical constraints, social isolation, and diminished adaptive capacity (Bukvic et al., 2018; Kinay et al., 2019). Older persons face increased exposure to climate-related stressors, requiring specific treatments that tackle both environmental and health vulnerabilities (Oven et al.,

2019). These findings highlight the significance of age-sensitive adaptation frameworks in public health planning.

Alongside age-related concerns, those with impairments constitute another significantly neglected demographic in climate adaptation discussions. Structural impediments, such as restricted mobility, insufficient access to information, and exclusion from policy formulation, intensify their vulnerability to climate threats (Gaskin et al., 2017). Rectifying these disparities necessitates the formulation of inclusive policies that guarantee accessibility and the active engagement of excluded groups in adaptation planning initiatives.

Global studies highlight that the effects of climate change are intricately linked to ecological and socioeconomic systems. The Intergovernmental Panel on Climate

Change emphasizes that vulnerability results from the interplay of exposure, sensitivity, and adaptation capacity, all of which are shaped by structural inequalities (IPCC, 2014). Moreover, the susceptibility of species and ecosystems to climate change might trigger cascade impacts on human health and livelihoods, especially in populations reliant on natural resources (Pacifici et al., 2015). These interdependencies underscore the imperative of implementing integrated, systems-oriented strategies for climate adaptation/ppRecent scholarship advocates for a more explicit incorporation of health equality into climate research and policy frameworks. There is growing acknowledgment that adaptation programs must tackle fundamental structural determinants of health, like poverty, education, and healthcare access, to attain significant and durable results (Pearson et al., 2023). This corresponds with the overarching aim of guaranteeing that climate action fosters both environmental sustainability and social fairness.

The data indicates that equitable climate adaptation necessitates a multifaceted approach that integrates traditional knowledge systems, inclusive governance, and interdisciplinary collaboration. By addressing the distinct needs of vulnerable people and acknowledging the interrelated nature of climate hazards, policymakers can formulate more effective solutions that enhance both resilience and health equality.

Conclusion

Climate change adaptation is crucial for mitigating the disproportionate health impacts faced by marginalized populations. This review highlights the intersection of climate adaptation strategies and health equity, emphasizing the need for inclusive and community-driven approaches. While adaptation measures such as climate-resilient healthcare systems and disaster preparedness initiatives have demonstrated benefits, disparities persist due to socioeconomic barriers, inadequate policy integration, and limited access to resources. Addressing these challenges requires interdisciplinary collaboration, evidence-based policymaking, and sustained investment in equitable adaptation frameworks. Future research should focus on refining targeted interventions and ensuring the voices of vulnerable communities are central to adaptation planning. Policy and practitioners can foster resilience and create sustainable, just outcomes for all populations by integrating health equity into climate adaptation strategies.

References

Abbass, K., Qasim, M. Z., Song, H., et al. (2022). A review of the global climate change impacts, adaptation, and sustainable mitigation measures. *Environmental Science and Pollution Research, 29*, 42539–42559. https://doi.org/10.1007/s11356-022-19718-6

Al-Falahi, A. H., Barry, S., Gebrechorkos, S. H., Spank, U., & Bernhofer, C. (2023). Potential of traditional adaptation measures in mitigating the impact of climate change. *Sustainability, 15*(21), 15442. https://doi.org/10.3390/su152115442

Al-Jundi, A., & Sakka, S. (2017). Critical appraisal of clinical research. *Journal of Clinical and Diagnostic Research, 11*(5), Je01–je05. https://doi.org/10.7860/jcdr/2017/26047.9942

APA. (2019). *Sustainability, climate, and the natural environment*. https://www.apa.org/topics/climate-change

Bekkar, B., Pacheco, S., Basu, R., & DeNicola, N. (2020). Association of air pollution and heat exposure with preterm birth, low birth weight, and stillbirth in the US: A systematic review. *JAMA Network Open, 3*(6), e208243. https://doi.org/10.1001/jamanetworkopen.2020.8243

Bolte, G., Dandolo, L., Gepp, S., Hornberg, C., & Lumbi, S.L. (2023). Climate change and health equity: A public health perspective on climate justice. *Journal of Health Monitoring, 8*(Suppl 6), 3–35. https://doi.org/10.25646/11772. PMID: 38105794; PMCID: PMC10722520.

Borg, M. A., & Bi, P. (2021). The impact of climate change on kidney health. *Nature Reviews Nephrology, 17*(5), 294–295. https://doi.org/10.1038/s41581-020-00365-4

Bukvic, A., Gohlke, J., Borate, A., & Suggs, J. (2018). Aging in flood-prone coastal areas: Discerning the health and well-being risk for older residents. *International Journal of Environmental Research and Public Health, 15*(12), 2900. https://doi.org/10.3390/ijerph15122900

D'Amato, G., Vitale, C., Lanza, M., Molino, A., & D'Amato, M. (2016). Climate change, air pollution, and allergic respiratory diseases: An update. *Current Opinion in Allergy & Clinical Immunology, 16*(5), 434–440. https://doi.org/10.1097/ACI.0000000000000301

Das, M. M., Kalyan, S., Kumari, A., Prakash, M., & Sharma, T. (2026). Climate disasters and financial resilience among women—Case studies of India. In M. Bhattacharyya & R. Shaw (Eds.), *Gender, climate change and disaster resilience* (pp. 353–371). Springer Nature Singapore. https://doi.org/10.1007/978-981-95-6188-9_18

Epstein, Y., & Yanovich, R. (2019). Heatstroke. *New England Journal of Medicine, 380*(25), 2449–2459. https://doi.org/10.1056/NEJMra1810762

Gaskin, C. J., Taylor, D., Kinnear, S., Mann, J., Hillman, W., & Moran, M. (2017). Factors associated with the climate change vulnerability and the adaptive capacity of people with disability: a systematic review. *Weather, Climate, and Society, 9*(4), 801–814.

Haigh, F., Harris-Roxas, B., & de Leeuw, E. (2020). Understanding vulnerability in the context of climate change—A health impact assessment framework. *European Journal of Public Health, 30*(Supplement_5), ckaa165.513. https://doi.org/10.1093/eurpub/ckaa165.513

Intergovernmental Panel on Climate Change (IPCC). (2014). Climate change 2014: impacts, adaptation, and vulnerability. Part A: global and sectoral aspects. In C. B. Field, V. R. Barros, D. J. Dokken, K. J. Mach, M. D. Mastrandrea, T. E. Bilir, M. Chatterjee, K. L. Ebi, Y. O. Estrada, R. C. Genova, B. Girma B, E. S. Kissel, A. N. Levy, S. MacCracken, P. R. Mastrandrea, & L. L. White (Eds.), *Contribution of working group II to the fifth Assessment report of the intergovernmental panel on climate change*. Cambridge University Press, UK.

Intergovernmental Panel on Climate Change (IPCC). (2018). Global warming of 1.5°C. Special Report Intergovernmental Panel on Climate Change.

John, S. (2024). The unequal burden: Climate change and vulnerable populations. In S. Chakraborty & S. S. Satapathy (Eds.), *Advances in religious and cultural studies* (pp. 359–380). IGI Global. https://doi.org/10.4018/979-8-3693-6069-9.ch018

Khanal, S., Ramadani, L., & Boeckmann, M. (2023). Health equity in climate change and health policies: A systematic review. *Sustainability, 15*(13), 10653. https://doi.org/10.3390/su151310653

Kinay, P., Morse, A. P., Villanueva, E. V., Morrissey, K., & Staddon, P. L. (2019). Direct and indirect health impacts of climate change on the vulnerable elderly population in East China. *Environmental Reviews, 27*(3), 295–303.

Kosanic, A., Petzold, J., Martín-López, B., & Razanajatovo, M. (2022). An inclusive future: Disabled populations in the context of climate and environmental change. *Current Opinion in Environmental Sustainability, 55*, 101159. https://doi.org/10.1016/j.cosust.2022.101159

Lame, G. (2019). *Systematic literature reviews: An introduction*. Paper presented at the Proceedings of the design society: international conference on engineering design

Malak, M. A., Sajib, A. M., Quader, M. A., & Anjum, H. (2020). We are feeling older than our age: Vulnerability and adaptive strategies of aging people to cyclones in coastal Bangladesh. *International Journal of Disaster Risk Reduction, 48*, Article 101595.

Martinez, E. C., Valdés, J. R. F., Castillo, J. L., Castillo, J. V., Montecino, R. M. B., Jimenez, J. E. M., & Diarte, E. (2023). Ten steps to conduct a systematic review. *Cureus, 15*(12).

Oven, K., Wistow, J., & Curtis, S. (2019). Older people and climate change. *People and Climate Change: Vulnerability, Adaptation, and Social Justice, 68.*

Pacifici, M., Foden, W. B., Visconti, P., Watson, J. E., Butchart, S. H., Kovacs, K. M., Scheffers, B. R., Hole, D. G., Mar-tin, T. G., Akçakaya, H. R., & Corlett, R. T. (2015). Assessing species vulnerability to climate change. *Nature Climate Change, 5*(3), 215–224.

Page, M. J., McKenzie, J. E., Bossuyt, P. M., Boutron, I., Hoffmann, T. C., Mulrow, C. D., Shamseer, L., Tetzlaff, J. M., Akl, E. A., Brennan, S. E., Chou, R., Glanville, J., Grimshaw, J. M., Hróbjartsson, A., Lalu, M. M., Li, T., Loder, E. W., Mayo-Wilson, E., McDonald, S., & Moher, D. (2021). The PRISMA 2020 statement: An updated guideline for reporting systematic reviews. *BMJ, 372*, 71. https://doi.org/10.1136/bmj.n71

Parmar, A., & Chan, K. K. W. (2020). Prioritising research into cancer treatment delays. *BMJ*, m4261. https://doi.org/10.1136/bmj.m4261

Pearson, A. R., White, K. E., Nogueira, L. M., Lewis, N. A., Jr., Green, D. J., Schuldt, J. P., & Edmondson, D. (2023). Climate change and health equity: A research agenda for psychological science. *American Psychologist, 78*(2), 244–258.

Peters, A., & Schneider, A. (2021). Cardiovascular risks of climate change. *Nature Reviews Cardiology, 18*(1), 1–2. https://doi.org/10.1038/s41569-020-00473-5

Pollock, A., & Berge, E. (2018). How to do a systematic review. *International Journal of Stroke, 13*(2), 138–156.

Rao, S. C. (2023). An introduction to systematic reviews and meta-analyses. *APIK Journal of Internal Medicine, 11*(3), 141–146.

Romanello, M., McGushin, A., Di Napoli, C., Drummond, P., Hughes, N., Jamart, L., Kennard, H., Lampard, P., Solano Rodriguez, B., Arnell, N., Ayeb-Karlsson, S., Belesova, K., Cai, W., Campbell-Lendrum, D., Capstick, S., Chambers, J., Chu, L., Ciampi, L., Dalin, C., & Hamilton, I. (2021). The 2021 report of the Lancet Countdown on health and climate change: Code red for a healthy future. *The Lancet, 398*(10311), 1619–1662. https://doi.org/10.1016/S0140-6736(21)01787-6

Schmidt, L., Finnerty Mutlu, A. N., Elmore, R., Olorisade, B. K., Thomas, J., & Higgins, J. P. T. (2021). Data extraction methods for systematic review (semi)automation: Update of a living systematic review. *F1000Res, 10*, 401. https://doi.org/10.12688/f1000research.51117.2

Stoll, C. R. T., Izadi, S., Fowler, S., Green, P., Suls, J., & Colditz, G. A. (2019). The value of a second reviewer for study selection in systematic reviews. *Research Synthesis Methods, 10*(4), 539–545. https://doi.org/10.1002/jrsm.1369

Turek-Hankins, L. L., Hino, M., & Mach, K. J. (2020). Risk screening methods for extreme heat: Implications for equity-oriented adaptation. *PLoS ONE, 15*(11), Article e0240841. https://doi.org/10.1371/journal.pone.0240841.PMID:33147245;PMCID:PMC7641348

Turner, G., Kovats, S., Brisley, R., Landeg, O., Brown, S., & Connor, L. O. (2024). Can we adapt fairly? Systematic review of health equity implications of climate change in coastal communities in the UK. *European Journal of Public Health, 34*(Supplement_3), ckae144.285. https://doi.org/10.1093/eurpub/ckae144.285

Wilkerson, I. (2020). *Caste: The origins of our discontents* (1st ed.). Random House.
World Health Organization. (2023). *Climate change and health.* https://www.who.int/news-room/fact-sheets/detail/climate-change-and-health
Yadav, M., Chandel, A., Agrawal, H., & Quttainah, M. (2025). Climate change and global inequality: how does climate change exacerbate existing global inequalities and its implications. In B. Sevinçli & D. Savaş (Eds.), *Effects of climate change on social and economic factors* (pp. 21–48). IGI Global Scientific Publishing. https://doi.org/10.4018/979-8-3693-5792-7.ch002

Chapter 33
Generational Shifts in Sustainable Built Environment Investment: The Influence of Gen-Z and Millennials on Green Real Estate and Smart Urban Development

Jacinta Dsilva and Sahil Buch

Abstract This research investigates the perspectives of Gen-Z and Millennials on sustainable living, with a specific focus on their attitudes toward future real estate investments in sustainability focused communities. The study highlights a preference for sustainable living environments, with respondents consistently emphasizing the importance of energy-efficient appliances and renewable energy sources. However, significant barriers such as financial limitations and restricted access to sustainable products are identified as key obstacles hindering widespread adoption of sustainable real estate. The study further examines the primary motivators behind these age groups' interest in sustainable investments, revealing that economic considerations and personal values are central to their decision-making processes. The findings underscore the growing significance of sustainability within the context of real estate, suggesting that these generations not only prioritize environmental responsibility but also view sustainable investments as aligned with their long-term economic and ethical goals. This research provides valuable insights for developers, policymakers, and investors aiming to cater to the evolving demands of Gen-Z and Millennials in the real estate sector.

Keywords Sustainable living · Responsible investing · Sustainable urban developments · Eco-consumer behaviour

J. Dsilva (✉) · S. Buch
CEO and Co-Founder, InLabs Group, Sharjah Media City, Sharjah, UAE
e-mail: inlabsgroup@outlook.com

S. Buch
MICE, Tourism & Visitor Economy Specialist, RISA Consultancy, Dubai, United Arab Emirates

J. Dsilva (ed.), *Sustainability Education, Youth and Innovation*,
https://doi.org/10.1007/978-3-032-21948-0_33

Introduction

In recent years, sustainability has shifted from a niche topic to a central global concern, influencing lifestyle choices, well-being, and consumption habits (Faber, 2024). At the forefront of this transformation are Gen-Z and Millennials often called the "Sustainability Oriented Generations". These groups are increasingly vocal about environmental issues, making sustainability a defining factor in how they live and what they buy. As their influence grows, understanding their preferences is becoming essential for real estate developers, policymakers, and investors (Lee & Thompson, 2022). This insight is vital not only for market alignment but also for advancing climate action and sustainable practices (Coughlin & Dorner, 2023). Gen-Z (born 1997–2012) and Millennials (born 1981–1996) are more aware of climate risks than previous generations. Experiencing rising global temperatures and environmental instability, they feel compelled to adopt practices like energy-efficient living, waste reduction, and renewable energy use (Sengupta et al., 2024; Le et al., 2022). For them, sustainability is not just ethical, it guides their lifestyle, investment, and consumption decisions (Miller & Carter, 2021).

As sustainability becomes a growing priority, it's critical to understand how these generations make decisions and how their concerns can drive broader systemic change (Lopes et al., 2024). This study therefore aims to investigate: *"In what ways do Gen-Z and Millennials prioritize sustainability in their lifestyle choices, and how can these priorities be integrated into urban development, housing, and public policy?"* Addressing this will offer insights for designing sustainable communities, eco-friendly housing, and responsive policy frameworks.

Methodology

This study aims to explore the perspectives of Gen-Z and Millennials on sustainability and their future real estate investments. Specifically, it examines the importance of sustainable features in real estate properties, the barriers to investing in sustainable real estate, and the economic and personal drives influencing their decisions. The research question addresses the growing significance of sustainability in real estate investment decisions among younger generations, which has both theoretical and practical implications. Theoretically, it contributes to the expanding body of knowledge on sustainability and consumer behavior, while practically, it informs real estate developers and policymakers about the preferences and challenges faced by these generations when considering sustainable investments (Consumer Behaviour Research, 2024).

The research adopted a multi-method approach (Dawadi et al., 2021) to provide a comprehensive understanding of the sustainability preferences and barriers faced by Gen-Z and Millennials in investing in sustainable living as a future choice of investment in real estate. The framework integrates quantitative data collection through structured survey questions and qualitative data through open-ended responses, allowing for the identification of key themes and insights. This approach is particularly suitable for the research question as it enables the collection of both statistical trends and in-depth qualitative insights into the attitudes, beliefs, and behaviors of participants (Raman et al., 2024; Rosdiana et al., 2026).

The research study, particularly the survey, was designed to address the purpose of the study by collecting data on participants' demographics, current sustainability practices, preferences for sustainable features in real estate, and investment considerations. Key focus included participants' commitment to sustainability, their desired sustainable features in real estate (e.g., energy-efficient appliances, renewable energy sources), and the financial or logistical challenges that might hinder their interest in sustainable real estate. These questions were intended to capture not only the participants' general attitudes towards sustainability but also specific challenges they perceived in investing in sustainable real estate.

- Participant Demographics

A total of 226 participants were surveyed, and the sample included individuals across various age groups, educational backgrounds, and income levels to provide a diverse representation of the target demographic. Age distribution was as follows: 4.6% were aged 15–17, 23.7% were aged 18–20, 12.8% were aged 21–23, 21.0% were aged 24–27, 7.3% were aged 28–30, and 30.6% were aged 30 or older (maximum 43). In terms of education, 22.8% held a high school diploma, 40.6% had a bachelor's degree, and 35.2% had a master's degree or higher. The income distribution showed that 18.7% earned less than $5,000 annually, while 9.1% earned between $5,000 and $9,999, and 15.5% earned more than $40,000. This diverse demographic ensured that the research captured a wide range of perspectives on sustainable real estate investment, providing robust and representative data.

- Data Collection Methods

Data was collected through an online survey administered via email invitations and social media channels, an efficient method that ensured broad participation from a geographically and socioeconomically diverse sample. This format facilitated data collection across various locations, increasing representativeness and external validity. The online survey method is particularly well-suited to the research scope as it allows for the efficient capture of a large sample, ensuring that trends can be extracted that are applicable to the wider Gen-Z and Millennial population on a global scale.

Findings

This section presents the key findings derived from the survey data and provides a comprehensive analysis of participants' views, behaviours, and the factors influencing their decisions. The findings are divided into four main areas: current sustainable living practices, challenges to sustainable living, influences on investment decisions, and preferences for sustainable transportation.

1. Current Sustainable Living Practices.

Respondents were asked to evaluate the sustainability of their communities, revealing a mixed perception. 48% of respondents consider their community to be sustainable, while 34% disagree and 18% remain unsure. Among those who identified sustainable aspects, the most frequently mentioned elements include Waste Management and Recycling (54%), Energy Efficiency (50%), Water Conservation (44%), Food Sustainability (approximately 38%), Transport Sustainability (around 42%), and the Use of sustainable materials (roughly 30%). These findings suggest that while communities exhibit notable sustainable features, their application and effectiveness may vary, leading to differing perceptions among residents. This variation could perhaps stem from differences in local policies, infrastructure, and community engagement, highlighting the need for more consistent and widespread sustainability efforts to ensure long-term impact.

The survey also highlights varied personal engagement and participation from respondents in sustainability behaviors. Many respondents actively participate in conservation efforts, while others show room for growth in certain areas:

- Energy Consumption Reduction: 45% frequently or always reduce their energy use, demonstrating a commitment to energy conservation within their communities.
- Waste Reduction: 44% actively minimize waste, reflecting strong engagement in sustainability.
- Recycling: 57% regularly participate, making it the most widely adopted practice.
- Sustainable Food Consumption: 42% engage in sustainable food practices, indicating potential for growth.
- Renewable Energy Use: 22% rarely and 20% sometimes use renewable energy, suggesting barriers like cost or accessibility.

Challenges to Sustainable Living

Despite the enthusiasm for sustainability shown by the respondents, several barriers were cited as hindrances in the widespread adoption of sustainable behavior. Financial constraints were the most frequently mentioned challenge, with 55% of respondents identifying this as a significant obstacle. Limited access to sustainable options was another common issue, with 48% of respondents acknowledging this difficulty.

Additionally, 51% of respondents cited inconvenience as a major barrier to adopting more sustainable practices, while 28% mentioned a lack of knowledge. The data highlights that the financial constraint to afford sustainable alternatives, that are usually more expensive and not easily available, is more commonly seen in the bracket of individuals with lower disposable incomes, which further highlights the need for more affordable and accessible sustainable options (Dragolea et al., 2023). Addressing these barriers could involve providing more economical and convenient sustainable products and services to ensure that sustainability is accessible to all income groups.

2. Influences on Investment Decisions

When asked about the factors influencing their decision to invest in sustainable real estate, respondents identified economic considerations (25%) as the most significant factor, followed by personal values and beliefs (21%) and government policies or incentives (18%). Other influences included educational background (12%), peers and friends (10%), social media (10%), and media coverage (4%), suggesting that financial and personal values drive decisions more than external influences. These results also indicate that while economic and ideological factors play a primary role, social influences like peers, social media, and media coverage still contribute to a lesser extent, however.

3. Preferences for Sustainable Transportation

Transportation preferences were also examined, with respondents showing a strong preference for low-carbon mobility infrastructure in sustainable communities. The most popular options included bike lanes (61%), electric vehicle (EV) charging stations (51%), electric community shuttle buses (53%), and walkability (58%). These preferences highlight the importance of integrating sustainable transportation options into community designs. The high demand for bike lanes and walkability indicates a desire for easily accessible, low-impact modes of transport, while EV charging stations and electric shuttle buses reflect an increased interest in electric vehicle adoption.

Discussion

Implications for Sustainable Real Estate Investment

The survey data revealed strong demand among Gen-Z and Millennials for sustainable real estate features (Choudhary, 2020). Respondents consistently emphasized the importance of energy-efficient appliances, renewable energy sources, and smart home technologies. These preferences align with existing research showing younger generations prioritize sustainability and tech integration in housing (Jones et al., 2021). Their interest in energy-efficient homes reflects broader trends in environmental awareness and cost-conscious decision-making (Smith & Green, 2020). Developers looking to meet these expectations should incorporate renewable energy systems, efficient appliances, and smart technologies in new residential projects. These features resonate with environmentally conscious buyers and increase long-term property value (Brown, 2019).

Impact of Influencing Factors

The influence of sustainability on real estate decisions is shaped by both values and economics. Survey respondents cited financial viability, long-term savings, and environmental impact as key motivators. For Gen-Z and Millennials, sustainability is no longer a niche concern but a practical and ethical standard for investment. Developers who ignore these factors risk losing relevance in a market that increasingly prioritizes green living.

Addressing Challenges

Despite growing interest, significant barriers limit the adoption of sustainable practices. Financial constraints (55%), lack of access to sustainable options (48%), and inconvenience (51%) were among the most cited challenges. To address these, policymakers should consider:

- Financial Incentives: Subsidies or rebates can reduce cost barriers and make sustainable living more accessible (Pulido-Fernandez et al., 2024).
- Improved Access to Eco-Friendly Products: Expanding product availability and service reach addresses both convenience and accessibility (Pradeep, 2023).
- Public Awareness Campaigns: Educational initiatives can shift behaviors and improve understanding of sustainability benefits (Pulido-Fernandez et al., 2024).

Conclusion

This study highlights the central role of sustainability in the choices of Gen-Z and Millennials, particularly in real estate. While many individuals' practice sustainability through actions like recycling and energy conservation, barriers such as affordability and access limit broader adoption. Financial support mechanisms and improved product availability are essential to closing this gap.

Economic and ethical drivers influence investment in sustainable real estate. Energy-efficient, tech-enabled homes that reflect environmental values are increasingly favored. Developers must respond by integrating renewable energy, smart technologies, and sustainability features into design and planning. The rising demand for green transportation infrastructure—bike lanes, EV charging stations, and walkable neighborhoods—also highlights the need for integrated urban development.

By addressing these challenges and aligning with the values of younger generations, developers and policymakers can accelerate the shift toward sustainable living and build resilient, future-ready communities.

References

Brown, M. (2019). Sustainable housing: The future of real estate development. *Green Building Journal, 14*(2), 45–59.

Choudhary. (2020). Generation green is leading the sustainability agenda. www.capgemini.com/2020/08/generation-green-is-leading-the-sustainabilityagenda

Consumer Behavior Research. (2024, February 13). NIQ. https://nielseniq.com/global/en/info/consumer-behavior-research/

Coughlin, A., & Dorner, E. (2023). *Shaping tomorrow's sustainability: Unraveling Gen Z's decision-making journey for sustainable consumption.* https://www.diva-portal.org/smash/get/diva2:1777795/FULLTEXT01.pdf

Dawadi, S., Shrestha, S., & Giri, R. A. (2021). Mixed-methods research: A discussion on its types, challenges, and criticisms. *Journal of Studies in Education, 2*(2), 25–36. https://www.researchgate.net/publication/349663713_Mixed-Methods_Research_A_Discussion_on_its_Types_Challenges_and_Criticisms

Dragolea, L.-L., Butnaru, G. I., Kot, S., Zamfir, C. G., Nuţă, A.-C., Nuţă, F.-M., Cristea, D. S., & Ştefănică, M. (2023). Determining factors in shaping the sustainable behavior of the generation Z consumer. *Frontiers in Environmental Science, 11*. https://doi.org/10.3389/fenvs.2023.1096183

Faber, E. (2024, May 15). Gen Zs and millennials seek purpose and progress in a changing world: Insights from deloitte's 2024 Gen Z and millennial survey. Forbes. https://www.forbes.com/sites/deloitte/2024/05/15/gen-zsand-millennials-seek-purpose-and-progress-in-a-changing-world-insights-from-deloittes-2024-gen-z-andmillennial-survey/

Jones, P., Comfort, D., & Hillier, D. (2021). Smart and sustainable homes: A critical review. *Journal of Sustainable Development, 29*(3), 101–115.

Le, T. H. M., Tran, H. V., & Hoang, T. T. H. (2022). Ethically minded consumer behavior of Generation Z in Vietnam: The impact of socialization agents and environmental concern. *Cogent Business & Management, 9*(1). https://doi.org/10.1080/23311975.2022.2102124

Lee, T., & Thompson, R. (2022). The role of economic incentives in sustainable housing investment. *International Journal of Housing Research, 18*(1), 22–37.

Lopes, J. M., Gomes, S., Suchek, N., et al. (2024). The hidden reasons behind generation Z's green choices. *Discover Sustainability, 5*, 520. https://doi.org/10.1007/s43621-024-00764-8

Miller, S., & Carter, J. (2021). Financial and environmental benefits of sustainable homes: A market analysis. *Real Estate Economics Review, 32*(4), 78–94.

Pradeep, S., & Pradeep, M. (2023). Awareness of sustainability, climate emergency, and generation Z's consumer behaviour in UAE. *Cleaner and Responsible Consumption, 11*, 100137. https://doi.org/10.1016/j.clrc.2023.100137

Pulido-Fernández, J. I., Casado-Montilla, J., López-Sánchez, Y., Durán-Román, J. L., & Carrillo-Hidalgo, I. (2024). Gen Zers' behaviour towards sustainability in a mature coastal destination. *International Journal of Tourism Research, 26*(5). https://doi.org/10.1002/jtr.2758

Raman, R., Das, P., Mandal, S., Vijayan, V., AR, A., & Prema, N. (2024). The impact of Gen Z's pro-environmental behavior on sustainable development goals through tree planting. *Sustainable Futures, 8*, 100251–100251. https://doi.org/10.1016/j.sftr.2024.100251

Rosdiana, R., Hariani, S., Pramudena, S. M., Bakti, I. G. M. Y., Syahlani, N., Utomo, A. W., Sumaedi, S., Widyanty, W., Prasetya, P., & Shiratina, A. (2026). Understanding the future of investment: an analysis of millennials and generation Z. *Future Business Journal, 12*(1). https://doi.org/10.1186/s43093-026-00724-w

Sengupta, D., Mathews, M., Bridges, L., D'Costa, R., & Bastian, B. L. (2024). Sustainability orientation of generation Z and its role in their choice of employer—A comparative qualitative inquiry of India and United States. *Administrative Sciences, 14*(10), 249–249. https://doi.org/10.3390/admsci14100249

Smith, K., & Green, R. (2020). Energy efficiency in residential buildings: Consumer preferences and policy implications. *Environmental Policy Journal, 27*(1), 55–72.

Chapter 34
Summar Welfare Organization—A Vision of Healing and Hope

Sara Ray and Samra Rai

Abstract Summar Welfare Foundation is launching a 200-bed charitable hospital in an underserved region of Pakistan to improve healthcare access for low-income families, women, and children (World Health Organization. (2022). Primary health care and health systems. https://www.who.int/news-room/fact-sheets/detail/primary-health-care). It aligns with UN SDG 3 – Good Health and Well-being, aiming for long-term impact through sustainable, inclusive healthcare services (United Nations. (2015). Transforming our world: The 2030 Agenda for Sustainable Development. https://sdgs.un.org/goals/goal). The Emergency Department will be dedicated to His Highness Sheikh Zayed bin Sultan Al Nahyan, honoring his legacy of compassion and humanitarian service (Al Nahyan Foundation. (2020). Legacy of Compassion: Sheikh Zayed's Humanitarian Vision. UAE Publications.). This initiative also creates local jobs, empowers medical professionals, and promotes health equity across communities. Through global partnerships and collaboration, the project envisions a scalable healthcare model rooted in dignity and care. Access to quality healthcare remains a major challenge in many regions of Pakistan, particularly for low-income families, women, and children (Government of Pakistan. (2021). National Health Vision Pakistan 2016–2025. Ministry of National Health Services, Regulations & Coordination.).

Keywords Summar welfare trust hospital · Sustainable healthcare · Underserved communities · SDG 3-good health and well-being · Sheikh Zayed humanitarian legacy

---

S. Ray (✉) · S. Rai
Summar Welfare Foundation Pakistan, Lahore, Pakistan
e-mail: Sararay786@gmail.com

J. Dsilva (ed.), *Sustainability Education, Youth and Innovation*,
https://doi.org/10.1007/978-3-032-21948-0_34

Introduction

Summar Welfare Foundation, a registered nonprofit since 2015, is working to close this gap through the development of a 200-bed charitable hospital in a medically underserved area. This initiative reflects a strong commitment to humanitarian values and sustainable development, providing equitable access to essential medical services and promoting long- term community health. Summar Welfare Foundation envisions a future where healthcare is not a privilege but a basic human right. In pursuit of this mission, the Foundation is establishing the Summar Welfare Trust Hospital. This hospital project addresses the critical gap in healthcare infrastructure in remote and low-resource areas, where thousands lack access to timely medical intervention.

Methodology

Summar Welfare Trust project is based on a needs-based assessment aimed at addressing the critical healthcare gaps in underserved regions of Pakistan. Recognizing that access to quality healthcare remains a major challenge, particularly for low-income families, women, and children, the Summar Welfare Foundation has identified the urgent requirement for a healthcare facility in a medically underserved area. To respond to this need, the project involves the establishment of a 200-bed charitable hospital designed to provide free or low-cost medical services. The hospital will include essential facilities such as a 24/7 Emergency Unit, OPD, Maternity Ward, NICU, Nursery, and an in-house Pharmacy, ensuring comprehensive and accessible care. The design adopts a patient-centered approach and is structured to expand as community needs evolve, thereby ensuring long-term sustainability. Furthermore, the project aligns with the United Nations Sustainable Development Goal (SDG) 3—Good Health and Well-being, focusing on improving maternal and child health outcomes, addressing emergency medical needs, and strengthening local healthcare infrastructure. By selecting a location with limited healthcare resources and tailoring services to the needs of vulnerable populations, the methodology ensures that the project directly addresses existing disparities and contributes to equitable and inclusive healthcare delivery.

Project Vision and Mission

Its mission is to offer free and affordable treatment to the most vulnerable—especially women, children, and low-income families—ensuring dignity, equity, and care.

Hospital Design and Services

The Summar Welfare Trust Hospital will launch with essential departments including a 24/7 Emergency Unit, OPD, Maternity Ward, NICU, Nursery, and an in-house Pharmacy. The design prioritizes accessible, patient-centered care and is structured to expand as community needs grow. All services will be offered free or at low cost, focusing on underserved families, women, and children.

Sustainability and SDG 3 Alignment

The Summar Welfare Trust Hospital is grounded in the principles of sustainable healthcare delivery and directly aligns with United Nations Sustainable Development Goal (SDG) 3—Good Health and Well-being (United Nations, 2015).

The hospital is designed to operate as a sustainable health ecosystem, prioritizing health equity, preventive care, and clean energy.

By improving maternal and child health outcomes, addressing emergency medical needs, and empowering local healthcare infrastructure, the hospital will make a measurable contribution to SDG 3 targets.

Tribute to Sheikh Zayed

This initiative reflects a model of compassion-driven sustainability, inspired by the humanitarian legacy of His Highness Sheikh Zayed bin Sultan Al Nahyan (Al Nahyan Foundation, 2020).

The Summar Welfare Trust Hospital honors the legacy of His Highness Sheikh Zayed bin Sultan Al Nahyan, the founding father of the United Arab Emirates, whose humanitarian vision and commitment to global well-being continue to inspire generations.

Sheikh Zayed's legacy represents dignity, empathy, and action for the underserved.

By dedicating this life-saving wing to him, the Foundation affirms its mission to serve with integrity and ensure that no life is left behind due to lack of medical care.

Community Benefits and long-term impact.

The hospital aims to evolve into a community-owned health hub, strengthening public trust and reducing long-term healthcare disparities (Health Policy Forum, 2023).

Summar Welfare Trust Hospital is a long-term vision to bridge healthcare gaps in underserved communities. We invite collaborators who share our commitment to sustainable, equitable healthcare to join us in bringing this vision to life.

Conclusion and Future Steps

The Summar Welfare Trust Hospital represents a compassionate step toward equitable healthcare in underserved communities. Moving forward, the Foundation seeks strategic partnerships and sustainable funding to complete construction and expand services. Our goal is to build a model of community-based healthcare that inspires long-term impact and global collaboration.

References

Al Nahyan Foundation. (2020). *Legacy of compassion: Sheikh Zayed's humanitarian vision*. UAE Publications.

Government of Pakistan. (2021). *National Health Vision Pakistan 2016–2025*. Ministry of National Health Services, Regulations & Coordination.

Health Policy Forum. (2023). *Community health hubs: Reducing disparities through local empowerment. international health review*.

United Nations. (2015). *Transforming our world: The 2030 Agenda for sustainable development*. https://sdgs.un.org/goals/goal

World Health Organization. (2022). *Primary health care and health systems*. https://www.who.int/news-room/fact-sheets/detail/primary-health-care

Part VI
Sustainability and Climate Resilience in Tourism, Hospitality, and Events

Chapter 35
Sustainable Tourism and Educational Travel: Expanding. Net-Zero Concepts Through Dubai's Sustainable City

Keigo Nagatani

Abstract Sustainable tourism and educational travel play a crucial role in fostering environmental awareness and responsible citizenship. This study examines how net-zero concepts can be effectively introduced through structured educational programs, using Dubai's Sustainable City as a case study. In August 2024, 80 high school students from the Tokyo Metropolitan area participated in an immersive sustainability-focused study tour, exploring key aspects of urban planning, renewable energy, and resource-efficient food production. Through on-site observations, student reflections, and academic presentations, this research analyzes the impact of experiential learning on students' understanding of sustainability and its potential applications in Japan. Findings indicate that well-structured educational travel programs significantly enhance knowledge retention and engagement with sustainability concepts, particularly in urban development and resource management. The study also highlights the lack of domestic sustainability education opportunities in Japan and recommends academic collaboration, interdisciplinary curriculum reform, and teacher training as strategies to advance long-term environmental consciousness.

Keywords Sustainable tourism · Experiential learning · Net-zero · Urban sustainability · Educational · Travel

Introduction

As the world faces pressing environmental challenges, sustainability education has become increasingly vital. Climate change, resource depletion, and urbanization require innovative solutions that future generations must be prepared to implement. However, despite Japan's technological advancements, there remains a lack of structured sustainability education, particularly at the secondary and university levels (United Nations, 2023). This research was inspired by participation in the Global

K. Nagatani (✉)
Aoyama Gakuin University, Tokyo, Japan
e-mail: keigo.nagatani@gmail.com

J. Dsilva (ed.), *Sustainability Education, Youth and Innovation*,
https://doi.org/10.1007/978-3-032-21948-0_35

Sustainable Tourism Council (GSTC) Sustainable Tourism Training Program in Kyoto in 2022, which emphasized the potential role of tourism in fostering sustainability awareness (GSTC, 2022). Building upon this foundation, the study focuses on a case where 80 Tokyo Metropolitan high school students visited Dubai's Sustainable City in August 2024. Through experiential learning, students engaged with net-zero urban planning, renewable energy integration, and sustainable food production. The primary objectives of this research are (1) to analyze how international study tours influence students' understanding of sustainability and (2) to explore the integration of sustainability education into Japan's curriculum at both secondary and tertiary levels.

Methodology

- This research adopts a qualitative approach, employing multiple methods to analyze the impact of experiential learning on sustainability education. The methodology includes:
- On-site observations: Direct assessments of students' engagement and learning during their visit to Dubai's Sustainable City.
- Student reflections: Post-tour written feedback to evaluate changes in understanding and attitudes.
- Academic presentations: Student-led sessions held at their respective schools and at Aoyama Gakuin University to share learnings and stimulate peer discussion.
- Comparative analysis: Examination of sustainability education models in Japan and Dubai to identify best practices. This research aligns with GSTC principles, ensuring that findings adhere to globally recognized sustainability standards (GSTC, 2022).

Findings

Experiential Learning Enhances Engagement

The immersive nature of the study tour significantly enhanced students' understanding of sustainability concepts. Many students reported that physically experiencing sustainable infrastructure—such as solar-powered homes and water-efficient urban systems—provided a deeper learning experience than traditional classroom education. According to experiential learning theories, hands-on experiences enhance knowledge retention and engagement, making them effective tools for sustainability education (SEE Institute, 2024).

Key Sustainability Concepts Explored

Renewable Energy Integration Dubai's Sustainable City operates entirely on solar energy, demonstrating how urban areas can transition to clean power sources. Students observed photovoltaic panel installations and smart grid technologies designed to optimize energy distribution (Dubai Sustainable City, 2024). Comparing these findings with Japan's energy policies, students reflected on the potential for greater solar energy adoption in Japanese cities.

Sustainable Agriculture Visits to vertical farms and hydroponic systems highlighted innovative methods for food production in resource-constrained environments. Students learned about the water efficiency of hydroponics and the potential for urban agriculture to enhance food security. In contrast, Japan's agricultural sector remains dependent on rural farming, suggesting a need for increased urban farming initiatives (SEE Institute, 2024).

Waste and Water Management Students observed recycling programs and grey-water reuse systems that emphasized circular economy principles. Dubai's Sustainable City incorporates advanced waste management techniques, such as composting and bio-digesters, which could serve as models for Japanese municipalities (Dubai Sustainable City, 2024).

Application of Net-Zero Strategies to Japan

Although Japan is technologically advanced, many net-zero strategies observed in Dubai remain underutilized domestically.

Urban Planning Adopting smart city designs similar to Dubai's Sustainable City could improve energy efficiency and reduce Japan's carbon footprint.

Food Security Expanding hydroponic farming and vertical agriculture within urban centers could increase Japan's food self-sufficiency.

Education System Introducing structured sustainability curricula at the high school and university levels would foster greater environmental awareness among future generations.

Educational Impact and Student Reflections

Post-tour reflections indicated that students not only acquired technical knowledge but also developed a stronger sense of agency. Many expressed a desire to study environmental science at university or engage in sustainability-related careers. Academic presentations reinforced this impact, as students translated their experiences into advocacy and action within their school communities.

Barriers to Sustainability Education in Japan

Despite increasing awareness of climate change and global sustainability efforts, Japan faces several institutional and cultural barriers to integrating sustainability education across all levels of the education system. One significant challenge is the rigid national curriculum, which leaves little room for interdisciplinary or project-based learning. Subjects such as science, geography, and civics are taught separately, which often prevents students from seeing the interconnectedness of environmental, social, and economic issues. Another obstacle is the lack of trained educators who specialize in sustainability-related topics. Even at the university level, few programs provide students with practical, systems-based approaches to learning about sustainability. This creates a gap in the academic pipeline, where students interested in sustainability may struggle to find appropriate guidance or career pathways. Compared to other developed countries, Japan has a lower number of environmental studies programs per capita, particularly in regional universities. Moreover, there is a cultural tendency to avoid controversial or globally political topics in schools, such as climate justice or environmental ethics. While this promotes harmony in the classroom, it can also suppress open discussion and critical thinking about Japan's role in global sustainability. As a result, students may graduate with technical knowledge but without the mindset required to act as sustainability leaders. Without strategic reforms, these barriers will continue to prevent the widespread implementation of sustainability education in Japan. Greater flexibility in curriculum design, teacher development, and institutional support are essential to overcome these limitations.

Discussion

The findings of this study demonstrate that immersive, sustainability-focused study tours can yield significant cognitive and affective gains among participating students. Beyond merely conveying factual knowledge about net-zero technologies and urban design, the Dubai Sustainable City experience fostered deeper reflective processes, resulting in measurable shifts in attitudes toward environmental stewardship. Students not only reported heightened awareness of renewable energy systems and resource management but also articulated concrete personal and collective aspirations for engaging with sustainability in their home communities. The presentation session held upon return served as both a pedagogical reinforcement and a catalyst for peer-to-peer learning. Through this collective debrief, initial individual insights were translated into communal discourse, suggesting that short-term international experiences, when coupled with structured follow-up activities, can ripple outward to influence broader school culture. Moreover, the framework employed by the SEE Institute—integrating expert lectures, on-site technical demonstrations, and guided reflection—proved effective in scaffolding student learning. This multi-modal approach aligns with constructivist principles, whereby learners actively construct new understanding through direct engagement and social negotiation of meaning. In the Japanese context, where classroom instruction often relies on rote learning, such experiential models present a valuable complement. They can serve to bridge theory and practice, thereby strengthening students' self-efficacy and sense of agency. The fact that several participants expressed interest in pursuing sustainability-related fields of study underscores the potential long-term impact on career trajectories and professional identity formation.

Implications for Policy and Curriculum

Educational policymakers and curriculum designers should recognize the value of embedding international, practice-oriented modules within existing science and social studies programs. First, formalizing partnerships between Japanese schools and global sustainability hubs—such as the SEE Institute or similar centers of excellence—can facilitate regular exchange programs. Funding mechanisms, including government grants and private sponsorships, could be established to subsidize travel costs and offset logistical barriers. Second, national curriculum frameworks should incorporate explicit learning objectives related to net-zero urban planning, renewable energy systems, and circular economy principles. These objectives can be achieved through locally coordinated field projects, virtual collaborations with overseas institutions, or blended-learning modules that precede and follow in-person study tours. By situating sustainability education within core subject areas, policymakers can ensure that such content is not marginalized but rather woven into the overarching educational mission.

Pedagogical Implications

For classroom practitioners, the results suggest that experiential learning—when integrated with reflective practice and peer teaching—can amplify both intellectual and emotional engagement. Teachers should be encouraged to adopt a "flipped" approach, wherein foundational knowledge (e.g., basic principles of net-zero architecture) is delivered asynchronously through readings or multimedia, freeing up face-to-face time for hands-on activities, site visits, and group reflection. In addition, establishing mentorship linkages with sustainability professionals can enrich the learning environment and provide students with role models in emerging green industries. Assessment strategies should also evolve to capture not only cognitive mastery but also dispositional shifts, such as increased environmental concern and willingness to engage in collective action. Portfolios, reflective journals, and student-led presentations—similar to the post-tour session described in this study—can serve as robust, authentic assessment tools that honor the holistic nature of sustainability learning.

Limitations and Future Directions

Although this research provides promising evidence of the benefits of educational travel, several limitations warrant consideration. First, the sample was limited to a single cohort of 80 students from the Tokyo Metropolitan area, which may not be representative of the broader student population in Japan. Future studies should include multiple cohorts across diverse geographic and socioeconomic contexts to assess generalizability. Second, data collection relied primarily on self-reported reflections and a single post-tour presentation; longitudinal tracking of actual behavioral change over one to two years would yield stronger evidence of lasting impact. Third, the study tour experience involved substantial resource investment—financial, logistical, and human—which may pose challenges for large-scale implementation. Research on cost-effective adaptations, including virtual reality site simulations or domestic field trips, could help identify scalable models. Building on these limitations, future research might explore comparative analyses between different types of experiential programs (e.g., domestic vs. international, short-term vs. longterm) and their respective influence on student learning outcomes. Investigating the role of individual difference variables—such as prior environmental attitudes, socioeconomic background, and learning style—could illuminate which students benefit most from various program designs. Finally, collaboration with educational technology developers to create hybrid learning environments that blend in-person tours with immersive digital platforms may offer a pathway to broaden access while maintaining pedagogical fidelity.

Conclusion and Recommendations

6.1. Summary of Key Findings.

This study underscores the importance of experiential learning in sustainability education. By participating in a structured study tour, students gained firsthand exposure to net-zero practices, fostering a deeper appreciation for sustainable living. The research findings highlight the benefits of integrating sustainability education into Japan's academic system.

Recommendations for Enhancing Sustainability Education in Japan

Expansion of Sustainability Study Tours Increasing international study programs would expose more students to global best practices in sustainability.

Strengthening Japan–UAE Academic Collaboration Deepening academic partnerships could enhance knowledge exchange and foster crosscultural environmental innovation.

Integration of Sustainability Education into Curricula Structured courses at secondary and tertiary levels should address renewable energy, urban planning, and sustainable agriculture.

Government–Private Sector Collaboration Public–private partnerships can provide funding and resources to support sustainability education and infrastructure.

Enhancing Teacher Training and Capacity Building Japan has limited opportunities for educators to gain expertise in net-zero concepts. Investment in teacher development through expert partnerships and workshops is crucial. This is particularly important given the scarcity of university programs specializing in sustainability. Empowered educators will play a central role in transforming school culture and student outcomes.

Establishing Interdisciplinary Learning Hubs To address the shortage of academic opportunities in sustainability fields, Japanese universities should consider establishing interdisciplinary learning hubs focused on environmental challenges. These hubs could serve as collaborative spaces where students from different faculties—engineering, business, sociology, and education—can work together on real-world sustainability projects. The integration of such hubs into existing institutions would not only broaden academic access but also foster innovation by promoting diverse perspectives. These learning environments would also benefit from partnerships with local governments, NGOs, and private sector actors. For instance, students could engage in urban design projects for carbon–neutral districts, or develop food waste reduction initiatives tailored to their communities. Such practical, community-embedded learning models have shown to improve both student motivation and long-term impact. Furthermore, hosting public forums, exhibitions, and workshops through these hubs would extend the reach of sustainability education beyond enrolled students. Creating an ecosystem of academic and civic engagement around sustainability is essential for building societal resilience in the face of environmental challenges.

Expanded Future Research Directions

To advance the field of sustainability education in Japan, several research directions should be pursued. One area is the longitudinal tracking of students who participate in experiential programs, such as the Dubai study tour, to assess how early exposure influences academic choices, values, and career development. This kind of follow-up research can provide evidence of long-term effectiveness and justify greater investment in similar programs. Another valuable focus would be comparative studies between Japanese institutions and their counterparts in countries with more developed sustainability curricula. Investigating how curriculum frameworks, teaching methodologies, and evaluation systems differ can help identify adaptable strategies for Japan. Finally, it would be useful to explore the role of digital technology in scaling sustainability education. Virtual reality simulations, online exchange programs, and interactive platforms could democratize access to high-quality learning experiences, especially for schools in rural or resource-constrained settings. As sustainability is a global challenge, international collaboration through digital education should also be considered a vital component of future strategies.

References

Dubai Sustainable City. (2024). Net-zero urban development strategies. https://www.thesustainablecity.ae

Global Sustainable Tourism Council. (2022). *GSTC sustainable tourism training program.* https://www.gstcouncil.org

SEE Institute. (2024). *Sustainability research and education.* https://www.seeinstitute.ae

United Nations. (2023). *Sustainable development goals.* https://sdgs.un.org/goals

Chapter 36
Tourism Beyond Cities: Advancing and Embracing Sustainability in the UAE's Natural and Cultural Landscapes

Sahil Buch

Abstract This research study examines the UAE's transition from urban-centric tourism to sustainable eco-tourism, focusing on emerging destinations such as Ras Al Khaimah, Al Ain, and Khorfakkan. Utilizing a mixed-methods approach—combining in-person site visits, qualitative observations, and secondary data analysis—the study explores how these regions integrate environmental conservation, cultural heritage preservation, and community engagement. The scope includes assessments of eco-tourism initiatives aligned with the United Nations Sustainable Development Goals (SDGs). The findings demonstrate how eco-tourism supports biodiversity, fosters local economic growth, and positions the UAE as a regional leader in sustainable tourism development.

Keywords Regenerative economy · Local visitor economy · Eco-tourism · Ecological conservation · Urban tourism

Introduction

Tourism has been a key driver of economic growth and urbanization in the GCC, transforming once-arid landscapes into thriving cities. For decades, tourism in the United Arab Emirates (UAE) has been synonymous with grand architectural feats and cutting-edge urban developments that have captivated the world. Cities like Dubai and Abu Dhabi positioned themselves as global tourism hubs by constructing iconic landmarks such as the Burj Khalifa, the Palm Jumeirah, the Yas Island, Saadiyat Cultural District and the Sheikh Zayed Grand Mosque—structures that not only showcased engineering excellence but also symbolized economic prosperity and modernity. These ambitious projects have long transformed the UAE into a premier travel destination, attracting millions of visitors annually and significantly contributing to GDP

S. Buch (✉)
MICE, Tourism & Visitor Economy Specialist, Dubai, UAE
e-mail: sahil.buch@risaconsultancies.com

J. Dsilva (ed.), *Sustainability Education, Youth and Innovation*,
https://doi.org/10.1007/978-3-032-21948-0_36

growth, job creation, and international inward investments. The UAE's strategic focus on tourism has yielded substantial economic benefits. In 2024, the travel and tourism sector contributed approximately AED 236 billion to the Visitor Economy (VE) of the UAE. Projections for 2025 indicate a further 5 to 10% growth on the VE of the UAE upwards from 2024 figures. This robust growth underscores the sector's pivotal role in the UAE's economic diversification and its emergence as a global tourism leader.

Dubai and Abu Dhabi have leveraged their architectural marvels by collaborating with globally renowned engineering and design firms, to create record-breaking structures and luxury tourism developments. These strategic partnerships have helped cement their status as premier global tourism destinations, attracting millions of visitors seeking cutting-edge urban experiences. However, this rapid urbanization has had a detrimental impact on the environment, leading to land reclamation, habitat destruction, and significant stress on marine ecosystems, including coral reefs and marine life due to coastal development, man-made islands and other mega-infrastructure projects. The extensive use of natural resources to support these developments has also contributed to ecological imbalances in both land and sea.

In response to these environmental concerns, a new wave of competition is emerging from the Northern Region—primarily Ras Al Khaimah, Khorfakkan and Al Ain, which are shifting the narrative away from conventional, city-centric tourism by focusing on the more natural eco-tourism. These regions are emphasizing nature-based experiences, cultural heritage, and sustainable practices, offering visitors an opportunity to connect with the natural landscapes and local communities in a way that traditional urban tourism cannot. This shift reflects a broader movement towards responsible tourism that prioritizes environmental preservation and socio-cultural sustainability over luxury and modernity.

In light of this evolving landscape, this study seeks to explore the Research Question—"In what ways are the transition towards eco-tourism reshaping the UAE's tourism sector to align with the Sustainable Development Goals". Investigating this shift is significant for understanding how the UAE can balance economic diversification with the preservation of its natural and cultural heritage, ensuring long-term sustainability and resilience within its tourism industry.

Methodology

This study adopts a qualitative, exploratory research design, combining secondary data analysis with primary observational research to examine the UAE's evolving eco-tourism landscape. The qualitative approach allows for an in-depth understanding of the complex relationship between sustainable tourism initiatives, economic development, and environmental preservation. The research focuses on emerging eco-tourism destinations such as Ras Al Khaimah, Al Ain, and Hatta, which are repositioning the UAE's tourism sector towards greater alignment with Sustainable Development Goals.

- Research Framework and Design

The research is guided by established sustainable tourism and eco-tourism frameworks, particularly the Ecotourism Principles and Standards Model (Blamey, 2001) and the Global Sustainable Tourism Council (GSTC) Criteria. These frameworks emphasize the integration of environmental conservation, socio-cultural respect, and local community engagement as core components of eco-tourism. Applying these models provides a structured lens through which to assess how UAE destinations are balancing ecological protection with tourism-driven economic growth.

The Ecotourism Principles and Standards Model (Blamey, 2001) focuses on three main pillars: Minimizing environmental impact; Generating financial benefits for conservation and providing meaningful experiences that raise environmental awareness. The GSTC Criteria, adopted globally as a benchmark for sustainable tourism, is used to evaluate the sustainability efforts of the destinations studied, including governance, socio-economic benefits, cultural heritage preservation, and environmental conservation.

This multi-framework approach ensures the study captures both theoretical and applied perspectives on eco-tourism, assessing how global sustainability principles are interpreted and implemented within the UAE's local context.

- Research Method

An extensive review of existing literature and data was conducted, including tourism articles, government policy documents, and industry reports from reputable organizations such as the Global Destination Sustainability (GDS) Index, AESG, and the World Tourism Organization (UNWTO). Academic publications on eco-tourism, urban tourism impacts, and sustainable destination management were systematically analysed to establish the theoretical foundation for the study. To complement secondary desk research, in-person site visits were undertaken to emerging eco-tourism spots in Ras Al Khaimah, Al Ain and Khorfakkan. These visits involved direct observation of eco-tourism initiatives, conservation projects, and visitor experiences. Attention was given to environmental management practices, infrastructure that supports sustainability (e.g., eco-lodges, renewable energy use), and community involvement in tourism activities. Informal conversations with local stakeholders, including Tourism Operators and Destination Management Organizations, were also conducted to gain qualitative insights into the practical applications and challenges of eco-tourism in these regions.

Data Analysis

The collected data is analysed thematically, guided by the key principles outlined in the selected eco-tourism frameworks. Themes such as environmental sustainability, community empowerment, cultural preservation, and economic diversification were identified and explored. The thematic analysis helped in understanding patterns, trends, and best practices that could inform the broader implications for sustainable tourism development in the UAE.

Findings and Discussion

This study explored the UAE's evolving tourism landscape, focusing on the strategic transition from urban-centric tourism models to nature-based, sustainable alternatives in destinations such as Ras Al Khaimah, Al Ain, and Hatta. The findings are derived from thematic analysis of secondary research, academic literature, industry reports, and first-hand observations from eco-tourism sites. The analysis was guided by key eco-tourism frameworks, including the Ecotourism Principles of The International Ecotourism Society and the Global Sustainable Tourism Council (GSTC) Criteria, which emphasize sustainability in environmental, socio-cultural, and economic dimensions.

1. Transition from Urban-Centric Tourism to Eco-Tourism

For decades, Dubai and Abu Dhabi have served as the epicenters of the UAE's tourism sector, driven by ambitious urban development projects, iconic architectural landmarks, and large-scale events. These cities have positioned themselves as global hubs for luxury tourism, retail, and MICE (Meetings, Incentives, Conferences, and Exhibitions), collectively generating over 75% of the UAE's tourism GDP in 2024. However, this rapid urban expansion has placed significant pressure on the natural environment, contributing to habitat destruction, land reclamation, and increased marine pollution (Epure et al., 2011). While Dubai and Abu Dhabi have implemented sustainability certifications, such as LEED and local green building codes, their environmental footprints remain substantial due to the high-intensity nature of urban tourism (Gössling & Higham, 2021).

In contrast, emerging destinations like Ras Al Khaimah, Khorfakkan and Al Ain are pioneering a shift towards eco-tourism, offering a more sustainable and environmentally conscious tourism model. These regions focus on nature-based experiences, cultural heritage preservation, and adventure tourism that actively minimize ecological impact. They prioritize conservation efforts, community involvement, and the promotion of responsible tourism through global sustainability standards such as EarthCheck and the Global Sustainable Tourism Council (GSTC) certifications (RAKTDA, 2023). These destinations provide an alternative to urban-centric tourism by leveraging their natural landscapes, biodiversity, and traditional cultural practices to create meaningful, low-impact visitor experiences. Table 1 highlights the different focus between Urban and Eco-Centric Tourism based on different categories.

2. Environmental Conservation and Education

The findings presented in this section are informed by in-person site visits to key eco-tourism destinations in the UAE, including Ras Al Khaimah, Al Ain, and Wadi Shees. These observations, supplemented by secondary data, highlight how these

Table 1 Urban-centric tourism versus eco-tourism in the UAE

Category	Urban-centric tourism	Eco-tourism
Tourism focus	Luxury, architectural wonders, MICE	Nature-based, heritage, adventure, community
Environmental impact	High: Land reclamation, marine degradation	Lower: Conservation-focused initiatives
Community involvement	Limited to absent	Active participation and local entrepreneurship
Sustainability certifications	LEED, BREEAM, Estidama	Earth check, LEED, green key, green globe
Primary attractions	Downtown Dubai, Palm Jumeirah, Yas Island, Saadiyat District	Wadi Shawkah, Green Mubazzarah Plains, Wadi Shees, Jebel Jais, Al Wadi Desert Reserve
Economic focus	Luxury tourism, retail, events	Community-based tourism, adventure tourism

initiatives integrate environmental conservation, education, and sustainable tourism practices to foster responsible visitor engagement and contribute to the local visitor economy, also summarized in Table 2.

- Sheikh Zayed Desert Learning Centre—Al Ain

A defining feature of eco-tourism initiatives is the integration of environmental education and conservation. One exemplary initiative is the Sheikh Zayed Desert Learning Centre (SZDLC) in Al Ain, part of the Al Ain Zoo complex and the broader Al Ain UNESCO World Heritage Site. The SZDLC is the UAE's first building to receive a Platinum rating on all three global green building systems: LEED, Estidama, and BREEAM. The Centre serves as an educational and interpretive hub, promoting knowledge about the UAE's unique desert ecosystem, conservation challenges, and sustainable living practices. It attempts to align with SDG 4 (Quality Education) and SDG 15 (Life on Land) by fostering environmental awareness and stewardship among visitors.

- Green Mubazzarah Hot Springs—Jebel Hafeet

Located at the foot of Jebel Hafeet in Al Ain, Green Mubazzarah is an eco-tourism destination that blends natural beauty with wellness tourism and outdoor recreation. This lush oasis, developed as part of Al Ain's sustainable tourism initiatives, features hot springs, man-made lakes, and green open spaces that provide a stark contrast to the surrounding desert landscape. The area supports biodiversity conservation through the preservation of indigenous flora and fauna, while offering sustainable infrastructure, including eco-lodges and environmentally friendly facilities. Visitors can enjoy activities such as hiking, cycling, and picnicking in a landscape that highlights Al Ain's commitment to balancing leisure tourism with ecological preservation. Green Mubazzarah also plays a role in reinforcing the region's cultural identity by showcasing traditional Emirati agricultural practices and land use within a modern, sustainable tourism framework.

- Al Wadi Desert Reserve—Ras Al Khaimah

Similarly, the Al Wadi Desert Reserve in Ras Al Khaimah—home to the Ritz-Carlton Ras Al Khaimah, Al Wadi Desert—offers a model of conservation-focused luxury tourism. This private nature reserve spans over 500 hectares and is dedicated to the preservation of native species, such as the Arabian oryx, sand viper, gazelles, and desert foxes. Visitors engage in guided nature walks, bird watching, and wildlife conservation activities that align with Earth Check sustainability standards and promote low-impact tourism. The Ritz-Carlton Ras Al Khaimah, Al Wadi Desert, is certified under the Responsible RAK program, recognizing its commitment to sustainable tourism. The resort implements eco-friendly practices such as renewable energy use, water recycling, and conservation initiatives. It actively protects native species within the Al Wadi Desert Reserve while offering educational programs for visitors. These efforts align with Ras Al Khaimah's vision for responsible, low-impact tourism development.

- Wadi Shees—Khorfakkan

Wadi Shees, located on the East Coast of the UAE, in the Emirate of Sharjah (Khorfakkan), has recently undergone significant ecological regeneration, making it a model for sustainable eco-tourism in the UAE. In collaboration with Emirates Nature-WWF and the Ministry of Climate Change and Environment, the site has been restored to enhance its natural biodiversity and support climate resilience. A key feature of this project is the restoration of the traditional falaj irrigation system, which plays a vital role in preserving both the ecological and cultural heritage of the area. By improving the water supply and supporting traditional agricultural practices, the restored falaj system helps maintain the sustainability of local ecosystems while familiarizing visitors with the UAE's rich heritage and traditions. The initiative highlights the importance of productive public-private partnerships in promoting environmental stewardship and ensuring the well-being of the UAE's people. Additionally, the development of the Shees Nature Trail offers visitors an immersive, educational experience, showcasing the successful integration of conservation and cultural preservation in eco-tourism.

3. Community Empowerment—Creating a Local Visitor Economy (LVE)

The Local Visitor Economy (LVE) within the broader tourism industry is commonly referred to the economic benefits generated by tourism within specific regions, including visitor spending on accommodation, food services, cultural activities, and locally produced goods. Focusing on the eco-tourism destinations discussed in this paper, they have strategically leveraged this model to diversify their economies and foster community-based development. Insights from the Ras Al Khaimah Tourism Development Authority (RAKTDA) highlight that eco-tourism and adventure tourism have contributed to a 17% increase in visitor numbers in 2023, with

Table 2 Summary of salient eco-tourism features of selected UAE destinations

Location	Key eco-tourism features	Sustainability initiatives	Cultural/environmental significance
Sheikh Zayed Desert Learning Centre - SZDLC AL AIN	– UAE's first building with platinum LEED, Estidama, and BREEAM ratings – Part of Al Ain UNESCO world heritage site – Environmental education exhibitions and interactive displays	– Renewable energy and water conservation systems – Sustainable building materials – Educational programs aligning with SDG 4 & SDG 15	– Promotes understanding of desert ecology – Enhances awareness of conservation challenges and sustainable practices
Green Mubazzarah Hot Springs AL AIN	– Geothermal hot springs and wellness tourism – Verdant parklands at the foothills of Jebel Hafeet – Popular destination for eco-recreation and nature-based activities	– Sustainable landscaping and water conservation practices – Promotion of geothermal energy use	– Historically significant site for traditional Emirati wellness practices using natural hot springs. – Gathering place for local communities, fostering social interaction and cultural celebrations
Al Wadi Desert Conservation Reserve RAS AL KHAIMAH	– Private nature reserve spanning over 500 hectares – Wildlife conservation experiences (Arabian oryx, gazelles, desert foxes) – Guided nature walks, bird watching	– EarthCheck-certified sustainability practices – Renewable energy and water recycling – Responsible RAK certification for eco-tourism excellence	– Protects native species and habitats – Demonstrates conservation-focused luxury tourism and low-impact travel
Wadi Shees KHORFAKKAN	– Ecological regeneration and biodiversity enhancement – Restored falaj irrigation system – Shees Nature Trail for immersive experiences	– Collaboration with Emirates Nature-WWF and Ministry of Climate Change and Environment – Climate resilience initiatives – Promotion of traditional agricultural practices	– Preserves cultural heritage through traditional irrigation – Supports local biodiversity and sustainable farming

nearly 40% of tourism spending directly benefiting local businesses and service providers. Similarly, in Al Ain, the Sheikh Zayed Desert Learning Centre (SZDLC) attracted over 500,000 visitors in 2023, generating significant revenue for local tour operators, cultural sites, and traditional markets and retail destinations in the vicinity of Al Ain.

These eco-tourism initiatives have also driven employment opportunities for local populations. In Ras Al Khaimah, the Al Wadi Desert Reserve supports over 300 direct and indirect jobs, including roles in conservation, hospitality, and local artisanal production. Meanwhile, Wadi Shees in Khorfakkan has seen increased community participation since its ecological regeneration, with approximately 60% of vendors and service providers being local residents, contributing to a sustainable and inclusive visitor economy.

By prioritizing local sourcing, community engagement, and cultural heritage experiences, these destinations highlight how eco-tourism can create resilient local economies while promoting environmental stewardship and cultural preservation.

Conclusion

The Sustainable Eco-tourism initiatives observed across Ras Al Khaimah, Al Ain, and Khorfakkan increasingly demonstrate the UAE's strategic shift toward sustainable and responsible tourism development. These destinations collectively highlight how environmental conservation, cultural preservation, and community engagement can be integrated into successful tourism models that align with international sustainability frameworks such as the Sustainable Development Goals (SDGs). The findings underline the critical role of public-private partnerships, environmental education, and low-impact infrastructure in promoting eco-tourism that benefits both local communities and urban ecosystems. As the UAE continues to diversify its tourism offerings beyond urban-centric experiences, these emerging eco-tourism hubs set a precedent for sustainable growth.

References

Blamey, R. (2001). Principles of ecotourism. In *The encyclopedia of ecotourism* (pp. 5–22). EarthCheck. (2022). Certification Programs and Case Studies. https://earthcheck.org

Epure, M., Barna, C., & Vasilescu, R. (2011). Ecotourism—Conservation of the natural and cultural heritage, 87–96.

Gössling, S., & Higham, J. (2021). The low-carbon imperative: Destination management under urgent climate change. *Journal of Travel Research, 60*(6), 1167–1179. https://doi.org/10.1177/0047287520933679

Honey, M. (2008). *Ecotourism and sustainable development: Who owns paradise?* (2nd ed.). Island Press.

RAKTDA (Ras Al Khaimah Tourism Development Authority). (2023). Annual tourism report.

Chapter 37
The Role of Mega-Events in Promoting National Commitment to Sustainability: Insights from the UAE's Capital City

Ana Stojković

Abstract This study investigates the influence of hosting mega-events on promoting governments' commitment to sustainability. Despite the significant number of studies focusing on the contribution of mega-events to a city's political, economic, and cultural development, there remains a lack of research exploring how these events can support the host city in positioning itself as a global leader in sustainability. Given the growing international concern regarding climate change, examining the role of events like Coldplay concerts in Abu Dhabi in promoting sustainability is crucial. Through qualitative content analysis, this study examines a range of primary and secondary sources, including government policy documents, press releases, and media reports. The findings suggest that mega-events can be a powerful tool for governments to frame their global agendas, such as the commitment to sustainability. In the case of the examined case study, the event was found to contribute to a broader strategic objective: to position the city as a leader in impactful, sustainable events and to solidify the legacy of Environment Vision 2030. These findings provide valuable insights into the potential influence of mega-events on promoting environmentally sustainable behaviours, with implications for event organisers and policymakers aiming to achieve sustainable development objectives.

Keywords Mega-events · Sustainability · Abu Dhabi

Introduction

Mega-events serve as powerful platforms through which cities assert their global presence and frame their global agendas. Roche defines mega-events as "large-scale cultural (including commercial and sporting) events, which have a dramatic character, mass popular appeal, and international significance" (Roche, 2000). Over the past

---

A. Stojković (✉)
Faculty of Political Sciences, University of Belgrade, Belgrade, Serbia
e-mail: stojkovi@eim.ae

J. Dsilva (ed.), *Sustainability Education, Youth and Innovation*,
https://doi.org/10.1007/978-3-032-21948-0_37

decade, the global visibility of mega-events has increased (Ollion, 2010), capturing the attention of city authorities and researchers alike.

Previous studies have primarily examined the economic costs and benefits associated with hosting mega-events (Kirton et al., 2010). Over time, sustainable development has become increasingly integrated into the objectives of hosting mega-events. It is now nearly inconceivable to discuss the planning or impact of such events without referencing sustainability in some form (Wolfe et al., 2024). The concept of sustainability has become a central part of common policy language and increasingly features in the discourse surrounding mega-events. It is now embedded in the policies of governments at all levels, industry bodies, private firms, and non-governmental organisations. It also plays an integral role in shaping the policy frameworks within which mega-events are planned and evaluated, particularly in terms of their contributions to sustainability.

Despite the growing emphasis on sustainability within the event industry (Whitfield et al., 2025), significant gaps persist in the academic literature, with important implications for both host cities and event organisers. In light of the escalating climate crisis, it is critical to prioritise the environmental sustainability of mega-events. This research seeks to address this gap by examining how mega-events can enhance a city's reputation as a leader in sustainability. In doing so, the study contributes to a deeper understanding of the role of mega-events in advancing sustainable development agendas.

The paper is organised thematically. It begins by defining mega-events and outlining their key dimensions, before examining their potential to serve as instruments for advancing sustainable practices. The discussion then turns to the broader implications for host cities. An extensive literature review follows, surveying existing research at the intersection of mega-event hosting and sustainability promotion. This section identifies dominant scholarly trends and proposes a preliminary evaluative framework through which to interpret the Abu Dhabi case study. The paper concludes with a detailed analysis of this case, offering insight into the practical application of sustainability principles in the context of a contemporary mega-event.

Mega-Events

Mega-events constitute influential platforms through which cities assert their global presence. Müller noted that the term mega-event was first introduced during the 37th Congress of the Association Internationale d'Experts Scientifiques du Tourism, which was organized under the theme "The role and impact of mega-events and attractions on national and regional tourism development" (Müller, 2015). The conference proceedings indicated that the purpose of the mega-event is primarily to attract tourists. Similarly, Wang and Jin observed that recent mega-events have increasingly focused on attracting attendees, with event size often measured by ticket sales (Wang and Jin, 2018). As Azzali noted, although the concept of mega-events has existed since the 1960s, such events have increasingly become central to strategies aimed at urban transformation, tourism promotion, and efforts to enhance international reputation (Azzali, 2019).

Over the past decade, the global visibility of mega-events has intensified (Ollion, 2010), drawing heightened attention from municipal authorities and scholars alike. Previous studies concerning the impacts of hosting international mega-events have primarily focused on their economic costs and benefits (Kirton et al., 2010). The typical economic benefits from hosting these events are (1), short-term stimulus of higher spending at hotels, restaurants and shops and the creation of temporary jobs; (2) longer-term economic benefits such as increased tourist traffic and investment due to media; (3) new, permanent, public infrastructures and upgrades and (4) training for security forces to prevent mass emergency events (Guebert & Tanna, 2010).

Among the major non-economic benefits of hosting these events are boosting favourable publicity for the host (Giffard & Rivenburgh, 2000), strengthening its global engagement and improving its international image (Cherry & Dobson, 2012). Additionally, mega-events can bring unparalleled international public attention to host cities. They provide 'vehicles for nation-states to construct or reshape specific brand identities and achieve certain foreign policy objectives' (Brannagan & Giulianotti, 2015), though there is no guarantee this will succeed.

It is reasonable to conclude that economic imperatives often motivate cities to host mega-events (Raj & Musgrave, 2009), and economic impacts remain one of the most extensively studied aspects of such events (Hall, 2012; Koba et al., 2019; Minnaert, 2012). Additionally, much of the existing research focuses on single case studies of sport mega-events, such as the Olympic Games and the FIFA World Cup (Brannagan & Giulianotti, 2015; Preuss, 2004; Manzenreiter, 2010; Hiller, 2000; Roche, 2000; Short, 2008).

However, as sustainability has become increasingly embedded in event-related discourse—particularly regarding the policy contexts in which such events operate and their contribution to broader sustainability objectives—this study examines how mega-events can enhance a city's reputation as a leader in sustainability. In doing so, it aims to deepen understanding of the role of mega-events in advancing sustainable development agendas.

Mega-Events and Sustainability

Sustainable development has become increasingly integrated into the objectives of hosting mega-events. For example, with the adoption of the Olympic Movement's Agenda 21 in 1999, the International Olympic Committee (IOC) incorporated sustainability principles into the organisation and implementation of the Games (International Olympic Committee, 1999). This institutionalisation has led mega-event organisers to embed sustainability into their planning and operations through various reform documents and progress reports (FIFA, 2015, 2020; International Olympic Committee, 2018; International Olympic Committee, 2021).

Similarly, Union of European Football Associations (UEFA) touted its commitments to sustainable planning and implementation (UEFA, 2021), and the Commonwealth Games Foundation emphasized sustainable cities and communities as part of its strategic priorities, harmonizing with the framework of the United Nations Sustainable Development Goals (Commonwealth Games Federation, 2022).

Thus, from a certain perspective, the modern era of mega-events can be understood as one of increasing concern with sustainability. Ideas of sustainable development have gradually become mainstream, including their integration into the planning and delivery of mega-events. These efforts gained traction in 2012 with the creation of the ISO 20121 Sustainable Event Management standard, developed as a legacy of the London 2012 Olympic Games. ISO 20121 has emerged as a key framework for sustainable event management, enabling organisations to embed sustainability into all aspects of event planning and execution.

These are just a few examples that illustrate the growing prominence of sustainability-oriented discourses in the hosting of mega-events. Indeed, it has become nearly inconceivable to engage in discussions about the planning or impacts of such events without incorporating considerations of sustainability.

Research Methodology

This research uses a case study approach, drawing on primary and secondary sources to examine how mega-events promote sustainable behaviour. Primary data consists of policy documents and official statements from government bodies and affiliated organisations. Only documents published by state authorities or their subsidiaries were included. These are analysed through content analysis and critically evaluated against secondary sources to contextualise claims and better understand the role of mega-events in promoting sustainable behaviour.

The research also draws on secondary sources, including journal articles, academic books, and media reports. Relevant literature was identified via Google Scholar and Scopus using terms like "mega-events" and "sustainability," with selection based on relevance and academic quality. Secondary data were analysed through content analysis to contextualise findings and explore how governments use mega-events to advance sustainability agendas.

Data were subjected to a thematic analysis through repeated readings, coding key content, identifying relationships among codes, and refining distinct themes. Each theme was then named to reflect its overarching narrative, informed by a review of relevant literature.

Historical Overview of the UAE'S Commitment to Sustainability

Sustainability is deeply embedded in the UAE's national identity, and this commitment originates from the legacy of the UAE's Founding Father, the late Sheikh Zayed bin Sultan Al Nahyan, and continues to guide current leadership in environmental stewardship (Sustainability in the United Arab Emirates, 2015). As stated by Dr. Amna bint Abdullah Al Dahak Al Shamsi, the UAE's Minister of Climate Change and Environment:

"Sustainability is part of the UAE DNA, and a legacy that continues to guide our path with clear vision and objectives. We consider sustainability a core value that not only shapes domestic policies but also positions the UAE as a global leader" (Emirates News Agency—WAM, 2014). Supporting this claim, the UAE recognises the urgent need for climate action and has demonstrated consistent leadership - most notably as the first Gulf Cooperation Council country to ratify the Paris Agreement, thereby setting a regional precedent for climate action (United Nations, 2015). The country has since formalised its commitment through instruments such as the National Climate Change Plan 2017–2050 (UAE Ministry of Climate Change & Environment, 2015) and the establishment of the National Committee on SDGs in January 2017 (The United Arab Emirates' Governmental portal, 2017). Further advancing its environmental agenda, the UAE launched the Net Zero by 2050 Strategic Initiative in 2021—the first of its kind in the MENA region— (The United Arab Emirates' Governmental portal, 2021), and formed the UAE Council on Climate Action in 2022 to ensure coherent governance across federal and emirate levels (UAE Ministry of Climate Change & Environment, 2022). Strategic initiatives like Abu Dhabi Vision 2030 (Abu Dhabi Urban Planning Council, 2007), Environment Vision 2030 (The United Arab Emirates' Government portal, 2024), and Estidama Pearl Rating System (Municipality of Abu Dhabi City, 2011) further reaffirm the UAE's long-standing commitment to sustainable development and bolster its regional leadership in this domain.

In the next paragraph, it will be further examined how mega-events are reinforcing the UAE's position as a leader in sustainable development.

Sustainable Development through Mega-Events: Insights from the UAE

The UAE, and particularly the Emirate of Abu Dhabi, strategically leverages mega-events as platforms to embed and showcase sustainability within its national branding agenda. A salient example of this strategy is the Coldplay Music of the Spheres World Tour, which took place in Abu Dhabi in January 2025. Drawing over 200,000 attendees, it became the UAE's largest music event to date (Mair, 2025), generating significant economic impact across the tourism, retail, and hospitality sectors. The event also enhanced cultural tourism, as international visitors engaged with iconic local attractions such as the Sheikh Zayed Grand Mosque, the Louvre Abu Dhabi, and Qasr Al Watan. As Florek et al. argue, the integration of mega-events with destination-specific cultural assets can amplify place branding outcomes by fostering deeper visitor engagement (Florek et al., 2008).

The economic and reputational impacts of such events are magnified through the so-called "halo effect," wherein positive spillovers are observed across multiple sectors. In certain Middle Eastern contexts, this effect has resulted in increases of up to 30% in tourism and hospitality activity (Hymers, 2022). For the UAE, this underscores the efficacy of mega-events as multi-dimensional policy tools that advance economic diversification while reinforcing global narratives around sustainability and innovation.

Coldplay's tour represents a distinctive convergence of entertainment and environmental values. The band committed to reducing the tour's direct emissions by over 50% compared to their 2016–17 tour and partnered with local stakeholders to mitigate environmental impact further (Coldplay, 2024). Sustainability measures included the deployment of solar power stations, kinetic dance floors that generate electricity through audience movement, and the provision of free water to reduce single-use plastics. The tour's routing strategy prioritized ecological considerations over commercial convenience, using ground transport powered by biofuels and electric vehicles where possible and adopting a regional "hub" approach to minimize air travel (Saeed, 2025).

Abu Dhabi's selection as the regional anchor for the tour was influenced not only by its infrastructure but also by its geographic accessibility. Positioned within a six-hour flight of nearly two-thirds of the global population, the Emirate offered an ideal location to reduce the tour's carbon footprint. The alignment between Coldplay's environmental goals and Abu Dhabi's Climate Change Strategy (Government of Abu Dhabi, 2023) provided a coherent framework for the event's sustainability agenda.

Beyond infrastructural capacity, the concert incorporated a wide range of sustainability practices: gamified sustainability dashboards, accessible concert experiences such as sensory bags and touch tours, recyclable wristbands, and real-time data on energy generation at each location.

The coherence between Coldplay's brand identity and Abu Dhabi's environmental aspirations reflects a broader strategic alignment. Nevertheless, the complexities of achieving fully sustainable events remain considerable. Challenges include emissions from international logistics, material sourcing, and stakeholder coordination. However, this example illustrates how mega-events, when carefully integrated with sustainability strategies, can serve as effective instruments for environmental leadership.

Findings

The study demonstrates that mega-events serve as a powerful instrument for governments to advance their global agendas, particularly concerning sustainability. In the case examined—Coldplay's concerts in Abu Dhabi—the event contributed to a broader strategic objective: positioning Abu Dhabi as a leader in hosting sustainable events and reinforcing the legacy of Environment Vision 2030. Coldplay's concerts illustrate the potential of mega-events to model sustainable practices—through emissions reduction, renewable energy use, eco-conscious routing, and inclusive audience engagement—while simultaneously supporting the UAE's climate goals. The event also contributed to the local economy and cultural diplomacy by attracting international visitors and enhancing Abu Dhabi's reputation as a global sustainability hub.

In conclusion, the study argues that while sustainability in mega-events remains an aspirational goal, such events—when strategically aligned with broader policy frameworks—can serve as effective platforms for reinforcing environmental commitments, shaping global narratives, and catalyzing systemic change in host cities. These findings yield important insights for the fields of sustainability and event management and offer practical implications for integrating sustainable behaviour into urban policy agendas.

Conclusion

This study aims to enhance understanding of how mega-events can contribute to the fulfilment of governmental sustainability commitments. The Coldplay concerts held in Abu Dhabi serve as a case study to examine the potential of such events to advance sustainability agendas.

Findings indicate that this event contributes to a broader strategic mission: positioning Abu Dhabi as a leader in hosting impactful, sustainable events and reinforcing the legacy of Environment Vision 2030. Furthermore, hosting mega-events alongside investments in urban infrastructure designed to accommodate a sustainability-oriented "visitor class" can enable cities to enhance their leadership in sustainability. These insights offer valuable guidance for event organizers and policymakers striving to align mega-event planning with sustainable development goals.

The study also enriches the broader scholarly discourse on mega-events and sustainability, providing a foundation for future research. It holds benchmarking potential, encouraging comparative analyses of similar strategies across different national contexts. While this research deepens understanding of mega-events' role in promoting sustainability commitments, it does not address other critical impacts of hosting such events, including urban security challenges, cultural benefits and drawbacks, and social tensions. Consequently, further research is necessary to comprehensively assess these multifaceted effects.

References

Abu Dhabi Urban Planning Council. (2007, January). *Plan Abu Dhabi 2030: Urban structure framework plan*. https://andp.unescwa.org/sites/default/files/2020-10/Plan%20Abu%20Dhabi%202030-Urban%20Structure%20Framework%20plan.pdf.

Azzali, S. (2019). Challenges and key factors in planning legacies of mega sporting events. *International Journal of Architectural Research Archnet-IJAR, 14*(2), 203–218. https://doi.org/10.1108/ARCH-04-2019-0093

Brannagan, P. M., & Giulianotti, R. (2015). Soft power and soft disempowerment: Qatar, global sport and football's 2022 world cup finals. *Leisure Studies, 34*(6), 703–719.

Cherry, J., & Dobson, H. (2012). Seoul-searching: The 2010 G-20 Seoul summit. *Global Governance A Review of Multilateralism and International Organizations, 18*(3), 363–381. https://doi.org/10.1163/19426720-01803007

Coldplay. (2024). Retrieved January 28, 2025, from https://sustainability.coldplay.com

Commonwealth Games Federation. (2022). *Commonwealth sport strategy 2023–2034*. https://resources.cwg-qbr.pulselive.com/qbr-commonwealth-games/document/2025/01/14/d1d43b92-b657-4b12-b787-ce8bf2531367/CGF-Strategic-Plan-2023-2034.pdf

Emirates News Agency–WAM. (2014, November 16). *NYU Abu Dhabi to hold conference on mega sports events in UAE*. https://www.wam.ae/en/article/hsz97x4j-nyu-abu-dhabi-hold-conference-mega-sports-events

FIFA. (2020, August). *FIFA world cup Qatar 2022 sustainability strategy*. https://digitalhub.fifa.com/m/5adbe651c67c78a3/original/o2zbd8acyiooxyn0dwuk-pdf.pdf

FIFA. (2015, July). *Sustainability strategy 2018 FIFA world cup*. https://www.icsspe.org/system/files/FIFA%20-%20Sustainability%20Strategy%20for%202018%20FIFA%20World%20Cup.pdf

Florek, M., Breitbarth, T., & Conejo, F. (2008). Mega event=mega impact? Travelling fans' experience and perceptions of the 2006 FIFA world cup host nation. *Journal of Sport and Tourism, 13*(3), 199–219.

Giffard, A. C., & Rivenburgh, N. K. (2000). News agencies, national images, and global media events. *Journalism and Mass Communications Quarterly, 77*(1), 8–21. https://doi.org/10.1177/107769900007700101

Government of Abu Dhabi. (2023). *Abu Dhabi climate change strategy*. https://www.abudhabi.gov.ae/en/programmes/abu-dhabi-climate-change-strategy

Guebert, J., Tanna, S. (2010, June 14). *Economic benefits of hosting G8 and G20 summits. University of Toronto*. Retrieved March 4, 2025, from https://g7.utoronto.ca/evaluations/factsheet/factsheet_ecobenefits.pdf

Hall, C. M. (2012). Sustainable mega-events: Beyond the myth of balanced approaches to mega-event sustainability. *Event Management, 16*(2), 119–131. https://doi.org/10.3727/152599512X13343565268294

Hiller, H. H. (2000). Mega-events, urban boosterism and growth strategies: An analysis of the objectives and legitimations of the Cape Town 2004 Olympic bid. *International Journal of Urban and Regional Research, 24*(2), 449–458.

Hymers, S. H. (2022, September 29). *The region's sports tourism industry could generate as much as US$600bn*. Connecting Travel. Retrieved January 27, 2025, from https://www.connectingtravel.com/news/middle-east-becoming-the-worlds-fastest-growing-sports-tourism-destination.

International Olympic Committee. (2018, February). *Olympic agenda 2020. Olympic Games: The New Norm*. https://stillmed.olympic.org/media/Document%20Library/OlympicOrg/News/2018/02/2018-02-06-Olympic-Games-the-New-Norm-Report.pdf

International Olympic Committee. (2021, March). *Olympic Agenda 2020+5: 15 recommendations*. https://stillmed.olympics.com/media/Document%20Library/OlympicOrg/IOC/What-We-Do/Olympic-agenda/Olympic-Agenda-2020-5-15-recommendations.pdf

International Olympic Committee. (1999, October). *Olympic movement's Agenda 21: Sport for sustainable development*. https://stillmed.olympic.org/media/Document%20Library/OlympicOrg/Documents/Olympism-in-Action/Environment/Olympic-Movement-s-Agenda-21.pdf
Kirton, J., Guebert, J. & Tanna, S. (2010, July 5). *G8 and G20 summit costs*, Munk school for global affairs, University of Toronto. Retrieved March 2, 2025, from https://g7.utoronto.ca/evaluations/factsheet/factsheet_costs.pdf
Koba, T., Gong, H., Ross, W. J., & Grady, J. (2019). Sustainable Olympic development: A proposed benchmark for managing economic outcomes. *Journal of Global Sport Management, 6*(4), 1–21. https://doi.org/10.1080/24704067.2019.1565925
Mair, F. (2025, January 15). Coldplay in Abu Dhabi officially the biggest concert series in UAE history. *TimeOut*. https://www.timeoutdubai.com/news/coldplay-biggest-concert
Manzenreiter, W. (2010). The Beijing games in the Western imagination of China: The weak power of soft power. *Journal of Sport and Social Issues, 34*(1), 29–48. https://doi.org/10.1177/0193723509358968
Minnaert, L. (2012). An Olympic legacy for all? The non-infrastructural outcomes of the Olympic games for socially excluded groups (Atlanta 1996–Beijing 2008). *Tourism Management, 33*(2), 361–370. https://doi.org/10.1016/j.tourman.2011.04.005
Müller, M. (2015). What makes an event a mega-event? Definitions and sizes. *Leisure Studies, 34*(6), 627–642. https://doi.org/10.1080/02614367.2014.993333
Municipality of Abu Dhabi City. (2011, June 27). *Estidama @ Municipality of Abu Dhabi City*. https://www.dmt.gov.ae/adm/-/media/Project/DMT/ADM/E-Library/Estidama.pdf
Ollion, E. (2010). Summits of information: Advocacy officers, journalists, and newsmaking at international summits. *Qualitative Sociology, 33*(3), 211–227. https://doi.org/10.1007/s11133-010-9163-z
Preuss, H. (2004). *The economics of staging the Olympics: A comparison of the games, 1972–2008*. EE Publishing
Raj, R., & Musgrave, J. (2009). The economics of sustainable events. In R. Rak & J. Musgrave (Eds.), *Event management and sustainability* (Vol. 1, pp. 56–65). CABI Digital Library. https://doi.org/10.1079/9781845935245.0056
Roche, M. (2000). *Mega-events and modernity: Olympics and expos in the growth of global culture*. Routledge.
Saeed, S. (2025, January 09). *Coldplay's leaner and cleaner tour: Why band chose to come to Abu Dhabi*. The National. https://www.thenationalnews.com/arts-culture/music-stage/2025/01/09/coldplay-abu-dhabi-tour-routes-sustainability/
Short, J. R. (2008). Globalization, cities and the summer Olympics. *City: Analysis of Urban Change, Theory, Action, 12*(3), 321–340.
Sustainability in the United Arab Emirates. (2015). *Founding father's sustainability legacy*. https://www.sustainableuae.ae/en/founding-fathers-legacy.html
The United Arab Emirates' Government portal. (2024, May). *Environment Vision 2030 (Abu Dhabi)*. https://u.ae/en/about-the-uae/strategies-initiatives-and-awards/strategies-plans-and-visions/environment-and-energy/environment-vision-2030-abu-dhabi
The United Arab Emirates' Governmental portal. (2017). *The UAE'S National Committee on SDGs*. https://u.ae/en/about-the-uae/leaving-no-one-behind/uae-and-the-sdgs/the-uae-national-committee-on-sdgs
The United Arab Emirates' Governmental portal. (2021). *The UAE'S net zero 2050 strategy*. https://u.ae/en/about-the-uae/strategies-initiatives-and-awards/strategies-plans-and-visions/environment-and-energy/the-uae-net-zero-2050-strategy
UAE Ministry of Climate Change & Environment. (2015). *National Climate Change Plan of the United Arab Emirates 2017–2050*. https://www.moccae.gov.ae/assets/download/d3d90eba/National%20Climate%20Change%20Plan.pdf.aspx

UAE Ministry of Climate Change & Environment. (2022, March 24). *UAE Council for Climate Change and Environment Holds First Meeting of 2022.* https://www.moccae.gov.ae/en/media-center/news/24/3/2022/uae-council-for-climate-change-and-environment-holds-first-meeting-of-2022.aspx#page=1

UEFA. (2021). *UEFA football sustainability strategy 2030.* https://editorial.uefa.com/resources/0270-13f888ffa3e5-931c597968cb-1000/uefa_football_sustainability_strategy.pdf

United Nations. (2015, December 12). *Paris Agreement.* https://unfccc.int/sites/default/files/resource/parisagreement_publication.pdf

Wang, Y., & Jin, X. (2019). Event-based destination marketing: The role of mega-events. *Event Management, 23*(1), 109–118. https://doi.org/10.3727/152599518X15378845225384

Whitfield, J., Gouthro, M. B., & Moital, M. (Eds.). (2025). *The Routledge handbook of events and sustainability.* Routledge.

Wolfe, S. D., Gogishvili, D., & Muller, M. (2024). Mega-events and triple-baseline (un)sustainability. In H. A. Solberg, R. K. Storm, & K. Swart (Eds.), *Research handbook on major sporting events* (pp. 399–414). Edward Elgar Publishing. https://doi.org/10.4337/9781800885653.00039

Chapter 38
Greenwashing in Corporate Communications in the UAE: Uncovering the Dynamics of a Non-Regulated Space

Noushin Bagheri and Nada Sayarh

Abstract Greenwashing is increasingly becoming a global concern, as corporations make misleading environmental claims to attract eco-friendly consumers. This study aims to explore the prevalence and impact of greenwashing in corporate communications with the UAE, a country with a diverse consumers base and growing sustainability awareness. The research employs a qualitative approach using semi-structured interviews with sustainability experts across industries in the UAE. The findings indicate that green washing is driven by incorporate transparency, and educating consumers are essential to mitigate the issue. The study concluded with recommendations for addressing greenwashing through policy and consumer driven initiatives.

Keywords Greenwashing · Communications · Sustainability · Environmental

Introduction

Greenwashing is defined as "communication that misleads people regarding environmental performance by disclosing negative information and disseminating positive information about an organization, service, or product" Tateishi (2017). The increasing emphasize on sustainability reporting and the impact of corporate actions on environmental, social and governance (ESG) criteria's have contribute to the rise of misleading information, commonly referred to as greenwashing (Freitas et al., 2020). As highlighted by Longoni and Cagliano (2015) sustainability is the greatest challenge confronting today's manufacturing firms. The adverse market reaction should be worrying, as nearly 90% of investors view a sustainability strategy as essential for firms to remain competitive (Unruh et al., 2016).

N. Bagheri · N. Sayarh (✉)
SP Jain School of Global Management, Dubai, UAE
e-mail: Nada.sayarh@spjain.org

N. Bagheri
e-mail: Noushin.bagherin@spjain.org

J. Dsilva (ed.), *Sustainability Education, Youth and Innovation*,
https://doi.org/10.1007/978-3-032-21948-0_38

The UAE is home to more than 200 nationalities, each bringing diverse cultural perspectives on sustainability. In the UAE, sustainability is highly valued, with 81% of consumers recognizing the impact of climate change and 65% willing to adopt eco-friendly practices (Press Release, 2022). Many consumers prefer eco-friendly products due to their upbringing in environmentally conscious cultures, further driving the demand for genuine green initiatives. While this encourages corporations to adopt green strategies and provide eco-friendly products, it also raises concerns that companies may exaggerate their sustainability claims to appeal to consumers. According to YouGov (2022) survey results, 55% of consumers are skeptical of most brand's sustainability claims, aligning with global concerns about greenwashing. This skepticism highlights the urgent need for stronger corporate transparency and regulatory oversight in the UAE. Moreover, the country's position as global business hub makes it a crucial case for examining how regulatory framework influence corporate sustainability efforts.

Research Objectives

The main research objectives of this study are:

To uncover the elements of Greenwashing with the context of the UAE across industries.

To Identify the key strategies that can effectively reduce greenwashing in the UAE.

Literature Review

Institutional Theory and Corporate Legitimacy

Institutional theory posits that corporations operate within a larger institutional framework shaped by social and cultural influences (Scott, 1987). As dynamic entities, businesses continuously adapt to external expectations to establish legitimacy within their respective industries. Legitimacy is seen as "a process whereby organizations seek approval for their acts from groups in society" (Kaplan & Ruland, 1991, pp. 320). From this perspective, corporations seeking legitimacy tend to conform to prevailing environmental norms and societal expectations, resulting in structural and operational similarities across industries (Starik & Marcus, 2000).

Greenwashing and its Impact on Consumer Trust

Greenwashing refers to deceptive corporate practices that mislead consumers regarding a company's environmental performance. When corporations engage in greenwashing, they obscure transparency with their customers, making it difficult for them to make well-informed purchasing decisions, particularly regarding the environmental impact of products or services (Horiuchi et al., 2009). Moodaley and Telukdarie (2023) highlighted that greenwashing undermines sustainability efforts, drawing attention from industries and regulatory agencies as they seek to address its consequences. Additionally, Nyilasy et al. (2014) stated that greenwashing has negatively impacted consumer trust, further reinforcing skepticism toward corporate environmental claims.

Environmental Communication and Transparency

Environmental communication refers to the sharing of information about the impact of a company's operations on the natural or physical environment (Wilmshurst & Frost, 2000). Transparency in environmental communication is crucial, as it helps organizations respond to institutional pressure while strengthening their legitimacy, reputation, and stakeholder trust (Patten, 2002). Positive environmental disclosure can reduce consumer skepticism and mitigate corporate liability concerns (Cho et al., 2009). However, misleading or exaggerated sustainability claims can have the opposite effect, damaging corporate credibility and fostering distrust among stakeholders.

The Role of Social Media in Greenwashing Accountability

The effectiveness of corporate environmental disclosures presents both opportunities and challenges. While stakeholders may appreciate genuine sustainability efforts, they also remain skeptical of potential greenwashing motives (Vries et al., 2015; Yoon et al., 2006). The rise of social media has created a double-edged communication platform where greenwashing messages can be amplified, but consumers can also hold corporations accountable for their sustainability claims. Social media allows educated consumers to scrutinize brands more closely, demanding greater transparency and ethical responsibility in corporate sustainability initiatives.

Methodology

This exploratory study employs a qualitative research approach, utilizing semi-structured interviews to gather insights on greenwashing in the UAE. Data was collected from 10 sustainability experts across various industries in the UAE. Each interview lasted between 45 min to 1 h, with a set of structured questions guiding the discussion. Follow-up questions were asked as needed to gain deeper insights. The respondents were analyzed using thematic analysis to identify key patterns and themes.

Findings

Based on findings from the thematic analysis shown in (Table 38.1), we found that greenwashing remains a significant challenge in the UAE, driven by inconsistent regulations and poor consumer awareness. While sustainability is culturally valued, weak enforcement allows misleading claims to persist. Strengthening regulatory frameworks, enhancing transparency in corporate communication, and educating consumers are key to mitigating greenwashing.

Although consumers recognize the term "greenwashing", many lack a clear understanding of how it occurs, which creates disparities in verified communication at the corporate level. The government is perceived as a key player that should establish standardized regulations, conducting regular audits, and offering incentives for responsible corporate behavior. These measures should be widely communicated through various channels and integrated into corporate policies from the outset to ensure long-term accountability and trust.

Conclusions

The increasing demand for sustainability in the UAE has prompted corporations to adopt more sustainable products. However, this rising emphasis on sustainability and green practices has led some corporations to engage in greenwashing, in an attempt to retain their customers and enhance their brand image. This situation underscores the urgent need for action from researchers and policy-makers and desicion-makers. The findings of this study emphasize the need for stronger regulatory frameworks, corporate accountability and consumer education to reduce greenwashing. To address these challanges, policy makers should implement standardized regulations, conduct frequent audits, and introduce penalties for misleading environmental claims. Corporations must adopt transparent sustainability communication, while consumers should be empowered with knowledge to critically assess green claims.

Table 38.1 Thematic analysis

Theme	Description
Trust	Trust is key to building long-term consumer relationships, especially through transparency and ethical practices.
Corporate motivations for green washings	Companies may engage in greenwashing due to competitive pressure, cost saving, consumer demand for sustainability and brand images.
Consumer awareness	Educated consumers are scrutinizing brands more closely and demanding transparency, using tools like social media to hold companies accountable.
Greenwashing	Greenwashing tactics are widespread, and there's a growing need for stronger regulations to hold brands accountable for misleading claims.
Transparency in advertisement	Honest and open communication from brands is essential in building consumer trust and avoiding reputational damage.
Environmental impact and sustainability	Collaborations between brands and environmental organizations are pushing for real sustainability innovation and environmental action.
Regulatory influence	Effective regulatory frameworks are crucial for ensuring accountability and curbing greenwashing. Quick legislative actions, especially in regions like the UAE, are seen as advantageous.
Consumer-drive market change	Consumer have the power to influence corporate behavior by choosing ethical and sustainable brands.

The results of this study confirmed that transparency and accountability, business in the UAE can build stronger consumer trust and contribute to genuine sustainability effort, ensuring long term credibility. This study focused on industry experts rather than consumers, which is a limitation of this study. Interviewing consumers would provide a deeper understanding of how greenwashing influences their purchasing decisions and explore consumer awareness, expectations, and the criteria they use to identify misleading sustainability claims. Understanding how consumers recognize and respond to greenwashing could provide more actionable insights for policymakers and businesses.

References

De Vries, G., Terwel, B., Ellemers, N., & Daamen, D. (2015). Sustainability or profitability? How communicated motives for environmental policy affect public perceptions of corporate greenwashing. *Corporate Social Responsibility and Environmental Management, 22*(3), 142–154.

Freitas Netto, S. V., Sobral, M. F. F., Ribeiro, A. R. B., & Soares, G. R. L. (2020). Concepts and forms of greenwashing: A systematic review. *Environmental Sciences Europe, 32*, 19.
Horiuchi, R., Schuchard, R., Shea, L., & Townsend, S. (2009). *Understanding and preventing greenwash: A business guide*. Futerra Sustainability Communications.
Kaplan, S. E., & Ruland, R. G. (1991). Positive theory, rationality and accounting regulation. *Critical Perspectives on Accounting, 2*(4), 361–374.
Longoni, A., & Cagliano., R. (2015). Environmental and social sustainability priorities: Their integration in operations strategies. *International Journal of Operations & Production Management*.
Moodaley, W., & Telukdarie, A. (2023). Greenwashing, sustainability reporting, and artificial intelligence: A systematic literature review. *Sustainability, 15*, 1481.
Nyilasy, G., Gangadharbatla, H., & Paladino, A. (2014). Perceived greenwashing: The interactive efects of green advertising and corporate environmental performance on consumer reactions. *Journal of Business Ethics, 125*(4), 693–707. https://doi.org/10.1007/s10551-013-1944-3
Patten, D. M. (2002). Media exposure, public policy pressure, and environmental disclosure: An examination of tri data availability. *Accounting Forum, 26*(2), 152–171.
Press Release. (2022, November 30). *GCC survey: Consumers willing to pay more for sustainability-minded companies*. ACE Times. https://www.zawya.com/en/press-release/research-and-studies/gcc-survey-consumers-willing-to-pay-more-for-sustainability-minded-companies-vgmjmv6x
Scott, R. (1987). The adolescence of institutional theory. *Administrative Science Quarterly, 32*(4), 493–511.
Starik, M., & Marcus, A. A. (2000). Introduction to the special research forum on the management of organization in the natural environment: A field emerging from multiple paths, with many challenges ahead. *Academy of Management Journal, 43*(4), 539–546.
Tateishi, E. (2017). Craving gains and claiming "green" by cutting greens? An exploratory analysis of greenfield housing developments in Iskandar Malaysia. *Journal of Urban Affairs, 40*(3), 370–393. https://doi.org/10.1080/07352166.2017.1355667
Unruh, G., Kiron, D., Kruschwitz, N., Reeves, M., Rubel, H., Felde, M. Z., & A. (2016). Investing for a sustainable future: Investors care more about sustainability than many executives believe. *MIT Sloan Management Review, 57*, 1–29.
Wilmshurst, T. D., & Frost., G. R. (2000). Organizational environmental reporting: A test of legitimacy theory. *Accounting, Auditing & Accountability Journal, 13*(1), 10–26.
Yoon, Y., Gürhan-Canli, Z., & Schwarz, N. (2006). The effect of corporate social responsibility (CSR) activities on companies with bad reputations. *Journal of Consumer Psychology, 16*(4), 377–390.
YouGov. (2022, November 21). https://business.yougov.com/content/46714-more-than-half-of-global-consumers-are-skeptical-of-a-brands-sustainability-claims

Part VII
Sustainability Education for All

Chapter 39
Education for Sustainability: The Case Study of Nepal

Filipe Duarte and Pradipta Kadambari

Abstract The challenges of achieving sustainability are substantial. While political and societal will are essential, meaningful progress also requires significant financial investment and a strong educational commitment. This study explores the role of education in advancing sustainability by presenting findings from a two-year research project on the engagement of international and national non-governmental organizations (INGOs and NGOs) with the Sustainable Development Goals (SDGs) in Nepal. Using an interpretative research design, the study involved 20 semi-structured interviews with heads of INGOs/NGOs whose work is aligned with the SDGs. The findings reveal a notable gap between the participants' professional identity (educational/training background), their professional role and area(s) of intervention, their understanding of the United Nations (UN) SDGs, and the education/training needs related to them. Three key themes emerged: (1) challenges and opportunities in engaging with the SDGs, (2) the need for targeted SDG localization training, and (3) the urgency of integrating the SDGs into local higher education curricula. The study highlights the importance of transformative educational strategies to advance sustainability learning and practice. It offers valuable insights for scaling sustainability education not only in Nepal but also in global contexts, including in emerging knowledge hubs such as the United Arab Emirates (UAE).

Keywords Education · SDGs · Sustainability · Sustainable development · Nepal

F. Duarte (✉)
Universidade Católica Portuguesa, Lisbon, Portugal
e-mail: cfduarte@ucp.pt

P. Kadambari
Kadambari Memorial College, Kathmandu, Nepal
e-mail: pradiptakadambari@gmail.com

J. Dsilva (ed.), *Sustainability Education, Youth and Innovation*,
https://doi.org/10.1007/978-3-032-21948-0_39

Introduction

In 2015, all United Nations (UN) member states adopted the 2030 Agenda for Sustainable Development (UN, 2015). The sustainable development goals (SDGs) are a blueprint comprising 17 goals and 169 targets to be met over 15 years by 2030. The SDGs offer a holistic vision of society and a normative and ethical perspective, identifying goals for what a "good society" should be, a society that should be prosperous, inclusive, environmentally sustainable and well-governed (Sachs, 2015). They represent a global political mapping of sustainability and sustainable development (Le Blanc, 2015). The SDGs were, in fact, a transition from the Millennium Development Goals (MDGs) to sustainable development through a common global agenda with a set of universally applicable goals to commit rich and poor countries alike (Sachs, 2012, 2015).

The challenges for sustainability are enormous. The ties to investment make progress towards sustainability and sustainable development a very ambitious target (SDSN, 2024). Political and societal will is key, however, progress towards the SDGs requires both substantial financial investment and educational commitment from higher-education institutions (Aramburuzabala & Cerrillo, 2023; UN, 2024; Žalėnienė & Pereira, 2021).

This global call to action offers a unique and coordinated policy framework to address both global and local challenges faced by societies (Holden, Linnerud, & Banister, 2017). Linking the UN's global goals to national and local-driven agendas and engaging local/national stakeholders' collaboration is imperative to achieve the goals and targets envisioned by 2030 (Kawakubo & Murakami, 2020). Higher education institutions play a crucial role in embedding sustainability principles into curriculum design, stakeholder engagement, and knowledge transfer (Leal Filho et al., 2018b; Žalėnienė & Pereira, 2021). These institutions also serve as hubs for innovative pedagogies, such as service-learning, which foster civic-driven learning and practical sustainability outcomes (Aramburuzabala & Cerrillo, 2023).

The Sustainable Development Goals (SDGs) offer a global framework for collective action "for people, planet, prosperity, peace, and partnerships" (UN, 2015). As Sachs (2012, 2015) argues, the transition from the Millennium Development Goals to the SDGs reflects a shift toward a more integrated and multidimensional understanding of sustainable development. This research adopts the SDGs as its primary entry point for addressing the principle of Leaving No One Behind (Kharas, McArthur and Ohno, 2020; UN, 2024), particularly within contexts marked by structural and educational inequalities. Grounded in systems thinking and place-based approaches (Brychkov et al., 2024; Grenni, Soini and Horlings, 2020; Meadows and Wright, 2008; Stroh, 2015; Wilbanks, 2015), this study investigates how INGOs and NGOs in Nepal engage with the SDGs. Beyond treating the SDGs as aspirational goals, this research follows Leal Filho et al. (2018b) in framing them as practical instruments for institutional transformation. The study focuses specifically on the educational and training dimensions of sustainability implementation in development practice. Findings reveal significant capacity-building gaps and highlight the need for transformative educational strategies. These insights are relevant not only for development practitioners but also for higher education institutions aiming to strengthen their role in sustainable development through curriculum innovation, partnerships, and civic engagement.

Higher education institutions, particularly in low- and middle-income countries, are increasingly recognized as key actors in driving sustainability transitions. The Nepal case study contributes valuable insights by emphasizing the role of education in fostering not just awareness of the SDGs, but also local ownership and the capacity to implement them effectively. These findings also offer relevance for other rapidly developing contexts, such as the United Arab Emirates (UAE), where higher education plays an expanding role in shaping sustainability governance and advancing national development priorities.

Methodology

Drawing from systems thinking and placed-based approaches, given the exploratory and interpretative nature of this inquiry, a qualitative study design was adopted to flesh out the understanding and engagement experiences of participants with the SDGs, representing INGOs and NGOs in Kathmandu, Nepal, at the time of data collection (Grenni Soini & Horlings, 2020; Meadows & Wright, 2008; Miles & Huberman, 1994; Srivastava & Hopwood, 2009; Stroh, 2015; Walsham, 2006; Wilbanks, 2015). Through an interpretative approach, the authors aimed to have in-depth access to key participants, issues and data related to the intervention/practices and education/training related to the SDGs. Interpretative research allows one to construct a particular reality through the eyes of other people's constructions (Walsham, 2006). Thus, the qualitative and interpretative approach informed the data collection method, procedures and techniques (Bazeley, 2009; Miles & Huberman, 1994). Data collection consisted of 20 in-depth interviews with heads of INGOs/NGOs. The selection criteria was based on areas of intervention within eight selected SDG areas (poverty, zero hunger, good health and well-being, quality education, gender equality, reduced inequalities, sustainable cities and communities, and climate change). Ethical issues were taken into account. Ethics approval was granted in 2022 by a university research ethics board from one of the authors at the time of data collection. In terms of confidentiality and anonymity, both authors were not involved in any stage of data collection to protect participants from any risks arising from their voluntary participation in the research project.

Findings

All 20 participants expressed a strong interest in expanding their knowledge and skills related to sustainability and sustainable development. As professionals in the international development sector, they emphasized the critical role of higher education institutions in advancing transformative learning—an observation well supported in the literature (Aramburuzabala and Cerrillo, 2023; Leal Filho et al., 2018a; Žalėnienė and Pereira, 2021). Participants consistently called for SDG localization training, identifying it as key to adapting global sustainability goals to Nepal's diverse development contexts. Several also stressed "the importance of reflective and analytic education to work more professionally towards the SDGs." Many highlighted the urgent need to embed sustainability principles into school and university curricula, with one participant noting that "education develops and builds professional behaviour for social change." These findings align with recent research highlighting the institutional impact of SDG-driven education (Leal Filho et al., 2025b), particularly within higher education settings that engage multiple stakeholders and foster place-based learning (Leal Filho et al., 2018a). This Nepal-based case study underscores the link between development practice, SDG program delivery, and the transformative potential of sustainability education. It provides timely insights for universities aiming to deepen their engagement with the SDGs through curriculum design, professional training, and civic-oriented pedagogies (Aramburuzabala and Cerrillo, 2023; Brychkov et al., 2024; Leal Filho et al., 2025a).

Conclusion

Integrating sustainability and sustainable development into higher education curricula is a global priority aligned with the ambitions of the UN 2030 Agenda. This study reveals that professionals already engaged in development practice demonstrate considerable pedagogical needs, particularly in acquiring the knowledge and skills necessary to effectively implement sustainability principles. While case study research carries inherent limitations—such as geographic scope, sample size, and cultural specificity—systems thinking offers a flexible framework for exploring diverse methodologies and pedagogical strategies in sustainability education. To deepen our understanding, future research should investigate the specific competencies required for sustainability across varied geographic, educational, economic, cultural, and religious contexts. Such comparative insights will be essential for informing curriculum design and advancing transformative education for sustainability and sustainable development.

References

Aramburuzabala, P., & Cerrillo, R. (2023). Service-learning as an approach to educating for sustainable development. *Sustainability, 15*(14), 11231. https://doi.org/10.3390/su151411231

Bazeley, P. (2009). Analysing qualitative data: More than 'identifying themes'. *The Malaysian Journal of Qualitative Research, 2*(1), 6–22.

Brychkov, D., McKeown, P. C., Domegan, C., Spillane, C., & Brychkova, G. (2024). "Connect the circle" systems thinking tool for postgraduate sustainability education: Case study. *International Journal of Sustainability in Higher Education, 25*(9), 437–454. https://doi.org/10.1108/IJSHE-10-2023-0507

Grenni, S., Soini, K., & Horlings, I. (2020). The inner dimension of sustainability transformation: How sense of place and values can support sustainable place-shaping. *Sustainability Science, 15*(2), 411–422. https://doi.org/10.1007/s11625-019-00743-3

Holden, E., Linnerud, K., & Banister, D. (2017). The imperatives of sustainable development. *Sustainable Development, 25*(3), 213–226. https://doi.org/10.1002/sd.1647

Kharas, H., McArthur, J. W., & Ohno, I. (2020). Getting specific to leave no one behind on sustainable development. In H. Kharas, J. W. McArthur, & I. Ohno (Eds.), *Leave no one behind: Time for specifics on the sustainable development goals* (pp. 1–19). Brookings Institution Press.

Kawakubo, S., & Murakami, S. (2020). Development of the local SDGs platform for information sharing to contribute to achieving the SDGs. IOP conference series. *Earth and environmental. Science, 588*(2), 22019. https://doi.org/10.1088/1755-1315/588/2/022019

Leal Filho, W., Raath, S., Lazzarini, B., Vargas, V. R., de Souza, L., Anholon, R., Quelhas, O. L. G., Haddad, R., Klavins, M., & Orlovic, V. L. (2018). The role of transformation in learning and education for sustainability. *Journal of Cleaner Production, 199*, 286–295. https://doi.org/10.1016/j.jclepro.2018.07.017

Leal Filho, W., Sigahi, T. F. A. C., Anholon, R., Rebelatto, B. G., Schmidt-Ross, I., Hensel-Börner, S., Franco, D., Treacy, T., & Brandli, L. L. (2025). Promoting sustainable development via stakeholder engagement in higher education. *Environmental Sciences Europe, 37*(1), 1–20. https://doi.org/10.1186/s12302-025-01101-0

Leal Filho, W., Tripathi, S. K., Andrade Guerra, J. B. S. O. D., Giné-Garriga, R., Orlovic Lovren, V., & Willats, J. (2018). Using the sustainable development goals towards a better understanding of sustainability challenges. *International Journal of Sustainable Development & World Ecology, 26*(2), 179–190. https://doi.org/10.1080/13504509.2018.1505674

Leal Filho, W., Viera Trevisan, L., Sivapalan, S., Mazhar, M., Kounani, A., Mbah, M. F., Abubakar, I. R., Matandirotya, N. R., Pimenta Dinis, M. A., Borsari, B., & Abzug, R. (2025). Assessing the impacts of sustainability teaching at higher education institutions. *Discover Sustainability, 6*(1), 227. https://doi.org/10.1007/s43621-025-01024-z

Le Blanc, D. (2015). Towards integration at last? The sustainable development goals as a network of targets. *Sustainable Development, 23*(3), 176–187. https://doi.org/10.1002/sd.1582

Meadows, D. H., & Wright, D. (2008). *Thinking in systems: A primer*. Chelsea Green Publishing.

Miles, M. B., & Huberman, A. M. (1994). Qualitative data analysis an expanded sourcebook. *Sage*. Publications.

Sachs, J. (2012). From millennium development goals to sustainable development goals. *The Lancet, 379*(9832), 2206–2211. https://doi.org/10.1016/S0140-6736(12)60685-0

Sachs, J. (2015). *The age of sustainable development*. Columbia University Press.

SDSN. (2024). *Part 1–recommendations of the UN sustainable development solutions network for the summit of the future*. SDSN.

Srivastava, P., & Hopwood, N. (2009). A practical iterative framework for qualitative data analysis. *International Journal of Qualitative Methods, 8*(1), 76–84. https://doi.org/10.1177/160940690900800107

Stroh, D. P. (2015). *Systems thinking for social change: A practical guide to solving complex problems, avoiding unintended consequences, and achieving lasting results*. Chelsea Green Publishing.

UN (2015). Transforming our world—The 2030 agenda for sustainable development. https://sdgs.un.org/2030agenda
UN. (2024). *Pact for the future that includes a global digital compact and a declaration on future generations*. United Nations.
Walsham, G. (2006). Doing interpretive research. *European Journal of Information Systems, 15*(3), 320–330. https://doi.org/10.1057/palgrave.ejis.3000589
Wilbanks, T. (2015). Putting "place" in a multiscale context: Perspectives from the sustainability sciences. *Environmental Science & Policy, 53*, 70–79. https://doi.org/10.1016/j.envsci.2015.04.009
Žalėnienė, I., & Pereira, P. (2021). Higher education for sustainability: A global perspective. *Geography and Sustainability, 2*(2), 99–106. https://doi.org/10.1016/j.geosus.2021.05.001

Chapter 40
Green Minds, Bright Futures: Sustainable Quality Practices in Academia

Connie Mitchell

Abstract Sustainable quality practices in the context of higher education refer to strategies, principles, and actions that educational institutions adopt to simultaneously promote quality in education and sustainability in environmental, social, and economic dimensions. Given the increased demands placed on societies and the environment due to, among other factors, increased human migration, increased urbanization, and industrialization, as well as the ongoing depletion of non-renewable resources, it is clear that global action is needed to create a more sustainable future. This is why higher education institutions must actively modify their approach to education, including all stakeholders, like faculty, staff, students, and cleaners. The question is what actions are being taken to promote change that goes beyond the going green concept. Sustainability goes beyond planting trees and recycling. It is about changing mindsets that support sustainable practices for long-term growth and impact. Aligning the curricula with the SDGs is the first step in this process. Incorporating critical thinking, problem-solving, cultural awareness, and reflective activities can support the alignment and can lead to student empowerment through the SDGs. Prince Sultan University has taken measures to align the curricula and offers programs and initiatives focused on promoting the SDGs and sustainability as a way of life.

Keywords Sustainability · Quality assurance · Higher education · SDGs · Climate change

Introduction

Higher education plays a pivotal role in educating individuals about sustainability (UNESCO, 2020). The concept often evokes thoughts of recycling and environmental conservation, but sustainability extends beyond these notions. In 2025 and beyond,

---

C. Mitchell (✉)
Prince Sultan University, Riyadh, Saudi Arabia
e-mail: cmitchel@psu.edu.sa

J. Dsilva (ed.), *Sustainability Education, Youth and Innovation*,
https://doi.org/10.1007/978-3-032-21948-0_40

sustainable quality practices in academia focus on shifting mindsets, ensuring long-term growth for sustainable systems, and fostering a lasting impact that benefits future generations (Leal Filho et al., 2015).

Defining sustainable quality practices in higher education requires embedding environmental sustainability into educational processes, infrastructure, and curricula while maintaining high academic standards. Global challenges—including climate change, resource depletion, and socioeconomic disparities—necessitate a strategic transformation of higher education institutions. Climate action, particularly addressing climate change through institutional initiatives, is integral to this transformation. Universities must go beyond traditional green initiatives by actively incorporating sustainability into educational frameworks, research, and operational policies. Integrating environmental sustainability into higher education curricula is essential for addressing global challenges such as climate change, resource depletion, and socioeconomic disparities. This integration not only enhances academic standards but also equips students with the knowledge and skills necessary to tackle pressing environmental issues. For instance, a study examining Imam Abdulrahman bin Faisal University's curricula in Saudi Arabia found a low degree of sustainability integration, highlighting the need for curriculum reforms to promote sustainability learning (Bataeineh & Aga, 2023). Sustainable quality practices in the context of higher education refer to strategies, principles, and actions that educational institutions adopt to simultaneously promote quality in education and sustainability in environmental, social, and economic dimensions. Given the increased demands placed on societies and the environment due to, among other factors, increased human migration, increased urbanization, and industrialization, as well as the ongoing depletion of non-renewable resources, it is clear that global action is needed to create a more sustainable future. This is why higher education institutions must actively modify their approach to education, including faculty, staff, and students. Taking climate action that results in climate change is what is needed. The question is what actions are being taken to promote change that goes beyond the going green concept. Sustainability goes beyond planting trees and recycling. It is about changing mindsets that support sustainable practices for long-term growth and impact. Aligning the curricula with the SDGs is the first step in this process. Incorporating critical thinking, problem-solving, cultural awareness, and reflective activities can support the alignment and can lead to student empowerment through the SDGs. Therefore, Universities must play a crucial role in promoting sustainability through research, teaching, and operations. They are responsible for developing future leaders who will drive sustainable development.

Research Questions

(i) What actions have helped establish sustainable quality practices at Prince Sultan University?

(ii) How has Prince Sultan University tackled the issue of the sustainable development goals to support the Quality practices of incorporating sustainability into its infrastructure to foster Green minds and bright futures?

Literature Review

Sustainable quality practices in the context of higher education refer to strategies, principles, and actions that educational institutions adopt to simultaneously promote quality in education and sustainability in environmental, social, and economic dimensions (Leal Filho et al., 2015). Given the increased demands placed on societies and the environment due to, among other factors, increased human migration, urbanization, and industrialization, as well as the ongoing depletion of non-renewable resources, it is clear that global action is needed to create a more sustainable future (UNESCO, 2012). This is why higher education institutions must actively modify their approach to education, including all stakeholders—faculty, staff, and students—to meet these challenges (UNESCO, 2020). Climate action, which is central to addressing the effects of climate change, is crucial in this context. However, the question remains: What actions are being taken to promote change that goes beyond the "going green" concept and addresses the broader dimensions of sustainability? (Sachs, 2015).

Prince Sultan University has taken measures to align the curricula and offers programs and initiatives that are focused on promoting the SDGs and sustainability as a way of life. Revising the curricula to prepare professionals and citizens to be able to meet the current and future challenges of converging global environmental problems through increasing levels of "ecological literacy" or "environmental literacy" is a necessary action that needs to happen. The university has created a well-thought-out infrastructure that complies with UNESCO's Education for Sustainable Development (ESD) concept paper.

ESD sees education as the key to unraveling progress in all global sustainable development goals. It teaches individuals to make informed decisions and act individually and collectively to change society and protect the planet. It equips people of all ages with the knowledge, skills, values, and ability to tackle issues such as climate change, biodiversity loss, overuse of resources, and inequality that impact the well-being of people and the planet.

ESD advocates for learning that is:

(i) Cognitive: Improving how we think and understand information.
(ii) Socio-emotional: Building social skills, empathy, and emotional intelligence.
(iii) Behavioral: Encouraging positive actions and behaviors. (UNESCO, 2024, para 2)

Greening education is a comprehensive approach that equips learners with the knowledge, skills, and attitudes needed to take meaningful action on climate change mitigation, adaptation, and resilience (UNESCO, 2024).

Nogueiro and Saraiva (2023) argue that Higher Education Institutions (HEIs) play a vital role in societal development, as they are key players in training professionals who will shape labor markets and societal trends. These institutions must serve as foundations for knowledge dissemination and the promotion of sustainable practices. The Times Higher Education (THE) Impact Rankings, which are the only global rankings linking universities to the United Nations' Sustainable Development Goals (SDGs), assess universities' progress in achieving these goals. The third edition of the rankings, released in 2021, included 1118 universities from 94 countries and regions. This study focuses on a qualitative analysis of the sustainability practices and SDG-related content presented on the websites of four selected Portuguese HEIs. The authors aim to contribute to the ongoing discussion about HEIs' roles in sustainability and the 2030 Agenda, highlighting the institutions' commitment to fostering sustainability and achieving the SDGs, ensuring that no one is excluded or discriminated against in the process (Nogueiro & Saraiva, 2023).

Education for sustainable development (ESD) is a key component of SDG target 4.7, offering a comprehensive approach that equips learners of all ages with the knowledge, skills, values, and agency to make informed decisions that foster environmental protection, equitable economic growth, and social justice. Climate change education is a crucial aspect of ESD, focusing on how climate change and its impacts are taught in various educational settings. The concept of "greening education" expands on this by not only addressing climate change content but also considering factors such as the learning environment, teacher training, and community involvement. This approach ensures that learners gain the necessary knowledge, skills, and attitudes to effectively engage in climate change mitigation, adaptation, and resilience efforts. As the climate crisis becomes increasingly urgent, there is growing demand from young people for this more comprehensive form of education, which the Greening Education Partnership supports through a whole-institution approach (UNESCO, 2024).

Methodology

This research utilizes a qualitative approach, combining narrative inquiry with a case study approach about one university. The research framework centers on evaluating the effectiveness of sustainability-focused educational practices. The study also examines the impact of quality assurance practices and accreditation processes on fostering a sustainable campus culture. At Prince Sultan University, multiple centers (e.g., Center for Sustainability and Climate, centers under the Deanship of Quality and Development, as well as the Evaluation and Academic Accreditation Center) collectively contribute to reviewing strategic plans, accreditation procedures and sustainability projects integrated into the curriculum. Furthermore, student clubs and extracurricular activities, along with integrating sustainability principles into courses, offering interdisciplinary courses, providing experiential learning opportunities, and encouraging project-based learning, have progressed the sustainable culture. The relevance of the research questions lies in their potential to inform both theory and practice by providing actionable insights into effective sustainability education.

Findings

Prince Sultan University (PSU) is dedicated to fostering a deep understanding of sustainable quality practices in higher education among its stakeholders, including faculty, staff, students, and external partners. PSU encourages stakeholders to recognize the importance of integrating sustainability principles into both the educational framework and the operational aspects of the university (Abowardah et al., 2024). By offering various training and development programs, the university helps participants explore strategies for incorporating sustainability into quality assurance processes, ensuring that sustainability is considered in all aspects of academic and administrative practices. This includes the identification of specific strategies to integrate sustainability into the curriculum, enabling students to gain the knowledge and skills needed to contribute to sustainable development in their future careers.

PSU also emphasizes the need to recognize different strategies for measuring and evaluating sustainability performance within higher education. The university supports stakeholders in understanding how to assess sustainability efforts, from curriculum content to campus operations (Javed & Alenezi, 2023). In addition, PSU equips participants with the tools necessary to create actionable plans for implementing sustainable quality practices within their own institutions. This holistic approach ensures that sustainability is not just a theoretical concept but a practical, measurable goal that is integrated across all levels of the institution.

Key aspects of PSU's sustainable quality practices include fostering a sustainability culture that engages all members of the university community, from students to administrators. Employees and students are strongly committed to actively participating in sustainability initiatives (Hoke, 2024). The university promotes community and stakeholder engagement by collaborating with external organizations and industry partners while also encouraging student engagement through sustainability-focused extracurricular activities and projects. Faculty and staff receive regular training to stay informed on sustainability trends and best practices, ensuring they are equipped to integrate these principles into their teaching and administrative responsibilities (Javed & Alenezi, 2023).

PSU also focuses on curriculum integration, where sustainability topics are embedded in various academic programs, preparing students to address complex global challenges. In the realm of research and innovation, the university supports projects that contribute to sustainable solutions, particularly in areas such as renewable energy, environmental conservation, and social responsibility. Effective resource management practices ensure that PSU operates efficiently, minimizing waste and optimizing energy use, while sustainable campus operations promote a green and energy-efficient physical environment.

The university's green campus initiatives reflect its commitment to sustainability, incorporating eco-friendly infrastructure, renewable energy sources, and waste reduction strategies. PSU also fosters a sense of global citizenship, encouraging students and faculty to think beyond national borders and address global sustainability challenges. Transparency and reporting are essential to PSU's sustainability practices, with regular updates on sustainability efforts shared with internal and external stakeholders. Ethical leadership is evident in the university's decision-making processes, which prioritize sustainability and long-term societal benefits.

Through its rigorous quality assurance and assessment processes, PSU continuously evaluates its sustainability efforts to ensure they align with the highest standards. The university is committed to continuous improvement, regularly reviewing and enhancing sustainability practices to adapt to emerging challenges. External recognition, such as awards and accreditation for sustainability, further reinforces PSU's commitment to sustainable quality practices in higher education. These comprehensive strategies position PSU as a leader in integrating sustainability into higher education, ensuring that the institution remains at the forefront of educational and environmental stewardship.

1. What actions have helped establish sustainable quality practices at Prince Sultan University?

    (a) Institutionalizing Continuous Improvement: PSU has implemented a robust system of continuous improvement that spans all academic programs and operational functions. This includes regular self-assessments, feedback loops from students, faculty, and stakeholders, and the integration of data-driven decision-making processes. Faculty development programs are regularly organized to keep educators up to date with best practices, ensuring that quality teaching and learning practices are sustained across all disciplines.
    (b) Accreditation and International Standards: Accreditation and International Standards: PSU has actively pursued and achieved various prestigious accreditations such as AABI (Accreditation Board for Engineering and Technology), ABET, AACSB, HCERES, and NCAAA, among others, ensuring that academic and administrative operations adhere to internationally recognized standards. This commitment to global standards promotes a culture of excellence and establishes long-term quality sustainability across its programs.
    (c) Technology-Driven Innovation: By adopting advanced learning technologies, PSU has ensured that educational methods remain cutting-edge and aligned with global trends. The integration of Learning Management Systems (LMS), online learning tools, and digital platforms like Coursera has enhanced teaching, learning, and assessment, ensuring quality education is accessible to all students regardless of their location. This technological integration helps sustain the delivery of quality education and support services.
    (d) Collaboration with Industry and Research Initiatives: PSU has established strong relationships with industry partners to bridge the gap between

academia and the workforce. Through internships, collaborative research, and industry advisory boards, PSU ensures its programs remain relevant, maintaining high academic and practical standards. These partnerships also enable PSU to introduce innovative practices and solutions that ensure the university's education remains of the highest quality.

2. How has Prince Sultan University tackled the issue of the sustainable development goals to support the quality practices of incorporating sustainability into its infrastructure to foster Green minds and bright futures?

    (a) Integration of Sustainability into Curriculum: PSU has embedded sustainability principles across its academic programs, particularly in disciplines such as engineering, architecture, business, and environmental sciences. Courses on sustainable development, climate change, and renewable energy technologies are integrated into the curriculum, fostering a deep understanding of sustainability issues among students. By doing so, PSU ensures that its graduates are not only skilled professionals but also conscious global citizens who can contribute to sustainable development (Hoke, 2024).
    (b) Green Campus Initiatives: PSU has made significant strides in transforming its physical infrastructure into a more sustainable and eco-friendly environment. The campus boasts energy-efficient buildings, the use of renewable energy sources such as solar panels, and the promotion of waste management and recycling programs. These green initiatives are part of PSU's broader commitment to creating a sustainable campus that aligns with the United Nations' Sustainable Development Goals (SDGs).
    (c) Research and Innovation for Sustainability: Research at PSU focuses on tackling real-world sustainability challenges. Faculty and students are involved in research projects related to sustainable energy solutions, water conservation, environmental conservation, and smart cities. The university actively collaborates with governmental and private sector organizations to drive innovation in these areas, contributing to sustainable development goals such as affordable and clean energy (SDG 7), climate action (SDG 13), and sustainable cities (SDG 11).
    (d) Green Mindset for Students: PSU actively promotes sustainability as part of its student engagement and extracurricular activities. The university organizes awareness campaigns, workshops, and sustainability-themed events, encouraging students to adopt sustainable lifestyles. Student-led environmental clubs, sustainability challenges, and collaboration with NGOs provide opportunities for students to take an active role in creating a greener future (Abowardah, Labib, Aboelnagah, & Nurunnabi, 2024).
    (e) Community Engagement and Outreach: PSU has partnered with local and international organizations to promote sustainability in the wider community. Through outreach programs, community service initiatives, and public awareness campaigns, PSU students and faculty work to address local sustainability challenges. By engaging with the community, PSU helps to

spread knowledge of sustainable practices and inspires others to take action toward environmental conservation and social responsibility.

(f) Commitment to SDGs in Governance: PSU's strategic planning and governance structures are deeply aligned with the Sustainable Development Goals. The university's leadership has incorporated sustainability principles into the institution's long-term vision, policies, and decision-making processes, ensuring that all actions undertaken by PSU contribute to the SDGs. This holistic approach to governance ensures that sustainability becomes an integral part of PSU's mission and operations.

Conclusions

In conclusion, while adopting sustainable practices at Prince Sultan University (PSU) presents challenges such as significant financial investment, infrastructure changes, and the ongoing need for training and stakeholder engagement, these obstacles are not insurmountable. The opportunities for PSU to lead as a model of sustainability within the region are vast. By collaborating with international universities, PSU can enhance its role as a thought leader in sustainability, fostering global partnerships that elevate its impact on higher education. As a core value, sustainability is not just an option but an essential commitment for the future. By embedding sustainable quality practices into its operations and curriculum, PSU can make a profound impact on environmental preservation, societal well-being, and academic excellence. In this way, PSU has the potential to position itself as a leader in sustainability within the Middle East, contributing to the development of future generations of green scholars dedicated to creating a sustainable world. Through these efforts, PSU can ensure that sustainability becomes deeply ingrained in the educational experience, advancing both local and global sustainability goals for years to come (Nogueiro & Saraiva, 2023).

The study highlights the broader implications of sustainable quality practices in academia. Other institutions can adopt PSU's model by embedding sustainability into their curricula, research initiatives, and campus operations. This fosters a global network of higher education institutions committed to sustainability, ultimately contributing to long-term environmental preservation, social equity, and economic resilience. As sustainability becomes a fundamental pillar of education, universities worldwide must embrace their role in shaping future leaders who are equipped to address global sustainability challenges.

References

Abowardah, E. S., Labib, W., Aboelnagah, H., & Nurunnabi, M. (2024). Students' perception of sustainable development in higher education in Saudi Arabia. *Sustainability, 16*(4), 1483. https://doi.org/10.3390/su16041483

Bataeineh, M., & Aga, O. (2023). Integrating sustainability into higher education curricula: Saudi vision 2030. *Emerald Open Research, 1*(3). https://doi.org/10.1108/eor-03-2023-0014

Hoke, T. (2024). AI with a heart: Integrating emotional intelligence and sustainability in smart education for 21st-century learners. *Springer Singapore.* https://doi.org/10.1007/978-981-97-8348-9_56

Javed, Y., & Alenezi, M. (2023). A case study on sustainable quality assurance in higher education. *Sustainability, 15*(10), 8136. https://doi.org/10.3390/su15108136

Leal Filho, W., Londero, L. M., & Brandli, L. L. (2015). *Sustainability in higher education: A global perspective.* Springer.

Nogueiro, T., & Saraiva, M. (2023). Quality and practices for sustainability in higher education—An impact ranking approach. In A. Mesquita, A. Abreu, J. V. Carvalho, & C. H. P. de Mello (Eds.), *Perspectives and trends in education and technology* (Vol. 320, pp. 245–261). Springer Singapore. https://doi.org/10.1007/978-981-19-6585-2_27

Sachs, J. D. (2015). *The age of sustainable development.* Columbia University Press.

UNESCO. (2012). Education for sustainable development and global citizenshipds. http://unesdoc.unesco.org

UNESCO. (2020). *Higher education and sustainable development goals: A global overview.* United Nations Educational. http://unesco.org

UNESCO. (2024). What you need to know about education for sustainable development. United Nations. https://www.unesco.org/en/sustainable-development/education

United Nations. (2015). Transforming our world: The 2030 agenda for sustainable development. United Nations. https://sdgs.un.org/2030agenda.

Chapter 41
Localizing Global SDGs: A Comparative Analysis of University-Led Sustainability Initiatives in the Global South and North

Victor Lobo

Abstract Universities worldwide are important players in turning the global Sustainable Development Goals (SDGs) into something that works locally, but their approaches split sharply between the Global North and South based on what they've got to work with and what matters most where they are. This paper digs into how higher education spots in these regions make SDGs their own through sustainability efforts. This paper presents a comparative analysis of university-led sustainability initiatives across institutions in the Global South and North, exploring how universities in these regions localize the SDGs to address their specific challenges. Drawing on case studies from the University of Leeds (UK), Aalborg University (Denmark), Stanford University (USA), University of Cape Town (South Africa), University of São Paulo (Brazil), University of Indonesia (Indonesia), Makerere University (Uganda), University of Delhi (India). The paper employs a mixed-methods approach through data collection, case study selection, and analysis. The findings suggest that while universities in the Global North tend to focus on technological innovation and research-driven sustainability, institutions in the Global South emphasize community-based approaches and local partnerships. Despite these differences, both regions demonstrate a commitment to advancing the SDGs within the context of their unique challenges. The study concludes with policy recommendations for universities to enhance cross-regional collaboration in achieving the SDGs and fostering sustainable development at local, national, and global levels.

Keywords Sustainable development goals · Higher education · Global South · Global North · Sustainability initiatives · University governance

V. Lobo (✉)
St Joseph's University, Bengaluru, India
e-mail: victorlobo@sju.edu.in

J. Dsilva (ed.), *Sustainability Education, Youth and Innovation*,
https://doi.org/10.1007/978-3-032-21948-0_41

Introduction

The Sustainable Development Goals (SDGs), instituted by the United Nations in 2015, set a global program for sustainability, encouraging universities to integrate these targets into their education, research, and operations (United Nations, 2015). However, decoding these collective goals into actionable, locally relevant initiatives varies widely between the Global North (developed countries) and Global South (developing countries), due to economic capacity, technological access, and societal demands (Filho et al., 2021). In the Global North, educational institutions like those in Europe and North America often localize SDGs using extensive resources and modern tools to address global challenges (Findler et al., 2019), whereas at Global South the universities, those in Africa and Latin America, fit them to tackle immediate local issues with limited means (Mawonde & Togo, 2019).

This variation reflects broader global disparities, raising questions about how universities can locate SDGs to fit their needs (Bull & McNeill, 2021). Global Northern universities often channel SDGs into research and technological innovation, supported by robust partnerships (Lozano et al., 2013), while Southern ones prioritize practical, community-focused adaptations despite financial limitations (Salvia et al., 2019). For example, the University of British Columbia aligns SDG 13 (Climate Action) with carbon neutrality in infrastructure, while the University of Cape Town adjusts SDG 6 (Clean Water) to address the issues of regional water scarcity (Leal Filho et al., 2021). The aim of this study is to explore how do the universities in the Global North and South localize SDGs through their sustainability initiatives along with the analysis on what challenges arise in this process? By analyzing these efforts, the research aims to inform strategies for aligning global goals with local needs, building on work by Filho et al. (2019) and Verhulst and Lambrechts (2015).

Literature Review

Research on sustainable initiatives through education examines how institutions adapt global SDGs to local and regional settings, revealing regional contrasts. In the Global North, universities like the University of Leeds (UK) and Stanford University (USA) localize SDGs through research-intensive projects and operational targets, such as net-zero emissions by 2030 (Leal Filho et al., 2021). The Times Higher Education (THE) Impact Rankings often highlight their focus on innovation and international collaboration to advance goals like SDG 13 (Climate Action) (Times Higher Education, 2023). Findler et al. (2019) note that Northern institutions use ample funding and infrastructure to adapt SDGs with technologies like renewable energy systems. Aalborg University (Denmark), for example, embeds SDG 4 (Quality Education) into its problem-based learning approach (Lozano et al., 2013).

In the Global South, universities such as the University of Cape Town (South Africa) and the University of São Paulo (Brazil) localize SDGs by addressing community needs and resource constraints, like water scarcity or poverty (Filho et al., 2021). The UI GreenMetric World University Rankings praise their efforts in waste and energy management, though fewer participate due to data limitations (Peirchinunno & Cazzolle, 2020). Mawonde and Togo (2019) describe how the University of South Africa adapts SDG 12 (Responsible Consumption) through recycling and local education. Alcántara-Rubio et al. (2022) find that Latin American universities align SDG 11 (Sustainable Cities) with local partnerships, a trend less prominent in the North.

Comparisons show distinct localization strategies. Northern universities leverage research and technology to adapt SDGs globally, supported by policies and networks (Verhulst & Lambrechts, 2015), as seen in the University of British Columbia's SDG 9 (Industry, Innovation) efforts with green buildings (Salvia et al., 2019). Southern institutions focus on local relevance, with Makerere University (Uganda) tailoring SDG 2 (Zero Hunger) to agroecology (Adomßent et al., 2014). Funding disparities hinder Southern scalability, with budgets estimated at 10–20% of Northern levels (Filho et al., 2019).

Challenges persist across regions. Northern efforts to localize SDGs face high per-capita emissions despite tech advances (Bull & McNeill, 2021), while Southern adaptations struggle with infrastructure and instability (Omrcen et al., 2018). Partnerships (SDG 17) are key to localization, but Northern dominance may sideline Southern priorities (Ramos et al., 2015). Barth et al. (2014) and Sedlacek (2013) suggest education adapts SDGs universally, yet Northern curricula emphasize global scope while Southern ones target regional issues. This review underscores the need to compare how universities localize SDGs effectively.

Methodology

This study employs a mixed-methods framework to examine how universities in the Global North and South localize SDGs through sustainability initiatives, ensuring a transparent and repeatable process. It proceeds in three phases: data collection, case study selection, and analysis.

Data Collection

Two datasets were compiled. First, sustainability reports, strategic plans, and policies were sourced from university websites and repositories, outlining how institutions adapt SDGs to their contexts. Second, quantitative data came from the 2024 THE Impact Rankings, assessing SDG alignment, and the 2024 UI GreenMetric World University Rankings, evaluating campus sustainability metrics like energy and waste.

These provide qualitative insights and measurable outcomes, though Southern data was sometimes less complete.

Case studies.

Ten universities—five per region—were selected to represent diverse contexts and SDG localization efforts. For the Global North:

University of Leeds (UK): Aligns SDG 13 with a net-zero target by 2030 and climate research.

Aalborg University (Denmark): Adapts SDG 4 through sustainable education models.

Stanford University (USA): Localizes SDG 7 (Affordable Energy) with campus solar systems.

University of British Columbia (Canada): Targets SDG 9 with green infrastructure.

ETH Zurich (Switzerland): Advances SDG 13 via climate research and facilities.

For the Global South:

University of Cape Town (South Africa): Adjusts SDG 6 to address water scarcity.

University of São Paulo (Brazil): Aligns SDG 11 with community urban farming.

University of Indonesia (Indonesia): Adapts SDG 12 through waste management.

Makerere University (Uganda): Localizes SDG 2 with agroecology programs.

University of Delhi (India): Targets SDG 7 with energy efficiency measures.

These were chosen for regional significance, ranking participation, and available data, reflecting trends from Filho et al. (2021) and Mawonde and Togo (2019).

Analysis

Analysis combined qualitative and quantitative methods. Policy documents were examined via content analysis, coding for SDG-specific adaptations (e.g., energy, community, research) to identify localization patterns. Quantitatively, THE Impact Rankings scores (out of 100) for SDGs 4, 11, 13, and 17, and UI GreenMetric scores (out of 10,000) were averaged across each region's five universities. A t-test evaluated significant score differences. This dual approach illuminated how SDGs are tailored regionally.

The ten-university sample limits generalizability, and incomplete Southern data may underrepresent efforts. Reporting inconsistencies and language barriers added challenges. Nonetheless, this method offers a clear lens on SDG localization.

Results and Discussions

The analysis showed how universities in the Global North and South localize SDGs differently, according to their own capacities and settings.

The Global North universities designed SDGs through technology and research. They averaged 85/100 on SDG 13 (Climate Action), with Stanford's solar system (50% campus energy) and Leeds' 2030 net-zero goal as key examples. Policies showed 70% focus on research, like ETH Zurich's SDG 13-aligned climate models with global impact. Funding averaged $10–20 million annually, often from public and private sources, as at UBC.

Global South universities localized SDGs via community engagement and resource efficiency. They scored 78/100 on SDG 11 (Sustainable Cities), with Cape Town cutting water use by 40% (SDG 6) and Makerere recycling 60% of waste (SDG 12). Policies indicated 80% community focus, such as São Paulo's SDG 11 farming partnerships. Funding averaged $1–3 million yearly, often supplemented by international support, as at Indonesia.

Comparatively, technology use was 90% in the North (e.g., Aalborg's grids) versus 40% in the South (e.g., Delhi's solar). Community initiatives were 60% in the South versus 30% in the North. Both averaged 70/100 on SDG 17 (Partnerships), with Northern universities leading globally.

The findings reveal two approaches to localizing SDGs, connected to the introduction's focus on contextual adaptation and the literature's regional divide (Filho et al., 2021; Findler et al., 2019). Northern universities, like Stanford and Leeds, use their resources to adapt SDGs with technology, aligning with Lozano et al. (2013). Their high SDG 13 scores reflect a global climate focus, enabled by significant funding and research capacity. However, their 30% community engagement suggests a weaker local lens, echoing Bull and McNeill's (2021) critique of Northern trade-offs. UBC's SDG 9 green buildings impress, but high energy use per person shows limits in local impact.

Southern universities, such as Cape Town and Makerere, localize SDGs by addressing immediate needs, supporting Mawonde and Togo (2019). Their SDG 11 and 6 efforts—like São Paulo's farming—tailor global goals to local crises (Alcántara-Rubio et al., 2022). Yet, with 40% technology adoption and low funding, expansion is tough, as Filho et al. (2019) note. Indonesia's SDG 7 solar steps are modest compared to Stanford, underscoring resource gaps.

SDG 17 (Partnerships) highlights collaboration's role in localization, averaging 70/100 (Ramos et al., 2015). Northern leadership, as in ETH Zurich's networks, contrasts with Southern reliance on external aid, like Delhi's, raising equity issues (Bull & McNeill, 2021). This ties to the introduction's call for balanced progress.

Challenges reflect the literature (Omrcen et al., 2018; Verhulst & Lambrechts, 2015). Northern tech-driven localization doesn't always shift habits—Stanford's solar gains are offset by consumption. Southern efforts face infrastructure limits—Makerere's SDG 12 recycling lacks scale. These insights, drawn from case studies and rankings, suggest Northern universities enhance local ties, while Southern ones need resources to broaden SDG impact.

Conclusions

According to this study, universities in the North use technology and research to localize SDGs, but those in the South adapt them through efficiency and community. Although both approaches are essential, unequal access to resources and capacity hinder the advancement of the SDGs. Increasing Southern support through partnerships and promoting North–South exchanges are among the recommendations.

Although colleges in the Global South and North are dedicated to the SDGs, the comparative research shows that because of the diverse socioeconomic and environmental settings, their methods to localization differ greatly. While the Global North prioritizes innovation, technology, and research-driven solutions, the Global South often concentrates on community-based, socially inclusive alternatives.

The study's limitations include its narrow emphasis on a few universities and the availability of publicly available data. Future studies might increase the sample size and concentrate on more diverse universities, especially those from underrepresented Global South regions.

Cross-regional cooperation can help universities address global sustainability issues by utilizing community expertise and technology innovation.

References

Adomßent, M., Fischer, D., Godemann, J., Herzig, C., Otte, I., Rieckmann, M., & Timm, J. (2014). Emerging areas in research on higher education for sustainable development—Management education, sustainable consumption and perspectives from Central and Eastern Europe. *Journal of Cleaner Production, 62*, 1–7. https://doi.org/10.1016/j.jclepro.2013.09.045

Alcántara-Rubio, L., Valderrama-Hernández, R., Solís-Espallargas, C., & Ruiz-Morales, J. (2022). The implementation of the SDGs in universities: A systematic review. *Environmental Education Research, 28*(11), 1585–1615. https://doi.org/10.1080/13504622.2022.2063798

Barth, M., Adomßent, M., Fischer, D., Richter, S., & Rieckmann, M. (2014). Learning to change universities from within: A service-learning perspective on promoting sustainable consumption in higher education. *Journal of Cleaner Production, 62*, 72–81. https://doi.org/10.1016/j.jclepro.2013.04.006

Bull, B., & McNeill, D. (2021). SDG partnerships may perpetuate the global North-South divide. *Scientific Reports, 11*, 1–10. https://doi.org/10.1038/s41598-021-01563-7

Filho, W. L., Amaro, N., Avila, L. V., Brandli, L., Damke, L. I., Vasconcelos, C. R. P., & Salvia, A. (2021). Mapping sustainability initiatives in higher education institutions in Latin America. *Journal of Cleaner Production, 315*, Article 128093. https://doi.org/10.1016/j.jclepro.2021.128093

Filho, W. L., Shiel, C., Paço, A., Mifsud, M., Ávila, L. V., Brandli, L. L., & Salvia, A. L. (2019). Sustainable development goals and sustainability teaching at universities: Falling behind or getting ahead of the pack? *Journal of Cleaner Production, 232*, 285–294. https://doi.org/10.1016/j.jclepro.2019.05.309

Findler, F., Schönherr, N., Lozano, R., Reider, D., & Martinuzzi, A. (2019). The impacts of higher education institutions on sustainable development: A review and conceptualization. *International Journal of Sustainability in Higher Education, 20*(1), 23–38. https://doi.org/10.1108/IJSHE-07-2017-0114

Leal Filho, W., Salvia, A. L., & Pires Eustachio, J. (2021). The role of universities in accelerating the sustainable development goals in Europe. *Scientific Reports, 11*, 1–12. https://doi.org/10.1038/s41598-021-92991-8

Lozano, R., Lukman, R., Lozano, F. J., Huisingh, D., & Lambrechts, W. (2013). Declarations for sustainability in higher education: Becoming better leaders, through addressing the university system. *Journal of Cleaner Production, 48*, 10–19. https://doi.org/10.1016/j.jclepro.2011.10.006

Mawonde, A., & Togo, M. (2019). Incorporation of the SDGs in university operations: A case study of the University of South Africa. *Sustainability, 11*(23), 1–15. https://doi.org/10.3390/su11236789

Omrcen, E., Lundgren, U., & Dalhberg, S. (2018). Integrating sustainability in higher education: A Swedish case. *International Journal of Sustainability in Higher Education, 19*(6), 1131–1150. https://doi.org/10.1108/IJSHE-10-2017-0168

Peirchinunno, P., & Cazzolle, M. (2020). Re-evaluating the UI Green Metric: A global sustainability ranking for universities. *Sustainability, 12*(15), 1–18. https://doi.org/10.3390/su12156123

Ramos, T. B., Caeiro, S., van Hoof, B., Lozano, R., Huisingh, D., & Ceulemans, K. (2015). Experiences from the implementation of sustainable development in higher education institutions: Environmental management for sustainable universities. *Journal of Cleaner Production, 106*, 3–10. https://doi.org/10.1016/j.jclepro.2015.05.110

Salvia, A. L., Leal Filho, W., Brandli, L. L., & Griebeler, J. S. (2019). Assessing research trends related to sustainable development goals: Local and global issues. *Journal of Cleaner Production, 208*, 841–849. https://doi.org/10.1016/j.jclepro.2018.09.242

Sedlacek, S. (2013). The role of universities in fostering sustainable development at the regional level. *Journal of Cleaner Production, 48*, 74–84. https://doi.org/10.1016/j.jclepro.2013.01.029

Times Higher Education. (2023). *Impact rankings 2023.* Accessed https://www.timeshighereducation.com/impactrankings

United Nations. (2015). *Transforming our world: The 2030 agenda for sustainable development.* United Nations General Assembly. https://sdgs.un.org/2030agenda

Verhulst, E., & Lambrechts, W. (2015). Fostering the incorporation of sustainable development in higher education: Lessons learned from a change management perspective. *Journal of Cleaner Production, 106*, 189–204. https://doi.org/10.1016/j.jclepro.2014.09.049

Chapter 42
Framework for Promoting School Leadership for Climate Education

Swetal Sindhvad

Abstract School-based climate education entails community-centered learning experiences linked to real-world environmental challenges that give students insight into decision-making processes. However, school leadership prioritizes curriculum delivery that limit climate education to science class rather than embedding it in an interdisciplinary learning experience. School leadership frameworks that guide the development of climate education initiatives in alignment with broader school improvement strategies is few and far between. Moreover, school leaders often operate as school managers rather than instructional leaders that lead them to disregard opportunities in driving systemic change through their own climate-focused leadership. This paper proposes a framework that positions school leaders as central agents in advancing climate education. Using the core disciplines of schools as learning organizations, it identifies three key leadership strategies towards adopting a systems thinking approach to integrate climate education across the whole school while rooted in sustainable development principles: (i) fostering a shared vision that enables collaboration among teachers, students, and the community; (ii) strengthening collaboration by building partnerships within and beyond the school; and (iii) promoting teachers' personal mastery to enhance their capacity to deliver climate education effectively. This framework provides a structured approach for school leaders to rethink traditional practices and incorporate climate education into their broader school improvement efforts. This paper contributes to Sustainable Development Goal (SDG) 13: Climate Action, specifically Target 13.3, which aims to "*improve education, awareness-raising and human and institutional capacity on climate change mitigation, adaptation, impact reduction and early warning*."

Keywords Climate education · Education for sustainable development · School leadership · School improvement · Learning organization

S. Sindhvad (✉)
i3Development, Dubai, UAE
e-mail: ssindhvad@i3development.org

J. Dsilva (ed.), *Sustainability Education, Youth and Innovation*,
https://doi.org/10.1007/978-3-032-21948-0_42

Introduction

Around the world, climate education has transitioned from being largely ignored to becoming an integral part of curricula, policy discussions, and research (Stevenson et al., 2024; Gan, 2021). Initially identified as environmental education and associated with science and geography, climate education now incorporates interdisciplinary approaches aimed to foster climate literacy and climate action among today's youth (Gan, 2021; Kwak, 2020). Case studies of climate education identify efforts as being community-centered and linking students' learning experiences to real-world environmental challenges that give them insight into decision-making processes (Bauman-Hill et al., 2022; Cebrián et al., 2022; Ceblano et al., 2023). School leadership plays a crucial role in integrating climate education into curricula and ensuring teachers are well equipped to deliver it (Cebrián et al., 2022).

While some school leaders are proactive in promoting climate education, others face institutional and resource constraints such as competing policy priorities, overcrowded curriculum, overworked teachers, and insufficient parent and community involvement (Cebrián et al., 2022; Shabalala, 2024). Also, research points to school leaders' lack of understanding for climate education and its place in the school curriculum. Al'Abri et al. (2024) indicate that school leaders may be more inclined to promote values associated with climate education such as global citizenship than the knowledge, skills, and attitude needed to act as change agents in climate adaptation and mitigation.

The objective of this paper is to introduce a framework for promoting school leadership for climate education through the concept of school as learning organization (Kools & Stoll, 2016; Senge, 2006; Senge 2012). By applying Senge's concept of learning organization, school leaders can leverage continuous learning, collaboration, and adaptation as drivers of climate education through 5 core disciplines: (i) systems thinking; (ii) personal mastery; (iii) mental models; (iv) shared vision; and (v) team learning. Systems thinking can enable school leaders to design and implement climate education from a holistic perspective through an interconnected approach rather than through a focus on isolated factors such as students, teachers, and curriculum. Personal mastery relates to lifelong learning and a growth mindset which school leaders can prioritize for themselves, students, and teachers. Priority for developing personal mastery throughout the school can foster a culture motivating students and teachers to innovate approaches toward climate adaptation and mitigation by way of curriculum and experiential learning. Mental models can help school leaders to challenge traditional assumptions that may prevent climate education from being implemented such as the role of stakeholders. The development of a shared vision for climate education can allow school leaders to align students, teachers, parents, and the community at large towards common goals related to climate adaptation and mitigation within the school and community. Team learning would encourage school leaders to foster interdisciplinary, community-based collaboration to promote student-led initiatives in climate adaptation and mitigation for the school community and beyond.

Methodology

Development of the framework will be dependent on integrative literature review addressing learning organization and school improvement as applicable to climate education. Key research questions guiding the integrative literature review include: (i) How is the concept of learning organization applied to schools? (ii) What are the essential components of a learning organization in the context of promoting climate education? (iii) How can existing models for learning organizations inform a framework for school-based climate education? Data sources include peer-reviewed journal articles, books, policy reports, and case studies on learning organization and school improvement with a focus on climate education and associated education for sustainable development and environmental education. Data sources accessed through Google Scholar will meet the following search criteria: (i) published between 2015 and 2025; and (ii) empirical and theoretical studies on the role of school leadership in schools as learning organizations, climate education and school improvement, and school leadership for education for sustainable development. Thematic analysis applied to identify patterns and framework.

Findings

Schools implementing education for sustainable development are effective when they function as learning organizations and prioritize structured policies, strategic visioning, and professional development (Gericke & Torbjörnsson, 2022). School leaders can engage in systems thinking by employing a whole school approach integrating principles of sustainable development into school vision, mission, routines, pedagogical practices, and teacher professional development highlighted in Fig. 42.1 (Mogren et al., 2018). The success of environmental education initiatives is linked to teacher training, school-wide collaboration, and trust-building (Gericke & Torbjörnsson, 2022). For school principals to promote personal mastery, teacher professional development for climate change education implementation will be critical (Novawan & Aisyiyah, 2020). Schools with high levels of trust are more successful in integrating sustainability practices (Gericke & Torbjörnsson, 2022). For school leaders to promote shared vision for climate education, collaboration among students, teachers, and the administration would be a critical enabler. Initial findings do not relate approaches towards mental models for school leaders.

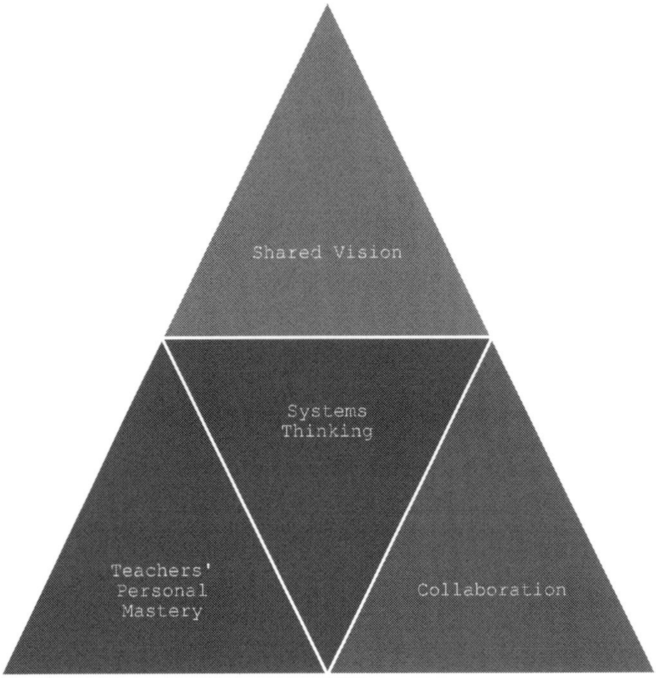

Fig. 42.1 Framework for school leadership for climate education (*Note* Framework is undergoing development based on further research)

Conclusions

A whole school approach grounded in systems thinking is essential for integrating climate education into the fabric of schools. It should be recognized that school leaders have a significant role to enable climate education by dually aligning their schools' vision, mission, pedagogical practices, and teacher professional development with the principles of sustainable development together with systems thinking. By doing so, a foundation for climate education can thrive as part of school culture.

Empowering teachers through personal mastery is a critical step. However, its impact may be limited if school-wide collaboration and trust-building are not operationalized. Both can help to propel teachers' sharing of knowledge and their collective engagement. A school principal's shared vision for climate education has the potential to catalyze collaboration not only between teachers but among students, teachers, and the wider community.

However, the lack of attention to school leaders' mental models suggests that traditional assumptions about education remain in operation which makes the adoption of climate education a challenge. For climate education to be fully integrated into schools, school leaders must question and rethink outdated norms about what and how students should be learning. Overcoming this challenge requires support to school leaders to be bold enough to redefine priorities and positioning climate education among school improvement efforts. Future research needs to focus on school leaders as change agents in reforming learning environments that prepare students to tackle climate challenges faced within the school environment and the community-at-large.

References

Al'Abri, K., Al-Ghaithi, B., Al-Ghafri, A., & AlBarashdi, H. (2024). Achieving sustainable development by promoting global citizenship education (GCED): The role of school principals. *Globalisation, Societies and Education, 1*(1), 1–14. https://doi.org/10.1080/14767724.2024.2330365

Bauman-Hill, K., Dai, S., & Payne, A. (2022). Creating brighter futures: Building climate leaders in the United States through a community-focused curriculum. In F. M. Reimers, U. Amaechi, A. Banerji, & M. Wang (Eds.), *Education to build back better*. Springer. https://doi.org/10.1007/978-3-030-93951-9_8

Ceblano, M. V., Canlas, I. P., & Japay, R. (2023). Leadership for sustainable development in public rural schools. In *SDGs in the Asia and Pacific region* (pp. 1–20). Springer International Publishing. https://doi.org/10.1177/17411143212438220

Cebrián, G., Mogas, J., Palau, R., & Fuentes, M. (2022). Sustainability and the 2030 agenda within schools: A study of school principals' engagement and perceptions. *Environmental Education Research, 28*(6), 845–866. https://doi.org/10.1080/13504622.2022.2044017

Gan, D. (2021). Environmental education leadership: The perceptions of elementary school principals as expressed in their drawings and explanations. *Environmental Education Research, 27*(10), 1440–1466. https://doi.org/10.1080/13504622.2021.1959899

Gericke, N., & Torbjörnsson, T. (2022). Identifying capital for school improvement: Recommendations for a whole-school approach to ESD implementation. *Environmental Education Research, 28*(6), 803–825. https://doi.org/10.1080/13504622.2022.2045256

Kools, M., & Stoll, L. (2016). What makes a school a learning organisation?.

Kwauk, C. (2020). *Roadblocks to quality education in a time of climate change* (Brief). Center for Universal Education at the Brookings Institution.

Mogren, A., Gericke, N., & Scherp, H. Å. (2018). Whole-school approaches to education for sustainable development: A model that links to school improvement. *Environmental Education Research, 25*(4), 508–531. https://doi.org/10.1080/13504622.2018.1455074

Novawan, A., & Aisyiyah, S. (2020). The role of leadership in education for sustainable development curriculum reform in Indonesian higher education. In *Introduction to sustainable development leadership and strategies in higher education* (pp. 145–159). Emerald Publishing Limited. https://doi.org/10.1108/S2055-364120200000022014

Shabalala, N. P. (2024). Distributed leadership as a potential strategy for the management of environmental education curriculum in secondary schools. *Pedagogical Perspective, 1*(X), 255–272.

Stevenson, R. B., Whitehouse, H., & Field, E. (2024). Teaching climate justice education holistically in schools. *Oxford Research Encyclopedia of Education*. https://doi.org/10.1093/acrefore/9780190264093.013.1791

Senge, P. M. (2012). Creating schools for the future, not the past for all students. *Leader to leader, 2012*(65), 44–49. https://doi.org/10.1002/ltl.20035

Senge, P. M. (2006). *The fifth discipline: The art and practice of the learning organization*. Broadway Business.

Chapter 43
Fostering Sustainability and Innovation in Education: An Overview of the Foundational Pillars of the School of Climate Action at Mohawk College

Amanda Smits and Lorraine McKellar

Abstract Postsecondary institutions play a critical role in leading global solutions to climate change and sustainability, and are uniquely positioned to influence local, regional, and global communities through the education of future generations of students, the workforce, and leaders. Drawing on a comprehensive literature review, this extended abstract outlines best practices from sustainability education literature that informed the design of the foundational pillars for Mohawk College's new School of Climate Action. The literature review underscores the need for leadership in sustainability, interdisciplinary and action-oriented climate and sustainability education, in addition to an emphasis on the development of core competencies through diverse teaching methods. The foundational pillars developed based on this comprehensive literature review include Climate Leadership, Academic Programming, Research and Innovation, and Partnerships and Initiatives. This extended abstract highlights the integration of Indigenous Knowledge Systems as fundamental to the school's framework and explores its integration within the college's overall strategic vision. Through additional discussion, the authors provide tangible examples of how the newly formed School of Climate Action may serve as a model for post-secondary institutions seeking to lead in sustainability education, research, community engagement and partnership.

Keywords Sustainability education · Climate action · Climate education · Indigenous knowledge

A. Smits (✉) · L. McKellar
School of Climate Action, Mohawk College, Hamilton, Canada
e-mail: amanda.smits@mohawkcollege.ca

L. McKellar
e-mail: lorraine.mckellar@mohawkcollege.ca

J. Dsilva (ed.), *Sustainability Education, Youth and Innovation*,
https://doi.org/10.1007/978-3-032-21948-0_43

Introduction

Postsecondary institutions play a critical role in leading toward global solutions to climate change and are uniquely positioned to impact local, regional and global communities by educating future generations of students, the workforce, and leaders (Aung & Hallinger, 2022; NASEM, 2020). Mohawk College is taking a leadership role through the creation and launch of the School of Climate Action. This extended abstract explores the emerging themes related to addressing sustainability and climate action, particularly in a postsecondary context through a review of the literature. Beginning with the exploration of Mohawk College's current state, including the connection to strategic priorities and Indigenous Knowledge Systems, the authors will then describe the alignment between emerging themes within literature and the foundational pillars of the School of Climate Action.

Embedded as strategic and academic priorities, Mohawk College has clearly articulated its commitment to sustainability and climate action within the Strategic Plan (Mohawk College, 2022), the Academic Plan (Mohawk College, 2024), and its Climate Action Plan (CAP) (Mohawk College, 2023). More specifically, the College's Strategic and Academic Plans outline several aspirations including making a measurable impact on climate change and launching a School of Climate Action. The CAP commits to decarbonization by setting targets for the college to reach by 2030, as it progresses towards its 2035 goal.

An authentic commitment to the inclusion of Indigenous Knowledge Systems and ways of knowing underpins both Mohawk College and the School of Climate Action. Mohawk College's Centre for Indigenous Relations, Knowledge and Learning (CIRKL) have developed and are implementing Indigenous Knowledge Learning Outcomes (IKLOs) across the institution, with the School of Climate Action collaborating directly with CIRKL. The IKLOs "are an opportunity to decolonize curriculum and education" by meaningfully embedding Indigenous knowledge into college wide curriculum (Mohawk College, n.d.-b). This commitment is also demonstrated through the influence of the Two Row Wampum (Haudenosaunee) and Two-Eyed Seeing (Mi'kmaq) teachings in the School of Climate Action's design and academic programs. The Two Row Wampum teaching "generate[s] respect between [Indigenous and non-Indigenous] groups so they can share the river that sustains all life" (Hill & Coleman, 2019, p. 345), with the two rows as "exactly identical, with no indications of superiority, hierarchy, or preference between them" (Hill & Coleman, 2019, p. 349). Similarly, Two-Eyed Seeing refers to "learning to see from one eye with the strength of Indigenous ways of knowing and from the other eye with the strengths of Western ways of knowing and to use both of these eyes together" (Hatcher et al., 2009, p. 146). By combining these teachings, the College endeavours to raise understanding of Indigenous ways of knowing. Leddy and Miller (2024) argue that "Indigenous ways of knowing are inherently interdisciplinary" (p. 8), which aligns seamlessly with the college's approach to sustainable education.

Methodology

This extended abstract provides an overview of a literature review which was undertaken to inform the development of the four pillars of the School of Climate Action: Climate Leadership, Academic Programs, Research and Innovation, and Partnerships and Initiatives. Leveraging academic search engines, including One Search and Google Scholar, the researchers used a series of key words to identify articles for consideration. Articles were reviewed for relevance to the research topic. In addition to peer reviewed scholarly articles, key resources including sectoral based information and Mohawk College commitments were included for additional context. Ultimately, this comprehensive literature review considered best practices in sustainability education and leadership, ensuring a strong foundation from which to launch the school.

Findings

School of Climate Action's Foundational Pillars

Climate Leadership

The American Colleges & Universities President's Climate Commitment (ACUPCC) requires institutions to create comprehensive plans with immediate and long-term goals for tracking GHG emissions, integrating sustainability into academics, expanding research, and engaging the community (Dyer & Dyer, 2017). A key requirement is publicly sharing plans and progress to ensure accountability. Launched in 2023, Mohawk College's Climate Action Plan (CAP) and reporting on the plan closely reflect ACUPCC priorities, including transparent reporting.

With a focus on integrating sustainability as a social norm within postsecondary institutions, Aung and Hallinger (2022) recommend that sustainability-focused leadership must target the institutional culture as a key leverage point. In addition, sustainability leadership practices include vision, transforming facilities and operations, and integrating sustainability into research and academic programming to develop a culture of sustainability. Senior leadership at Mohawk College have clearly articulated a vision through the Strategic Plan, Academic Plan, and CAP, which are now being actioned.

Beyond the campus, Mohawk College has proven leadership in climate action demonstrated through partnership and collaborations internally and externally with industry and community-based partnerships and initiatives. Nationally, Mohawk College has led a pan-Canadian sectoral and industry partnership with the development and launch of the Canadian Colleges for Resilient Recovery, focused on upskilling and reskilling the workforce for a green and low carbon economy (C2R2, 2024) and is well positioned to show similar leadership around climate action through this new school. Regionally, Mohawk College's Centre for Climate Change Management, an applied research centre, is focused on co-developing decarbonization solutions with partners from industry, government, and within the regional community.

Based on the literature reviewed around sustainability leadership, climate leadership was selected as a pillar of the School of Climate Action as there is a clear need for institutional commitment to climate action, and recommendations around the required to actions needed to create changes in areas such as institutional culture. The School of Climate Action aims to act as a conduit for climate leadership on campus and beyond.

Sustainability-Focused Academic Programs

The academic pillar of this new school underscores the vital role that education plays in equipping future leaders with the knowledge and skills necessary to drive climate action through sustainability education. Sustainability-focused curricula are essential in addressing the wicked problems associated with a changing climate. Designing academic programs with a sustainability lens requires an interdisciplinary approach, with a focus on systems thinking, critical thinking, and the problem-solving skills necessary.

Wiek et al. (2011) define core competencies, as they relate to sustainability, to include "complexes of knowledge, skills, and attitudes that enable successful task performance and problem solving with respect to real-world sustainability problems, challenges, and opportunities" (p. 204). Diverse teaching methods can include incorporating project-based learning, experiential education, as well as work-integrated learning. Ensuring core competencies are well-defined and aligned across curriculum and program development, research, policy, communication and pedagogical approaches at academic institutions is key to ensuring successful program outcomes (NASEM, 2020). The Consensus Study Report published by the National Academy of Science, Engineering, and Medicine (2020) further highlights the critical importance of "incorporate[ing] the principles of diversity, equity, and inclusion in sustainability education programs" (p. 7). It is through this inclusion of diverse experiences, perspectives and knowledge systems that wholistic and interdisciplinary programs are being created and developed within the new school.

These approaches to sustainability education are foundational to the curriculum development process currently being undertaken within the School of Climate Action. The new School of Climate Action will offer a diverse set of in-demand credential programs, including micro credentials, diplomas and degrees, designed using a Two-Eyed Seeing approach, focused on meeting work force requirements, and supporting the incorporation of sustainability-focused learning outcomes across the college-wide curriculum. To ensure learners in these programs are equipped with the skills necessary to solve the complex sustainability challenges a focus has been placed on utilizing diverse teaching methods to develop core competencies can enhance the effectiveness of sustainability programs in higher education (NASEM, 2020).

Research and Innovation

Mohawk College ranks in Canada's top ten research colleges, with applied research being foundational to its mission (Coffey, 2024). Applied research is focused on solving real-world problems with practical solutions. As previously noted, leadership in sustainability is closely tied to several factors, including integration of sustainability into research (Aung & Hallinger, 2022). It is also widely recognized that to solve complex problems, such as climate and sustainability challenges, postsecondary education should be undertaking interdisciplinary approaches and incorporating systems perspectives (Dyer & Dyer, 2017).

The college has placed a strong emphasis on addressing climate-related challenges through several applied centres which bring together academic faculty, students, and industry. Over the last two years, the Centre for Climate Change Management has established partnerships with 25 organizations, undertaken 25 projects, and secured $1.061M in research funding. Additionally, the Energy and Power Innovation Centre offers specialized expertise and facilities to support Canadian companies in addressing operational challenges, with a particular emphasis on energy and resource management. Mohawk College has also been a leader in developing sustainable infrastructure on campus through the construction of buildings such as the Joyce Centre for Partnership and Innovation, Hamilton's first zero-carbon institutional building, and one of the nations largest (Mohawk College, n.d.-c). By implementing and testing climate solutions on campus, leading in applied research and leveraging industry connections, Mohawk College is committed to "influencing local communities and other sectors…to demonstrate that climate action is a legitimate and necessary priority for all organizations" (Dyer & Dyer, 2017, p. 115).

Building on this foundation in applied research, the research and innovation pillar of the School of Climate Action is focused on creating a collaborative hub for climate action research and innovation across the college, with direct connections to industry and workforce partners.

Partnerships and Initiatives

Partnerships and initiatives play a pivotal role in the School of Climate Action and represent the fourth pillar of the school. Azizi's (2022) research suggests that postsecondary institutions "advance the sustainable development of surrounding communities" (p. 46) through curriculum, research and operational perspectives. The Consensus Report (NASEM, 2020) put forth a series of recommendations to strengthen sustainability education that specifies the importance of collaboration with professional, public and private sector partnerships, and cross-sectoral employment tracking.

Considering sustainability education specifically, transdisciplinary partnerships are imperative. Plummer et al.'s (2022) research on transdisciplinary partnerships for sustainability suggests that bringing together academics and non-academics from a variety of disciplines offers the platform to creatively solve sustainability related problems by leveraging diverse knowledge and experiences. Transdisciplinary partnerships directly target sustainability challenges and offer rich learning environments for students. Robinson et al. (2022) suggest that the "inclusion of transdisciplinary experiences will allow students to build the competencies required for solving real-world problems that cannot be learned in a classroom setting" (p. 174).

Mohawk College currently partners with ECO Canada (national) and the SEE Institute in Dubai, UAE (global). Future partnership development will explore connections with other educational institutions, industry leaders, financial institutions, government and non-profit organizations as well as Indigenous communities in collaboration the College's Centre for Indigenous Relations, Knowledge and Learning. These formal and informal partnerships will be reciprocal and interdisciplinary in nature and aim to support learning and developmental opportunities like work integrated learning and industry projects for learners.

Conclusions

In conclusion, the School of Climate Action at Mohawk College can continue to evolve by drawing upon peer reviewed literature focused on the importance of sustainability education, leadership, partnerships and Indigenous Knowledge Systems. Continuing to develop interdisciplinary teaching competencies should remain a focus to guide the development of academic programs that foster essential competencies in sustainability and climate leadership, with a specific focus on work force development and needs. Furthermore, by engaging in activities such as partnership building, the school can expand its impact, connecting diverse knowledge systems from academia, industry, and Indigenous communities. These ongoing collaborations will not only enhance program offerings but also strengthen the college's role in leading local, national, and global solutions to climate and sustainability challenges. Through sustained engagement in both research and partnerships, the School of Climate Action can ensure that it remains at the forefront of climate leadership, education, innovation, and action.

References

Aung, P. N., & Hallinger, P. (2022). Research on sustainability leadership in higher education: A scoping review. *International Journal of Sustainability in Higher Education, 24*(3), 517–534. https://doi.org/10.1108/IJSHE-09-2021-0367

Azizi, L. (2022). Which leadership processes encourage sustainable transitions within universities? *International Journal of Sustainability in Higher Education, 24*(1), 46–68. https://doi.org/10.1108/IJSHE-12-2021-0510

Canadian Colleges for a Resilient Recovery (C2R2). (2024). About Us. https://resilientcolleges.ca/about/

Coffey, S. (2024, December 10). Mohawk College retains top 10 ranking in applied research Mohawk College Newsdesk. https://www.mohawknewsdesk.ca/mohawk-college-retains-top-10-ranking-in-applied-research/

Dyer, G., & Dyer, M. (2017). Strategic leadership for sustainability by higher education: The American college & university presidents' climate commitment. *Journal of Cleaner Production, 14*, 111–116. https://doi.org/10.1016/j.jclepro.2015.08.077

Hatcher, A., Barlett, C., Marshall, A., & Marshall, M. (2009). Two-eyed seeing in the classroom environment: Concepts, approaches and challenges. *Canadian Journal of Science, Mathematics and Technology Education, 9*(3), 141–153.

Hill, R. W., & Coleman, D. (2019). The two row wampum-covenant chain tradition as a guide for indigenous-university research partnerships. *Cultural Studies—Critical Methodologies, 19*(5), 339–359.

Leddy, S., & Miller, L. (2024). *Teaching where you are.* University of Toronto Press.

Mohawk College. (n.d.-a) Land Acknowledgement. https://www.mohawkcollege.ca/land-acknowledgement

Mohawk College. (n.d.-b) Indigenous Knowledge Learning Outcomes (IKLOs). https://www.mohawkcollege.ca/centre-for-indigenous-relations-knowledge-and-learning/learning-education/indigenous-knowledge

Mohawk College. (n.d.-c). The Joyce Centre for Partnership & Innovation. https://www.mohawkcollege.ca/sustainability/buildings-and-energy/joyce-centre-for-partnership-and-innovation

Mohawk College. (2022). Strategic Plan 2022–2027. https://www.mohawkcollege.ca/about-mohawk/leadership-and-administration/mission-vision-and-values/strategic-plan

Mohawk College. (2023). Climate Action Plan 2023–2030. https://climateaction.mohawkcollege.ca/

Mohawk College. (2024). Academic Plan 2024–2029. https://www.mohawkcollege.ca/about-mohawk/leadership-and-administration/mission-vision-and-values/academic-plan-2024-2029

National Academies of Sciences, Engineering, and Medicine (NASEM). (2020). *Strengthening sustainability programs and curricula at the undergraduate and graduate levels.* The National Academies Press. https://doi.org/10.17226/25821

Plummer, R., Blythe, J., Gurney, G. G., Witkowski, S., & Armitage, D. (2022). Transdisciplinary partnerships for sustainability: An evaluation guide. *Sustainability Science, 17*, 955–967. https://doi.org/10.1007/s11625-021-01074-y

Robinson, J., Ariga, A., Cameron, S., & Wang, R. (2022). Reaching the rest: Embedding sustainability in undergraduate student learning. *Journal of Integrative Environmental Sciences, 19*(1), 171–187. https://doi.org/10.1080/1943815X.2022.2131829

Wiek, A., Withycombe, L., & Redman, C. L. (2011). Key competencies in sustainability: A reference framework for academic program development. *Sustainability Science, 6*, 203–218. https://doi.org/10.1007/s11625-011-0132-6

Chapter 44
Environmental Citizenship in the UAE: An Analysis of the Educational System

Tilde Rosmer and Magdalena Karolak

Abstract The aim of this study is to measure the progress of the UAE towards fostering environmental citizenship through its educational system. Using a triangulation of policy and teaching material analyses, as well as interviews with educators, this research aims to be the first of its kind, providing a comprehensive overview of the UAE in this regard. Although not all findings have been achieved yet, when completed, this research will serve as a basis for providing recommendations to other countries in the region and beyond on how to prepare the young generation for a sustainable future. The latter is especially important in the GCC region, which is one of the highest carbon-emitting areas given its dependence on oil extraction.

Keywords UAE · Sustainability · Education · Environmental citizenship · Middle school

Introduction

This study aims to analyze the role of national Emirati and international curricula in the UAE in fostering environmental citizenship and is supported by a Research Incentive Grant from Zayed University. The acceleration of global warming and its dire consequences already felt around the world have put sustainability at the center of planning for societies globally. One of the key factors in creating such preparedness and changing mindsets towards conservation, away from over-consumption, is the educational system (UNESCO, 2016). The European Network for Environmental Citizenship (ENEC) put forward the concept of Education for Environmental Citizenship (EEC), defined as "the type of Education which promotes Environmental Citizenship" (ENEC, n.d.).

There is still a shortage of studies that provide evidence of the specifics of education in the formation of youth into environmental citizens. This is especially true for the Middle East, and particularly for the GCC. Alzeer and Rosmer (2023) assessed the

T. Rosmer (✉) · M. Karolak
Zayed University, Dubai Campus, UAE
e-mail: Tilde.Rosmer@zu.ac.ae

J. Dsilva (ed.), *Sustainability Education, Youth and Innovation*,
https://doi.org/10.1007/978-3-032-21948-0_44

attitudes of Emirati students in tertiary education towards sustainability and found that the level of environmental awareness is generally low. This confirms earlier observations by Ouis (2002) and Luomi, who in 2012 concluded that there is "… no profound environmental awareness among most UAE citizens" (Luomi, 2012:342).

The UAE is at high environmental risk due to its geographic location, and it is one of the world's highest carbon-emitting countries per capita, currently ranked 40th in the Global Sustainability Index (Earth.org, 2025). However, in recent years, the UAE has committed to several strategies to promote sustainable development (DEWA, 2020; Utilities ME, 2022) and has taken a leadership role in organizing the COP28 talks in 2023. Within this context, it is necessary to evaluate the approaches to national education in environmental education. Indeed, education has been identified as a vital element in the global development agenda and lies at the heart of the roadmap to achieve the Sustainable Development Goals (UNESCO, 2016).

Emirati society is unique in the make-up of its population, where the locals are outnumbered ten to one by migrant workers. According to governmental sources, the "… population of the UAE grew to about 10 million by the end of 2021. Out of this, the number of UAE citizens is around 1 million." (U.A.E. 2022). Therefore, the potential environmental citizens are a minority, and this will also impact how sustainable education should be taught most effectively so that it may be put into action. Consequently, in the UAE, sustainability should be a duty of both nationals and expatriates, and it should be fostered in schools that serve multiple communities. Nonetheless, to date, no studies have been conducted in the UAE to assess the role of the educational system in this regard. This study will be the first evaluation of this kind.

Literature Review

So far, most studies that assess questions pertaining to climate change and a green shift in the Middle East and North-African region have focused on technical and economic possibilities of or limitations to renewable energy initiatives (Arouri et al., 2012; Nader, 2009; Sgouridis et al., 2013), as well as on policy, economic development, and international branding (Alhoweish & Orujov, 2016; Gelil, 2016; Luomi, 2012). Few studies, however, have evaluated sustainability in the educational context in the MENA region. The present study will contribute to the emerging field of environmental humanities—an endeavor to embed environmental issues within "human ways of being in the world, and broader questions of politics and social justice" (Rose et al., 2012: 1). As such, this research will move beyond the policy focus of many previous studies on systems and institutions to capture the importance of the interplay between the many actors involved in communicating and shaping perceptions and practices regarding climate change and sustainable energy (Devine-Wright et al., 2017), focusing specifically on the role of the education in achieving a sustainable future.

Methodology

The appreciation for a contextualized approach when studying climate change and sustainability is not new (Adger et al., 2009; Wilhite, 2014; Hanson-Easey et al., 2015). It is especially significant when assessing the local understanding and teaching of environmental and sustainable adaptation to pupils. This research will compare the approaches in governmental and international schools in the UAE, thereby providing a comprehensive overview of the population through a triangulation analysis that combines policy comparisons, educational materials, and interviews with educators. This research is devoted to middle school education, given its foundational role in fostering environmental citizenship (Taylor et al., 2019), specifically in the early years of high school, where pupils become increasingly aware of the world around them and can take actions on both individual and collective levels. Thus, the present study excludes primary schools, high schools, and universities, which could be included in future research. The study's outcomes will provide a comprehensive overview of the UAE's progress in fostering environmental citizenship. Our qualitative approach will proceed from content analysis of educational strategies and curricula mandates, textbook materials, and interviews with educators. As such, this study aims to become the first fully comprehensive research that compares general educational policies and their application in practice within the UAE context. Additional data will be collected from official websites of the relevant organizations and other study materials used for these curricula and courses.

Findings

This study is currently underway, and as it is in its early stages, it is not yet possible to present conclusive findings. However, thus far, we have identified that the many school frameworks operating in the UAE are both a strength and a challenge when it comes to implementing national requirements. In Dubai alone, there are over fifteen international school curricula supervised by the Dubai Knowledge and Human Development Authority (KHDA), in addition to UAE government schools for citizen children overseen by the Ministry of Education. In practice, this means that there are more than 16 different approaches to addressing and incorporating sustainability topics into school curricula in the UAE. In most cases, environmental issues, climate change, and sustainability are taught in an interdisciplinary manner and often in science subjects and geography classes. This is experienced as both an opportunity and a challenge simultaneously.

To illustrate, in British international schools, which constitute the majority of international schools in Dubai, they are free to design their own curriculum for these topics, as long as they address them. Some of the schools we interviewed focus on relevant events in the local context when teaching these topics, such as EXPO 2020 and natural disasters in the region (for example, recent floods in Pakistan). They use interactive methods to teach these topics, including workshops with invited speakers, group work with presentations, and model projects such as a Model COP, providing pupils with an interactive opportunity to engage creatively with the subjects. This also relates the teaching of these issues to current debates about the relevant topics.

However, most teachers interviewed expressed facing challenges and saw room for improvement, particularly due to insufficient coordination between subject departments on what they teach and how, such as the overlap in topics taught in science and geography lessons. Another related challenge identified is the overlap and/or gaps in coverage of these topics due to a lack of coordination. It seems to us that what is needed is to have one administrator responsible for the overall teaching of environmental issues, climate change, and sustainability, to ensure that all necessary topics are covered duly and none are duplicated.

In addition, we observed a lack of mention of school/class visits to local places of relevance, such as the solar plantations in Dubai and/or Abu Dhabi, the Mohammed bin Rashid al Maktoum Solar Park & Sustainable City, and Noor Abu Dhabi, as well as visits to or by local organizations working on sustainability, such as Azraq, EEG, WWW local chapter, Goumbook to mention a few. As part of the contextualization of the teaching of these topics addressed above, we suggest utilizing the local knowledge and engagement for sustainability in the teaching of these subjects.

Regarding governmental schools, according to our interviews with representatives from the Ministry of Education, they have recently revised their policy on teaching environmental issues, climate change, and sustainability. However, they are not yet able to provide us with any outputs from the new curricula and teaching experience. Previously, they had relied on extracurricular programs, such as the Sustainable School Initiative, which provided project-based learning in schools; however, it was up to the individual school to sign up for this, and thus it was not a nationwide experience.

Thus far, we can summarize that there is no concise approach or understanding of how to teach environmental issues, climate change, and sustainability across schools in the UAE. This is why we suggest implementing a new national curriculum to streamline the teaching of these subjects. This should include a bottom-up, whole-school approach that addresses climate change and sustainability in schools, making this topic a practical reality in schools for pupils to experience, rather than simply a theoretical issue they are taught in class. Furthermore, this curriculum should be based on local research and case studies so that the theory is relevant to the reality of the pupils. Such a curriculum would connect the sustainable efforts and implementations of the UAE leaders with the educational reality on the ground among all UAE residents.

Conclusions

This study presents a novel approach to examining the intersections of education strategies and environmental citizenship through a triangulation of policy content analysis, educational materials content analysis, and interviews with educators in middle schools in the UAE. Given this methodological approach, it should be one of the most comprehensive on the topic in general and the only study of this scope conducted to date in the UAE. When finalized, this study is expected to reveal how national strategies are implemented in educational systems, thereby preparing future generations to become environmentally responsible citizens. Based on the findings, it will be possible to provide recommendations to local authorities, schools, and other countries in the GCC and beyond.

References

Adger, N. W., Lorenzoni, I., & O'Brien, K. (2009). Adaptation now. In N. W. Adger, I. Lorenzoni & K. O'Brien (Eds.), Adapting to climate change. Thresholds, values, governance (pp. 1–23). Cambridge University Press.

Alhoweish, B., & Orujov, C. (2016). Promoting an effective energy efficiency programme in Saudi Arabia: Challenges and opportunities. In M. A. Raouf & M. Luomi (Eds.), *The green economy in the Gulf* (pp. 100–123). Routledge.

Alzeer, G., & Rosmer, T. (2023). Emirate youth's perceptions of green branding. In S. Wippel, D. R. Bell (Eds.), Branding the Middle East, Gruyter, Berlin & Boston. Studies on Modern Orient Series (2005). *Liberal Environmental Citizenship, 14*, 179–194.

Arouri, M. E. H., Youssef, A. B., M'henni, H., & Rault, C. (2012). Energy consumption, economic growth and CO_2 emissions in Middle East and North African countries. *Energy Policy, 45*, 342–349.

Devine-Wright, P., Batel, S., Aas, O., Sovacool, B., LaBelle, M. C., & Ruud, A. (2017). A conceptual framework for understanding the social acceptance of energy infrastructure: Insights from energy storage. *Energy Policy, 107*, 27–31.

DEWA (Dubai Electricity and Water Authority). (2020, Jan 5). Renewable energy shaping the future of sustainability. https://www.dewa.gov.ae/en/about-us/media-publications/latest-news/2020/01/renewable-energy-shaping-the-future-of-sustainability. Accessed 27 Apr 2020.

ENEC (n.d.). Defining environmental citizenship. Retrieved April 31, 2023, from https://enec-cost.eu/our-approach/enec-environmental-citizenship/

Gelil, I. A. (2016). Barriers to greening the energy sector in the Gulf Cooperation Council. In M. A. Raouf & M. Luomi (Eds.), *The green economy in the Gulf* (pp. 25–53). Routledge.

Hanson-Easey, S., Williams, S., Hansen, A., Fogarty, K., & Bi, P. (2015). Speaking of climate change: A discursive analysis of lay understandings. *science Communication, 37*(2), 217–239.

Luomi, M. (2012). *The Gulf Monarchies and climate change: Abu Dhabi and Qatar in an era of natural unsustainability*. London: Hurst.

Nader, S. (2009). Paths to a low-carbon economy—the Masdar example. *Energy Procedia, 1*(1), 3951–3958.

Ouis, P. (2002). 'Greening the Emirates': The modern construction of nature in the United Arab Emirates. *Cultural Geographies, 9*(3), 334–347. https://doi.org/10.1191/1474474002eu252oa. Accessed 18 Feb 2021.

Rose, D. B., van Dooren, T., Chrulew, M., Cooke, S., Kearnes, M., & O'Gorman, E. (2012). Thinking through the environment, unsettling the humanities. *Environmental Humanities, 1*(1), 1–5.

Sgouridis, S., Griffiths, S., Kennedy, S., Khalid, A., & Zurita, N. (2013). A sustainable energy transition strategy for the United Arab Emirates: Evaluation of options using an integrated energy model. *Energy Strategy Reviews, 2*(1), 8–18.

Taylor, N., Quinn, F., Jenkins, K., Miller-Brown, H., Rizk, N., Prodromou, T., Serow, P., & Taylor, S. (2019). Education for sustainability in the secondary sector—a review. *Journal of Education for Sustainable Development, 13*(1), 102–122. https://doi.org/10.1177/0973408219846675

UNESCO. (2016). *Education 2030: Incheon declaration and framework for action—towards inclusive and equitable quality education and lifelong learning for all.* UNESCO.

Wilhite, H. (2014). The problem of habits for a sustainable transformation. In K. V. L. Syse & M. L. Mueller (Eds.), *Sustainable consumption and the good life: Interdisciplinary perspectives.* Routledge.

Chapter 45
Integrating Sustainability into the Core of University Governance: Challenges and Opportunities for Transforming Institutional Practices

Victor Lobo

Abstract Higher education institutions play a pivotal role in advancing global sustainability by shaping the knowledge, values, and actions of future leaders. However, despite their significant potential, many universities still approach sustainability as an ancillary concern rather than an integral part of their governance. This paper examines the challenges and opportunities of integrating sustainability into the core of university governance. Specifically, it explores how universities can align their institutional practices—ranging from decision-making processes to resource management—with the principles of sustainability. The research draws on case studies of leading universities worldwide, analysing their governance structures, sustainability policies, and engagement with the wider community. Through qualitative interviews with administrators, faculty, and students, the paper identifies key barriers such as resistance to change, institutional inertia, and limited financial resources, while also highlighting successful strategies for overcoming these challenges. The findings suggest that while significant hurdles remain, the integration of sustainability into university governance offers transformative potential, not only for institutional operations but also for the broader societal impact of higher education. Recommendations are provided for universities seeking to transition to sustainable governance models, emphasizing the importance of leadership commitment, stakeholder engagement, and long-term planning.

Keywords Sustainability governance · Higher education · Institutional change · University leadership · Transformative practices

V. Lobo (✉)
St Joseph's University, Bengaluru, India
e-mail: victorlobo@sju.edu.in

J. Dsilva (ed.), *Sustainability Education, Youth and Innovation*,
https://doi.org/10.1007/978-3-032-21948-0_45

Introduction

The urgency of addressing climate change and social inequality has made sustainability a critical issue across all sectors. Higher education institutions are crucial in addressing the global sustainability challenges, as influential institutions that shape future policymakers, business leaders, and educators, must lead by example in integrating sustainability into their core operations (Shih, 2025; Lozano, 2006). They are also imperative in achieving the United Nations Sustainability Development Goals (UNSDGs).

Universities have been at the forefront in creating and breaking paradigms, and educating the future decision-makers, entrepreneurs, and leaders (Cortese, 2003). But universities remains traditional (Elton, 2003) which remains as a real challenge in incorporating sustainability into institutional governance. The objective of this paper is to explore the opportunities and challenges associated with embedding sustainability in the governance frameworks of universities. While many universities have adopted sustainability as part of their curriculum or research agendas, integrating sustainability into institutional governance remains a less explored but vital frontier (Lozano, 2006).

The role of higher education institutions in achieving sustainability is critical, since they serve as pivotal drivers in educating and developing future leaders equipped to implement the United Nations Sustainable Development Goals (SDGs). It influence the education sector and broader societal dimensions, shaping policies, technologies, values, and mind-sets essential for global sustainability. Despite this significant potential, challenges remain in how such can effectively contribute to sustainability transitions (Shih, 2025). A key issue mentioned by Wals et.al, 2017 is the geographic heterogeneity in the implementation of SDGs, where varying regional contexts necessitate tailored approaches in implementation (Arjen & Wals, 2017).

This research investigates how universities can redesign governance structures, policies, and strategic goals to foster sustainability at an organizational level.

Literature Review

A growing body of literature emphasizes the role of higher education in sustainable development. Studies by Svensson & Wagner highlight the challenges universities face in implementing comprehensive sustainability practices at the governance level. These include bureaucratic resistance, lack of financial incentives, and insufficient cross-departmental collaboration. Conversely, research by Wright & Wilton suggests that universities that successfully integrate sustainability into their governance structures often achieve long-term operational efficiencies, improved reputation, and greater student and staff satisfaction.

Research Questions:

1. What are the primary barriers to integrating sustainability into university governance?
2. What best practices can universities adopt to overcome these barriers?
3. How can university leadership foster a culture of sustainability across all levels of governance and operations?

Methodology

This research employs a qualitative case study approach to analyze the governance structures of leading universities that have made significant strides in integrating sustainability into their operations. The study was carried out in four stages. The first stage involves an extensive literature review to gather and understand the existing studies, policy frameworks and documentations including annual reports on sustainability in higher education. This resulted in identifying the focus area for each institutions and in general the research gaps.

The second stage was to conduct stakeholder mapping and sampling to ensure a diverse collection of respondents. Followed by this in the third phase data collection was conducted through semi-structured interviews with the identified stakeholders. Data was collected through semi-structured interviews with 30 stakeholders including university administrators, sustainability officers, faculty and students from ten universities globally, including institutions in Europe, North America and Asia. Additionally, an institutional document analysis was conducted, reviewing sustainability policies, strategic plans, annual reports and governance frameworks from these universities.

In the fourth phase a detailed analysis of collected resources and data obtained through interviews and case studies was undertaken to draw conclusions.

The research framework draws on institutional theory, which proposes that organizational behaviour is shaped by societal norms and external pressures. This theoretical lens helps to explore how universities governance structures are shaped by internal and external pressures related to sustainability. The framework also includes change management principles to assess how universities can effectively implement sustainability into their governance systems, ensuring that it is not just a superficial add-on but embedded in institutional culture and practices. Change management principles were also incorporated to analyse how sustainability can be incorporated into governance systems.

Data analysis allowed for the identification of recurring patterns, such as barriers and best practices, across the case studies and interviews. Document analysis complemented this by providing contextual evidence of institutional commitments.

Case studies included sustainability achievements and governance strategies of global universities. Semi-structured interviews were conducted with 30 stakeholders from universities recognized for their sustainability leadership. Key documents including sustainability policies, strategic planning documents, and annual reports were analyzed for insights into institutional commitments to sustainability.

Findings and Discussions

The current study shed light on the grey areas which challenge the incorporation of sustainability into education. In most cases, sustainability is treated as a stand-alone subject rather than being incorporated into the major courses.

(i) Institutional Inertia: Many universities have deeply entrenched governance structures that prioritize academic and administrative functions over sustainability initiatives. This institutional inertia can make it difficult to integrate sustainability into the heart of institutional decision-making
(ii) Less acceptance to changes: Despite the growing importance of sustainability, many university leaders remain focused on traditional academic priorities, such as research output and institutional rankings, rather than on sustainability outcomes.
(iii) Lack of Financial Incentives: Universities often struggle to secure the necessary funding for sustainability initiatives due to budget constraints, leading to prioritization of short-term goals over long-term sustainability projects.

On the other hand, the research also identifies several successful strategies for overcoming these barriers:

(i) Leadership Commitment: Institutions with strong, top-down leadership commitment have been more successful in embedding sustainability within governance frameworks. For example, The University of Cambridge has incorporated sustainability into its strategic goals, with the Vice-Chancellor leading sustainability efforts through a Sustainability Leadership Team that coordinates initiatives across all departments.
(ii) Cross-Departmental Collaboration: Universities like The University of California, Berkeley have formed interdisciplinary sustainability committees to ensure that all faculties are involved in sustainability governance, from the humanities to the natural sciences.
(iii) Stakeholder Engagement: Successful institutions actively involve students, faculty, staff, and external stakeholders in decision-making processes regarding sustainability, fostering a sense of shared ownership. The University of Edinburgh has established a Sustainability Advisory Group, which includes representatives from all university sectors, ensuring broad engagement in sustainability decisions.
(iv) Integrating Sustainability into Governance Frameworks: Universities such as Monash University have created a Sustainability Governance Model that ties sustainability goals to key performance indicators (KPIs) in the university's annual review process, making sustainability a central aspect of institutional governance.

Table 45.1 Governance practices for integrating sustainability

University name	Key governance practice	Outcome
University of Cambridge	Sustainability leadership team	Enhanced integration of sustainability across departments
University of California, Berkeley	Cross-departmental sustainability committees	Improved coordination and policy implementation
University of Edinburgh	Sustainability advisory group	Increased stakeholder engagement and transparency
Monash University	Sustainability governance model	Stronger alignment between sustainability goals and institutional KPIs

The Role of Sustainability in Institutional Reputation

Integrating sustainability into governance has also been shown to significantly enhance the reputation of universities as shown in Table 45.1. Research institutions that embed sustainability within their governance structures are increasingly viewed as leaders in corporate social responsibility (CSR), attracting students, faculty, and funding from sustainability-conscious stakeholders. According to a report by Universitas Indonesia, universities with clear sustainability policies are ranked higher in global sustainability indices, which in turn improves their global positioning and influence.

Conclusion

This research underscores the importance of integrating sustainability into the core governance structures of higher education institutions. Universities that take proactive steps to embed sustainability in decision-making processes, leadership strategies, and operational frameworks are better positioned to meet the challenges of climate change and social inequality. However, several barriers—such as institutional inertia, resistance from leadership, and limited financial resources—must be overcome to achieve lasting change.

The study is limited by its focus on a relatively small number of global universities, which may not fully represent the diversity of challenges and opportunities faced by universities in different regions. Further research is needed to explore the specific needs of universities in the Global South and how they can integrate sustainability in resource-constrained environments.

For universities looking to integrate sustainability into governance, the research highlights several key actions:

(i) Establish clear leadership structures that prioritize sustainability at the highest levels.

(ii) Foster interdisciplinary collaboration across departments and faculties.
(iii) Engage stakeholders across all university sectors to create a shared vision of sustainability.
(iv) Align sustainability with institutional KPIs to ensure it is prioritized in strategic planning.
(v) Universities should develop clear, actionable policies that define sustainability goals and outline steps for their implementation
(vi) Universities must allocate financial and human resources to sustainability initiatives, recognizing them as long-term investments.
(vii) Engage students actively in sustainability governance through leadership roles in sustainability committees and task forces.
(viii) Universities should seek partnerships with local governments, businesses, and NGOs to amplify the impact of their sustainability efforts.

References

Arjen, E. J., & Wals, A. B. (2017). Can we meet the sustainability challenges? The role of education and lifelong learning. *European Journal of Education*, 404–413.

Cortese, A. (2003). The critical role of higher education in creating a sustainable future. *Elsiever, planning for Higher Education*, 15–22.

Elton, L. (2003). Dissemination of innovations in higher education: A change theory. *Tertiary Education and Management*, 199–214.

Lozano, R. (2006). Incorporation and institutionalization of SD into universities: Breaking through barriers to change. *Journalof Cleaner Production*, 787–796.

Shih, Y. H., & Hsu, M. C. (2025). Sustainability transformations in higher education: Global perspectives on the challenges and solutions. *International Journal of Education and Humanities (IJEH)*, 126–139.

Chapter 46
How Educational Institutions Promote Sustainability Knowledge, and Innovation

Kaunain Shahidi

Abstract This paper explores how educational institutions can drive sustainability through integrated curricula, interdisciplinary research, and leadership development while leveraging behavioural science and cultural relevance. It emphasizes the importance of embedding sustainability across all academic disciplines, fostering experiential learning, and promoting active student engagement through nudge theory, neuroscience, and behavioural psychology. Special attention is given to Islamic values such as stewardship and moderation, highlighting their alignment with modern sustainability principles, especially in the context of the UAE. By drawing on successful international case studies and proposing actionable strategies, including technological innovations, community collaboration, and government support, the study outlines a holistic model to cultivate a culture of sustainability among students and academic leaders. The paper advocates for systemic, culturally grounded, and evidence-based reforms in sustainability education that support national strategies like UAE Vision 2031 and Net Zero 2050.

Keywords Sustainability education · Behavioural insights (Nudge Theory & Neuroscience) · Islamic environmental stewardship · Interdisciplinary research · UAE Vision 2031 · Net Zero 2050

Introduction

Educational institutions play a crucial role in addressing global sustainability challenges by equipping students with the necessary knowledge, skills, and mindset to drive sustainable development. Recent advances in neuroscience are highlighting connections between emotion, social functioning, and decision making that have the potential to revolutionize our understanding of the role of affect in education (Immordino-Yang & Damasio 2007). Theoretical concepts such as nudge theory,

K. Shahidi (✉)
Sustainability and Packaging Expert, Sharjah, United Arab Emirates
e-mail: kaunain.shahidi@almarai.com

J. Dsilva (ed.), *Sustainability Education, Youth and Innovation*,
https://doi.org/10.1007/978-3-032-21948-0_46

neuroscience, and behavioural psychology offer valuable insights into fostering pro-sustainability behaviours. Research suggests that integrating behavioural insights into sustainability education enhances student engagement and helps them retain sustainability principles more effectively. Additionally, aligning sustainability initiatives with cultural values, such as the Islamic principles of moderation and environmental stewardship, strengthens their relevance in regions like the Middle East. This paper explores how educational institutions incorporate sustainability into curricula, research, leadership, and technology, shaping a more sustainable future, with a specific focus on the UAE education framework.

Methodology

This study adopts a qualitative approach, drawing from literature, case studies, and policy reports to assess how effectively sustainability education is integrated into academic institutions. Data sources include peer-reviewed journals, university sustainability reports, and government policy documents. The study focuses on leading institutions worldwide that have successfully embedded sustainability into their curricula and operations. A purposive sampling method was used to highlight best practices. The research follows a thematic content analysis approach to identify key trends and strategies employed by these institutions, with a specific lens on how such models can be tailored for the UAE's education system.

Significance and Contribution

Unlike studies that primarily emphasize policy and curriculum changes, this paper highlights the integration of behavioural science and cultural values into sustainability education. It underscores how neuroscience and psychology shape attitudes toward sustainability, offering actionable insights for policymakers, educators, and institutional leaders aiming to cultivate lasting behavioural change among students to align with UAE's national strategies (e.g., UAE Net Zero 2050, Vision 2031).

Uniqueness

This paper stands out by bridging behavioural science, cultural values, and sustainability education in a comprehensive manner tailored to the UAE's needs getting insights from neuroscience and psychology to enhance engagement. Additionally, it integrates Islamic environmental stewardship principles, making sustainability education more culturally relevant in the UAE. The paper also emphasizes technological advancements such as AI and gamification as tools to make sustainability learning more interactive and impactful. Its recommendations move beyond theoretical discourse, offering practical, research-backed strategies for educational institutions, policymakers, and industry leaders to foster a sustainability mindset among students.

Limitations

This paper relies on secondary sources rather than primary field data. Future research could involve direct engagement with students and educators through surveys or interviews to gain first-hand empirical insights.

Sustainability in Curriculum Design

Integrating sustainability into the curriculum is a foundational step for educational institutions to instil environmental and social consciousness. Sustainability-focused courses can encourage critical thinking, problem-solving, and decision-making skills. Curricula that emphasize real-world applications can prepare students to address global challenges such as climate change, resource depletion, and social inequities. Research suggests that interdisciplinary learning plays a key role in building a holistic understanding of sustainability.

1. Integrated Learning: Embedding sustainability topics across disciplines can ensures a comprehensive understanding. Sustainability should be infused across various disciplines. For instance, business courses can introduce circular economy concepts, while engineering programs can focus on renewable energy innovations tailored to UAE's unique climate.
2. Active Learning: Experiential learning through projects, fieldwork, and internships would allow students to engage with sustainability challenges first-hand. Examples include participation in clean energy projects or urban greening initiatives or beach cleaning etc.
3. Skill-Based Learning: Curriculum modules can integrate technical skills such as life cycle assessment, carbon footprint analysis, and sustainable design principles, equipping students to understand the importance and innovate effectively and prepare them for real-world challenges in the UAE's growing green economy.

Interdisciplinary Research for Sustainability

Collaboration across disciplines is essential for addressing the multifaceted nature of sustainability. Educational institutions are uniquely positioned to foster interdisciplinary research that combines science, technology, social sciences, and humanities to create innovative solutions they can establish research centres dedicated to sustainable development.

1. Collaborative Research Centres: Establishing sustainability research hubs within institutions can promotes synergy between departments. For example, partnerships between environmental science and business schools can lead to sustainable business models.
2. Community Engagement: Research initiatives involving local communities can create impactful solutions tailored to specific regional challenges. For instance, water conservation projects in arid regions can combine hydrology, engineering, and cultural studies.
3. Technology and Data: Leveraging artificial intelligence, big data, and IoT for sustainability research can provide powerful tools for addressing complex environmental issues. Students and researchers can work on predictive models for climate adaptation or resource optimization.

The Role of Schools and Educational Institutions in Advancing Sustainability Leadership

Educational institutions must nurture future sustainability leaders by instilling a sense of responsibility, innovation, resilience and ethical decision-making. Leadership programs and mentorship initiatives can empower students to champion sustainability within their communities and beyond. Education for Sustainable Development (ESD) learning is sometimes interpreted as the process of gaining knowledge, values and theories related to sustainable development but it also prioritises the changing of mindsets and active engagement of the learner in matters relating to more sustainable futures. (Tilbury 2011).

1. Leadership Development: Programs emphasizing sustainability leadership can equip students with soft skills such as communication, collaboration, and ethical decision-making.
2. Nudge Theory in Leadership Training: Nudge theory suggests that recognition-based incentives can encourage sustainable behaviours among students. Subtle behavioural interventions, such as recognition for sustainable practices or incentives for participation in green initiatives, would encourage students to embrace leadership roles in sustainability.
3. Student-Led Initiatives: Encouraging students to lead sustainability clubs or organize events would fosters a sense of ownership and responsibility.

Insights from Nudge Theory, Neuroscience and Behavioural Psychology

Behavioural psychology and neuroscience can enhance sustainability education. The integration of behavioural insights shall be looked into to strengthen the efficacy of sustainability education by aligning it with the way humans learn, think, and act. Learning from others is how individuals and societies develop. Humans are easily influenced by the statements and actions of others because we naturally like to conform. One of the most effective ways to encourage change is through social influence. (Thaler & Sunstein 2008)

1. Nudge Theory: Small, strategic interventions can encourage sustainable behaviours. For instance, placing recycling bins in visible locations on campus nudges students toward waste segregation.
2. Neuroscience: Emotional engagement enhances learning. Storytelling, such as sharing success stories of sustainable innovations, stimulates empathy and inspiration, engaging the brain's reward systems.
3. Behavioural Psychology: Techniques like positive reinforcement and habit formation can create lasting behavioural changes. For example, offering certificates and wider publication for completing sustainability workshops reinforces participation.

Islamic Cultural Influence in the UAE and Middle East

Islamic values and cultural practices in the UAE and Middle East provide a strong foundation for promoting sustainability. Principles of stewardship (Khalifa), moderation (wasatiyyah), and conservation resonate deeply with modern sustainability goals.

1. Stewardship of Resources: The Quran emphasizes the responsibility to protect and preserve the Earth. Integrating these teachings into education fosters a moral obligation toward sustainability.
2. Community and Generosity: Islamic traditions of collective well-being encourage societal efforts toward sustainable development. For instance, community-based recycling or water conservation projects align with cultural practices of mutual support. The holy month of Ramadhan can be effectively used to inculcate the spirit of fasting (consuming less).
3. Minimalism and Moderation: The Middle Eastern lifestyle, historically shaped by resource scarcity, emphasizes minimalism and frugality. Educational initiatives can draw upon these values to promote a lifestyle of reduced consumption and waste.

Creating Sense of Responsibility and Passion

Engaging youth and society in sustainability requires fostering a sense of ownership and enthusiasm. Educational institutions can employ the following methods:

1. Interactive and Experiential Learning: Immersive activities, such as sustainability challenges or eco-tours, make learning engaging and impactful.
2. Role Models and Mentorship: Showcasing sustainability leaders and providing mentorship opportunities inspire students to follow similar paths.
3. Community Service: Encouraging participation in local sustainability initiatives cultivates a sense of responsibility and connection.
4. Positive Social Norms: Highlighting sustainable behaviours within peer groups reinforces their importance and normalizes them.
5. Recognition and Rewards: Acknowledging efforts, such as awarding sustainability champions, motivates continued engagement.

Collaborative Approach of Society and Educational Institutions

A collaborative approach between society and educational institutions is essential for creating sustainable futures. Schools and universities must lead by example through green campus initiatives, partnerships with industries, and community outreach programs.

1. Green Campuses: Implementing eco-friendly practices on campuses, such as renewable energy usage and waste reduction, demonstrates sustainability in action.
2. Community Collaboration: Partnering with local organizations and governments on sustainability projects amplifies impact and promotes collective responsibility.
3. Industry Partnerships: Collaborations with industries provide students with exposure to real-world sustainability challenges and solutions, fostering practical learning and innovation.

Case Studies Applying Nudge Theory in Education

1. Japan—Recycling and Waste Management: Japanese schools have successfully integrated waste segregation into daily routines. By using visual nudges, such as color-coded and transparent bins and signage, students are subtly encouraged to sort their waste correctly. This approach, combined with an educational curriculum on environmental stewardship, has resulted in a high level of recycling compliance and awareness among young learners.

2. USA—Sustainable Cafeterias: Universities like the University of California have implemented nudge strategies in campus dining facilities to promote sustainable food choices. Simple interventions, such as placing plant-based options at the start of buffet lines and using appealing descriptions, have increased the consumption of sustainable meals, reducing the campus's carbon footprint.
3. Denmark—Energy Conservation: Danish schools use energy dashboards that display real-time energy consumption in classrooms. This visual feedback acts as a nudge, encouraging students and teachers to switch off lights and devices when not in use, fostering a culture of energy conservation.

Changing Lifestyles, Habits, and Attitudes

Encouraging sustainable living involves reshaping lifestyles, daily habits and attitudes. Human brains are highly cooperative systems. Nonetheless, they are not uniformly structured networks, for they consist of many networks that are themselves connected in various ways. (Varela et al., 2020). Educational programs can focus on:

1. Minimalism: Promoting reduced consumption and mindful purchasing habits.
2. Eco-Friendly Practices: Encouraging behaviours such as using reusable items, conserving water, and minimizing energy use.
3. Long-Term Thinking: Instilling the importance of considering future generations in decision-making processes.
4. Collaborative Efforts: Facilitating group activities that emphasize collective responsibility, such as community clean-ups or tree planting drives.

Leveraging Technology for Sustainability Education

1. Gamification: Interactive apps and games can teach sustainability principles in an engaging manner, fostering healthy competition and learning.
2. Virtual Reality (VR): VR experiences that simulate real-world environmental challenges can immerse students in sustainability scenarios, enhancing their understanding and empathy.
3. Online Learning Platforms: Massive Open Online Courses on sustainability topics ensure wider accessibility and continuous learning opportunities.

The Role of Governments and Policy Support

Governments play a critical role in enabling educational institutions to prioritize sustainability. Policymaking that incentivizes green campuses, funds sustainability programs, and integrates sustainable practices in education creates a robust framework for long-term impact. Collaboration between government bodies and educational institutions ensures alignment with national and regional sustainability goals.

Recommendations for Implementation in the UAE:

1. Policy & Institutional Commitment–Establish a nationwide sustainability education policy, embedding sustainability in curricula and campus operations, with mandatory reporting by educational institutions.
2. Curriculum Enhancement–Integrate sustainability modules into all academic programs and train educators in sustainability-focused teaching. Including a module with focus of 'Sustainable Lifestyles' (as way of living, social behaviours and choices to minimize environmental degradation)
3. Research & Industry Collaboration–Promote university-business partnerships for applied research and support green start-ups with funding and job security.
4. Student Engagement–Launch sustainability competitions and student councils for green initiatives. Encourage with awards and recognition and publish on social media.
5. Impact Measurement–Develop KPIs and use AI-driven analytics to track progress and environmental impact.

Conclusion

Educational institutions serve as powerful agents of change in promoting sustainability knowledge and innovation. By incorporating sustainability into curricula, fostering interdisciplinary research, and leveraging behavioural insights, they prepare students to tackle global challenges effectively.

Drawing upon Islamic cultural values and integrating them with modern educational strategies further enriches this mission, ensuring alignment with regional traditions and lifestyles. Through targeted efforts to inspire responsibility, change behaviours, and cultivate a passion for sustainability, schools, educational institutions and universities can empower youth to lead sustainable lifestyle. Future studies should focus on empirical validation through primary research to further refine and strengthen these educational strategies.

References

Immordino-Yang, M. H., & Damasio, A. (2007). We feel, therefore we learn: The relevance of affective and social neuroscience to education. *Mind, Brain, and Education, 1*(1), 3–10.

Thaler, R. H., & Sunstein, C. R. (2008). *Nudge: Improving decisions about health, wealth, and happiness*. Yale University Press.

Tilbury, D. (2011). *Education for sustainable development: An expert review of processes and learning*. UNESCO.

Varela, F. J., Thompson, E., & Rosch, E. (2020). The embodied mind: Cognitive science and human experience.

Chapter 47
Empowering Sustainability Engagement in Higher Education: A Competency-Based Approach to Advancing SDG 4

Lana Hiasat, Wahida Dastakeer, and Christine Coombe

Abstract Youth today are more aware of climate issues than ever, yet this often does not lead to meaningful action which is seen as a gap between values and behavior. This study explores how competency-based learning (CBL) can close that gap by giving university students the skills and mindset to engage with sustainability in real ways. Focusing on Sustainable Development Goal 4.7, the researchers of this study used surveys, focus groups, and capstone project analysis to assess Emirati students in sustainability and future foresight courses. Comparing cohorts from 2022 and 2025, we examined the impact of major shifts like COP28 and Generative AI in education. Results show that CBL, when linked to real-world events and digital tools, builds systems thinking, agency, and global citizenship. The 2025 cohort showed stronger future-readiness and creative problem-solving. The findings offer a roadmap for designing education that prepares students for complex, fast-changing futures.

Keywords Sustainability engagement · Competency-based learning · Higher education · Future readiness · SDG 4 · Classrooms of tomorrow · Student empowerment

Introduction

Despite growing environmental awareness among youth, higher education institutions still struggle to convert that concern into sustained action, a phenomenon known as the value–action gap (Chaplin & Wyton, 2014). This paper explores how competency-based learning (CBL) can bridge that gap by equipping students with future-ready skills for sustainability leadership, in line with SDG 4.7's call for education that fosters knowledge, values, and global citizenship (UNESCO, 2017).

L. Hiasat (✉) · W. Dastakeer · C. Coombe
Higher Colleges of Technology, Dubai, United Arab Emirates
e-mail: lhiasat@hct.ac.ae

J. Dsilva (ed.), *Sustainability Education, Youth and Innovation*,
https://doi.org/10.1007/978-3-032-21948-0_47

Recent studies highlight the importance of interdisciplinary and justice-oriented pedagogies that go beyond content delivery, instead promoting systems thinking, agency, and real-world engagement (Bartlett et al., 2022; Membrillo-Hernández et al., 2021; Conroy et al., 2024). However, in the UAE and broader Gulf region, sustainability education often remains fragmented and overly focused on environmental practices, lacking holistic integration (Al Mahameed et al., 2023). This study investigates how a CBL framework, which was enhanced by generative AI tools and contextualised by real-world events like COP28, can empower Emirati university students to take meaningful sustainability action.

Using a mixed-method approach, we surveyed students, facilitated focus groups, and analysed capstone projects from two cohorts (2022 and 2025) enrolled in sustainability and foresight courses. Findings indicate that the 2025 cohort, exposed to CBL and Generative AI tools, demonstrated stronger systems thinking, greater confidence in sustainability action, and increased global orientation.

Methodology

This study employed a mixed-methods approach to investigate the impact of competency-based sustainability applied projects in two courses (Introduction to Sustainability and Applied Skills capstone) on student learning and engagement. The intervention took place at a major Emirati university during the Spring 2024 semester, coinciding with national efforts to raise climate change awareness after the successful Dubai COP28 event and the country's focus on the year of sustainability from 2023 to 2024. The projects were designed around real-world sustainability challenges aligned with SDGs (e.g. clean energy, sustainable cities) and required students to demonstrate defined competencies such as systems thinking, problem-solving, and community engagement. Following a competency-based learning framework, students worked on team projects addressing local sustainability issues (such as campus waste reduction and energy efficiency), reflecting the pedagogical shift towards experiential and problem-oriented learning (Membrillo-Hernández et al., 2021). To ensure cultural relevance, project topics were grounded in UAE's environmental context and national sustainability initiatives, echoing recommendations that local context and socio-cultural factors should be incorporated into sustainability curricula (Al Mahameed et al., 2023).

Researchers in this study have observed several problems in relation to sustainability and attitudes towards climate issues. The first problem is related to the observed attitudes and behaviours toward climate and sustainability issues. Mindsets, attitudes and behaviours of students in this region regarding sustainability issues do not appear to be altered using the current traditional strategies applied in higher education, such as the delivery of mandatory sustainability courses in the curriculum, running environmental campaigns, holding talks and webinars. It is observed that students tend to revert to their old habits once they are done with the activity. The second observed problem appeared to be on the other spectrum of awareness, where sustainability attitudes are already ingrained in society, and there is a sense of helplessness towards climate action. Scientists have observed these experiences as grief towards sustainability issues. A high level of awareness is causing anxiety or what is known as climate change anxiety and grief (Thompson, 2024).

We explored the following research questions:

RQ 1: *What are the observed factors that appear to impact the attitudes and behaviours toward climate and sustainability issues?*

RQ 2: *What are the stages of climate shock?*

RQ 3: *What are the proven effective tools in dealing with each stage of climate shock?*

These research questions were investigated through mixed-method surveys and focus group interviews with Emirati undergraduate students enrolled in the sustainability and applied skills capstone courses, drawn from diverse disciplines, including engineering, business, computer sciences and applied media. This multidisciplinary cohort was intentional, as recent studies highlight the importance of engaging students from all academic backgrounds in sustainability learning, not just those in environmental majors (Mokski et al., 2023). By involving a broad range of majors, the module aimed to foster interdisciplinary collaboration and mirror real-world diversity in problem-solving teams.

Findings

Several key findings resulted from quantitative survey analysis and thematic analysis of focus group interviews.

Key Finding 1: Enhanced Sustainability Competencies

Students reported an improved understanding of interdisciplinary aspects of sustainability; they recognised connections between environmental, social, and economic issues more clearly after working on their projects, especially evident in the applied skills capstone projects. This aligns with the notion that broad, "big tent" frameworks in sustainability education can widen participation and understanding across diverse topics (Conroy et al., 2024). Moreover, students' systems thinking abilities, which were assessed through scenario-based questions in the survey, showed marked improvement. Many students moved from a narrow focus on environmental issues (e.g., recycling or energy saving) to a more holistic view that included community and policy dimensions, addressing a common gap identified in regional studies (Al Mahameed et al., 2023).

Qualitative feedback underscored this shift. Several students mentioned in the reflective open-ended questions that initially, they thought sustainability was "*mostly about environmental conservation,*" but by the end, they saw it as "*a broad approach involving society, economy, and environment together.*" This reflects a deeper internalisation of the multi-dimensional nature of sustainability, which is a key learning outcome for transformative sustainability education (Walsh et al., 2020).

Key Finding 2: Empowerment and Self-Efficacy

The CBL approach, coupled with the use of Generative AI, appeared to empower students significantly. The 2025 cohort results indicated an increase in students' self-efficacy scores related to taking sustainability action. Students felt more confident in their ability to contribute to solutions for climate and sustainability challenges, which the researchers believed was primarily due to exposure to the COP28 events and the country's strong emphasis on the Year of Sustainability. Many of our students had opportunities to participate in the COP28 events, especially those that were held in our city of Dubai.

This finding resonates with Bartlett et al. (2022), who found that pedagogies emphasising justice and practical action can boost students' confidence to engage in climate action. In our focus groups, participants frequently mentioned that completing a real project (such as designing a campus recycling campaign or proposing a green transportation plan) made them feel "*like we can actually make a difference.*" One student noted, "*Instead of just learning theory, we applied it. Now I know how to start a sustainability initiative on campus, and that makes me feel capable.*" Such expressions of agency are indicative of transformative learning, wherein students shift from passive learners to active change agents.

The integration of the AI assistant also contributed to student empowerment. The teacher for the applied skills capstone developed a custom AI GPT to support students' systems thinking and problem-solving skills while working on their applied projects. About 80% of students reported that the AI tool was helpful for brainstorming and providing quick access to information. One focus group participant mentioned, "*Using the AI felt like having a tutor available 24/7. It guided us when we were stuck, which gave us the confidence to try more creative solutions*". This reflects how technology, when thoughtfully integrated, can enhance learning by building students' confidence to tackle complex problems (Mokski et al., 2023). However, a few students (approximately 15%) cautioned that they were initially unsure how to use the AI effectively, or they found some of its suggestions to be too generic and too complex to understand for second-language learners of English. This feedback indicates the necessity of proper guidance in using AI tools so that students benefit from them without becoming overly reliant or misled by less relevant outputs.

Key Finding 3: Student Engagement and Collaboration

The two courses of sustainability and applied skills capstone fostered high levels of engagement and collaboration. Class attendance and participation were notably high throughout the semester in both cohorts of 2022 and 2025, and most project teams completed their assignments with enthusiasm. Engaged students are known to play a crucial role in translating sustainability initiatives into positive outcomes for institutions (Al-Dmour, 2023), and our findings were consistent with this. For example, two project teams continued their initiatives even after the course ended; one team secured approval from the university administration to implement their waste reduction campaign campus-wide, and another team joined the student sustainability club to carry forward the momentum. Several students also participated in a notable national event, Sheikha Shamma competition for Intercollege Recycling Competition and won the award. These actions echo findings from Jordan, where student engagement in green campus efforts mediated improved institutional reputation and a culture of sustainability (Al-Dmour, 2023).

Collaboration across disciplines was another noted benefit. Engineering students, for instance, worked alongside business majors, as well as IT and applied media students, each contributing unique perspectives. This mirrors the call in the literature for interdisciplinary collaboration in sustainability initiatives, which can enrich problem-solving and ensure that diverse aspects of sustainability are considered (Mokski et al., 2023). Focus group discussions revealed that students valued learning from peers in other fields. Such peer-to-peer learning experiences are instrumental in breaking down the silos that often fragment sustainability education (Abo-Khalil, 2024).

Key Finding 4: Attitudinal Shifts and Behavioral Intentions

Beyond competencies and engagement, an important outcome was the shift in students' attitudes and intended behaviours regarding sustainability. The New Ecological Paradigm (NEP) scale in the survey showed a small but significant positive shift, indicating that students adopted more pro-environmental worldviews by the end of the courses in the 2022 cohort. This finding aligns with the conclusions by Wang et al. (2022) that developing sustainability competencies through education is directly related to students' pro-environmental attitudes and behaviours.

Developing a sense of global citizenship and responsibility are outcomes that global education scholars emphasise as critical for addressing worldwide challenges (Akçay et al., 2024), which was investigated in the focus group discussions with the 2025 cohorts. Our findings suggest that education can indeed play a transformative role in shaping such active and informed citizens.

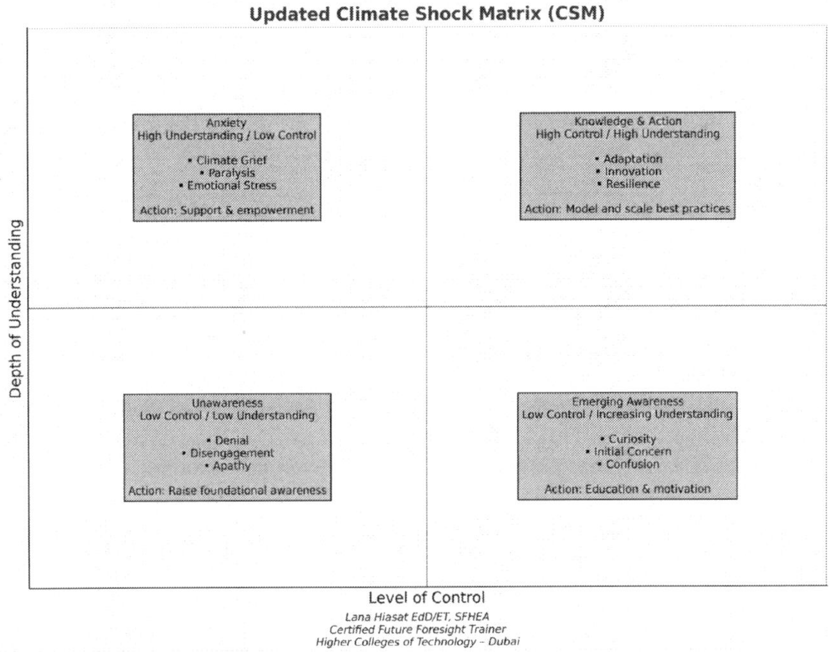

Chart 1 Stages of Climate Shock Matrix (CSM) developed by Hiasat, 2025

Key Finding 5: Stages of Climate Shock

Researchers hypothesised that two factors impacted the attitude and behaviour toward climate issues and sustainability. Chart 1 summarizes what the researchers of this study have coined 'climate shock' and the two factors. The stages of climate shock in this matrix include unawareness, emerging awareness, anxiety, and knowledge and action.

Stages of Climate Shock:

Conclusion

In conclusion, as the urgency for sustainable development intensifies, higher education has a pivotal role in equipping students with the competencies and mindset to drive change. This study contributed to the growing evidence that innovative pedagogical approaches such as blending competency-based learning, interdisciplinary projects, and incorporating Generative AI tools can create transformative educational experiences. Graduates of such programmes are more likely to emerge as informed, skilled, and motivated sustainability leaders capable of advancing the UAE's vision.

References

Abo-Khalil, A. G. (2024). Integrating sustainability into higher education challenges and opportunities for universities worldwide. *Heliyon, 10*(9), e29946. https://doi.org/10.1016/j.heliyon.2024.e29946

Akçay, K., Altinay, F., Altınay, Z., Daglı, G., Shadiev, R., Altinay, M., Adedoyin, O. B., & Okur, Z. G. (2024). Global citizenship for the students of higher education in the realisation of sustainable development goals. *Sustainability, 16*(4), 1604. https://doi.org/10.3390/su16041604

Al Mahameed, M., Riaz, U., Aldoob, M. S., & Halari, A. (2023). The implementation of sustainability practices in Arab higher education institutions. *Journal of Financial Reporting and Accounting*. https://doi.org/10.1108/JFRA-11-2022-0415

Al-Dmour, H. (2023). Green-smart university campuses: The mediating role of student engagement in enhancing corporate image. *SAGE Open, 13*(4). https://doi.org/10.1177/21582440231219591

Bartlett, M., Larson, J., & Lee, S. (2022). Environmental justice pedagogies and self-efficacy for climate action. *Sustainability, 14*(22), 15086. https://doi.org/10.3390/su142215086

Chaplin, G., & Wyton, P. (2014). Student engagement with sustainability: Understanding the value–action gap. *International Journal of Sustainability in Higher Education, 15*(4), 404–417. https://doi.org/10.1108/IJSHE-04-2012-0029

Conroy, M. M., Mansfield, B., Irwin, E., Jaquet, G., Hitzhusen, G., & Brooks, J. (2024). Six dimensions of sustainability: A framework for organising diverse university sustainability curricula at the Ohio State University. *International Journal of Sustainability in Higher Education, 25*(9), 316–332. https://doi.org/10.1108/IJSHE-08-2023-0344

Hiasat, L. (2025). Stages of climate shock matrix (CSM). Copyrighted Literary Work available at: https://app.copyrighted.com/work/ZpRWUCGFNmuPKmw9.

Membrillo-Hernández, J., Lara-Prieto, V., & Caratozzolo, P. (2021). Sustainability: A public policy, a concept, or a competence? Efforts on the implementation of sustainability as a transversal competence throughout higher education programs. *Sustainability, 13*(24), 13989. https://doi.org/10.3390/su132413989

Mokski, E., Leal Filho, W., Sehnem, S., & Andrade Guerra, J. B. S. O. de. (2023). Education for sustainable development in higher education institutions: An approach for effective interdisciplinarity. *International Journal of Sustainability in Higher Education, 24*(1), 96–117. https://doi.org/10.1108/IJSHE-07-2021-0306

Thompson, A. (2024). Coping with eco-grief and climate anxiety. Retrieved from https://catapult.co.nz/coping-with-eco-grief-and-climate-anxiety/#:~:text=This%20feeling%2C%20termed%20eco%20or,guilt%2C%20helplessness%2C%20and%20anxiety.

UNESCO. (2017). Education for sustainable development goals: Learning objectives. UNESCO Publishing. https://unesdoc.unesco.org/ark:/48223/pf0000247444.

Walsh, Z., Böhme, J., Lavelle, B. D., & Wamsler, C. (2020). Transformative education: Towards a relational, justice-oriented approach to sustainability. *International Journal of Sustainability in Higher Education, 21*(7), 1587–1606. https://doi.org/10.1108/IJSHE-05-2020-0176

Wang, Y., Sommier, M., & Vasques, A. (2022). Sustainability education at higher education institutions: Pedagogies and students' competences. *International Journal of Sustainability in Higher Education, 23*(8), 174–193. https://doi.org/10.1108/IJSHE-11-2021-0465

Chapter 48
Sustainability in Education Through Transforming Teaching, Learning, and Campus Practices

Nitin Patwa, Roshni Paul, and Pranab Kumar Pani

Introduction

Education is the systematic development of one's intellectual and practical capabilities and aims to broaden students' horizons in terms of knowledge, skills, attitudes, and appreciation for diversity. Integrating principles of environmental stewardship and economic and social sustainability brings a sense of responsibility, critical thinking, and motivation to innovate solutions for a sustainable future. Education's central purpose is to help students make sense of both the internal world they create and the external world they inhabit. It empowers to lead initiatives such as waste reduction campaigns, renewable energy projects, and advocacy for green policies (Miao & Nduneseokwu, 2025).

Adopting sustainability in all aspects of our education system is the need of the hour. Education for Sustainable Development refers to a pedagogical approach that integrates essential sustainable development concerns into the curriculum and all five pillars of sustainability. **The five key pillars identified are: (i) Curriculum, (ii) University global coalition. (iii) Extra-curricular activities (iv) Infrastructure (v) Faculty and student-driven research towards sustainability.**

Sustainability education is a foundation for achieving the United Nations Sustainable Development Goals (SDGs), from quality education and lifelong opportunities for all (SDG 4) to equipping youth with the knowledge and skills to address critical challenges like climate change (SDG 13), resource depletion (SDG 12), and social inequality (SDG 10). By embedding environmental stewardship (SDG 15), economic viability (SDG 8), and social equity (SDG 5) into curricula, sustainability education

N. Patwa · R. Paul (✉) · P. K. Pani
S P Jain School of Global Management, Dubai, UAE
e-mail: roshni.paul@spjain.org

N. Patwa
e-mail: nitin.patwa@spjain.org

J. Dsilva (ed.), *Sustainability Education, Youth and Innovation*,
https://doi.org/10.1007/978-3-032-21948-0_48

fosters critical thinking and systems-based understanding, inspiring innovation for a more sustainable future.

Strengthening the Five Core Pillars

Curriculum

Integrating sustainability into the curriculum is fundamental to fostering a generation of responsible global citizens. Education should be transformative rather than merely transmissive, shifting from rote learning to an active, experiential, and critical-thinking-based approach. A transformative education empowers students to question existing systems, develop sustainable solutions, and take action rather than passively receiving information (O'Brien et al., 2013). A well-structured curriculum emphasizing sustainability equips students with the knowledge, skills, and values to address environmental, social, and economic challenges. This transformation requires a shift in pedagogy, content, and assessment methods to embed sustainability across disciplines.

Curriculum for sustainable education is an approach to education that creates individuals, institutions, and societies equipped with the values and drive to work toward a more sustainable future, locally and globally. Sustainable education aims to increase students and institutions and foster the growth of critical thinkers, innovators, and problem solvers who can help society transition to more sustainable lifestyles. Research by Mohanty (2018) proposed a conceptual model—a sustainable education triangle for the Indian school education system by adopting the triple-bottom-line model of sustainable development where the three sustainable education goals were inclusive education, effective management, and community development (Mohanty, 2018).

Education should be transformative instead of transmissive; the transmission approach assumes that education transmits information, abilities, and values from a knowing source (teacher or textbook) to an unknowledgeable recipient. It emphasizes mastery of traditional school subjects through traditional teaching methods, especially textbook learning.

In transmissive education, the instructor controls the knowledge provided to students, and students show their understanding through testing. Transmissive techniques seem efficient since pupils are expected to retain more knowledge. The teacher's job is to devise exercises that deliver as much knowledge as feasible in the time allowed, while pupils are viewed as generic knowledge consumers based on their commonalities as learners (Gesthuizen et al., 2025).

Global University Coalition

As per the results of the 2020 statistics, 90% of the world's population had completed primary education; however, the result dropped to 66% in secondary education and even lower to only 40% in tertiary education. Although the literacy gender gap has shrunk recently, there is still a gap of around 7% between the two genders. The countries that rank high in tertiary education are Canada, Japan, Luxembourg, South Korea, Israel, United States, Ireland, United Kingdom, Australia, Finland, Switzerland, Norway, Sweden, Lithuania, Netherlands, Belgium, Estonia, Iceland, New Zealand, France, Spain, and Denmark.

Data analysis reflects that these countries have a very shared economic structure among themselves. Many of these countries are part of the OECD, comprising 38 countries, and are part of Economic Cooperation. These countries enjoy stable governance and part of democracy, and the transition of power does not affect the stability of their government.

Despite the varied population rates, the countries still demonstrate a very high per capita income. This very high gross enrollment ratio (GER) and tertiary education rate can be achieved by these countries' very high education budgets. These countries have world-class institutions that attract foreign talent to get educated in these countries. Their education budget for higher education is very high compared to the developing economies. Luxembourg ranks top in the list, spending 47,694 USD per student, followed by the USA, which spends around 34,036 USD per student. The budget of OECD countries is very high compared to that of other developing economies. Primarily, we may reason that a great deal of economic cooperation among these countries and the budget spent contributes to the high gross enrollment ratio (Charlton, 2019).

A World Bank Group report from 2017 argues that investing in higher education is essential for economic growth and the development of a knowledge-based economy. Increased production and growth, and people's ability to take part in that growth, are often attributed to higher levels of education. According to the study, "this has resulted in an increasing demand for and support to tertiary education."

The countries with better tertiary education rates positively correlate to the economy, quality of life, health, psychological well-being, health, happiness index, arts and culture, environment, etc. A stable economy attracts the global investor community, paving the way for youth employment (Erudera, 2022).

Sustainability in Extra-Curricular Activities

Academic institutions offer extra-curricular activities like access to academic clubs, sports participation, and creative arts pursuit. Extracurriculars are an integral element of a well-rounded education since they allow students to practice what they have learned in the classroom outside of the classroom. Participation in meaningful extra-curricular activities is crucial to a balanced education. It's an excellent tool for building community and teamwork skills among high school students. Students can benefit socially and emotionally from participating in extra-curricular activities. Kids learn about their strengths and areas for improvement through extra-curricular activities, and they also get to try out new interests. Students can engage in academic study, research, and a diverse range of extra-curricular activities, yet discovering a particular interest that evolves into a long-term passion often occurs unexpectedly. Empirical research indicates that institutions providing a broad spectrum of extra-curricular activities demonstrate higher levels of student engagement and overall academic motivation (Gordon, 2006).

Academic Clubs Fostering Sustainability: Current academic clubs in which institutions operate are paramount to STEM education and creative arts and drama. Although some extra-curricular activities offered by schools render some social services like the National Cadet Service and National Social Service, the institutions' goals are still narrowed to the said objective and do not germinate from the fundamental 17 UN sustainability goals. Many academic course structures were not so prevalent in the last quarter of the century in many countries, such as computer science education and, very recently, artificial intelligence and data science. By absorbing sustainability as a part of extra-curricular activity with dedicated class timetables set for the year, students will embrace sustainability as part of their culture since they will realize this is a fundamental need for the planet's future.

Schools partnering with non-governmental organizations: Institutions should incorporate a service component into their extra-curricular activities, and it should nurture their students to develop a passion element along with value, which will combine the service component to their area of interest. And there is more good news: volunteer opportunities tend to flourish in environmental research.

Infrastructure

As important as making our curriculum sustainable for the current and future generations, it is also vital to implement a sustainable design on the educational infrastructure. Currently, our school buildings remain dormant for a significant amount of time. The architecture of our educational infrastructure needs to be sustainable. In the broad sense, sustainable architecture refers to structures meant to reduce humanity's environmental impact. An eco-friendly approach to modern buildings includes all aspects of the entire planning and construction process. This includes everything from designing and implementing the heating, ventilation, and air conditioning system to waste management and plumbing systems to integrating the built environment into the natural landscape. Through good practices, one of the UN Sustainability we encourage would-be goal no.9: "Industry, innovation and infrastructure".

Faculty and Student-Driven Research Towards Sustainability

Higher education institutions (HEIs) have been instrumental in encouraging sustainable development in today's world, whether through education, research, operations, community partnerships, or other means. The importance of sustainability in today's world is only growing, and HEIs have been at the forefront of this movement. The term HEI refers to all higher education institutions, including colleges, universities, technical schools, community colleges, and law schools, among other educational facilities. A university has a unique potential to inspire and prepare its students, professors, and visitors to make a difference in the world because of its role as a usually credible organization that acts as a miniature version of society. There are many ways for universities to reach these goals, but the most common ways are for the administration, students, and faculty, or a mix of both, to push for change. These are called "top-down" and "bottom-up" methods, respectively.

Universities are frequently seen as having a "moral responsibility" to promote sustainable practices because universities educate the next generation of young adults who will enter the workforce. This is because universities provide education for the next generation of young adults. Graduates who know their world and the difficulties they will be confronted with in the next years become beneficial assets as environmental problems continue to grow more complex and deeply ingrained in our world daily. Therefore, educational institutions have a "moral responsibility" not only to strengthen environmentally friendly practices and economically and socially beneficial systems but also to ensure that their graduates are educated about issues related to sustainability and understand how they can act.

Challenges and Conclusion

Challenges

Several obstacles must be overcome for educational institutions to successfully include sustainable practices in their curricula, research, business practices, and community relationships. Nevertheless, doing so is highly advantageous. Students holding leadership positions at Canadian universities report that money is the most difficult obstacle to overcome when integrating sustainability into educational institutions. If only a small amount of funds are available from the government, then the university's budget will suffer, and with it will come a diminished capacity to make investments in environmentally friendly policies and procedures (Aleixo et al., 2018).

Researching and implementing sustainable practices can be challenging due to some colleges lacking specialized committees and offices.

In terms of education, one study found that sustainability and the SDGs were not always included in coursework because it was hard to match them up with the course content. Many professors didn't know much about this field or how to teach it in their classes best because they hadn't been trained in it (Leal Filho, 2019). Another possible barrier is that university departments and programs are often set up in a way that does not leave much room for interdisciplinary work. Even though research on sustainability has shown that interdisciplinary work is needed to educate students and prepare them to deal with sustainability issues, some faculty members think students should focus on one field of study at a time.

Conclusion

Governments and businesses worldwide are increasingly pushing for additional opportunities and investments in sustainability and sustainable development as the effects of the global climate crisis worsen. Many institutions, business owners, and government organizations have stated sustainability as one of their primary goals. However, it is hard to measure how sustainable an organization is. Historically, businesses just cared about their financial bottom lines. Today, businesses know that success isn't just measured by how much money they make and lose. Instead, organizations must go above and beyond compliance to fully account for all business costs to get a full, accurate picture of how their operations affect the environment, the community, and the economy. John Elkington, an author and business owner, built on the triple bottom line (TBL) idea to try to change the current business system, which is based on financial accounting, into one that measures impact and success in a broader way.

The triple bottom line theory expands how businesses measure their success to include how they help the environment, help people, and ensure the economy is fair. People, the planet, and Profit are often called the three "P's" of the bottom line. (The Triple Bottom Line, n.d.) The people's bottom line is a way to measure the good and bad effects of your business on the people directly involved with it. This includes the shareholders and the people who have a stake in the business.

With universities putting a lot of effort into sustainability, a new positive aspect, "placement," is coming to light in addition to people profit planet as defined by John Elkington. Sustainability and green jobs are growing industries that are taking the country by storm. The UN's adoption of the 2030 Agenda for Sustainable Development is a good example for groups and businesses worldwide. There has never been a better time to start a career in sustainability because jobs and positions related to sustainable development and reducing adverse environmental effects are becoming increasingly important to how businesses plan and run daily.

References

Aleixo, A. M., Leal, S., & Azeiteiro, U. M. (2018). Conceptualization of sustainable higher education institutions, roles, barriers, and challenges for sustainability: An exploratory study in Portugal. *Journal of Cleaner Production, 172*, 1664–1673.

Charlton, E. (2019, October 16). These countries spend the most on education. *World Economic Forum*. https://www.weforum.org/stories/education-and-skills/education-spending-highest-school-brazil-chile-italy-mexico/

Erudera. (2022, October 5). World's most educated countries and their main common characteristics. *Erudera*. https://erudera.com/resources/worlds-most-educated-countries-their-main-common-characteristics/

Gesthuizen, R., Tan, H., & Kidman, G. (2025). Inspired to build understanding: Learning about sustainability and environmental education from within a virtual world. *International Research in Geographical and Environmental Education, 34*(1), 25–42.

Gordon, G. (2006). *Building engaged schools: Getting the most out of America's classrooms*. Simon and Schuster.

Leal Filho, W. (2019). Viewpoint: accelerating the implementation of the SDGs. *International Journal of Sustainability in Higher Education, 21*(3), 507–511. https://doi.org/10.1108/IJSHE-01-2020-0011

Miao, Q., & Nduneseokwu, C. (2025). *Advancing environmental leadership: Education, development, and training* (pp. 467–538). Environmental Leadership in a VUCA Era.

Mohanty, A., & Dash, D. (2018). Education for sustainable development: A conceptual model of sustainable education for India. *International Journal of Development and Sustainability, 7*(9), 2242–2255.

O'Brien, K., Reams, J., Caspari, A., Dugmore, A., Faghihimani, M., Fazey, I., et al. (2013). You say you want a revolution? Transforming education and capacity building in response to global change. *Environmental Science & Policy, 28*, 48–59.

Chapter 49
Transformative Pedagogies for the Twenty-First Century: Reimagining Higher Education's Role in Driving Global Sustainability

Victor Lobo

Abstract In the context of a rapidly changing global environment, higher education must play a transformative role in addressing pressing sustainability challenges. This paper explores the evolving pedagogical approaches that are necessary for preparing students to tackle complex sustainability issues in the twenty first century. With climate change, resource depletion, and social inequalities escalating, traditional modes of teaching are no longer sufficient. This research examines how transformative pedagogies -which emphasize active learning, interdisciplinary approaches, and social responsibility-can reshape the way higher education institutions engage with sustainability. Drawing on case studies, scholarly articles, and interviews with educators, this paper proposes an integrative framework for sustainability education that blends critical thinking, collaborative problem-solving, and global-local engagement. Findings highlight the importance of empowering students not only with knowledge but also with the agency to drive systemic change. The paper argues for the incorporation of experiential learning, participatory action research, and sustainability-driven innovation into curricula. The results emphasize the need for institutions to embrace interdisciplinary learning and inclusivity to ensure that sustainability education is equitable and globally relevant. The paper concludes with recommendations for higher education institutions to prioritize pedagogical innovations that align with global sustainability goals.

Keywords Transformative pedagogy · Sustainability education · Higher education · Interdisciplinary learning · Climate change

V. Lobo (✉)
St Joseph's University, Bengaluru, India
e-mail: victorlobo@sju.edu.in

J. Dsilva (ed.), *Sustainability Education, Youth and Innovation*,
https://doi.org/10.1007/978-3-032-21948-0_49

Introduction

The imperative to drive global sustainability requires a radical shift in how higher education institutions approach teaching and learning. As the impacts of climate change, environmental degradation, and social inequalities become more pronounced, it is crucial that universities develop transformative pedagogies that empower students to understand and confront these issues. Traditional educational models, which prioritize rote learning and siloed disciplines, have proven inadequate in addressing the complexities of sustainability. This paper seeks to reimagine the role of higher education in fostering sustainability, proposing an integrative pedagogy that emphasizes critical thinking, collaborative problem-solving, and global-local engagement. The key objectives are;

(i) To explore the role of transformative pedagogies in higher education and their potential for driving sustainability outcomes.
(ii) To examine how interdisciplinary approaches can bridge the gap between scientific knowledge, social activism, and policy in sustainability education.
(iii) To identify best practices in sustainability education that have emerged in global institutions.

Literature Review

A growing body of literature stresses the importance of active learning and student-centered pedagogy in fostering sustainable development. Scholars like (Tilbury, 2011) and (Orr, 2004) argue that the traditional educational paradigms fail to equip students with the necessary skills to tackle global challenges. The need for interdisciplinary curricula, community-based learning, and transformative pedagogies is becoming increasingly apparent in sustainability education. In 2019, a global framework on "Education for Sustainable Development: towards achieving the SDGs was adopted by the United Nations Educational, Scientific and Cultural Organization (UNESCO), which underscores the importance of equipping learners with the values and skills required for sustainable futures. (UNESCO, 2019).

Education for Sustainable Development is intrinsic for achieving the Sustainable Development Goals (SDGs) by 2030 (O'Grady, 2023). Through transformative pedagogies for sustainable development, it results in changing the direction of development by enhancing the learner's awareness, understanding of complexities, empathy, compassion and empowerment (Arto & Salonen, 2019). However, integrating education for sustainable development into the current education system poses several difficulties (Franco & Saito 2019).This study builds insights from the literatures to propose for a holistic pedagogical model for achieving SDGs by 2030.

Methodology

This study adopts a qualitative research approach that integrates multiple data sources to explore innovative pedagogical practices in sustainability education. The findings from the study is derived from a systematic methodology which includes in-depth interviews, curriculum analysis and learnings from case studies from different regions.

(i) *Case Studies:* The research reviews sustainability education programs from universities around the world, with a focus on institutions that have integrated transformative pedagogies in their curricula. Institutions like The University of Queensland (Australia), University College London (UK) and University of California (Berkeley)were selected due to their reputation for pioneering sustainability-focused teaching methods.

(ii) *Interviews:* In-depth interviews were conducted with educators, policy makers, and students to gather insights into how pedagogy can be transformed to drive sustainability. Interviewees were selected from diverse disciplines, including environmental science, economics, and social sciences. The participants were carefully sampled for their expertise in sustainability education.

(iii) *Curriculum Analysis:* A comparative analysis of sustainability programs at several leading universities was carried out to identify the strategies fostering problem solving and critical thinking. This involved reviewing course syllabi, learning outcomes, and teaching methods related to sustainability.

Findings and Discussions

Interdisciplinary Approaches: The Need for Holistic Learning

Climate change, environmental degradation, and social inequality do not exist in isolation and require solutions that draw on multiple knowledge areas, such as ecology, economics, political science, and social justice. The case studies undertaken during the study revealed that interdisciplinary programs encouraged student engagement for the comprehension of system interconnections. The University of Queensland's Masters of Sustainable Studies employs problem based learning which will directly enhance the learner's skill in identifying and solving the local and global issues. Similarly, for the University College London, MSc in Environmental Policy is a proper blend of social, economic and environmental aspects. One of the primary findings of this research is the growing recognition that sustainability education must be interdisciplinary. Table 49.1 reveals that the institutions that have integrated interdisciplinary learning report higher levels of student engagement and a better understanding of the interconnectedness of global sustainability challenges.

Table 49.1 Interdisciplinary approaches to sustainability in higher education

University	Key interdisciplinary program	Key Findings
University of Queensland, Australia	Master of sustainability studies	Emphasis on problem-based learning with a focus on local and global sustainability challenges
University College London, UK	MSc environmental economics & policy	Integration of policy analysis, economics, and environmental science
University of California, Berkeley	Environmental science, policy & management	Collaboration between natural and social sciences to address climate policy and social impacts

Global-Local Engagement: Bridging the Gap

Sustainability education gain significance when global challenges are contextualized locally. The University of Cape Town's programs, for example, set in local environmental issues within global frameworks, enabling students to devise context-specific solutions. Interviewees emphasized that this dual focus enhances applicability and fosters a sense of responsibility, corroborating findings from (Arjen & Wals, 2002).

Conclusion

This research highlights the importance of transformative pedagogies in shaping the future of higher education and advancing global sustainability. The study demonstrates that universities can and must move beyond traditional educational paradigms to incorporate active, experiential, and interdisciplinary learning that addresses the multifaceted nature of sustainability challenges. By integrating these pedagogies into curricula, universities can not only enhance student learning but also contribute meaningfully to achieving global sustainability goals.

However limitations exist; the focus on Global North institutions may overlook resource constraints in the Global South, needing further research into region-specific adaptations. Further research is needed to examine how transformative pedagogies can be tailored to the unique challenges and opportunities of these regions.

This paper provides practical recommendations for higher education institutions, such as adopting interdisciplinary curricula, fostering community partnerships, and integrating experiential learning. These practices not only enhance sustainability education but also prepare students to become proactive agents of change in their communities and beyond. Universities shall adopt for;

(i) *Curriculum Reform:* Universities should integrate sustainability education across disciplines and prioritize real-world problem-solving through interdisciplinary approaches.
(ii) *Policy Advocacy:* Higher education institutions should advocate for national and international policies that promote education for sustainable development.
(iii) *Community Engagement:* Partnerships between universities and local communities should be strengthened to create contextual sustainability solutions that are applicable in diverse settings.

References

Arjen, E. J., & Wals, B. J. (2002). "Sustainability" in higher education: From doublethink and newspeak to critical thinking and meaningful learning. *Elsevier, Higher Education Policy*, 121–131.

Arto, O., & Salonen, J. S. (2019). *Transformative pedagogies for sustainable development*. Springer.

Franco, I., & Saito, O. (2019). Higher education for sustainable development: Actioning the global goals in policy, curriculum and practice. *Sustainability Science*, 1621–1642.

O'Grady, M. (2023). Transformative education for sustainable development: A faculty perspective. *Environment Development and Sustainability*.

Orr, D. W. (2004). *The nature of design: Ecology, culture, and human intention*. Oxford University Press.

Salonen, A., & Siirilä, J. (2019). Transformative pedagogies for sustainable development.

Tilbury, D. (2011). *Education for sustainable development—An expert review of processes and learning*. UNESCO.

UNESCO. (2019). Framework for the implementation of education for sustainable development (ESD) beyond 2019. Retrieved from https:// unesd oc. unesco. org/ ark:/ 48223/ pf000 03702 15. locale= en.

Part VIII
Youth Leadership in Sustainability

Chapter 50
Green Skills Transformation: The Present and Future of Sustainability in Education, Industry, and Labor Markets

Polen Bahçebaşı, Hannah Paulson, Pearl Massoudi Samandari, Dimitra Pateraki, Nethanya Saji, and Marian Fletcher

Abstract Industries worldwide are experiencing more frequent and severe climate disruptions. As a result, roles and responsibilities at all levels from interns to C-suite leadership are adjusting toward sustainability functions. Still, many industries struggle to align workforce planning with emerging green job requirements, creating a persistent skills gap. It remains unclear how different regions and industries are adapting to climate shifts and whether the workforce's skills align with evolving sustainability demands. This paper examines global and regional labor markets, through an analysis of employment patterns and green sector job growth. We evaluate job seeker competencies and industry preparedness to adapt to evolving skill demands across regions, with the Gulf Cooperation Council (GCC) presented as a case study. We consider the impacts of climate migration on skill distribution, and evaluate and identify misalignments between academic output and industry needs. This study uses research gathered from a series of interviews with job seekers, educators and industry professionals and prevailing literature. By connecting regional and sectoral data, this research provides insights into the future of green skills and a strategy to connect talent to industry roles to build resilience and capacity in all industries.

Keywords Green skills · Education · Workforce development · Green economy

Introduction

Addressing the green skills gap is key to a sustainable and equitable future. As demand for sustainability-focused work grows, investing in skills training enhances productivity and prepares workers for the green economy. A well-equipped green workforce

P. Bahçebaşı · H. Paulson (✉) · P. M. Samandari · D. Pateraki · N. Saji · M. Fletcher
The Climate Consultancy, Dubai, UAE
e-mail: hello@theclimateconsultancy.com

J. Dsilva (ed.), *Sustainability Education, Youth and Innovation*,
https://doi.org/10.1007/978-3-032-21948-0_50

is essential for implementing sustainability policies and innovations. The green transition will create both opportunities and disruptions, but a just transition ensures fairness, opportunity, and inclusivity. By tapping into the trends in green job opportunities and upskilling initiatives, barriers to work can be reduced or removed, and vulnerable communities can be both empowered and benefit from the evolving economy. SustainabiliChapty is no longer a niche concern but a key factor in organizational resilience, shaping how businesses operate and how labor markets evolve (Deloitte & IEMA, 2022; Tsironis, 2023; Economist Impact, 2024). Challenges remain around effectively placing workers into sustainability roles, adapting to evolving job market trends, and recognizing sustainability expertise (ILO, 2018). The global green skills gap is persistent and pervasive across industries. Between 2022 and 2023, job postings on LinkedIn requiring at least one green skill rose by 22.4%, while the share of professionals possessing such skills grew by 12.3% (LinkedIn, 2024). Further, the LinkedIn Green Gender Report finds only 1 in 10 women identify themselves as having at least one green skill, compared to 1 in 6 men. This gap has widened by 25% over the past 7 years, indicating a pressing need for inclusive upskilling initiatives (LinkedIn, 2023). As individuals and organizations reckon with shifts in desired skills, we aim to document experiences of industry experts, hiring managers, and job seekers across industries, with a specific consideration around the green workforce in the GCC region. Many prevailing global reports on green skills rely primarily on quantitative labor market data, such as job postings, skills taxonomies, and workforce projections (World Bank 2022, Economists Impact, 2024; LinkedIn, 2024). While these provide useful insights, they fail to capture the lived experiences of job seekers and employees attempting to transition into the green workforce. Our study bridges this gap by integrating qualitative insights from job seekers, industry leaders, and policymakers with quantitative labor market trends. This experiential knowledge reveals barriers not reflected in traditional labor market data, such as unclear job descriptions, credential mismatches, and employer biases. By tapping into this untapped source of data, our research provides a more human-centered understanding of green workforce gaps, ensuring that policy and training recommendations reflect real-life challenges and opportunities. One source of confusion is the lack of consensus on what qualifies as a "green job" or a "green skill". The World Economic Forum (2024), LinkedIn (2024), the World Bank (2022), and the European Commission (Cedefop & OECD, 2015) each highlight the lack of a standardized, universally accepted green skills taxonomy. Some frameworks define green jobs broadly, incorporating any role that can be made more sustainable, while others focus strictly on roles directly tied to climate action and environmental management (Tsironis, 2023). This inconsistency makes it difficult to define, measure, and align skills with industry demands and businesses struggle to determine which competencies are most critical for hiring and upskilling (Deloitte & IEMA, 2022; Tsironis, 2023). Based on existing literature and professional experience, we define green skills under three primary categories: Foundational or Educational skills obtained through higher education and affiliated research capstones, internships or experiences; Technical skills obtained through traditional work experience; and Transversal skills or

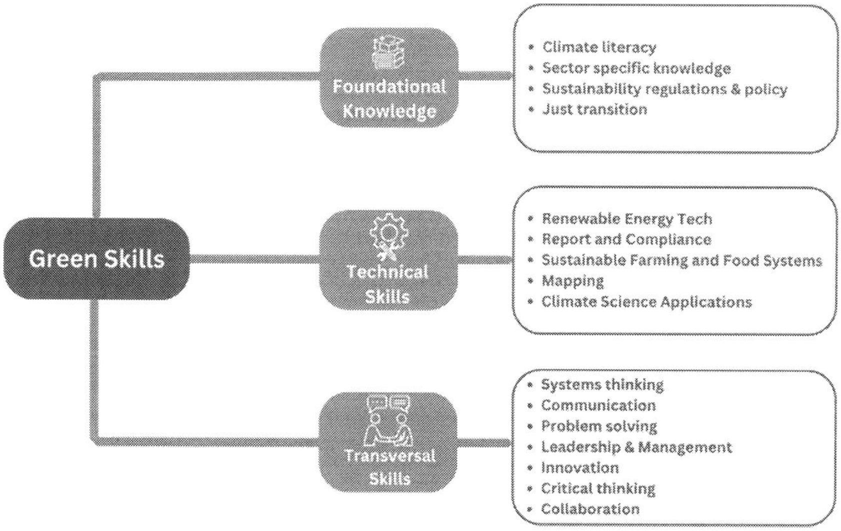

Fig. 50.1 Green skills categories defined by the research team. Adapted from Valente et al., 2015

"soft skills", which can be obtained a variety of ways, but most commonly constitute work related communication, problem-solving, and professionalism (Fig. 50.1).

Mismatches between academic training and employer expectations contribute to the green skills gap. While universities incorporate sustainability coursework, practicums and real-world case studies into curricula, hiring managers often fail to value these experiences in-line with traditional work experience (Valente et al., 2015). Many recent grads with advanced degrees resort to underpaid or unpaid roles to gain experience, where they may or may not receive mentorship, career counseling, or other advantageous personal growth opportunities that result in paid roles. Meanwhile, businesses struggle to predict, understand, and hire for green skills, with only 55% of business leaders actively implementing workforce transformation strategies (Economist Impact, 2024). This disconnect between education and industry needs hinders the transition to a sustainability-ready workforce (Poletaev & Robinson, 2008; Vona et al., 2015; Valente et al., 2015).

Methodology

In 2024, our team conducted a pilot study with green skills experts to explore the current and future trajectory of green skills. Insights from these interviews shaped our research questions, refined our investigative approach, and guided key areas of exploration. The pilot also identified emerging themes, allowing for more targeted interview questions and deeper insights in the main study. Here we employ a rigorous approach to ensure a balanced cross-section of interviewee profiles, geographies, and areas of expertise. In complement, we present a case study on the Gulf Cooperation Council (GCC) region. To determine the appropriate number of interviews to make narrative based claims, we consult existing literature and find saturation in interview-based qualitative research is generally agreed-upon to be achieved with sample sizes ranging from 12 to 20 (Guest et al., 2006; Francis et al., 2010; Hennink et al., 2017). In line with these recommendations, we have conducted 23 interviewees, in addition to 14 from the pilot study in July 2024 (Fig. 50.2). Interviewees in this main study were selected using purposive sampling, targeting individuals with specific expertise in various areas of green skills development, specific job roles in hiring and skills management, those with extensive experience in sustainability, or region-specific knowledge relevant to the study. Interviewees were sorted into the following profiles:

- Business Leaders: Executives from green technology and renewable energy sectors
- Job Seekers and Current Employees: Green sector professionals and career transitioners
- Climate Migration Specialists: Researchers studying global skill distribution
- Government Officials: Policymakers in sustainability and climate adaptation
- Experts and Academic Researchers: Researchers studying sustainability, green workforce shifts, green skills and other related topics.

Many participants were identified through our team's professional network, including The Climate Club, a global sustainability community managed by the author's place of business, The Climate Consultancy. This free global community of sustainability professionals includes entry-level, mid-level, and executive level sustainability professionals, as well as a key group of mid-career pivoters and first-time job seekers.

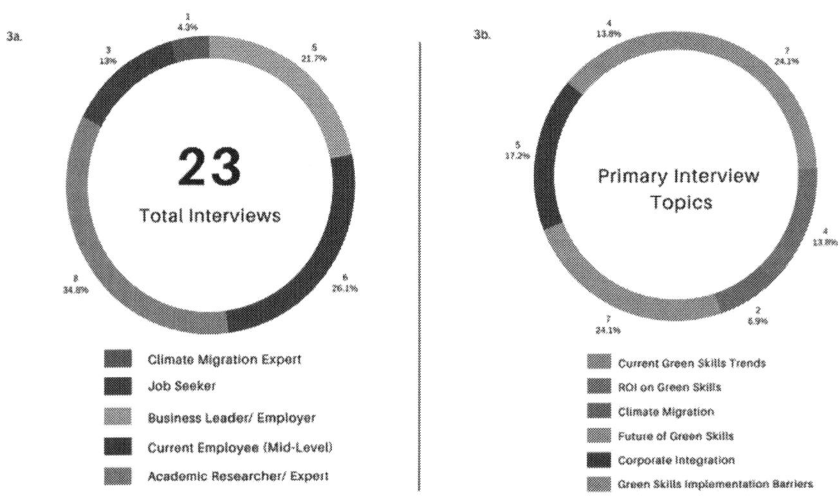

Fig. 50.2 Interviews and themes

Participation in both the group and this study was voluntary, with no compensation or incentives offered, ensuring no conflicts of interest. Additional interviewees with relevant expertise, industries, or geographic perspectives were identified via LinkedIn. Interviews were transcribed using Otter.ai and manually checked for accuracy by the research team. Subsequently, each transcription was manually tagged by topic and theme, and sorted into categories for analysis (Fig. 50.2).

Figure 50.2 Interviews conducted broken down by profile (**a**).Candidates often belong to multiple categories, such as Business Leader and Job Seeker, but are sorted into the most senior relevant category. Conducted Interviews sorted by interviewees area of expertise and primary topic discussed in interview, where some interviews cover multiple topics (**b**). 23 interviews have been conducted

In addition to quantitative interviews, we analyse labor market data, university learning outcomes, and demand projections for green jobs, both globally and in the GCC. The integration of these diverse data sources allows for a comprehensive assessment of workforce preparedness and the structural barriers hindering green job transitions. Expert surveys across industries, geographies, experience levels, and company roles will further broaden the study's scope. By incorporating a diverse range of perspectives, this research aims to capture the complexity of the green skills landscape and highlight key areas for intervention.

Findings and Arguments

*Cite transcriptions/interviews using this format (J. Smith, personal communication, March 10,

2025) Organize findings section by research question.

RQ1: How aligned are job seeker competencies with evolving sustainability skill demands across regions.and industries?

Themes:

- Presence of technical skills in job seeker experience (e.g., ESG reporting, carbon accounting)
- Mention of transversal/soft skills (e.g., communication, systems thinking)
- Mentions of misalignment between job requirements and actual competencies
- Employer perceptions of candidate preparedness

RQ2: What are the structural and educational barriers to entry into the green workforce?

Themes:

- Visa challenges or migration-related issues
- Lack of access to internships / paid roles
- Cost or confusion around credentials/certifications
- Reference to self-funding / unpaid work
- Lack of guidance or mentorship

RQ3: How do industry expectations around credentials, experience, and education shape workforce.

accessibility?

Themes:

- Employer prioritization of certifications vs degrees
- Emphasis on internships or paid experience over classroom learning
- Reference to specific credentials (e.g., LEED, CFA, GRI)
- Mismatch between education and actual job requirements

RQ4: In what ways do regional contexts, particularly in the GCC, affect the development and deployment.

of green skills?

Themes:

- Mentions of government policies (COP28, Vision 2030, UAE initiatives)
- References to local university programs
- Employer or skills demand in the GCC (vs globally)
- Role of expat labor / localization
- Industry-specific green job trends

RQ5: What role can integrated qualitative and quantitative data play in revealing hidden labor market.

barriers?

Themes:

- Mentions of lived experience contradicting job posting data
- Cases where real-world examples add nuance (e.g., job seeker stories)
- Absence of insight in quantitative-only data

Findings from interviews with 23 individuals across five stakeholder categories—business leaders & employers, government officials, experts & academic researchers, job seekers, and current employees in green roles—reveal significant challenges facing the green workforce, also summarized in (Table 50.1). Job seekers struggle with visa issues, unstable roles, and a mismatch between formal education and employer demands, particularly in business development and ESG reporting. While students in environmental sciences and STEM fields gain foundational knowledge in climate science, they often lack interdisciplinary systems thinking and business management skills, making it difficult to transition into industry roles. Employers frequently prioritize candidates with practical work experience, leaving graduates reliant on unpaid or underpaid positions to gain the necessary exposure—where they may or may not receive mentorship, career counseling, or clear pathways to paid employment (Fleacă et al., 2024). Certifications, often valued over degrees, present an additional barrier due to high costs and a lack of clarity on which credentials hold the most industry value. Additionally, two interviewees reported having to relocate from their home countries to find green roles, only to face visa complications and unstable employment. Given OECD's findings on cross-country disparities in green skills supply (2023), our research highlights the urgent need for tailored immigration policies to support green talent mobility. Transitioning from non-profit to for-profit sustainability roles poses additional obstacles, as employers often perceive non-profit experience as incompatible with the private sector's bottom-line mindset. This assumption leads to job seekers from non-profits being overlooked, despite possessing relevant sustainability expertise (Oelberger, 2024; Stirratt, 2019). At the same time, high turnover rates and economic instability in non-profit organizations further limit job security for green professionals, creating a bottleneck for their integration into the green workforce. For displaced workers affected by climate change, a lack of climate literacy and technical expertise poses another major barrier to green careers. Three green skills training providers interviewed specifically noted that state policies do not adequately support equitable workforce development for these individuals, aligning with Solis et al. (2024) findings. Employers also struggle to identify candidates with the right sustainability expertise, while job seekers lack clear guidance on which credentials to pursue. The demand for green skills is highly industry-specific, emphasizing the need for tailored education and training approaches that support both entry-level access and career mobility. Interview participants suggested that while universities incorporate real-world case studies and practical applications, employers do not always recognize these experiences as equivalent to work experience. Some job seekers noted that internships and paid roles tend to be valued more than soft and transversal skills developed through education. Additionally, interviewees indicated that students and graduates often emphasize their technical knowledge and coursework in job applications but struggle to effectively communicate their practical skills, further complicating their transition into the workforce.

GCC Findings

The UAE and broader Gulf Cooperation Council (GCC) region are experiencing a growing demand for green skills as sustainability gains prominence across industries. However, structural and policy challenges continue to shape the trajectory of green workforce development, affecting job seekers, employers, and policymakers alike. This study integrates qualitative insights from industry experts and job seekers in the UAE with empirical research to analyze the evolving landscape of green jobs, skill development, and labor market constraints in the region.

Green Skills Development and Industry Alignment

One of the key challenges facing job seekers in the GCC is the requirement for internships and work experience to secure entry-level positions. While universities are increasingly offering internship opportunities, dedicated placements for green jobs remain scarce, limiting students' ability to gain practical sustainability experience.

Table 50.1 Key findings

Stakeholder category	Key findings
Job seekers	• Job seekers face visa challenges, unstable roles, and a mismatch between formal education and employer demands, particularly in business development and ESG reporting. • In some cases, certifications are perceived to be preferred over specific degrees but are costly and difficult to navigate. • Transitioning from non-profit to for-profit roles is challenging due to employer biases.
Current employees in green roles	• Current employees recognize the gap between formal education and required skills. Many note that employers undervalue sustainability coursework as equivalent to work experience, leading to difficulties in career advancement. • High turnover rates in non-profit roles exacerbate job instability for green professionals.
Business leaders & Employers	• Employers struggle to identify candidates with appropriate sustainability expertise. Industry-specific green skills requirements complicate recruitment, and there is no consensus on which credentials indicate job readiness. Employers tend to prioritize paid experience over academic qualifications.
Government officials	• Government policies inadequately support equitable workforce development, particularly for displaced workers seeking green careers. There is an urgent need for tailored immigration policies to attract and retain green-skilled professionals.
Experts & academic Researchers	• Experts report that green skills curricula in formal education focus on broad environmental knowledge rather than practical skills development. The lack of standardized green skills taxonomies hinders workforce alignment with industry demands.

Some universities have sought to bridge this gap through eco-rep programs, which provide structured sustainability-focused roles for students. In the UAE, universities have also directly hired students for positions such as carbon accounting, offering a promising entry point into the green job market. This institutionalized model stands out as a best practice, as similar opportunities are less common in global labor markets. Despite these efforts, many graduates face immediate employment pressures to maintain their residency status in the UAE, underscoring the importance of clearer market signals regarding employer demand for green skills. Additionally, green job recruitment in the GCC remains largely focused on graduates from environmental science programs, while students in business and engineering fields often remain unaware of how their skills could translate into sustainability roles. Expanding awareness and interdisciplinary integration could unlock a broader and more diverse talent pool for the sustainability sector (AlNaqbi & Mohd Shamsudin, 2024).

Policy and Institutional Support for Green Workforce Development

The UAE government has played an increasing role in sustainability upskilling initiatives. The Environment Agency—Abu Dhabi offers green skills courses, yet these programs remain underutilized due to limited visibility among job seekers. The international spotlight on COP28 facilitated greater involvement from the Ministry of Education in global sustainability initiatives, such as UNESCO's Greening Education Partnership. This suggests that international collaborations can serve as catalysts for local green workforce development, but long-term policy implementation remains crucial for sustaining these efforts (Alherimi et al., 2024). A significant barrier for job seekers is the misalignment between international sustainability frameworks and GCC business practices. Many consultancies and multinational firms apply global sustainability standards without accounting for regional environmental, economic, and cultural contexts. This has resulted in a growing demand for green-skilled professionals who can translate global best practices into locally relevant business strategies. In Saudi Arabia, green job opportunities are expanding (Mulligan, 2024), but Arabic language proficiency remains a common requirement, creating barriers for expatriate workers and non-Arabic-speaking professionals seeking to enter the sector.

Barriers to Inclusion and Professional Advancement

Despite expanding sustainability roles, women in the GCC face challenges in accessing leadership positions within the green economy. This gender disparity mirrors broader trends in corporate leadership, limiting diversity in sustainability decision-making and slowing progress toward inclusive green workforce development. Additionally, GCC employers often place greater value on professional certifications over formal degrees, emphasizing the need for clearer credentialing pathways for green job seekers. This trend further highlights the importance of affordable, accessible certification programs to ensure equitable participation in the green labor market.

Future Considerations and Policy Implications

To create a more inclusive and responsive green workforce, the UAE and GCC nations must:

- Expand industry-integrated internship programs to provide structured pathways for sustainability careers.
- Improve awareness of sustainability roles across disciplines to attract business, engineering, and finance graduates to the sector.
- Align green skills training with local industry needs, ensuring practical applications of global sustainability frameworks.

- Promote gender-inclusive policies to support women in green leadership roles.
- Enhance access to professional certifications by standardizing credentialing frameworks and subsidizing certification costs for job seekers.

By addressing these structural gaps, the GCC can strengthen its sustainability workforce and build a resilient green economy that supports both national sustainability goals and global climate commitments.

Conclusion

As industries worldwide respond to increasing climate disruptions, the demand for green skills is growing rapidly. However, gaps in workforce preparedness, training effectiveness, and policy alignment continue to pose challenges for achieving a sustainable economy. By examining these issues through both global and regional lenses, with a particular focus on the GCC, this research seeks to provide a comprehensive analysis of labor market transitions, skill development needs, and workforce mobility trends.

Preliminary findings suggest a disconnect between how universities prepare students for the workforce and how employers evaluate job readiness. While academic programs are increasingly incorporating practical applications, case studies, and real-world problem-solving, these experiences are not always recognized by employers as valid indicators of professional capability. Further investigation into how different industries assess academic versus experiential learning will provide deeper insights into addressing this challenge.

Given these perceptions, students and graduates may benefit from developing clearer ways to articulate the practical and soft skills gained through their education. Exploring alternative approaches—such as framing coursework in terms of problem-solving, project management, and stakeholder engagement—could help bridge the gap between academic preparation and employer expectations.

Beyond individual skill development, this study aims to connect regional and sectoral data to provide a roadmap for addressing global and regional green skills gaps. Ensuring that businesses are prepared to transition to sustainable practices and workers are equipped with necessary skills is critical for the long-term success of green workforce development. The findings will support companies, policymakers, and educators in designing effective training programs and workforce development strategies. These strategies must extend beyond short-term compliance measures and focus on long-term capacity building that fosters innovation and resilience within the workforce.

This study also considers the policy implications of green skills development. Governments play a crucial role in shaping labor market dynamics through regulatory frameworks, incentives, and public-private partnerships. By evaluating existing policies and their effectiveness in promoting green workforce development, this research aims to provide recommendations for policymakers to create enabling environments that support sustainable labor markets. For example, in the GCC, national strategies such as Saudi Vision 2030 and the UAE's Green Economy Initiative have outlined ambitious sustainability targets, yet their success depends on a well-trained workforce capable of implementing them effectively.

Another key consideration is the role of technology in green skills development. Digital tools, artificial intelligence, and automation are reshaping industries, creating both challenges and opportunities for workforce adaptation. This study explores how emerging technologies can be leveraged to enhance green skills training, improve job matching, and streamline workforce transitions. For instance, online learning platforms and virtual reality simulations offer scalable solutions for upskilling workers in renewable energy, circular economy practices, and sustainable engineering. Understanding the potential of these technologies will help industries and governments design more effective workforce development programs.

Ultimately, a well-prepared workforce is essential for driving sustainable change across industries. The green transition is not only an environmental imperative but also an economic opportunity, and ensuring that workers are equipped with the right skills will be critical to realizing its full potential.

Acknowledgements The authors wish to thank the 23 individuals who generously gave their time as interview participants, and whose insights and experiences form the empirical foundation of this research. The authors are also grateful to the following experts whose thought leadership and collaboration shaped the conceptual framing of this work: Simon Bangs (Sustainable Recruitment Solutions), Romain Boitard (European Training Foundation), Efrem Bycer (LinkedIn), Ximena Játiva (World Economic Forum), Till Leopold (World Economic Forum), Hamid Samandari (McKinsey & Company, Emeritus), and Leila Takeh (Spiralforms Strategy and Sustainability Consulting).

Declarations Author Pearl Samandari is related to Hamid Samandari, who is acknowledged for professional input. This relationship did not influence the research or its interpretation. The other authors declare no competing interests.

References

Deloitte & IEMA. (2022). *A blueprint for green workforce transformation*. Institute of Environmental Management & Assessment. https://www.iema.net

Economist Impact. (2024). *Green skills: Driving the transition to a more sustainable future*. Iberdrola. https://www.iberdrola.com

Tsironis, A. (2023). *Preparing the workforce for the low-carbon economy: A closer look at green jobs and green skills*. Asian Development Bank. https://doi.org/10.22617/BRF230397-2

Cedefop & OECD. (2015). *Green skills and innovation for inclusive growth*. European Centre for the Development of Vocational Training. https://www.cedefop.europa.eu/en/publications/3079

International Labour Organization. (2018). *World employment and social outlook 2018: Greening with jobs*. ILO. https://www.ilo.org/global/publications/books/WCMS_628654/lang-en/index.htm

LinkedIn. (2024). *Global green skills report 2024*. LinkedIn Economic Graph. https://economicgraph.linkedin.com/research/global-green-skills-report

Poletaev, M., & Robinson, C. M. (2008). Human capital specificity: Evidence from the dictionary of occupational titles and displaced worker surveys, 1984–2000. *Journal of Labor Economics, 26*(3), 387–420.

Valente, A. C., Domingues, A. R., Soares, C., Prata Dias, G., Correa, I., Van Den Bossche, P., & Ramos, T. B. (2015). Assessment of life-cycle skills and training needs in the renewable and energy efficiency sectors: Wind energy, electric mobility, and smart grids sectors. In *Green skills and innovation for inclusive growth* (pp. 33–51). Publications Office of the European Union. Cedefop reference series.

Vona, F., Consoli, D., & Castellani, M. (2015). *Green skills* (NBER Working Paper 21116). National Bureau of Economic Research. https://www.nber.org/papers/w21116

World Bank. (2022). *The green transition: Opportunities and challenges for labor markets.* World Bank Group. https://documents.worldbank.org/en/publication/documents-reports/documentdetail/379301632114763321/the-green-transition-opportunities-and-challenges-for-labor-markets

Alherimi, N., Marva, Z., Hamarsheh, K., et al. (2024). Employees' pro-environmental behavior in an organization: A case study in the UAE. Scientific Reports, 14, Article 15371. https://doi.org/10.1038/s41598-024-66047-4

Alnaqbi, E., & Mohd Shamsudin, F. (2024). Greening the UAE workforce: Can green human resource management and domestic environmental regulations support green innovation at the workplace? Problems and Perspectives in Management, 22(1), 593–604. https://doi.org/10.21511/ppm.22(1).2024.47

Mulligan, C. (2024). Measuring green jobs in Saudi Arabia: Saudis in green occupations (Discussion Paper No. KS-2024-DP28). King Abdullah Petroleum Studies and Research Center. https://doi.org/10.30573/ks--2024-dp28

Fleacă, B., Militaru, G., & Fleacă, E. (2024). Reinforcement of workforce training programs—Insights from pilot testing process to enhance greening practices in enterprises. Sustainability, 16(23), 10377. https://doi.org/10.3390/su162310377

Oelberger, C. (2024). Beyond assumptions of altruism: Examining nonprofit work with a job fit framework and response surface analysis. Nonprofit Management & Leadership. https://doi.org/10.1002/nml.21610

Solis, M., Oden, M., Lieberknecht, K., & Liu, H. (2024). Labor lacuna: Disjunctures between local climate action and workforce development in advancing just transitions. *Journal of Urban Affairs*. https://doi.org/10.1080/07352166.2023.2291072

Stirratt, D. A. (2019). Nonprofit crossover leadership: A phenomenological study of nonprofit executives from for-profit backgrounds [Doctoral dissertation, University of St. Thomas]. https://ir.stthomas.edu/cgi/viewcontent.cgi?article=1130&context=caps_ed_lead_docdiss

LinkedIn Economic Graph. (2023). Green gender gap report 2023. LinkedIn. https://economicgraph.linkedin.com/content/dam/me/economicgraph/en-us/Green-Gender-Gap-Report.pdf

Willige, A. (2024). Green job vacancies are on the rise: But workers with green skills are in short supply. World Economic Forum. https://www.weforum.org/stories/2024/02/green-jobs-green-skills-growth/

Guest, G., Bunce, A., & Johnson, L. (2006). How many interviews are enough? An experiment with data saturation and variability. Field Methods, 18(1), 59–82. https://doi.org/10.1177/1525822X05279903

Francis, J. J., Johnston, M., Robertson, C., Glidewell, L., Entwistle, V., Eccles, M. P., & Grimshaw, J. M. (2010). What is an adequate sample size? Operationalising data saturation for theory-based interview studies. Psychology & Health, 25(10), 1229–1245. https://doi.org/10.1080/08870440903194015

Hennink, M. M., Kaiser, B. N., & Marconi, V. C. (2017). Code saturation versus meaning saturation: How many interviews are enough? Qualitative Health Research, 27(4), 591–608. https://doi.org/10.1177/1049732316665344

Chapter 51
Blueprint for Change: How a University is Contributing to the Green Initiatives

Connie Mitchell and Tahira Hoke

Abstract As global environmental challenges such as climate change and resource depletion intensify, higher education institutions are crucial in equipping future generations with the knowledge and leadership skills necessary to address these pressing issues. This paper examines the strategies implemented by Prince Sultan University (PSU) to integrate sustainability within its curriculum and extracurricular initiatives, thereby fostering a new generation of sustainability leaders. PSU has successfully embedded sustainability principles into its educational framework through innovative curriculum design, interdisciplinary research, and effective academia-industry collaborations. This case study presents an in-depth analysis of PSU's efforts, supported by institutional archives, accreditation reports, ranking data, and sustainability-related policies. By highlighting best practices, identifying challenges, and proposing strategies for enhancing sustainability education, this paper underscores the vital role of higher education in cultivating youth leadership for sustainability.

Keywords Sustainability · Green movement · Student-led · SDGs in higher education · Greening the curriculum

Introduction

Sustainability has become a cornerstone of higher education, reflecting a global shift toward addressing pressing environmental challenges through academic innovation and leadership. The concept of greening the curriculum emphasizes integrating sustainability principles across all academic disciplines to cultivate ecological literacy and empower students as future sustainability leaders (Haigh, 2005). As

C. Mitchell (✉) · T. Hoke (✉)
Prince Sultan University, Riyadh, Saudi Arabia
e-mail: cmitchel@psu.edu.sa

T. Hoke
e-mail: thoke@psu.edu.sa

J. Dsilva (ed.), *Sustainability Education, Youth and Innovation*,
https://doi.org/10.1007/978-3-032-21948-0_51

climate change, resource depletion, and social inequities intensify, universities play a critical role in fostering solutions by embedding sustainability into their curricula and extracurricular activities (UNESCO, 2023a).

Higher education institutions are increasingly adopting interdisciplinary approaches to sustainability education, moving beyond traditional environmental science courses to incorporate sustainability into business, engineering, architecture, and urban planning programs (UNESCO, 2023b). This shift reflects a broader understanding that sustainability challenges require cross-sector collaboration and innovative thinking. Recent studies highlight the importance of experiential learning, including project-based courses, internships, and real-world research initiatives, to bridge the gap between theory and practice (Mohammadi et al., 2023).

Student engagement also plays a pivotal role in advancing campus sustainability. Pro-environmental behavior among students, supported by institutional policies and green student organizations, significantly enhances universities' sustainability efforts (Mohammadi et al., 2023). However, challenges persist. Barriers such as limited faculty training, rigid academic structures, and insufficient collaboration between academia and industry continue to hinder the seamless integration of sustainability into higher education (Haigh, 2005; UNESCO, 2024).

This research paper aims to explore the role of higher education institutions in fostering youth leadership in sustainability, with a particular focus on curriculum development and extracurricular activities at Prince Sultan University (PSU). By examining how PSU integrates sustainability principles into its courses and supports student-led green initiatives, this study seeks to highlight best practices, identify challenges, and propose strategies for enhancing sustainability education.

Research Questions:

1. How do universities integrate sustainability leadership into their curricula?
2. What role do extracurricular activities (e.g., sustainability clubs, hackathons, incubators) play in fostering youth leadership in sustainability?
3. How do university-industry collaborations support students in launching sustainability-focused projects and startups?
4. What are the barriers students face in engaging with sustainability leadership, and how can higher education institutions address them?

Prince Sultan University (PSU) is actively integrating sustainability into its curriculum and fostering student engagement through various extracurricular activities.

Literature Review

In recent years, the urgency of addressing global environmental challenges has intensified, prompting higher education institutions to play a critical role in advancing sustainability. Universities worldwide are recognizing their responsibility to cultivate ecological literacy and empower future leaders to tackle climate change and environmental degradation (Haigh, 2005). The concept of "greening the curriculum" involves integrating sustainability principles across diverse academic disciplines to prepare students for the multifaceted demands of sustainable development (UNESCO, 2023a). This approach aligns with international frameworks, such as the United Nations Decade of Education for Sustainable Development (2005–2014), which advocates for embedding sustainability in all levels of education (Haigh, 2005).

The Global Movement Towards Greening the University Curriculum

The integration of sustainability into higher education curricula has been a growing international movement. Haigh (2005) provides a historical perspective on this shift, tracing its evolution from early environmental education efforts in the 1970s to the UNESCO-driven Decade of Education for Sustainable Development (2005–2014). His work highlights the increasing role of higher education institutions in fostering ecological literacy, addressing global sustainability challenges, and preparing future leaders to engage in sustainable decision-making. However, he also acknowledges barriers such as institutional resistance, lack of funding, and interdisciplinary challenges that slow down progress. He suggests that external pressures, including reforms in primary and secondary education, might drive higher education institutions to embrace curriculum greening more effectively.

Experiential learning opportunities, including project-based learning and internships, are becoming essential in fostering practical understanding and real-world application of sustainability practices (UNESCO, 2023b). However, despite the progress made, significant challenges remain. Institutional barriers, including bureaucratic inertia, disciplinary silos, and inadequate faculty training, continue to hinder the comprehensive integration of sustainability into academic programs (Haigh, 2005; UNESCO, 2023c).

University Strategies for Sustainability Implementation

Recent efforts have focused on embedding sustainability within curricula through diverse strategies:

Interdisciplinary Course Integration: UNESCO (2023a) and Mohammadi et al. (2023) emphasize that sustainability is no longer confined to environmental science programs but is integrated into business, engineering, architecture, and social sciences. Courses on green business practices, renewable energy technologies, and sustainable urban planning equip students with the knowledge to tackle climate challenges.

Experiential Learning Approaches: Universities are increasingly employing project-based learning, internships, and real-world sustainability projects (UNESCO, 2024). This hands-on approach helps students understand the practical implications of sustainability beyond theoretical knowledge.

The Role of Student Engagement and Institutional Support

Students play a pivotal role in driving sustainability within universities. According to Mohammadi et al. (2023), student-led pro-environmental behaviors, such as participation in campus sustainability initiatives and advocacy for policy changes, significantly contribute to the overall sustainability of institutions. However, strong institutional support is required, including funding for green student organizations, sustainability workshops, and collaborative research projects.

Greening Curriculum as Part of a Larger Sustainability Framework

The concept of "Greening Every Curriculum" (UNESCO, 2023b) highlights that sustainability must be embedded across all disciplines, moving beyond standalone environmental courses to a fully integrated approach. This requires faculty training, curriculum redesign, and collaboration with policymakers and industry leaders to ensure that students receive a holistic sustainability education.

Challenges and Opportunities in Greening Higher Education

Despite Significant Progress, Challenges Remain

Institutional Barriers: Haigh (2005) and UNESCO (2023c) highlight bureaucratic resistance, rigid academic structures, and lack of interdisciplinary collaboration as obstacles to fully integrating sustainability into university curricula.

Need for Faculty Training: Many faculty members lack the expertise or training to effectively teach sustainability concepts in their courses (UNESCO, 2024).

Bridging Academia and Industry: There is a growing need for stronger university-industry collaborations to provide students with internships, research funding, and employment pathways in sustainability-related fields (UNESCO, 2023c).

The literature underscores the importance of integrating sustainability into higher education curricula as a crucial step toward developing a generation of sustainability-conscious professionals. Universities must adopt a holistic approach that includes interdisciplinary learning, experiential education, student engagement, and institutional policy support. While challenges remain, increasing awareness, policy initiatives, and global frameworks offer significant opportunities for progress in greening university curricula. For example, the Framework for Education for Sustainable Development developed by Advance He breaks it down into three parts students, educators, and institutions serves as a reference to guide ESD initiatives in higher education (Advance HE, n.d.).

Methodology

This research employs a mixed-methods approach, incorporating both qualitative and quantitative data collection methods to provide a comprehensive analysis of Prince Sultan University's (PSU) sustainability initiatives. By combining document analysis with survey data, the study aims to assess PSU's sustainability strategies, their effectiveness, and student engagement in sustainability-related activities.

Qualitative Data Collection

The qualitative component of the study is based on document analysis of institutional records, policies, and external reports to examine PSU's strategic implementation of sustainability initiatives. Data sources include:

- Accreditation reports detailing PSU's compliance with sustainability standards.
- Memos from university leadership outlining directives for sustainability integration.
- Ranking data assessing PSU's performance in sustainability-related metrics.
- Institutional documents related to sustainability policies and strategic planning.
- Student reflections and feedback on sustainability projects as part of coursework.

A thematic analysis was conducted to identify key trends, patterns, and areas for improvement in PSU's sustainability efforts. The findings were triangulated with external reports, including Saudi Arabia's Voluntary National Reports (VNRs) to the United Nations High-Level Political Forum on Sustainable Development (HLPF), to provide a broader context for PSU's role in advancing sustainability in higher education.

Quantitative Data Collection

To supplement the qualitative findings, a quantitative survey was conducted to measure student awareness, engagement, and perceptions of sustainability initiatives at PSU. The survey was administered online and targeted a sample of approximately 50 participants, including students from various academic disciplines. The survey included:

- Demographic questions (e.g., major, year of study).
- Likert-scale items to assess students' awareness of PSU's sustainability programs.
- Multiple-choice questions measuring participation in sustainability-related extracurricular activities.
- Open-ended responses to gather insights into students' views on PSU's sustainability initiatives and areas for improvement.

The survey data were analyzed using descriptive statistics, such as frequencies, means, and standard deviations, to identify trends and patterns in student engagement and perceptions.

Data Integration & Analysis

The qualitative and quantitative findings were analyzed concurrently to provide a holistic understanding of PSU's sustainability framework. The qualitative insights from document analysis were used to contextualize the quantitative survey results, allowing for a more nuanced interpretation of how students perceive and interact with sustainability initiatives at PSU.

By employing a mixed-methods approach, this study ensures a comprehensive evaluation of PSU's sustainability initiatives, offering data-driven recommendations for enhancing sustainability education and student engagement.

Findings

Case Study: Sustainability in Action at Prince Sultan University

Since the launch of Saudi Vision 2030, Prince Sultan University (PSU) has taken significant steps to integrate sustainability into its curriculum and operational framework. Under the leadership of its President, PSU has actively encouraged faculty members to embed sustainability principles into their courses, fostering a culture of environmental and social responsibility among students and staff. This case study explores PSU's strategic initiatives, their impact on student learning, and the university's broader contributions to sustainability within higher education.

PSU's commitment to sustainability is evident through its Center for Sustainability and Climate (CSC), which aligns with the United Nations Sustainable Development Goals (SDGs) (Prince Sultan University, 2025a). The CSC promotes interdisciplinary research and training on sustainability and climate, emphasizing the role of humans in the environment. This initiative ensures that sustainability principles are embedded across various academic programs, preparing students to address environmental challenges effectively.

Sustainability in Education

A central component of PSU's sustainability initiative is the mandatory Personal Development Skills (PDS) course, which is required across all colleges. Annually, 500–600 students engage deeply with the Sustainable Development Goals (SDGs), reflecting on their significance in various industries and contemplating their roles in advancing sustainability within Saudi Arabia. This engagement culminates in student-created posters illustrating their understanding and commitment to sustainability.

Following this foundational exposure, students continue exploring sustainability within their majors. The integration of sustainability principles across various disciplines ensures that students receive a comprehensive and industry-relevant education.

This educational approach has yielded positive outcomes. During an accreditation review, a student from the Preparatory Year Program (PYP) shared her enhanced understanding of sustainability, emphasizing how project-based learning reinforced her knowledge and appreciation of its relevance to her field. Such testimonials underscore the effectiveness of PSU's sustainability-infused curriculum.

Institutional Commitment and Impact

Beyond curriculum integration, PSU has taken strategic steps to incorporate sustainability into its mission and operations. As a result of broad community engagement and feedback, sustainability was formally included in PSU's mission statement, reinforcing the university's dedication to sustainable development.

> *PSU is committed to effectively managing institutional resources to optimize its multiple roles as a catalyst for new learning opportunities, national and international partnerships, continuous studies, professional growth, community service, and diversity in educational horizons for the good of humanity and sustainability.*
>
> (PSU Website, Prince Sultan University, 2025b)

Beyond the institution's mission statement, all colleges and departments were requested to embed the concept of sustainability into their mission statements. This has created an underlying infrastructure that supports sustainable practices across the institution.

Further cementing its leadership in this area, the Center for Sustainability at PSU has gained international recognition. The appointment of its leader as the UNESCO Chair on Sustainability highlights the institution's role as a model for sustainability education in the region.

PSU has also embraced digital transformation to enhance sustainability. By reducing reliance on paper, the university has saved over 200,000 sheets, equivalent to approximately one metric ton of paper waste (Tomberlin et al., 2020). This transition has significantly reduced PSU's carbon footprint, lowering greenhouse gas emissions by approximately 942 kilograms in the past year.

PSU's Contributions to Regional and Global Sustainability Initiatives

Prince Sultan University's sustainability efforts align with broader national and international initiatives. From 2021 to 2023, Saudi Arabia's Voluntary National Reports (VNRs) to the United Nations High-Level Political Forum on Sustainable Development (HLPF) highlighted PSU's commitment to sustainability, including its ambition to become a net-zero campus. Similarly, other regional initiatives, such as the UAE's Sustainable Development Research Programs, demonstrate a growing emphasis on sustainability within Middle Eastern higher education institutions.

The Role of Higher Education Institutions in Sustainability

Higher education institutions (HEIs) play a crucial role in raising awareness and fostering sustainability both locally and globally. Within the MENA region, universities are addressing key sustainability challenges by equipping faculty and students with the necessary knowledge and skills. At PSU, sustainability education focuses on:

- Understanding environmental issues such as resource depletion and pollution.
- Developing the skills needed to make responsible choices that protect the environment.
- Connecting sustainability concepts to students' daily lives and future careers.
- Educating students about current sustainability challenges.
- Promoting critical thinking and problem-solving skills.
- Encouraging proactive steps at both individual and societal levels.

- Conducting research on innovative sustainability solutions and integrating them into curricula.
- Establishing quality measures for sustainability education within higher education.

Prince Sultan University's commitment to sustainability is evident in its curriculum, institutional policies, and broader community engagement. By embedding sustainability into educational frameworks, PSU ensures that students are well-equipped to address global sustainability challenges. Its initiatives serve as a model for higher education institutions aiming to integrate sustainability into academic and operational practices, contributing to national and international sustainability goals.

Prince Sultan University (PSU) has demonstrated a strong commitment to sustainability by embedding it into its mission statement, a strategic decision influenced by stakeholder feedback. This integration underscores PSU's dedication to fostering environmentally responsible practices across its academic and operational initiatives.

The Impact of Extracurricular Activities

To complement its academic offerings, PSU encourages student participation in sustainability through dedicated clubs and initiatives:

- PSU SDG Club: Launched in September 2022, the PSU SDG Club is the first student-led SDG club in Saudi Arabia. It organizes educational and social events, raises awareness, and advocates for sustainable initiatives within the university and the broader community. Activities include the "Go Green Initiative", "Zero Waste Campaign", "Earth Day" celebrations, and participation in the "UN SDG and Climate Week". The club also offers a "Youth Sustainability Leadership Certification" and provides volunteering opportunities at the CSC.
- Future Vision Club: This social club supports Saudi Arabia's Vision 2030 by focusing on human development. It aims to enhance students' skills and experiences through events, camps, lectures, and workshops related to future-oriented topics.

Furthermore, PSU's College of Engineering actively participates in the Shell Eco-Marathon competition, a prestigious event promoting energy-efficient vehicle innovation. The university has achieved remarkable success, securing first place for the second consecutive year, highlighting its excellence in sustainable engineering and innovation.

Furthering its global engagement, PSU participated in COP 16, reinforcing its role as a thought leader in sustainability. During fringe events, PSU's interdisciplinary approach to addressing sustainability challenges. Additionally, the President of PSU delivered a unique address to the UNESCO Chairs on sustainability, emphasizing the university's commitment to integrating sustainable practices across higher education.

PSU also played a pivotal role in the CHEA Annual Conference, where its delegation underscored the critical importance of sustainability in higher education. The university's efforts extend beyond its own campus, as it actively promotes the adoption of sustainability measures among other institutions, advocating for continuous improvement and accountability.

Aligned with its strategic plan, PSU systematically implements sustainability initiatives and monitors its performance through prestigious global rankings. The university has achieved a top ranking in the Middle East in the Times Higher Education Impact Rankings and earned a 5-star rating in the QS Sustainability Rankings' environment indicators, reflecting its leadership in advancing sustainability in higher education.

Through these curriculum developments and extracurricular activities, PSU not only educates students about sustainability but also empowers them to lead and implement sustainable practices in their communities.

'Prince Sultan University's Center for Sustainability was highlighted in the Voluntary National Report, emphasizing the continuous culture of sustainability cultivated last year and this year. The report provides detailed examples from every college, including classroom projects, exhibitions, SDG Clubs, Model United Nations (MUN) events, competitions, and start-up initiatives.

These activities showcase the dynamic collaboration between faculty and students aimed at fostering sustainable leadership. This collaboration has sparked a theory based on extracurricular activities that not only develop individual skills but also enhance the capacity for leading sustainability initiatives. This approach underscores the university's commitment to integrating sustainability into its educational framework and cultivating leaders equipped to face global challenges.

Long-Term Impacts of Sustainability Education

Sustainability education in higher education institutions, such as Prince Sultan University (PSU), has profound long-term impacts on students, industries, and global sustainability efforts. By integrating sustainability principles into curricula and extracurricular activities, universities equip students with the knowledge, critical thinking skills, and leadership abilities necessary to address pressing environmental challenges, such as climate change and resource depletion (Haigh, 2005; UNESCO, 2023a). Graduates who undergo sustainability-focused education are more likely to pursue careers in green industries, sustainable business practices, and policymaking, contributing to a workforce that prioritizes environmental responsibility (Mohammadi et al., 2023). Additionally, sustainability education fosters a lifetime commitment to sustainable behaviors, encouraging individuals to adopt eco-conscious practices both in their personal and professional lives. At an institutional level, PSU's sustainability initiatives, such as interdisciplinary learning, student-led projects, and academia-industry collaborations—serve as a model for other universities, promoting widespread adoption of sustainability frameworks across higher education (UNESCO, 2023b). Over time, these efforts contribute to national and global sustainability goals, such as Saudi Vision 2030 and the United Nations Sustainable Development Goals (High-Level Political Forum, 2023), ensuring that future generations are well-prepared to create lasting, positive environmental and social change.

Conclusion

Prince Sultan University's (PSU) comprehensive approach to sustainability education highlights the transformative role higher education institutions can play in fostering environmental responsibility and leadership. By integrating sustainability across its curriculum, promoting interdisciplinary research, and actively engaging students through extracurricular initiatives, PSU has established a model for embedding sustainability within academic institutions. The university's commitment to sustainability is further reflected in its alignment with national and global sustainability frameworks, such as Saudi Vision 2030 and the United Nations Sustainable Development Goals (SDGs).

Despite PSU's notable achievements, challenges such as faculty training, interdisciplinary collaboration, and industry engagement remain areas for continuous improvement. Addressing these challenges requires ongoing innovation, institutional support, and expanded partnerships to ensure that sustainability remains a core pillar of higher education. By fostering a culture of sustainability, PSU equips students with the skills and knowledge to drive meaningful change in their communities and professional fields.

As universities worldwide strive to integrate sustainability into their academic and operational frameworks, PSU's experience offers valuable insights into best practices and potential areas for enhancement. Moving forward, further research should explore the long-term impact of sustainability education on graduates' career trajectories and contributions to global sustainability efforts. Ultimately, higher education institutions must continue evolving as catalysts for sustainability, ensuring that future generations are prepared to tackle the environmental challenges of the twenty-first century.

References

Advance HE (n.d.). Education for sustainable development in higher education. https://www.advance-he.ac.uk/teaching-and-learning/education-sustainable-development-higher-education

Filho, W. L. (2019). Encyclopedia of sustainability in higher education. In *Springer eBooks*. https://doi.org/10.1007/978-3-030-11352-0

Haigh, M. (2005). Greening the university curriculum: Appraising an international movement. *Journal of Geography in Higher Education, 29*(1), 31–48. https://doi.org/10.1080/03098260500030355

High-Level Political Forum (HLPF). (2023). Voluntary national reviews. https://hlpf.un.org/vnrs.
Mohammadi, Y., Monavvarifard, F., Salehi, L., Movahedi, R., Karimi, S., & Liobikienė, G. (2023). Explaining the sustainability of universities through the contribution of students' pro-environmental behavior and the management system. *Sustainability, 15*(2), 1562. https://doi.org/10.3390/su15021562
Prince Sultan University. (2025a). *Prince Sultan university|center for sustainability and climate*. Prince Sultan University. https://www.psu.edu.sa/en/CSC
Prince Sultan University. (2025b). *Prince Sultan university|vision, mission, and values*. Prince Sultan University. https://www.psu.edu.sa/en/vision-mission-and-values
Tomberlin, K. E., Venditti, R., & Yao, Y. (2020). Life cycle carbon footprint analysis of pulp and paper grades in the United States using production-line-based data and integration. *BioResources, 15*(2), 3899–3914. https://doi.org/10.15376/biores.15.2.3899-3914
UNESCO. (2023a). Greening every curriculum. In United Nations educational, scientific and cultural organization (UNESCO). https://www.unesco.org/en/sustainable-development/education/greening-future/curriculum
UNESCO. (2023b). Greening curriculum guidance: Teaching and learning for climate action. In United Nations educational, scientific and cultural organization (UNESCO). https://www.unesco.org/en/articles/greening-curriculum-guidance-teaching-and-learning-climate-action
UNESCO. (2023c). UNESCO launches new publications for "greening education partnership". In *International association of universities–higher education for sustainable development (IAU-HESD)*. https://www.iau-hesd.net/news/unesco-launches-new-publications-greening-education-partnership
UNESCO. (2024). Greening curriculum guidance: Teaching and learning for climate action. United Nations educational, scientific and cultural organization (UNESCO). https://www.unesco.org/en/sustainable-development/education/greening-future/curriculum

Chapter 52
Empowering Youth Leadership in Sustainability through Toastmasters International: A Case Study from Al Yamamah University, Saudi Arabia

Georgios Kormpas, Joud Hakeem, and Ayham Zeidan

Abstract Sustainable development requires a new generation of youth leaders equipped with communication, leadership, and mentorship skills. As sustainability efforts increasingly rely on community engagement and advocacy, effective leadership training becomes essential in ensuring long-term impact (Lozano, R., Ceulemans, K., & Seatter, C. S., Journal of Cleaner Production 171:438-451, 2017). Toastmasters International at Al Yamamah University (YU) in Riyadh, Saudi Arabia has emerged as a leadership incubator, providing students with structured training in public speaking, critical thinking, and mentorship, all of which are fundamental to advancing sustainability advocacy and leadership. This study employs a qualitative case study approach to examine the impact of Toastmasters on youth empowerment in sustainability leadership. Through participant observations, structured interviews, and survey analysis, the research explores how participation in Toastmasters fosters leadership growth, enabling students to lead sustainability initiatives, mentor peers, and engage in community outreach. Key findings suggest that structured leadership training significantly enhances students' confidence, strategic thinking, and ability to drive sustainability efforts (Dugan, J. P., Leadership theory: Cultivating critical perspectives. Jossey-Bass, 2017). Specifically, mentorship and public engagement within the club ensure knowledge transfer, continuity of leadership, and long-term sustainability action (Komives, S. R., Owen, J. E., Longerbeam, S. D., Mainella, F. C., & Osteen, L., Journal of College Student Development 52:515-539, 2011). The study concludes with policy recommendations for integrating communication and leadership training into sustainability education programs at higher education institutions. By equipping students with effective public speaking, strategic leadership,

G. Kormpas (✉) · J. Hakeem · A. Zeidan
Al Yamamah University, Riyadh, Saudi Arabia
e-mail: g_kormpas@yu.edu.sa

J. Hakeem
e-mail: 202312207@yu.edu.sa

A. Zeidan
e-mail: 202111210@yu.edu.sa

J. Dsilva (ed.), *Sustainability Education, Youth and Innovation*,
https://doi.org/10.1007/978-3-032-21948-0_52

and mentorship skills, educational institutions can prepare future leaders to address global sustainability challenges through advocacy, community initiatives, and policy engagement (Youth engagement and sustainability education: Global best practices. UNESCO Publishing, 2022).

Keywords Youth leadership · Sustainability education · Mentorship · Public speaking · Leadership development

Introduction

Sustainability and leadership are interconnected, as effective environmental action requires individuals who can communicate clearly, inspire change, and make strategic decisions (United Nations, 2023). Youth leadership development plays a crucial role in achieving long-term sustainability goals, particularly in nations undergoing economic and social transformations. In Saudi Arabia, Vision 2030 underscores the importance of youth engagement, leadership training, and education in driving sustainable economic and social reforms (Ministry of Economy and Planning, 2022). However, while universities serve as incubators for future sustainability leaders, a critical gap exists in developing students' public speaking, mentorship, and leadership skills necessary to influence policy, community initiatives, and corporate sustainability practices (MacDonald & Shriberg, 2016).

One effective approach to addressing this gap is through structured leadership training programs, such as those provided by Toastmasters International—a global nonprofit dedicated to fostering communication, leadership, and mentorship skills (Dugan, 2017). The Al Yamamah University Toastmasters Club serves as a model for integrating leadership training with sustainability education, equipping students with the ability to mentor peers, advocate for sustainability, and engage in decision-making processes.

This paper investigates how structured leadership training through Toastmasters contributes to sustainable development by:

1. Evaluating the role of Toastmasters in youth leadership development for sustainability, examining how structured communication and leadership training equip students with essential skills for sustainability advocacy.
2. Assessing the impact of structured mentorship on long-term leadership continuity, analyzing how mentorship within Toastmasters fosters leadership growth and sustainability over time, aligning with Saudi Vision 2030's commitment to youth empowerment and leadership development as a means to build a thriving knowledge-based economy.
3. Exploring student-led sustainability initiatives inspired by Toastmasters training, highlighting case studies where students apply their leadership skills in sustainability-driven projects and community engagement, supporting Vision

2030's emphasis on integrating sustainability education into national development efforts and fostering a culture of innovation among young leaders.
4. Providing recommendations for integrating communication skills into sustainability education, proposing strategies to embed public speaking and leadership training in sustainability programs for enhanced long-term impact.

A qualitative case study approach was employed to examine the impact of Toastmasters leadership training at Al Yamamah University. Data collection methods included participant observations, structured interviews, content analysis of leadership programs, and survey responses from current and past Toastmasters members. The findings indicate that mentorship, leadership training, and sustainability-focused public speaking significantly enhance students' ability to lead sustainability initiatives.

This study contributes to the growing discourse on youth leadership in sustainability education by demonstrating the importance of communication and mentorship in long-term sustainability efforts. The paper concludes with policy recommendations for integrating communication training into sustainability education programs to equip the next generation of leaders with the skills needed to address global sustainability challenges.

Methodology

This study employs a qualitative case study approach to explore the impact of Toastmasters leadership training at Al Yamamah University (YU) in Riyadh, Saudi Arabia, in fostering youth leadership, mentorship, and sustainability advocacy. Given the study's focus on understanding how structured leadership programs contribute to sustainability education and align with Saudi Vision 2030, a qualitative approach was deemed most suitable to capture in-depth insights from participants.

Research Framework

The research framework is designed to investigate the relationship between leadership training, mentorship, and sustainability engagement. The study explores how participation in Toastmasters equips students with the communication and leadership skills necessary for sustainability advocacy and how this experience translates into long-term leadership continuity and real-world sustainability initiatives.

The research is guided by three research questions:

1. How does participation in Toastmasters International at Al Yamamah University contribute to the development of leadership and communication skills among youth in the context of sustainability advocacy?

2. What is the role of structured mentorship within Toastmasters in fostering long-term leadership continuity and knowledge transfer for sustainability initiatives?
3. How does the leadership training provided by Toastmasters International align with Saudi Arabia's Vision 2030 goals for youth empowerment, sustainability education, and community engagement?

Data Collection Methods

To comprehensively answer these research questions, the study utilized multiple data collection methods to ensure triangulation and validate findings.

- Participant Observations: Direct observations were conducted in Toastmasters meetings, leadership transitions, and public speaking workshops. This allowed for real-time insights into how leadership is developed, how mentorship occurs, and how sustainability issues are integrated into speeches and discussions.
- Semi-Structured Interviews: Interviews were conducted with past and current student leaders, including Toastmasters club officers and active members. These interviews explored participants' experiences with leadership training, mentorship, and their engagement in sustainability initiatives.
- Content Analysis: Club events, sustainability-themed speeches, and leadership mentorship practices were analyzed to examine patterns in leadership development and sustainability advocacy within the club.
- Survey Responses: A structured survey was distributed to Toastmasters members to gauge their perceptions of leadership growth, mentorship experiences, and the influence of Toastmasters on their sustainability engagement.

Data Analysis

Data were analyzed using thematic analysis, a method well-suited for identifying patterns and recurring themes in qualitative research. The analysis focused on three key dimensions:

1. Leadership Development: How participation in Toastmasters enhances leadership and communication skills relevant to sustainability.
2. Mentorship Effectiveness: The role of mentorship in ensuring leadership continuity and transferring sustainability knowledge among students.
3. Sustainability Initiatives: How students apply their leadership training to sustainability-related projects and community engagement efforts.

Relevance to Theory and Practice

This study is relevant to both academic theory and practical implementation. Theoretically, it contributes to the growing literature on leadership development, sustainability education, and experiential learning (Dugan, 2017). It examines how structured communication training, mentorship, and leadership roles influence sustainability engagement, particularly in the context of higher education institutions.

From a practical perspective, the findings will provide valuable insights for universities, educators, and policymakers on how structured leadership programs can be leveraged to enhance sustainability education. Given Saudi Arabia's Vision 2030, which emphasizes youth empowerment, leadership, and sustainability, this study provides a model for integrating leadership training into sustainability education programs at higher education institutions.

By combining qualitative methods with a strong research framework, this study aims to provide actionable recommendations for embedding leadership, communication, and mentorship training into sustainability curricula to better equip future leaders with the skills needed to address global sustainability challenges.

Findings

This section presents the key findings from the study, structured according to the research questions. The results highlight the role of Toastmasters International at Al Yamamah University in developing youth leadership, fostering mentorship, and supporting sustainability advocacy, all within the framework of Saudi Arabia's Vision 2030 objectives. Thematic analysis of participant observations, semi-structured interviews, content analysis, and survey responses revealed three core findings:

1. Toastmasters fosters leadership and communication skills critical for sustainability advocacy.
2. Mentorship within Toastmasters ensures leadership continuity and knowledge transfer in sustainability efforts.
3. Toastmasters aligns with Saudi Vision 2030 by empowering youth through leadership and sustainability engagement.

Toastmasters and Leadership Development for Sustainability Advocacy

The findings indicate that participation in Toastmasters significantly enhances students' leadership and communication skills, equipping them with the confidence to advocate for sustainability initiatives. Interviews with past and current club officers revealed that 90% of respondents felt their public speaking skills improved, and 85% reported increased confidence in leading sustainability-related discussions.

Observation data further demonstrated that students actively integrate sustainability themes into their speeches, debates, and leadership roles, reinforcing previous research that suggests communication training is a crucial component of effective sustainability advocacy (Lozano et al., 2017). Survey results also indicate that 78% of students believe their ability to inspire action in their communities has improved through Toastmasters training, aligning with the study's first research question.

These findings suggest that effective leadership in sustainability requires a combination of technical knowledge and strong communication abilities. Without the ability to articulate ideas clearly and persuasively, sustainability initiatives struggle to gain traction in policy discussions, community projects, and corporate sustainability programs (Dugan, 2017).

The Role of Structured Mentorship in Leadership Continuity

Mentorship emerged as a key driver of leadership sustainability within the Toastmasters framework. The study found that structured mentorship programs within the club ensure leadership continuity, allowing knowledge and experience to transfer from one cohort to the next. 82% of survey respondents who previously held leadership positions reported that they had mentored incoming officers, helping to maintain a sustainable leadership pipeline within the club.

Analysis of mentorship structures within Toastmasters at YU identified three primary forms of leadership transfer:

1. Direct mentoring: Outgoing leaders provide one-on-one coaching to their successors.
2. Group workshops: Experienced members lead training sessions for new officers.
3. Peer learning: Members mentor each other informally, reinforcing leadership values.

These results highlight the effectiveness of peer mentoring in sustaining long-term leadership capacity, which aligns with Komives et al.'s (2011) research on leadership identity development in higher education. By creating a culture of continuous mentorship, Toastmasters ensures that sustainability leadership does not end with a single cohort but remains an evolving, ongoing initiative.

Furthermore, participants emphasized that mentorship helped bridge the leadership gap in sustainability engagement, as mentors encouraged newer members to integrate environmental and social responsibility themes into their speeches and leadership projects. This reinforces the broader goal of building sustainable leadership frameworks that extend beyond the university setting into professional and civic domains.

Alignment with Saudi Vision 2030: Youth Empowerment and Sustainability.

The study's final research question explored how Toastmasters leadership training aligns with Saudi Arabia's Vision 2030 objectives, particularly in youth empowerment, sustainability education, and community engagement.

Survey results indicate that 76% of participants believe Toastmasters aligns with Vision 2030 by equipping them with leadership skills necessary for the workforce. Additionally, students actively engage in community-driven sustainability projects, such as:

- Sustainability Speech Series: Public speaking events addressing environmental concerns.
- Community Outreach Programs: Youth-led efforts to educate the community on sustainability.

These initiatives reflect Vision 2030's emphasis on youth leadership in national development (Ministry of Economy and Planning, 2022). Furthermore, Toastmasters fosters a culture of civic engagement and professional skill-building, which is essential for achieving the kingdom's long-term sustainability goals.

Participants also highlighted how public speaking and leadership training prepare them for future leadership roles in sustainability-focused industries. Interviews with student leaders revealed that several alumni pursued careers in corporate social responsibility, environmental policy, and education, further demonstrating the real-world applicability of Toastmasters training in advancing Saudi Arabia's sustainability agenda.

Implications for Leadership and Sustainability Education

The findings highlight the necessity of integrating communication and leadership training into sustainability education programs at universities. Effective sustainability leadership extends beyond technical expertise—it requires the ability to engage, persuade, and inspire action. Toastmasters provides a model that can be replicated across educational institutions to strengthen youth leadership pipelines and enhance long-term sustainability engagement.

Furthermore, the mentorship structures observed within Toastmasters suggest that similar peer-learning approaches should be incorporated into sustainability-focused leadership programs. Encouraging young leaders to mentor their peers ensures that sustainability efforts remain continuous and expand beyond individual projects into long-term community initiatives.

Finally, by aligning leadership training with Saudi Vision 2030 objectives, this study reinforces the importance of equipping youth with the skills necessary to drive sustainable development, not just at the university level, but across industries, government sectors, and global sustainability initiatives.

Conclusion

Principles and Generalizations Inferred from the Results.

The findings of this study highlight the crucial role of structured leadership training in fostering youth engagement in sustainability initiatives. The Toastmasters International program at Al Yamamah University has demonstrated its effectiveness as a leadership incubator, equipping students with the necessary communication, strategic thinking, and mentorship skills to drive sustainability efforts. The study confirms that leadership in sustainability requires not only technical knowledge but also strong advocacy, mentorship, and public engagement capabilities.

Three key principles emerge from this research:

1. Leadership training enhances sustainability advocacy.
2. Mentorship ensures leadership continuity.
3. Toastmasters aligns with Vision 2030's youth empowerment goals.

Exceptions, Problems, and Limitations

While the study provides valuable insights into leadership development through Toastmasters, several limitations must be acknowledged:

- Sample Size and Generalizability: The research focuses on Toastmasters at Al Yamamah University, which may limit the generalizability of findings to other institutions or regions with differing educational structures and leadership cultures.
- Self-Reported Data: Interviews and surveys rely on self-perception of leadership growth, which may introduce bias or overestimation of leadership competencies.
- Lack of Longitudinal Data: The study examines leadership development within a specific timeframe; however, long-term impacts of Toastmasters training on

career progression and sustainability leadership beyond university require further investigation.

Despite these limitations, the findings remain significant in demonstrating how structured leadership programs can contribute to sustainability leadership and mentorship development among university students.

Theoretical and Practical Implications

Theoretical Implications

This research contributes to existing leadership and sustainability education literature by:

- Reinforcing Komives et al.'s (2011) Leadership Identity Development Model, demonstrating how structured mentorship programs support long-term leadership sustainability.
- Expanding on Dugan's (2017) work on leadership development, emphasizing that public speaking and advocacy skills are crucial in sustainability education.
- Aligning with research on experiential learning models, illustrating that hands-on leadership training improves student confidence and engagement in sustainability efforts.

Practical Implications

The study's findings offer several practical recommendations for educators, policymakers, and sustainability advocates:

1. Integrating Toastmasters into Sustainability Curricula: Universities should incorporate public speaking and leadership training into sustainability education programs to better prepare students for advocacy and leadership roles.
2. Expanding Mentorship Programs: Institutions should develop structured mentorship models similar to Toastmasters, ensuring leadership continuity and sustained engagement in sustainability projects.
3. Encouraging Multi-Stakeholder Collaboration: Universities, businesses, and government entities should partner to support leadership training programs that align with Vision 2030's sustainability goals.
4. Leveraging Student-Led Initiatives: Higher education institutions should empower student-led sustainability initiatives by providing resources, recognition, and platforms for engagement.

Conclusions and Recommendations

This study underscores that leadership development is a key driver of sustainability advocacy. By equipping students with communication, mentorship, and leadership skills, Toastmasters fosters youth engagement in sustainability initiatives and aligns with Saudi Arabia's Vision 2030 goals.

To further enhance sustainability education, universities and policymakers should:

- Expand leadership training programs to a broader student base, ensuring that all students, regardless of discipline, receive communication and leadership development.
- Develop mentorship-driven sustainability leadership models, integrating peer learning and coaching frameworks into sustainability curricula.
- Support longitudinal studies to examine the long-term impact of Toastmasters leadership training on career progression and sustainability leadership beyond university settings.

By embedding leadership and communication training into sustainability education, institutions can cultivate a new generation of leaders capable of driving global sustainability transformations.

References

Dugan, J. P. (2017). *Leadership theory: Cultivating critical perspectives*. Jossey-Bass.

Komives, S. R., Owen, J. E., Longerbeam, S. D., Mainella, F. C., & Osteen, L. (2011). Developing leadership identity in college students: A grounded theory. *Journal of College Student Development, 52*(5), 515–539. https://doi.org/10.1353/csd.2011.0061

Lozano, R., Ceulemans, K., & Seatter, C. S. (2017). Teaching organisational change management for sustainability: Designing and delivering a course at the University of Leeds to better prepare future sustainability leaders. *Journal of Cleaner Production, 171*, 438–451. https://doi.org/10.1016/j.jclepro.2017.10.068

MacDonald, L., & Shriberg, M. (2016). Sustainability leadership programs in higher education: Alumni outcomes and impacts. *Journal of Environmental Studies and Sciences, 6*(2), 360–370. https://doi.org/10.1007/s13412-015-0344-7

Ministry of Economy and Planning. (2022). *Vision 2030: Sustainable development and youth empowerment*. Government of Saudi Arabia. https://www.vision2030.gov.sa/

United Nations. (2023). Empowering youth for sustainable development: A global framework for action. In *UN department of economic and social affairs*. https://www.un.org/development/desa/youth

Chapter 53
A Multi-Sensory Approach for Reviving 'Sense of Belonging' to the Surrounding Environment in the Design/Build Process (Education & Practice) to Address Sustainability Problem Routes

Eman Abdellatif

Abstract 'Sense of belonging", a term that self-drove sustainability over time. Architects/builders and community have become less connected (physically, mentally, and psychologically) with surroundings and the design/build process. It's time to revive this connection, addressing sustainability from its routes. A *need for a new approach in the design/build process that revives this connection.* Thriving after this through accumulative research and practice I have reached how different architectural compositions and their surroundings produce different vibration symphonies (unseen and unheard yet felt through our body) that positively or negatively shapes/impacts the user and his/her fauna, environment, and climate according to its intensity and frequency. My latest practical/theoretical research at UAL linked architecture to sound, soundart, music, choreography, emotions, motion/active manipulation through space, deep/spatial listening and embodiment, revealing intrinsic relations to *possibly stimulate a design/build process* (*in education and practice*) to *guide the shaping of the built environment for reviving connectivity with surroundings.* Progressing all this into a series of courses in co-ordination with UCL (University College London, UK), it was implemented in NGU (New Giza University, Egypt). They are built on the wholistic process of creativity and findings through my Ph.D., previous explorations, using 'listening' and 'sound' as a key to trigger embodiment and a multi-sensory design aiming to stimulate the students' connectivity to their surroundings and senses, producing positively stimulating environments to similarly provoke user's connectivity. The outcome of these courses showed promising changes within the students' engagement and commitment to user and environment, reflected in their designs and behaviour. Community exhibited emotional 'involvement' and raised awareness. This approach is very important to consider in an era that is heavily oriented towards using AI in education from early stages. It is important to know *when* and *how* to use to avoid further separation.

E. Abdellatif (✉)
Dubai, United Arab Emirates
e-mail: eman.ibro@gmail.com

J. Dsilva (ed.), *Sustainability Education, Youth and Innovation*,
https://doi.org/10.1007/978-3-032-21948-0_53

Keywords Sense of belonging · Connecting · Listening · Multi-sensory · Vibrations

Introduction

Sustainability is today's menu main dish! However, it fails to be addressed correctly, form routes. It can only happen by provoking emotional connection to the whole. Feeling surroundings, sense of belonging, ownership, commitment, and design impact. An ability many architects lack.

We are floating in a sea of vibrations that even if we can't hear or see, it's felt via our bodies. Motions and emotions reflect the first indicative. Embodiment is the art of connecting to surrounding energy and vibrations. Listening is the trigger key.

This paper *aims* to display and propose a process of composing and implementing/exploring a multi-sensory design approach built on sound/music, abstraction, motion/choreography and rhythm analysis targeting shaping architects connected to surroundings (community/environment) and their design process. Congruently producing a built environment with positive vibrations that induces community's connectivity in a loop action.

The methodology was based on a multi-sensory and disciplinary journey, divided into two main interactive phases; a practical/theoretical phase of forming the proposed process, followed by shaping/implementing the process through course series and practice.

The first phase (practical/theoretical research) was a lifetime progressive process, crowned at UAL through linking my previous research, courses and practical explorations/work (Fig. 53.1). Seismic microphones and choreography were used to detect vibrations due to three different architectural compositions and their impact; traditional, contemporary, and modern. It was found, unprecedently, qualitatively, and quantitatively how different architecture compositions produce different intensity and frequencies of vibrations, namely infrasound, and how it impacts emotions and motion. This also impacts our health, mental and psychological abilities, the fauna and induces climate change (Persinger, 2014; Sokol, et al., 2022; Abdelatif, et al., 2023).

It similarly affects our motion, emotion and behaviour which was why parallel to this choreography, shown in Fig. 53.2 (the art of inhaling the surrounding and exhaling this into a motion performance) was used. Linking to my Ph. D and practice, a hypothesis for reaching positively shaping compositions was explored practically through the use a multi-disciplinary team; sound/music composition and motion/choreography, through a sound/music installation and performance event in Dilston Grove London. The choreographer was to perform accordingly in reflection to the composed sound/music installation, space and time. The performance was later analysed by the author through a series of abstract art paintings and rhythm analysis.

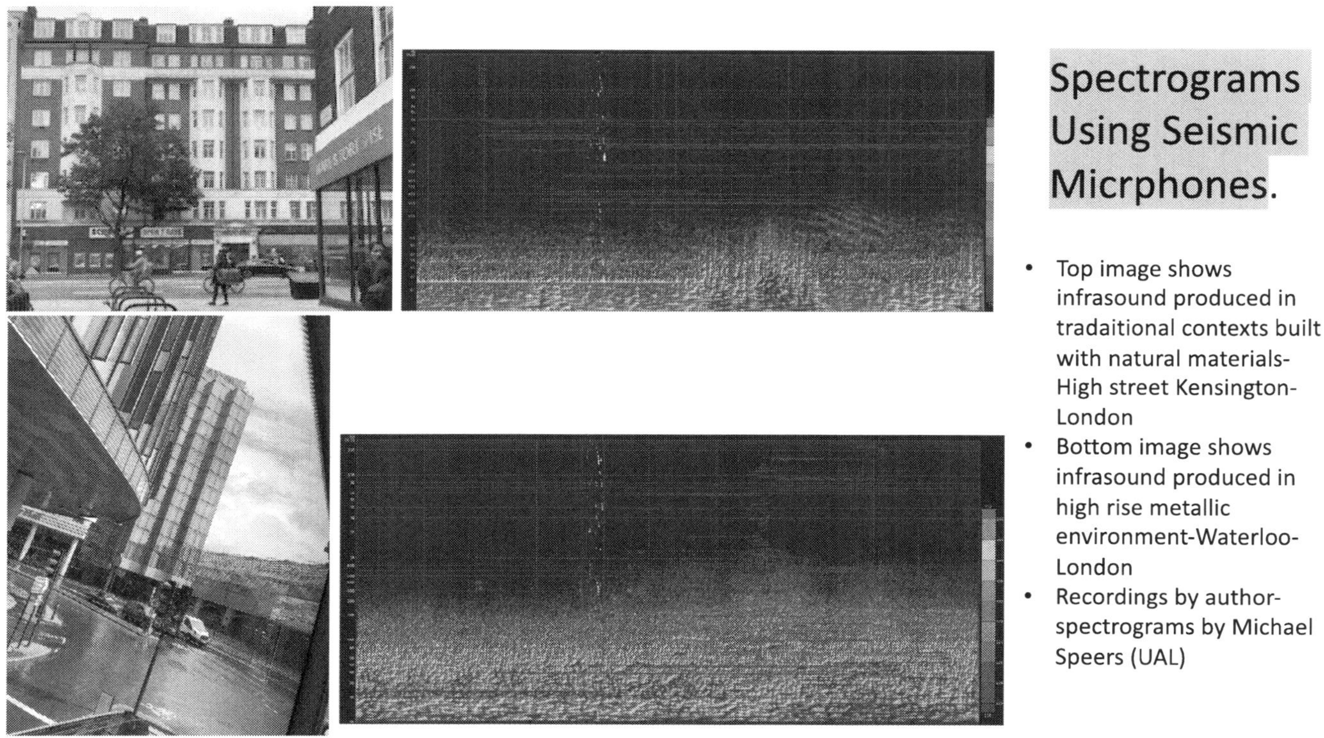

Fig. 53.1 Extract from my thesis at UAL "Listening to the Dance of the Built Environment" measuring infrasound (red indicating highest levels) in different architectural compositions (Abdellatif, 2022)–Photos and recordings by the author

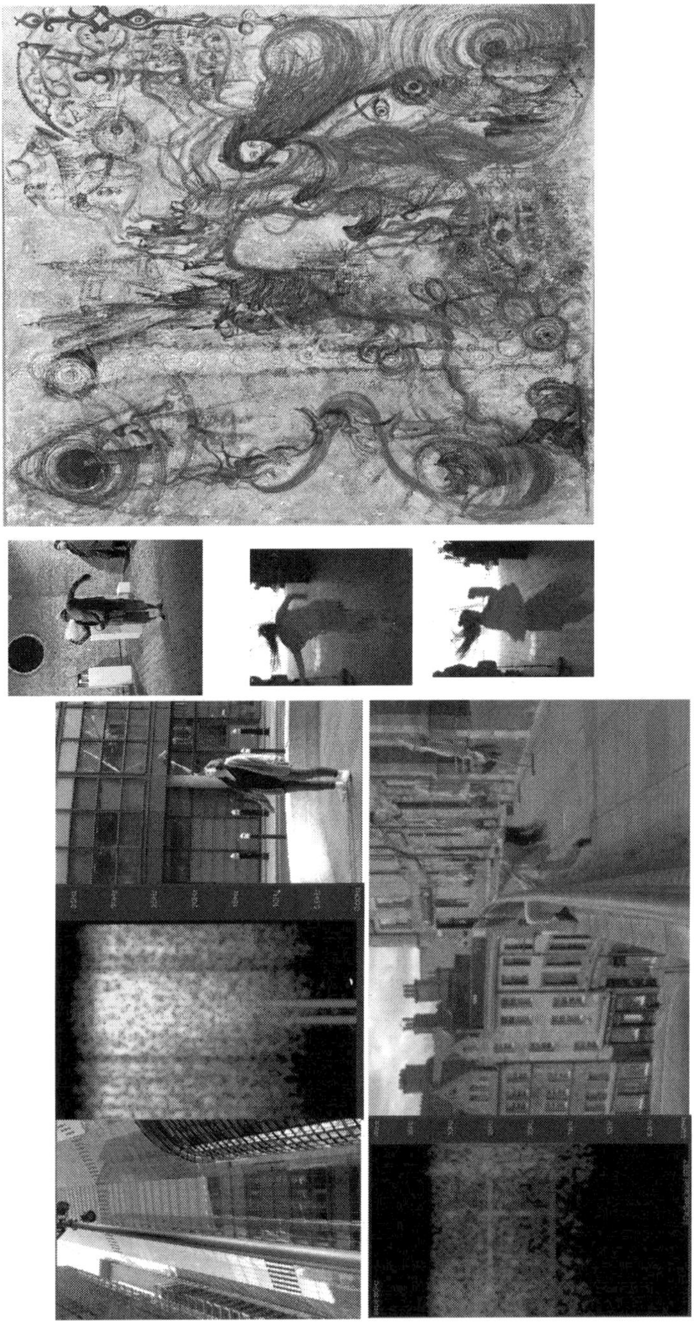

Fig. 53.2 Photographic display of the phase, exploring choreography/motion with different architectural composition and performance in Dilston Grove London, analysed into abstract paintings and rhythm analysis by the author. Photos by author. (Abdellatif, 2022)

The outcomes of this research along with further research, were developed into a series of courses in co-ordination with UCL, to be implemented in NGU, by the author as the main lecturer.

Findings for both parts proved how built environment compositions are literally a piece of music and hence can be designed as such. Sound/music and motion are key factors to use for designing built environment compositions and indicatives to its impact, whilst triggering an emotional link to the environment.

Methodology

The methodology uses a multi-disciplinary theoretical/practical approach, linking architecture with science and arts through two main phases in practice and education. The main question this paper focuses on; *What is the process that can enhance sustainability through reviving 'sense of belonging' and connection to the environment in the design/build process of shaping it in education & practice?*

To explore this the methodology used was a multi-disciplinary and sensory approach divided into two interconnected phases.

Phase1-Practical/theoretical research: This phase aimed to explore theoretically and practically my hypothesis that if architecture compositions are literally a symphony of vibrations, then using sound/music compositions and motions in the design process can help reach built environments compositions that are no longer frozen and reconnects people to surroundings. To explore the first part, the researcher selected three types of compositions: traditional, contemporary, and modern, in three different cities (London, Cairo, Dubai). The vibrations were measured quantitatively, using seismic microphones, and qualitatively using choreographers. To explore the second part the researcher composed a sound/music piece to test on the choreographer and audience how a targeted concept can be reached through the installation of a sound/music piece in a space, and how their motion can be analyzed to deduce the feeling and composition of the space required for a concept. First let's go through how the hypothesis was deduced.

Sustainability is continuity, it cannot happen one-sided. The current building process shows disconnection, between the architect, design, community, and environment, resulting in dead frozen built environments that further results in disconnected generations (Hirsch & Clark, 2019; Abdellatif, 2018; Pallasma, 2008; Alexander, 1979; Rasmussen, 1962). The flow needs to be reversed.

Finding solutions has been a lifetime journey, researching theoretically and practically the impact of the built environment. The built environment is the main shaper of our abilities, and we back in a continuous loop (Abdellatif, 2013). Awareness of its deep impact must be raised.

So, what is the process in education and design that can help regain this link to enhance connection?

Listening was proposed as a tool for architects/urban planners to improve quality of built environments (Oliveros, 2003). It is a means for embodiment (Carlyle,

2007). A key to all mental activities and sensors (Seth, 2012). Perception requires all sensors' involvement to be complete (Eagleman, 2015). 'Listening experiences' are the connector (LaBelle, 2021).

Sound holds meanings, memories, and emotions (Seth, 2012, and Blesser & Salter, 2006). It brings us into one continuum (Oliveros, 2003). It is used to inform architects about space's properties, culture, and psychophysics. (Blesser and Salter, 2006). Researchers and architects have relentlessly discussed sound's potential to inform and broaden architecture design (Sheridan, et al., 2003).

Music plays with emotions, motion, abilities, and behaviours (Levitin, 2006). It has the power of connecting and synchronizing people's motions and emotions (Abdellatif, 2018). It tells the story of a culture's history, values, and behaviours (Levitin, 2008). Architects have contemplated ways of integrating music into the architectural design process (Jencks, 2013).

Vibration is sound's partner we don't necessarily hear yet feel. It shapes and molds us (Seth, 2012). Everything around us is in constant vibration, constant motion (Hunt, 2018). Infrasound is a vibration below our hearing level (20 Hz), emitted by some built environment compositions and may cause health hazards, affect fauna and climate change (Persinger, 2014; Abdellatif, 2022). Different materials, geometries, and forms, etc., vibrate in different ways and synchronise in proximity. That is why different built environments produce different vibrational symphonies (Blesser and Salter, 2006 and Hunt, 2018 and Abdellatif, 2022).

Motion is the reason for sound and vibrations, as molecules are constantly in motion (Hunt, 2018). It creates emotions as an outcome. Music triggers both motion and emotions (Levitin, 2006), while active motion in a space bonds us with the space itself (Abdellatif, 2018). There is a link between motion, emotion and moving in a place through events which create this sense of belonging (LaBelle, 2021).

Rhythmanalysis was discussed by Henri Lefebvre who saw the unique role of architecture in relation to motion. He stated space and time needed to be connected, and motion in urban space did that (Lefebvre, 2004).

Dance and choreography in surroundings similarly require embodiment. Embodiments are defined as becoming tuned to surroundings and their vibrations (Callahan, 1995). There is a cross disciplinary link between movement, audio and visual (Jensenius et al., 2010).

Composition share boundaries between architecture design, choreography, music/sound composition, and film making. There is a cross modular phenomenon between the aural and visual. Coupling them shows deep value in relation to architecture (Abdellatif, 2018).

Hence, to conclude, through theory and practice, sound/music/vibration/motion holds potential to be a key for triggering this connection through a multi-sensory process, leading to the previously mentioned hypothesis.

Phase 2: design/implement personal courses: Outcomes of the previous phase were shaped into a proposed design process aiming to stimulate future architects capable of feeling 'sense of belonging', and commitment to environment and community, and impact of their designs. As shown in Fig. 53.3, this process was to be tested through educational channels.

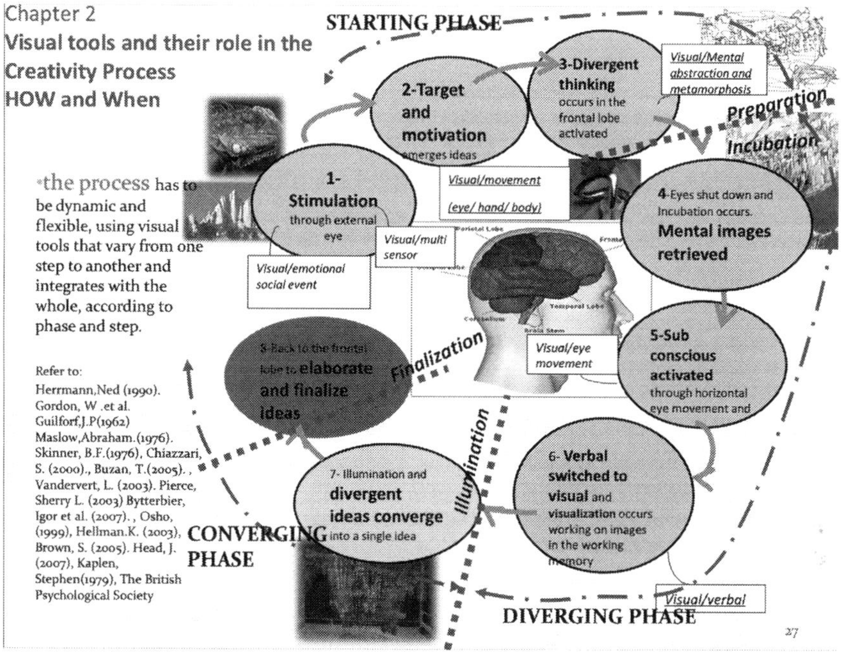

Fig. 53.3 An extract my Ph.D. showing the deduced 'wholistic creativity process' and the role of sensors (Abdellatif, 2013)

The process consisted of a three-course series to shape the above. Guided by the wholistic creativity process, in interaction with surroundings, the main theme revolved around using 'audio/visual musical abstraction' as a tool for enhancing creative abilities in a multi-sensory design/build approach (Abdellatif, 2013; Abdellatif, 2018). A fixed architecture student sample of 35 students was used throughout the course series. Their final project was to test courses' impact, along with events and activities, linked directly and indirectly with community and surroundings. The change was compared to their initial potential and my previous courses.

First course aims to enhance connection between individuals, their potentials, sensors, surrounding environment, and each other, as individuals and groups. It relies on deep listening, embodiment, and an audio/visual abstraction to trigger communication and engagement in a multi-sensory design/build process guided by the community and surroundings in a progressive event as their final participatory project.

Second course aims to shape architects that actively and creatively integrate the use of senses in their designs (sensuous urbanism) focusing on sound and natural light to emphasize concepts and experiencing architecture as a perceptual journey. It starts with the audio, shaping listening skills as the key to trigger other perceptual and analytical abilities required in creative architecture design. Light is understood as a creative tool to control the tempo of the journey and motion inside the space, shaping the mode and emotional melody of the concept using colours, shade and shadow, material, forms, and patterns in space. Sound is similarly understood as an integrative part of this musical symphony rather than a noise to avoid. The final years project links sound and light with the first course developing a design that reflects community and surroundings' need/want as a multi-disciplinary team in integration with community, surroundings in a real site.

Final course (yet to be implemented) is the crown. Students by now should be able to compose through their own senses and feelings, the built environment symphony/compositions that allows the users to stimulate personal connection and link to surrounding environment. It aims to revive urban and architectural environments, by retrieving missing emotional experiences in the physical context through coupling music/sound once again with the visual and motion in space. Students will learn to listen to the different layers in music, analyse them to rhythms, tempos, melodies, emotions, and motion, followed by shaping an ability to visualize music into lines, colours and forms and analyse motion through rhythmanalysis and abstraction. Finally translating into a creative design that triggers same emotions and values provoked through listening to their composed sound/music piece that acts as their design concept.

Findings

Theoretical/practical exploration of shaping the proposed process showed qualitatively and quantitively how different built environment compositions shape and impact the user shown in Fig. 53.4 (physically and psychologically), and that the built environment is literally a symphony of vibration that can be guided by sound/music motion and emotions. It may also produce vibrations that critically affect sustainability by impacting climate, surrounding environment, and health according to its vibrational intensity and frequency.

It also showed how when architecture students learn to connect with surrounding environment, feelings, sensors, vibrations, and community, it changes their visions towards architectural design and commitment to the environment and the design/build process. The design becomes through their sensors, heart, and soul in a continuum. Similarly, the community showed vibrant reactions and increased awareness towards environment.

Conclusion

This paper revealed both theoretically and practically that the sustainability problem if addressed from its roots, which is enhancing the 'sense of belonging' and connection to the environment, would result in deeper and more impactful results. It uncovered how the built environment around us is a symphony of vibration that if treated

Fig. 53.4 Shows how high rise vertical metallic composition showed repetitively high intensities of infrasound (below hearing level that impact climate, health and fauna) waves that limited motion of the choreographer, versus traditional low rise, natural material compositions that showed low levels of low frequency waves, giving freedom in motion and emotion-photos and measurements by author

as so through the architectural educational channels using multi-sensory and participatory design approaches that use sound/music and emotion as its base, can result in architects, community and a built environment that are naturally connected. Hence the natural flow of solving sustainability problems may be retrieved by reaching more profound results.

References

Abdelatif, E. M., & Doutsios, D. (2023). The impact of infrasound around us in our daily built environment. https://www.eumada.com/post/the-impact-of-infrasound-around-us-in-our-daily-built-environment

Abdellatif, E. M. (2013). *Visual tools as enhancers to a creative building process-towards a visually based teaching method in architecture.* Ph. D. Faculty of Engineering, Architecture Department, Cairo University. Egypt.

Abdellatif, E. M. (2018). Abstraction, music, art, emotions, visualization and creativity: A complex interconnected dynamic matrix for synergetic creative building process from education to implementation. In CAUMME conference, Istanbul, Turkey 2018. https://www.academia.edu/114240906/ABSTRACTION_MUSIC_ART_EMOTIONS_VISUALIZATION_AND_CREATIVITY_A_COMLEX_INTERCONNECTED_DYNAMIC_MATRIX_FOR_SYNERGETIC_CREATIVE_BUILDING_PROCESS_FROM_EDUCATION_TO_IMPLEMENTATION

Abdellatif, E. M. (2022). Listening to the dance of the built environment-multi-disciplinary exploration towards a process for composing built environments that enhances positive listeners in relation to culture through sound/music. M.Sc. in Soundarts, London College of Communication, University of Arts London, UK. https://artslondon.padlet.org/eabdellatif03202011/listening-to-the-dance-of-the-built-environment-4xvcekuad1x44m7u

Alexander, C. (1979). *Timeless way of building.* Oxford University Press.

Blesser, B., & Salter, L. R. (2006). *Spaces speak, are you listening? Experiencing aural architecture.* MIT Press.

Callahan, P. S. (1995), Paramagnetism–recovering nature's secret force of growth. Acres. http://www.richardmudhar.com/blog/wp-content/uploads/2015/07/acres_usa_callahan_paramag_secret_callahan.pdf

Carlyle, A. (2007). Autumn leaves-sound and the environment in artistic practice. Association Double–Entendre.

Eagleman, D. (2015). *The brain-the story of you.* Clays Ltd, Elsograf S.p.A. Great Britain.

Hirsch, J. L., & Clark, M. S. (2019). Multiple paths to belonging that we should study together. Perspectives on psychological science. *In Sage Journals, 4*(2) https://journals.sagepub.com/doi/full/10.1177/1745691618803629

Hunt, T. (2018December 5). *The hippies were right: It's all about vibrations, man! A new theory of consciousness.* Scientific American.

Jencks, C. (2013). Architecture Becomes Music. *Architectural Review.*

Jensenius, A. R., Wanderley, M. M., Godøy, R. I., & Leman, M. (2010). Musical gestures: Concepts and methods in research. In R. I. Godøy & M. Leman (Eds.), *Musical gestures: Sound, movement, and meaning* (pp. 12–35). Routledge.

LaBelle, B. (2021Jan). Acoustic spariality in infrasonic. January, 2021.https://infrasonica.org/en/sonic-realism-wave-3/nicole. (translated from Spanish to English).

Lefebvre, H. (2004). *Rhythmanalysis-space, time and everyday life (S. Elden & G. Moore, trans.). Continuum.* Cromwell Press.

Leivitin, D. (2006). *This is your brain on music-the science of the human obsession.* Dutton Penguin.

Levitin, D. (2008). *The world in six songs-how the musical brain created human nature.* Autmn Press.

Oliveros, P. (2003). *Deep listening–a composer's sound practice.* iUniverse, Inc.

Pallasma, J. (2008). *The eyes of the skin-architecture and senses.* TJ International LTD..

Persinger, M. A. (2014). Infrasound, human health, and adaptation: An integrative overview of recondite hazards in a complex environment. *Natural Hazards, 70,* 501–525. https://link.springer.com/article/10.1007/s11069-013-0827-3

Rasmussen, S. (1962). *Experiencing architecture.* MIT press.

Seth, H. (2012). *The universal sense: How hearing shapes the mind.* Bloomsbury.

Sheridan, T., et al. (2003). Hearing architecture: Exploring and designing the aural environment. *The Journal of Architectural Education, 57*(2) (Cambridge: The MIT Press, November 2003).

Sokol, G. I., et al. (2022). Impact of infrasound in the atmosphere of the earth. *New Frontiers in Physical Science Research, 1,* 37–51.

Chapter 54
New Ecological Paradigm in a Multicultural Context: Identity Integration, Global Mindset, and Values as Predictors of Ecocentrism

Magdalena Mosanya and Anna Kwiatkowska

Abstract In today's world, two pressing issues dominate the social sciences: sustainability and multiculturalism. These concepts are not just related but are deeply intertwined. The New Ecological Paradigm presents an ecocentric view that highlights the vital interdependence between humanity and nature, standing in stark contrast to anthropocentrism. This study investigated whether exposure to multiculturalism during early childhood can predict ecocentric worldviews among third culture kids (TCKs) (N = 399; mean age 21 years) through their multicultural identity, values, and global mindset. We hypothesized that different configurations of multicultural identity (integration, categorization, and compartmentalization) and value dimensions (self-transcendence, openness, and conservation), along with a global mindset, would influence ecocentric and anthropocentric worldviews. Our findings reveal a compelling trend: TCKs are notably more inclined toward ecocentrism. The path model illustrates that an integrated multicultural identity, combined with values of self-transcendence and openness, as well as a global mindset, strongly correlates with pro-ecological attitudes. Conversely, a tendency toward identity categorization and conservative values aligns more closely with anthropocentrism. We also uncovered intermediary effects that enrich our understanding of these dynamics.

Keywords Ecocentrism · Anthropocentrism · Multicultural identity · Third culture kids · Values

M. Mosanya (✉) · A. Kwiatkowska
Institute of Psychology, Polish Academy of Science, Warsaw, Poland
e-mail: Magdalena.Mosanya@murdoch.edu.au

A. Kwiatkowska
e-mail: akwiatkowska@psych.pan.pl

M. Mosanya
Murdoch University Dubai, Dubai, United Arab Emirates

J. Dsilva (ed.), *Sustainability Education, Youth and Innovation*,
https://doi.org/10.1007/978-3-032-21948-0_54

Introduction

In a world that is interconnected, environmental issues like global warming and the exhaustion of natural resources become widespread and intricate, necessitating cooperation on a global scale (Dunlap et al., 2000). The ecological harm recognized by scientists and world leaders is attributed to human actions, primarily linked to the advancements of industrialized societies (Jorgenson, 2006). Some people still lack concern about the ecological matters, despite an increasing popularity of the New Ecological Paradigm (NEP), which recognizes the interdependence between human well-being and nature while imposing restrictions on human activity and growth (Dunlap et al., 2000; Stern et al., 2008). Various authors attribute such neglect of environmental care to a widespread belief in human superiority over nature (Milfont et al., 2013). Social dominance theory explains such a mindset by suggesting that there are measurable differences in how much people prefer hierarchical social structures over egalitarian relationships (Sidanius & Pratto, 1999). Therefore, individuals vary in their positioning within the ecosystem. The inclination to isolate oneself from nature and exert dominance over it is associated with lower levels of environmentalism (Milfont et al., 2013). Thus, addressing environmental challenges may involve enhancing environmental awareness, promoting inclusivity, and valuing nature (United Nations, 2015). This can be facilitated by stimulating psychological factors such as self-transcendent and open-minded values along with a global perspective, which may mitigate dominant social influences and support inclusive consideration for non-human interests.

Transforming worldviews and attitudes from anthropocentric (focused on human progress through exploiting natural resources) to ecocentric (emphasizing the equilibrium between human needs and environmental requirements) presents difficulties. Studies highlight the tension between human demands and ecological necessities since environmentalism requires dedication and sacrifice (Kollmuss & Agyeman, 2002; Stern et al., 2008). When faced with a choice between personal advantages and environmental priorities, individuals often favor the former. Personal adherence to NEP may be rationalized if individuals internalize environmental arguments by connecting them with personal values and identity. Extensive research over the last few decades (e.g., Brieger, 2018; Duff et al., 2022; Ringov & Zollo, 2007) has underscored the influence of culture and identity on the development of environmental attitudes. Brieger (2018) highlighted how social identity plays a critical role in nurturing environmental awareness. Specifically, the inclination to safeguard the environment corresponds with the inclusiveness of the groups forming one's identity (community, nation, world). Identification with the global community, or world identity, proved to be especially advantageous for fostering environmental concern (Brieger, 2018).

This prompts inquiries regarding the impact of multicultural identities in individuals with extensive cross-cultural experiences, such as third culture kids (Pollock et al., 2017). Yampolsky et al. (2016) proposed that multicultural identities vary in their inclusiveness (integration versus categorization and compartmentalization). Consequently, environmental attitudes might also evolve in diverse manners. However, there are considerable deficiencies in the research concerning pro-environmental attitudes, particularly a significant absence of data from multicultural, non-Western, developing, and non-democratic settings (Tam et al., 2021). The Middle East, which contains multicultural populations, is particularly underrepresented and warrants further investigation. Furthermore, the current multicultural youth will encounter the consequences of environmental decline; their viewpoints on these issues should be acknowledged.

This study analyses whether multicultural individuals with mobile lifestyles, referred to as third culture kids (TCKs), could embrace ecocentrism through their prolonged, hybrid yet cohesive cross-cultural identity, global mindset, and self-transcendence and openness values. It also examines if specific strategies of multicultural identities, such as categorization or compartmentalization, encourage conservative beliefs and anthropocentric traditional social frameworks. In the twenty-first century, it has become clear that the survival and well-being of humanity hinge on global cooperation to guarantee environmental sustainability. Investigating third culture individuals is particularly pertinent, as their identities, attitudes, and values—reflecting post-modern adaptability and fluidity—are increasingly prevalent in a globalized context.

We assessed the presence of NEP factors among TCKs (RQ1). Additionally, we sought to analyze the relationships between multicultural identity configurations, value dimensions, and global mindset with respect to ecocentrism and anthropocentrism. We put forth the hypothesis that integration, self-transcendence, openness, and a global mindset correlate positively with ecocentrism (H1a), whereas categorization, compartmentalization, and conservation are positively linked to anthropocentrism (H1b). Lastly, appreciating the significance of integrative models, we proposed a framework illustrating the direct and indirect predictive impacts of multicultural identity configurations, value dimensions, and a global mindset on ecocentrism and anthropocentrism.

Metodology

Our participants were non-Western third culture kids (TCKs) ($N = 399$; average age 21). We used the Revised New Ecological Paradigm Scale (NEP-R) (Dunlap & Van Leire, 1978; Dunlap et al., 2000; Dunlap, 2008), Multicultural Identity Integration Scale (MULTIIS) (Yampolsky et al., 2016), Portrait Value Questionnaire—Short (PVQ-S) (Schwartz, 1996, 2008), and Global Mindset Scale (Den Dekker, 2011). This study was quantitative. We tested hypotheses using path analysis. Statistical tests were conducted using SPSS v.25 and Amos v.25.

Fındıngs

Overall, the findings revealed that Third Culture Kids (TCKs) typically lean towards ecocentrism. There were direct predictive effects of integrated multicultural identity and a global mindset on the ecological perspective. The conservative acted in the opposite direction (Fig. 54.1). Additionally, the indirect effects were observed. Integrated multicultural identity diminished the negative effect of compartmentalization, while a global mindset lessened the negative effect of conservative values on NEP. Our predictions that an integrated multicultural identity is associated with ecocentrism, while a categorical multicultural identity supports anthropocentrism, were hence validated. We therefore confirmed that identity and culture play crucial roles in shaping pro-environmental perspectives (Brieger, 2018), although only the integrated multicultural identity promotes ecocentrism among TCKs. Interventions that aim to encourage this strategy for dealing with multiculturalism could motivate TCKs to adopt ecocentrism. There are hence multiple advantages to promoting cultural integration.

We further observed a positive correlation between conservation values (tradition and conformity) and anthropocentrism (Fig. 54.2), which is consistent with Schwartz (2012), who highlighted that tradition is rooted in rigid, subordinate mindsets. Conservation values might support traditional social frameworks (anthropocentrism). Conservative values and anthropocentrism appear to conflict with the global mindset and the New Ecological Paradigm (Dunlap et al., 2000). Schwartz (2012) also proposed that values related to conservatism are associated with disruptions in intergroup relations. Numerous environmental issues today necessitate cooperation across groups and international collaboration, which traditional values might obstruct. In contrast, a global mindset that appreciates diversity and innovation (Den Dekker, 2011; Stokke, 2013) helps safeguard against these challenges, as it is positively linked with an eagerness to engage with outgroups (Bennett, 2017). Additionally, in our study, notable differences by gender and religion were found, with women, Hindus, and Buddhists exhibiting the strongest pro-ecological attitudes and inclinations.

Appendix

See Figs. 54.1 and 54.2.

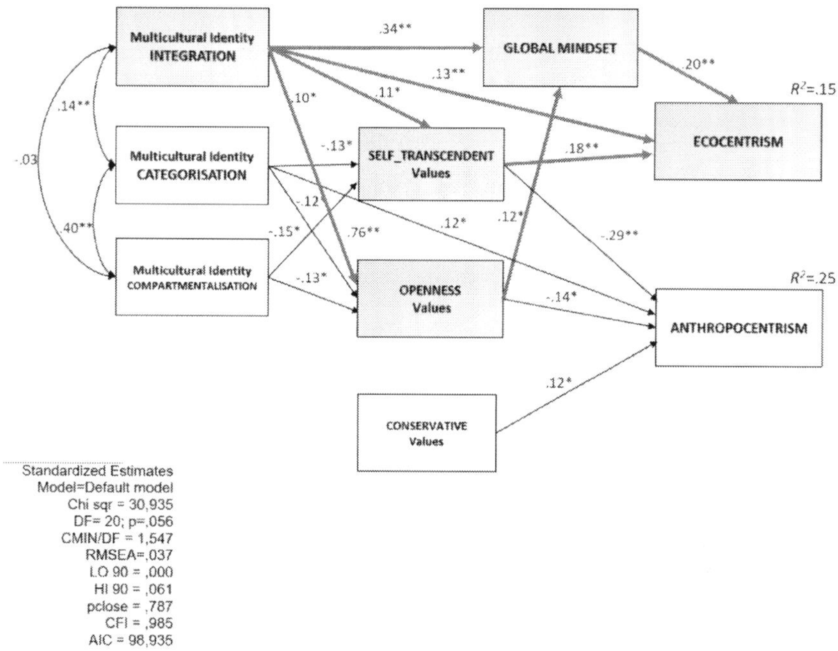

Fig. 54.1 Ecocentric pattern

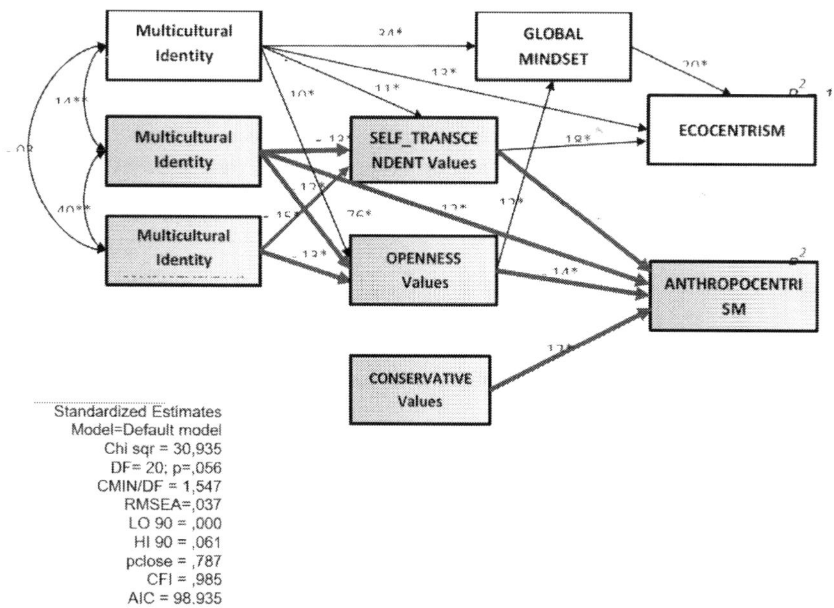

Fig. 54.2 Antropocentri pattern

References

Bennett, M. J. (2017). Developmental model of intercultural sensitivity. In Y. Kim (Ed.) *The International Encyclopedia of Intercultural Communication*. John Wiley & Sons. https://doi.org/10.1002/9781118783665.ieicc0182.

Brieger, S. A. (2018). Social identity and environmental concern: The importance of contextual effects. *Environment and Behavior, 51*(7), 828–855. https://doi.org/10.1177/0013916518756988

Den Dekker, W. (2011). *Global mindset. dimensions, measurement, and leadership effectiveness*. Doctoral dissertation. Amsterdam: Vrije Universiteit Amsterdam.

Duff, H., Vignoles, V. L., Becker, M., & Milfont, T. L. (2022). Self-construals and environmental values in 55 cultures. *Journal of Environmental Psychology, 79*, 101722. https://doi.org/10.1016/j.jenvp.2021.101722

Dunlap, R. E. (2008). The New Environmental Paradigm Scale: From marginality to worldwide Use. *The Journal of Environmental Education, 40*(1), 3–18. https://doi.org/10.3200/JOEE.40.1.3-18

Dunlap, R. E., & Van Liere, K. (1978). The new environmental paradigm. *The Journal of Environmental Education, 9*(4), 10–19. https://doi.org/10.1080/00958964.1978.10801875

Dunlap, R. E., Van Liere, K. D., Mertig, A. G., & Jones, R. E. (2000). Measuring endorsement of the new ecological paradigm: A revised NEP scale. *Journal of Social Issues, 56*(3), 425–442. https://doi.org/10.1111/0022-4537.00176

Jorgenson, A. (2006). Unequal ecological exchange and environmental degradation: A theoretical proposition and cross-national study of deforestation, 1990–2000. *Rural Sociology, 71*(4), 685–712. https://doi.org/10.1526/003601106781262016

Kollmuss, A., & Agyeman, J. (2002). Mind the gap: Why do people act environmentally and what are the barriers to pro-environmental behavior? *Environmental Education Research, 8*(3), 239–260. https://doi.org/10.1080/13504620220145401

Milfont, T. L., Richter, I., Sibley, C. G., Wilson, M. S., & Fischer, R. (2013). Environmental consequences of the desire to dominate and be superior. *Personality & Social Psychology Bulletin, 39*(9), 1127–1138. https://doi.org/10.1177/0146167213490805

Pollock, D., Van Reken, R., & Pollock, M. (2017). *Third culture kids. Growing up among worlds*. Nicholas Brealey Publishing.

Ringov, D., & Zollo, M. (2007). The Impact of national culture on corporate social performance. *Corporate Governance, 7*, 476–485. https://doi.org/10.1108/14720700710820551

Schwartz, S. H. (1996). Value priorities and behavior: Applying a theory of integrated value systems. In C. Seligman, J. M. Olson, & M. P. Zanna (Eds.), *The Psychology of Values: The Ontario Symposium* (Vol. 8, pp. 1–24). Erlbaum.

Schwartz, S. H. (2008). Values: cultural and individual. In F. J. R. van de Vijver, A. Chasiotis, & S. M. Breugelmans, (Eds.). *Fundamental questions in cross-cultural psychology* (pp. 463–493). Cambridge University Press. 2011

Schwartz, S. H. (2012). An overview of the Schwartz theory of basic values. *Online Readings in Psychology and Culture, 2*(1). https://doi.org/10.9707/2307-0919.1116

Sidanius, J., & Pratto, F. (1999). *Social dominance: An intergroup theory of social hierarchy and oppression*. Cambridge University Press.

Stern, P. C. (2000). New environmental theories: Toward a coherent theory of environmentally significant behavior. *Journal of Social Issues, 56*(3), 407–424. https://doi.org/10.1111/0022-4537.00175

Stern, M. J., Powell, R., & Ardoin, N. (2008). What difference does it make? Assessing outcomes from participation in a residential environmental education program. *The Journal of Environmental Education, 39*(4), 31–43. https://doi.org/10.3200/JOEE.39.4.31-43

Stokke, P. (2013). *Adult Third Culture Kids: Potential global leaders with global mindset*. Unpublished doctoral dissertation. Drexel University. https://doi.org/10.1108/JGM-12-2012-0021

Tam, K.-P., Leung, A.K.-Y., & Clayton, S. (2021), Research on climate change in social psychology publications: A systematic review. *Asian Journal of Social Psychology*, *24*, 117–143. https://doi.org/10.1111/ajsp.12477

Yampolsky, M. A., Amiot, C. E., & de la Sablonnière, R. (2016). The Multicultural Identity Integration Scale (MULTIIS): Developing a comprehensive measure for configuring one's multiple cultural identities within the self. *Cultural Diversity and Ethnic Minority Psychology*, *22*(2), 166–184. https://doi.org/10.1037/cdp0000043

Chapter 55
Fostering Inclusive and Sustainable Education: Enhancing Interdisciplinary Skills with AI-Driven Collaborative Learning

Senika Dewnarain

Abstract As artificial intelligence (AI) becomes increasingly embedded in educational practice, the need to develop interdisciplinary and collaborative skills in students has never been more critical. This paper explores how group-based learning using AI tools like ChatGPT can enhance student engagement, foster interdisciplinary understanding, and promote ethical AI literacy. By examining a real-world classroom intervention involving case study analysis on digital inequality in Uganda, the study demonstrates how ChatGPT supported collaborative learning, problem-solving, and critical thinking. Drawing on literature from interdisciplinary education, group learning, and AI ethics, the findings reveal that while ChatGPT facilitates immediate information access and reduces cognitive load, critical reflection remains essential to avoid surface-level learning. The paper concludes with recommendations for integrating AI in higher education to create sustainable, inclusive, and ethically grounded learning environments.

Keywords AI literacy · Interdisciplinary learning · ChatGPT · Higher education · Group-based learning · Digital inequality · Sustainable education

Introduction

Interdisciplinary education is increasingly essential in preparing students for the complexities of today's labour market. As societal problems grow more intricate, the limitations of single-discipline approaches have become evident (Borrego & Newswander, 2010). Higher education institutions must foster lifelong learners and adaptable professionals by emphasizing interdisciplinary skills. This shift in education moves away from passive knowledge acquisition to an active exploration of new concepts and competencies. Engaging with multiple disciplines helps students

S. Dewnarain (✉)
University of Birmingham Dubai, Dubai, United Arab Emirates
e-mail: s.dewnarain@bham.ac.uk

J. Dsilva (ed.), *Sustainability Education, Youth and Innovation*,
https://doi.org/10.1007/978-3-032-21948-0_55

develop critical thinking, holistic approaches, and metacognitive skills, which are essential for integrating diverse perspectives and understanding the interconnectedness of different fields (Klaassen, 2018). Real-world problems demand interdisciplinary solutions, making it imperative for education to promote demand-driven learning and transferable skills (Souppez, 2017).

The Evolution of AI and Its Implications for Education

Artificial intelligence (AI), since its inception in 1956, has rapidly progressed, incorporating machine learning, natural language processing, and autonomous systems. Innovations such as ChatGPT by OpenAI have brought AI into everyday applications, enabling sophisticated human-like interactions and offering potential benefits across various domains, including education (Crompton & Burke, 2023). However, this proliferation comes with challenges, such as employment disruptions and ethical implications (Cardon et al., 2023). The dual nature of AI, acting as both an economic driver and a labour market disruptor, necessitates a focus on AI literacy in higher education (Ng et al., 2021). This literacy involves both technical skills and critical thinking about AI's broader implications. Universities globally are adapting policies and training programs to prepare students for an AI-driven world, ensuring they possess the skills required to navigate this transformative era responsibly (Ng et al., 2023).

Advancing AI Literacy Through Interdisciplinary and Group Learning

AI literacy in higher education extends beyond technical skills to include ethical and applied thinking. This necessitates an interdisciplinary approach, integrating insights from various fields to navigate AI's complexities (Ng et al., 2023). Group-based learning, long valued for fostering transferable skills, offers significant potential in this regard. By emphasizing collaboration, communication, and teamwork, group learning enables students to engage with diverse perspectives and solve problems collectively (Volkov & Volkov, 2015). The integration of AI tools like ChatGPT into these group settings can enhance collaborative problem-solving and collective intelligence. ChatGPT's capabilities in real-time information sharing and discussion facilitation provide a unique opportunity to enrich learning experiences in interdisciplinary contexts (Crompton & Burke, 2023).

Research Aim and Objectives

This study aims to explore the potential of group-based learning using ChatGPT to foster AI literacy and enhance collaborative problem-solving skills in an interdisciplinary educational setting.

Main Research Questions:

1. How does group-based learning with ChatGPT influence student engagement, teamwork, and communication within an interdisciplinary context?
2. What are the strengths and limitations of ChatGPT in supporting case study analysis and fostering critical thinking?
3. How does interdisciplinary learning enhance students' understanding of AI ethics during AI-enhanced group activities?
4. How can instructional strategies be designed to integrate AI tools like ChatGPT in higher education to promote sustainability, inclusion, and critical engagement with learning materials?

Literature Review

The Role of Ethical Considerations in AI Education

The integration of AI in education introduces significant ethical considerations. ChatGPT's ability to generate human-like responses raises concerns about academic integrity, particularly in collaborative tasks where individual contributions may be difficult to discern (Ng et al., 2023). To address this, educators must establish guidelines for AI usage, fostering responsible practices and ensuring students understand the implications of AI-generated content. Ethical AI literacy involves educating students on issues such as bias, transparency, and data privacy. Teaching them to critically assess AI outputs and use such tools as supplements rather than substitutes for their efforts is crucial (Crompton & Burke, 2023). This instils a culture of ethical AI practices, equipping learners to navigate an AI-driven world with integrity and responsibility.

Evaluating the Impact of ChatGPT on Cognitive and Affective Learning Outcomes

The integration of ChatGPT into educational settings offers potential cognitive benefits, such as improved comprehension, enhanced problem-solving skills, and better knowledge retention. In group settings, ChatGPT facilitates shared cognition, enabling learners to collectively address complex tasks (Souppez, 2017). However, challenges remain. Over-reliance on ChatGPT can lead to surface-level learning if students fail to engage deeply with the material. The balance between reducing cognitive load and encouraging critical engagement is crucial. Instructional strategies must ensure that AI tools enhance learning without diminishing the depth of student understanding (Klaassen, 2018).

On an affective level, ChatGPT's interactive nature can increase motivation and engagement. The immediacy of AI responses reduces frustration and supports interdisciplinary learning by helping students feel more confident when tackling unfamiliar topics (Little & Hoel, 2011). However, over-reliance on AI can undermine intrinsic motivation and peer collaboration. Students' perceptions of ChatGPT's usefulness also influence their attitudes toward technology in education, highlighting the need for thoughtful integration to maintain engagement and foster positive learning experiences.

Methodology

An action research approach was employed to investigate student engagement with artificial intelligence (AI) in addressing issues related to digital inequalities. The study involved 30 students enrolled in the Consumer Insights module at the Master's level. Initially, students were divided into four groups of five for a group exercise, with 10 students additionally participating in a focus group discussion. The research began with two lectures: one on Ethical Consumption and Consumer Behaviour, delivered by the module lead in Dubai, and another on Socio-Economic Inequalities, led by an Economics lecturer based in the UK. Students were then given a case study on digital inequalities in the context of online purchases and accessibility in Uganda. Divided into four groups, they were allocated 45 min to address the case study questions using ChatGPT. After generating answers, students collaboratively reflected on and refined the AI-generated responses. To gain deeper insights, a focus group was conducted to explore the participants' learning experiences, their familiarity with AI tools, and their perspectives on using ChatGPT to address complex socio-economic issues. This methodology supported both individual and collaborative learning while examining the potential of AI to enhance critical thinking and problem-solving in educational contexts.

Findings and Discussions

RQ1–RQ1: The findings indicate that ChatGPT significantly enhanced student engagement during group activities. Students reported that the immediacy of AI-generated responses helped reduce frustration when encountering unfamiliar topics, particularly in interdisciplinary areas like ethical consumption and socio-economic inequalities. The tool also facilitated teamwork by acting as a neutral mediator, enabling students to focus on synthesizing diverse viewpoints rather than debating definitions or basic concepts (Berasategi et al., 2020). However, some participants noted that over-reliance on ChatGPT occasionally stifled deeper discussions, as groups tended to accept AI-generated responses without sufficient critical reflection.

RQ2-ChatGPT proved effective in supporting the initial stages of case study analysis by quickly generating ideas and organising information. This allowed students to efficiently address the case study on digital inequalities in Uganda. However, its strengths were accompanied by notable limitations. Students observed that while the AI provided well-structured responses, these often lacked depth and context-specific nuance. The collaborative reflection phase, where groups critically evaluated and refined the AI-generated answers, emerged as essential for fostering critical thinking.

RQ3-Interdisciplinary learning, supported by lectures on ethical consumption and socio-economic inequalities, enriched students' understanding of AI ethics. Group discussions and reflective activities encouraged students to critically evaluate AI-generated content through an ethical lens. Participants highlighted biases in ChatGPT's responses, such as oversimplifications of complex socio-economic dynamics, which spurred deeper discussions on the limitations and responsibilities associated with AI tools.

RQ4-To maximise the benefits of AI tools like ChatGPT, educators should adopt strategies that align with principles of sustainability and inclusion. Clear guidelines for AI usage are essential to ensure these tools complement, rather than replace, peer interaction and individual effort. Encouraging critical evaluation of AI responses can help students develop the analytical skills needed to navigate an AI-driven world. Positioning ChatGPT as a scaffolding tool rather than a primary source fosters deeper engagement with course materials while supporting diverse learning needs.

Conclusion

The integration of AI tools like ChatGPT into higher education represents both an opportunity and a challenge. When thoughtfully applied, these technologies can enhance interdisciplinary learning, foster critical and ethical thinking, and promote sustainability in education. However, their effective use requires a commitment to ethical practices, thoughtful instructional design, and continuous evaluation (Parsons & Curry, 2024). By addressing these considerations, higher education can equip students with the skills and knowledge needed to navigate the complexities of an AI-driven world while advancing a sustainable and inclusive educational paradigm.

References

Berasategi, N., Aróstegui, I., Jaureguizar, J., Aizpurua, A., Guerra, N., & Arribillaga-Iriarte, A. (2020). Interdisciplinary learning at university: Assessment of an interdisciplinary experience based on the case study methodology. *Sustainability, 12*(18), 7732. https://doi.org/10.3390/su12187732

Borrego, M., & Newswander, L. K. (2010). Definitions of interdisciplinary research: Toward graduate-level interdisciplinary learning outcomes. *The Review of Higher Education, 34*(1), 61–84. https://doi.org/10.1353/rhe.2010.0006

Cardon, P. W., Getchell, K., Carradini, S., Fleischmann, C., & Stapp, J. (2023). Generative AI in the workplace: Employee perspectives of ChatGPT benefits and organizational policies.

Crompton, H., & Burke, D. (2023). Artificial intelligence in higher education: The state of the field. *International Journal of Educational Technology in Higher Education, 20*, 8. https://doi.org/10.1186/s41239-023-00397-2

Klaassen, R. G. (2018). Interdisciplinary education: A case study. *European Journal of Engineering Education, 43*(6), 842–859. https://doi.org/10.1080/03043797.2018.1442417

Little, A., & Hoel, A. (2011). Interdisciplinary team teaching: An effective method to transform student attitudes. *Journal of Effective Teaching, 11*(1), 36–44.

Ng, D. T. K., Leung, J. K. L., Chu, S. K. W., & Qiao, M. S. (2021). Conceptualizing AI literacy: An exploratory review. *Computers and Education: Artificial Intelligence, 2*, Article 100041. https://doi.org/10.1016/j.caeai.2021.100041

Ng, D. T. K., Su, J., Leung, J. K. L., & Chu, S. K. W. (2023). Artificial intelligence (AI) literacy education in secondary schools: A review. *Interactive Learning Environments, 1–21*,. https://doi.org/10.1080/10494820.2023.2171771

Parsons, B., & Curry, J. H. (2024). Can ChatGPT pass graduate-level instructional design assignments? Potential implications of artificial intelligence in education and a call to action. *TechTrends, 68*(1), 67–78. https://doi.org/10.1007/s11528-023-00901-0

Souppez, J. B. (2017). Interdisciplinary pedagogy: A maritime case study. *Dialogue: Journal of Learning and Teaching, 1*(1), 37–44.

Volkov, A., & Volkov, M. (2015). Teamwork benefits in tertiary education: Student perceptions that lead to best practice assessment design. *Education + Training, 57*(3), 262–278. https://doi.org/10.1108/ET-10-2013-0133

Part IX
Sustainable Food Production, Security, and Systems

Chapter 56
Digital Agriculture for Sustainable Food Systems and Its Contested Futures

Anupam Kumar and Soutrik Basu

Abstract Over the past decade, the proliferation of ICT, IoT, big data analytics, digital platforms, and algorithms has opened new horizons for precision agriculture—commonly referred to as smart farming, digital agriculture, Agriculture 4.0, or the fourth agricultural revolution. Agricultural decision-making is increasingly framed as data-driven and digitally intermediated. Through a systematic literature review, this paper situates digitalization within four key promise areas aligned with sustainable food systems: (i) improving income, productivity, and farm life; (ii) enhancing resilience; (iii) promoting ecological sustainability; and (iv) empowering farmer agency. Drawing from agri-digital and sustainability transition literature, as well as critical social science and STS perspectives, the paper also examines the risks, uncertainties, and contested trajectories of digital agriculture. Finally, it explores how innovation pathways can be negotiated and governed to steer digitalization toward more inclusive, equitable, and sustainable agri-food futures.

Keywords Digital agriculture · Sustainability transition · Responsible innovation · ICT in agriculture

Introduction

In last one decade, proliferation of ICT, IoT, big data analytics, digital platforms, and Artificial Intelligence including Large Language Models have started giving new horizons to precision agriculture in form of digital agriculture or digitalization of agriculture. In a modernist sense, digitalization stands upon two vital traits—communication and comprehension. Technologies such as internet, ICT, IoT, and GenAI has enabled lightspeed communication between people, between things, and between people and things. At the same time, datafication of everything has exponentially enhanced our power of comprehension, often touted as 'disruptive revolution for decision-making processes (Sestino et al., 2020; p. 2).' Extent studies on

A. Kumar (✉) · S. Basu
School of Public Policy, Indian Institute of Technology Delhi, New Delhi, India
e-mail: anupam.kumar@sopp.iitd.ac.in

J. Dsilva (ed.), *Sustainability Education, Youth and Innovation*,
https://doi.org/10.1007/978-3-032-21948-0_56

role of decision support system, ICT, big data and AI/ML in agriculture have very well-articulated the wider implications of communication and comprehension for farming (Jakku & Thorburn, 2010; Mittal 2012; Aker et al., 2016; Wolfert et al., 2017; Zhai et al., 2020; Kumar & Basu 2022). All these have set the context for inventing the transitioning wheels of precision to decision. A summary of different technological bundles transforming agri-digitalization value chain (Porciello et al., 2022; Maffezzoli et al., 2022; Tseng et al., 2019; Usigbe et al., 2023) is provided in Table 56.1.

Decision making in agriculture, as AgTech players and Venture firms articulate, is poised to be data-driven, and digital innovation intermediated. This realm of digital technification is referred as smart farming, digital agriculture, 4th agriculture revolution, and agriculture 4.0.

Table 56.1 Digital intervention and its technology bundles

Sr	Digital intervention	Technology bundles	Application at different farm stage
1	Digital advisory & Extension	Big data analytics, AI & ML, GIS, Image processing, Digital apps, Large Language Model	Personalized packetization of extension—Crop selection, Input selection advisory; On farm agronomy, crop diagnostics and livestock management advisory; Weather alerts; Mandi prices
2	Digital farm-embedded tools	Autonomous drones, sensors and micro-controllers, IoT, Digital apps, AI/ML, Image processing; Cyber-physical system	Embodied technologies that generate data from the field and facilitates farm-centric actions such as selective spraying in diseased area, efficient seed sowing, digital soil testing and soil mapping, smart irrigation management, crop/livestock growth and health monitoring
3	Digitalized market bridging	Big data analytics, Digital apps, digital payment gateways	Digital platforms to facilitate forward and backward market linkages among farmers, intermediaries, market stakeholders etc. -facilitates purchase of inputs and new equipment, circular resale of equipment and livestock, supply chain management, food traceability
4	Digital finance services	Digital apps, Big data analytics, AI/ML	Platforms for crop insurance, farm credits, inclusion of farmers into commodity market

While many studies highlight the promises of agri-digitalization, few examine the emerging conflicts and contestations shaping its future. This paper addresses that gap through a systematic literature review, situating digitalization both as a paradigm aligned with sustainable food systems and as a site of risk and uncertainty. By synthesizing diverse narratives, we critically explore the contested nature of digital agri-food futures and ask: how can innovation be steered responsibly? We conclude by urging policymakers and innovators to shape agri-digital transitions with foresight, inclusivity, and reflexivity.

Methodology

Findings of this paper are based on the results of the systematic review of the scholarly literature on agriculture digitalization. A bibliography of Scopus indexed scientific paper was created based on a set of keywords: digital agriculture, smart farming, AI and agriculture, big data and agriculture, agriculture 4.0 and others, within the timeline of 2018–2024. As a first step, a preliminary review of abstracts and title of identified paper was taken to identify relevant literature and classify them under three cohorts: benefits and use cases of digitalization, risks and conflicts emerging from digitalization, and strategies to manage digitalization in agriculture. Selected papers were analysed to develop key themes emerging within these cohorts. Extent qualitative coding of articles led us to develop four themes on agri-digitalization's synergy with sustainable food futures and three themes around the conflict (see Fig. 56.1).

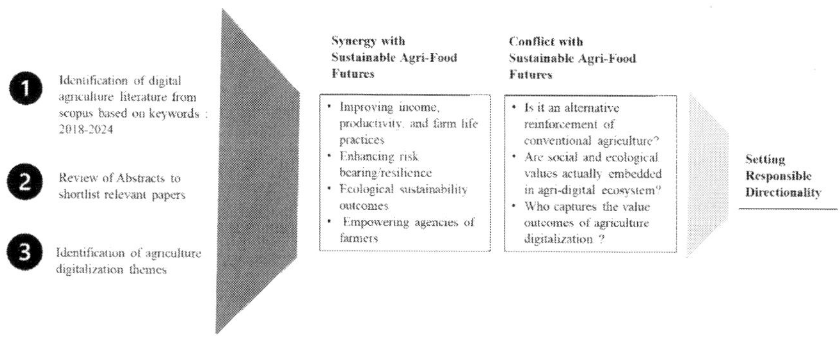

Fig. 56.1 Literature review framework and emergent themes

Findings

Based on the thematic analysis, as emergent from systematic literature review, findings of this paper reflect upon three key aspects of digital agriculture technologies and their role in sustainable food system. First, this article showcases role of digital agriculture in improving social, economic and ecological outcomes of agriculture ecosystem. Second, it further investigates different types of conflicts and uncertainties that may emanate as digital AgTech gets scaled and adopted at large. And thirdly, it illustrates the need for designing better policy instruments and governance frameworks for agriculture digitalization.

Synergy with Sustainable Agri-Food Futures

A series of studies have highlighted that digital technologies are increasingly seen as catalysts for transforming agriculture, offering multifaceted benefits that touch upon productivity, resilience, sustainability, and farmer empowerment.

Firstly, digitalization enhances productivity, income, and farm life by facilitating precision and timely decision-making; studies show a 4% yield increase through digital advisories (Fabregas et al., 2019), improved market access and transparency (Zhang & Fan, 2023), and significant yield gains with integrated digital-precision farming (Kayad et al., 2021). These tools also improve farm work-life conditions (through adoption of Autonomous Milking System—Vik et al., 2019), reduce everyday hassle (Abdulai et al., 2023), and help young farmers overcome rural isolation (Holton et al., 2023). For smallholders, smart farming apps have helped in streamlining advisory access, input procurement, and market linkages (Kumar & Basu, 2022).

Secondly, digital tools build rural resilience by enabling timely adaptation and knowledge access during crises (Ferdinand et al., 2021; Kumar et al., 2020), strengthening institutional capacity and social cohesion (Singh et al., 2023), and fostering social capital (Chaudhuri & Kendall, 2020; Baumüller et al., 2023). During the COVID-19 pandemic, digital platforms played an instrumental role in sustaining India's agri-value chains by enabling extension services, bridging producers and markets, and managing labour shortfalls (Kumar et al., 2020; Sridhar et al., 2022). Even different government's effort towards building Agri-stack and Digital Public Infrastructure for agriculture can empower farmers by providing integrated access to timely advisories, credit, markets, and entitlements (Kumar, 2025; Beriya, 2022).

Thirdly, digitalization contributes to ecological sustainability through resource-efficient precision tools, AI platforms for circular economy in agriculture (Ali et al., 2023), AI/IoT-based waste minimization (Myshko et al., 2024; Usigbe et al., 2023), and digital gamification for agroecological learning (Maurel & Huyghe, 2017; Jouan et al., 2021). At the governance level, digital public infrastructure fosters a new and emerging 'data-driven sustainability' frameworks.

And finally, underlying all these transformations is the empowerment of farmers as agents of change. Digital networks foster solidarity, peer learning, and the formation of virtual communities of practice (Slimi et al., 2023; Prost et al., 2022). Gender dynamics also shift, as technologies like precision livestock tools offer women greater visibility and decision-making roles (Hay & Pearce, 2014), reversing historically masculinized farm labour divisions (Saugeres, 2002). For youth, digitalization helps destigmatize farming and creates entrepreneurial pathways (Toumbourou et al., 2023).

Conflict with Sustainable Agri-Food Futures

As we articulate the techno-optimistic narratives in previous section, a set of studies have also highlighted that a socially-desired agri-digitalization is not a linear or uncontested paradigm—it is shaped by governance choices, ownership structures, and the socio-technical fabric of rural life. Scholars highlight three interlinked concerns that challenge the techno-optimistic narrative: digital solutionism, imbalanced value priorities, and unequal value capture.

Firstly, digital agriculture is increasingly critiqued as a form of digital solutionism—a tech-fix approach that oversimplifies complex, systemic agri-food challenges. Guthman and Butler (2023) argue that most AgTech solutions merely offer data-driven efficiencies—information provisioning, streamlined value chains, or repackaged tools—rather than addressing the deeply biophysical and ecological nature of agricultural crises. These solutions often fail to disrupt the existing industrial agriculture model, instead reinforcing it through digital means (Morozov, 2013). As Fraser (2022) and Carolan (2019) suggest, these technologies, often built by oligopolistic actors, are not emancipatory but paternalistic, crowdsourcing knowledge from farmers only to sell it back to them. This trend locks farmers into dependencies through proprietary platforms, raising serious concerns about data sovereignty and control.

Secondly, digital agriculture innovations often fail to embed or sustain the triple-bottom-line values of people, profit, and planet. While agri-digitalization is framed as a path to social and ecological sustainability, in practice, social value is frequently marginalized in policy, media, and design (Rose et al., 2021; Barrett & Rose, 2020). Bustamante's (2023) case of a Swedish food sector platform reveals how early commitments to transparency and farmer profitability were sidelined in favor of serving high-paying corporate clients. Similarly, Heimstädt (2023) shows how an AI-based ecological advisory app transformed into a data broker for agrochemical firms, betraying its original value proposition. Although ecological rationales are often used to legitimize digital farming (Clapp & Ruder, 2020), these are not always prioritized in development. Often digital tech-entrepreneurship turn anti-thetical (or to say violent) to lasting goals of sustainability transition (Agarwal, 2024).

Thirdly, the question of "digitalization for whom?" becomes central to understanding who benefits from these technologies. Evidence suggests that value capture is skewed toward tech firms, corporate actors, and higher nodes in the agri-food value chain. Duncan et al. (2021) note that AgTech often serves as an experimental arena for global tech giants to build future AI systems, using farmers' data as raw input. The real-time advisory primarily benefits service providers rather than farmers (Hackfort, 2023). Hidalgo et al. (2023), in their analysis of coffee value chains, show that while digital tools improve coordination, value is disproportionately captured by non-producer actors. Moreover, not all farmers are equally positioned to benefit. McCampbell et al. (2021) identify layers of exclusion, especially in the Global South, where high capital requirements, generic design choices, and dominant innovation trajectories lock out smallholders and digitally novice or vernacular farmers (as further highlighted by Kumar & Basu, (2022)). Even where adoption occurs, data inaccuracies and algorithmic limitations create risks. Visser et al. (2021) demonstrate how deviations in GPS and sensor data lead to a 10% error margin in yield maps, disproportionately affecting complex, diverse small farms. These precision traps exacerbate the "precision divide" between large monocultures and smallholders, shifting the burden of digital inaccuracies onto the most vulnerable.

All in all, digital agriculture risks reinforcing existing power asymmetries, marginalizing social and ecological goals, and concentrating value in the hands of a few. Recognizing and addressing these contested dimensions is essential if digital agriculture is to serve as a genuine lever for sustainable and equitable agri-food futures.

Setting Responsible Directionality

Given the contested nature of digital agriculture and its uncertain agri-food futures, a key research and governance imperative is setting responsible directionality for innovation. The framework of Responsible Research and Innovation (RRI) provides a valuable lens, emphasizing anticipation, inclusion, reflexivity, and responsiveness in innovation processes (Owen et al., 2013; Stilgoe et al., 2013). RRI promotes inclusive deliberation and mediating spaces to reconcile diverse stakeholder values and imaginaries (Regan, 2019), offering a path to navigate risks before socio-technical lock-ins emerge (Ingram & Maye, 2023). This is particularly urgent in the Global South, where digital agriculture is nascent and trajectories are still malleable (Rose & Chilvers, 2018).

Operationalizing RRI can be achieved through participatory co-design, farmer capacity-building, and investment in digital commons. Parallel perspectives call for attention to the socio-cyber-physical dynamics of digitalization (Rijswijk et al., 2021; Metta et al., 2022) and a shift from techno-optimistic, exclusionary "strong" data regimes to grounded, participatory "weak" data assemblages (Carolan, 2022). Similarly, the 4D framework—directionality, diversity, distribution, and democracy—articulates a pluralistic and inclusive innovation mission (Kok & Klerkx, 2023). Together, these approaches urge a deliberate, reflexive steering of agri-digital transitions toward socially desirable outcomes.

Conclusion

While much scholarly attention has focused on the scaler trajectory of agricultural innovation—how technologies scale and are adopted—this article argues for equal focus on the vector trajectory: the directions, lock-ins, and broader socio-technical consequences of agri-digital transitions. Given past unintended outcomes of agri-innovation, anticipating digital lock-ins early is vital. A nuanced understanding of both benefits and contestations can help shape inclusive, sustainable futures by identifying reflexivity (or its absence) among key actors and assessing the responsiveness of policies and innovations. Ultimately, steering digital agriculture requires deliberate governance that aligns innovation with equity, sustainability, and long-term agri-food system resilience.

References

Abdulai, A., Quarshie, P. T., Duncan, E., & Fraser, E. (2023). Is agricultural digitization a reality among smallholder farmers in Africa? Unpacking farmers' lived realities of engagement with digital tools and services in rural Northern Ghana. *Agriculture & Food Security, 12*(1). https://doi.org/10.1186/s40066-023-00416-6

Agrawal, N. (2024). Violent sustainability: Blitzscale and counteraccounting in an Indian agtech start-up. *Economic Anthropology*. https://doi.org/10.1002/sea2.12333

Aker, J. C., Ghosh, I., & Burrell, J. (2016). The promise (and pitfalls) of ICT for agriculture initiatives. *Agricultural Economics, 47*(S1), 35–48. https://doi.org/10.1111/agec.12301

Ali, Z. A., Zain, M., Pathan, M. S., & Mooney, P. (2023). Contributions of artificial intelligence for circular economy transition leading toward sustainability: An explorative study in agriculture and food industries of Pakistan. *Environment, Development and Sustainability*. https://doi.org/10.1007/s10668-023-03458-9

Barrett, H., & Rose, D. C. (2020). Perceptions of the fourth agricultural revolution: What's in, what's out, and what consequences are anticipated? *Sociologia Ruralis, 62*(2), 162–189. https://doi.org/10.1111/soru.12324

Baumüller, H., Ikpi, U., Jumpah, E. T., Kamau, G., Kergna, A. O., Mose, L., Nientao, A., Omari, R., Phillip, D., & Salasya, B. (2023). Building digital bridges in African value chains: Exploring linkages between ICT use and social capital in agricultural marketing. *Journal of Rural Studies, 100*, Article 103002. https://doi.org/10.1016/j.jrurstud.2023.03.010

Beriya, A. (2022). India Digital Ecosystem of Agriculture and Agristack: An Initial assessment. https://www.econstor.eu/handle/10419/250913

Bustamante, M. J. (2023). Digital platforms as common goods or economic goods? Constructing the worth of a nascent agricultural data platform. *Technological Forecasting & Social Change/ Technological Forecasting and Social Change, 192*, Article 122549. https://doi.org/10.1016/j.techfore.2023.122549

Carolan, M. (2019). Automated agrifood futures: Robotics, labor and the distributive politics of digital agriculture. *The Journal of Peasant Studies, 47*(1), 184–207. https://doi.org/10.1080/03066150.2019.1584189

Carolan, M. (2022). Digitization as politics: Smart farming through the lens of weak and strong data. *Journal of Rural Studies, 91*, 208–216. https://doi.org/10.1016/j.jrurstud.2020.10.040

Chaudhuri, B., & Kendall, L. (2020). Collaboration without consensus: Building resilience in sustainable agriculture through ICTs. *The Information Society, 37*(1), 1–19. https://doi.org/10.1080/01972243.2020.1844828

Clapp, J., & Ruder, S. (2020). Precision technologies for agriculture: Digital farming, gene-edited crops, and the politics of sustainability. *Global Environmental Politics, 20*(3), 49–69. https://doi.org/10.1162/glep_a_00566

Duncan, E., Glaros, A., Ross, D. Z., & Nost, E. (2021). New but for whom? Discourses of innovation in precision agriculture. *Agriculture and Human Values, 38*(4), 1181–1199. https://doi.org/10.1007/s10460-021-10244-8

Fabregas, R., Kremer, M., & Schilbach, F. (2019). Realizing the potential of digital development: The case of agricultural advice. *Science, 366*(6471). https://doi.org/10.1126/science.aay3038

Ferdinand, T., Illick-Frank, E., Postema, L., Stephenson, J., Rose, A., Petrovic, D., Migisha, C., Fara, K., Zebiak, S., Siantonas, T., Pavese, N., Chellew, T., Campbell, B., & Del Rio, C. R. (2021). A blueprint for digital climate-informed advisory services: Building the resilience of 300 million small-scale producers by 2030. *World Resources Institute.* https://doi.org/10.46830/wriwp.20.00103

Fraser, A. (2022). 'You can't eat data'?: Moving beyond the misconfigured innovations of smart farming. *Journal of Rural Studies, 91*, 200–207. https://doi.org/10.1016/j.jrurstud.2021.06.010

Guthman, J., & Butler, M. (2023). Fixing food with a limited menu: On (digital) solutionism in the agri-food tech sector. *Agriculture and Human Values, 40*(3), 835–848. https://doi.org/10.1007/s10460-023-10416-8

Hackfort, S. (2023). Unlocking sustainability? The power of corporate lock-ins and how they shape digital agriculture in Germany. *Journal of Rural Studies, 101*, Article 103065. https://doi.org/10.1016/j.jrurstud.2023.103065

Hay, R., & Pearce, P. (2014). Technology adoption by rural women in Queensland, Australia: Women driving technology from the homestead for the paddock. *Journal of Rural Studies, 36*, 318–327. https://doi.org/10.1016/j.jrurstud.2014.10.002

Heimstädt, C. (2023). The exploratory assetization of a crop protection app. *Environmental Science & Policy, 140*, 242–249. https://doi.org/10.1016/j.envsci.2022.12.014

Hidalgo, F., Quiñones-Ruiz, X. F., Birkenberg, A., Daum, T., Bosch, C., Hirsch, P., & Birner, R. (2023). Digitalization, sustainability, and coffee. Opportunities and challenges for agricultural development. *Agricultural Systems, 208*, 103660. https://doi.org/10.1016/j.agsy.2023.103660

Holton, M., Riley, M., & Kallis, G. (2023). Keeping on[line] farming: Examining young farmers' digital curation of identities, (dis)connection and strategies for self-care through social media. *Geoforum, 142*, Article 103749. https://doi.org/10.1016/j.geoforum.2023.103749

Ingram, J., & Maye, D. (2023). "How can we?" the need to direct research in digital agriculture towards capacities. *Journal of Rural Studies, 100*, Article 103003. https://doi.org/10.1016/j.jrurstud.2023.03.011

Jakku, E., & Thorburn, P. (2010). A conceptual framework for guiding the participatory development of agricultural decision support systems. *Agricultural Systems, 103*(9), 675–682. https://doi.org/10.1016/j.agsy.2010.08.007

Jouan, J., Carof, M., Baccar, R., Bareille, N., Bastian, S., Brogna, D., Burgio, G., Couvreur, S., Cupiał, M., Dufrêne, M., Dumont, B., Gontier, P., Jacquot, A. L., Kański, J., Magagnoli, S., Makulska, J., Pérès, G., Ridier, A., Salou, T., … Godinot, O. (2021). SEGAE: An online serious game to learn agroecology. *Agricultural Systems, 191*, 103145. https://doi.org/10.1016/j.agsy.2021.103145

Kayad, A., Sozzi, M., Gatto, S., Whelan, B., Sartori, L., & Marinello, F. (2021). Ten years of corn yield dynamics at field scale under digital agriculture solutions: A case study from North Italy. *Computers and Electronics in Agriculture, 185*, Article 106126. https://doi.org/10.1016/j.compag.2021.106126

Kok, K. P., & Klerkx, L. (2023). Addressing the politics of mission-oriented agricultural innovation systems. *Agricultural Systems, 211*, Article 103747. https://doi.org/10.1016/j.agsy.2023.103747

Kumar, A. (2025, May 15). Digital public Infrastructure—a catalyst for People-Centric development | RESO. Reso. https://resoinsights.com/insight/digital-public-infrastructure/

Kumar, A., & Basu, S. (2022). Can end-user feedback inform 'Responsibilisation' of India's policy landscape for agri-digital transition?*. *Sociologia Ruralis, 62*(2), 305–334. https://doi.org/10.1111/soru.12374

Kumar, A., Padhee, A. K., & Kumar, S. (2020). How Indian agriculture should change after COVID-19. *Food Security, 12*(4), 837–840. https://doi.org/10.1007/s12571-020-01063-6

Maffezzoli, F., Ardolino, M., Bacchetti, A., Perona, M., & Renga, F. (2022). Agriculture 4.0: A systematic literature review on the paradigm, technologies and benefits. *Futures, 142*, 102998. https://doi.org/10.1016/j.futures.2022.102998

Maurel, V. B., & Huyghe, C. (2017). Putting agricultural equipment and digital technologies at the cutting edge of agroecology. *Oilseeds and Fats, Crops and Lipids/OCL. Oilseeds & Fats Crops and Lipids, 24*(3), D307. https://doi.org/10.1051/ocl/2017028

McCampbell, M., Rijswijk, K., Wilson, H., & Klerkx, L. (2021). A Problematisation of inclusion and exclusion: Trade-offs and nuances in the digitalisation of African agriculture. In *The politics of knowledge in inclusive development and innovation* (pp. 199–213). Routledge. https://doi.org/10.4324/9781003112525

Metta, M., Ciliberti, S., Obi, C., Bartolini, F., Klerkx, L., & Brunori, G. (2022). An integrated socio-cyber-physical system framework to assess responsible digitalisation in agriculture: A first application with Living Labs in Europe. *Agricultural Systems, 203*, Article 103533. https://doi.org/10.1016/j.agsy.2022.103533

Mittal, S. (2012). *Modern ICT for agricultural development and risk management in smallholder agriculture in India*. CIMMYT.

Morozov, E. (2013). To save everything, click here: the folly of technological solutionism. In *Choice/Choice reviews* (Issue 01, pp. 51–0324). PublicAffairs.

Myshko, A., Checchinato, F., Colapinto, C., Finotto, V., & Mauracher, C. (2024). Towards twin transition in the agri-food sector? Framing the current debate on sustainability and digitalisation. *Journal of Cleaner Production, 452*, Article 142063. https://doi.org/10.1016/j.jclepro.2024.142063

Owen, R., Bessant, J., & Heintz, M. (2013). *Responsible Innovation: Managing the Responsible Emergence of Science and Innovation in Society* (Vol. 44, Issue 2, p. 127). https://doi.org/10.1002/9781118551424

Porciello, J., Coggins, S., Mabaya, E., & Otunba-Payne, G. (2022). Digital agriculture services in low- and middle-income countries: A systematic scoping review. *Global Food Security, 34*, Article 100640. https://doi.org/10.1016/j.gfs.2022.100640

Prost, M., Gross, H., & Prost, L. (2022). How could social media support farmers concerned with sustainability issues? *Journal of Agricultural Education and Extension/Journal of Agricultural Education and Extension, 1–23*,. https://doi.org/10.1080/1389224x.2022.2153888

Regan, Á. (2019). 'Smart farming' in Ireland: A risk perception study with key governance actors. *NJAS Wageningen Journal of Life Sciences, 90–91*(1), 1–10. https://doi.org/10.1016/j.njas.2019.02.003

Rijswijk, K., Klerkx, L., Bacco, M., Bartolini, F., Bulten, E., Debruyne, L., Dessein, J., Scotti, I., & Brunori, G. (2021). Digital transformation of agriculture and rural areas: A socio-cyber-physical system framework to support responsibilisation. *Journal of Rural Studies, 85*, 79–90. https://doi.org/10.1016/j.jrurstud.2021.05.003

Rose, D. C., & Chilvers, J. (2018). Agriculture 4.0: Broadening Responsible innovation in an era of smart Farming. *Frontiers in Sustainable Food Systems, 2*. https://doi.org/10.3389/fsufs.2018.00087

Rose, D. C., Wheeler, R., Winter, M., Lobley, M., & Chivers, C. (2021). Agriculture 4.0: Making it work for people, production, and the planet. *Land Use Policy, 100*, 104933. https://doi.org/10.1016/j.landusepol.2020.104933

Saugeres, L. (2002). Of tractors and men: Masculinity, technology and power in a French farming community. *Sociologia Ruralis, 42*(2), 143–159. https://doi.org/10.1111/1467-9523.00207

Sestino, A., Prete, M. I., Piper, L., & Guido, G. (2020). Internet of Things and Big Data as enablers for business digitalization strategies. *Technovation, 98*, Article 102173. https://doi.org/10.1016/j.technovation.2020.102173

Singh, N., Kumar, A., & Dey, K. (2023). Unlocking the potential of knowledge economy for rural resilience: The role of digital platforms. *Journal of Rural Studies, 104*, Article 103164. https://doi.org/10.1016/j.jrurstud.2023.103164

Slimi, C., Prost, L., Cerf, M., & Prost, M. (2023). The potential of community interactions as inducers of agroecological transition: The case of a digital agricultural community. *Journal of Agricultural Education and Extension, 1–17*,. https://doi.org/10.1080/1389224x.2023.2223576

Sridhar, A., Balakrishnan, A., Jacob, M. M., Sillanpää, M., & Dayanandan, N. (2022). Global impact of COVID-19 on agriculture: Role of sustainable agriculture and digital farming. *Environmental Science and Pollution Research International, 30*(15), 42509–42525. https://doi.org/10.1007/s11356-022-19358-w

Stilgoe, J., Owen, R., & Macnaghten, P. (2013). Developing a framework for responsible innovation. *Research Policy, 42*(9), 1568–1580. https://doi.org/10.1016/j.respol.2013.05.008

Toumbourou, T. D., Dressler, W. H., Sanders, A., Liu, E., Brown, T., & Utomo, A. (2023). Who are the future farmers? Media representations of youth in agriculture, food security and 'modern' farming in Indonesia. *Asia Pacific Viewpoint, 64*(2), 188–208. https://doi.org/10.1111/apv.12374

Tseng, F., Cho, H., & Wu, H. (2019). Applying big data for intelligent Agriculture-Based crop Selection analysis. *IEEE Access, 7*, 116965–116974. https://doi.org/10.1109/access.2019.2935564

Usigbe, M. J., Asem-Hiablie, S., Uyeh, D. D., Iyiola, O., Park, T., & Mallipeddi, R. (2023). Enhancing resilience in agricultural production systems with AI-based technologies. *Environment, Development and Sustainability.* https://doi.org/10.1007/s10668-023-03588-0

Vik, J., Stræte, E. P., Hansen, B. G., & Nærland, T. (2019). The political robot—The structural consequences of automated milking systems (AMS) in Norway. *NJAS Wageningen Journal of Life Sciences, 90–91*(1), 1–9. https://doi.org/10.1016/j.njas.2019.100305

Visser, O., Sippel, S. R., & Thiemann, L. (2021). Imprecision farming? Examining the (in)accuracy and risks of digital agriculture. *Journal of Rural Studies, 86*, 623–632. https://doi.org/10.1016/j.jrurstud.2021.07.024

Wolfert, S., Ge, L., Verdouw, C., & Bogaardt, M. (2017). Big data in smart farming—a review. *Agricultural Systems, 153*, 69–80. https://doi.org/10.1016/j.agsy.2017.01.023

Zhai, Z., Martínez, J. F., Beltran, V., & Martínez, N. L. (2020). Decision support systems for agriculture 4.0: Survey and challenges. *Computers and Electronics in Agriculture, 170*, 105256. https://doi.org/10.1016/j.compag.2020.105256

Zhang, X., & Fan, D. (2023). Can agricultural digital transformation help farmers increase income? An empirical study based on thousands of farmers in Hubei Province. *Environment, Development and Sustainability.*

Chapter 57
Understanding Sustainable Planet's Innovation: Cleantech Solutions for Better Governance and Communities Around the World

Hrushitaa Murali and Pradeep George

Abstract This paper delves into the current scenario of food security worldwide, while trying to understand the implications and applications of a revolutionary Water Lentil solution by Sustainable Planet, a global organization based in Europe. The research explores the varied themes interlinking food security (or food insecurity) to the various Sustainable Development Goals (SDGs) as dictated by the United Nations. These themes include, but are not limited to Gender, Land Usage and Food Security, Drought-induced and Flood-induced areas battling Food Insecurity and Economic Challenges amplified by impending Food Shortages and Changing Commodity Trade Policies Whilst exploring these multi-faceted challenges, this paper highlights the key areas that need immediate remediation and action. The aim of this paper is to highlight the interconnectedness of three Sustainable Development Goals, and to investigate the pipeline from cleantech innovations to more efficient modifications to policies enforced on a global basis and reflected as tangible impact in the form of better infrastructural and social development in urban and rural landscapes worldwide. Water lentils are a sustainable protein source, rich in complete amino acids and adaptable to all climates, making them a potential solution for global food security and environmental challenges. The UAE market presents significant opportunities for water lentil cultivation, driven by food security, sustainability, and economic diversification. However, challenges such as regulatory approval, market acceptance, and funding may hinder market penetration. To scale production, strong government policies, investment, and infrastructure support are essential. Additionally, geopolitical and economic factors, including trade, climate change, and conflicts, continue to impact food security, particularly in regions like MENA.

H. Murali
ESG Management Department, Step Change India, Bengaluru, India

P. George (✉)
Engineering, Infrastructure and Sustainability, De Montfort University Dubai, Dubai, United Arab Emirates
e-mail: Pradeep.george2@dmu.ac.uk

J. Dsilva (ed.), *Sustainability Education, Youth and Innovation*,
https://doi.org/10.1007/978-3-032-21948-0_57

Keywords Cleantech innovations · Sustainable agriculture · Water lentils · Food Security · Climate change

Introduction

Water lentils are a culturally significant, GMO plant with the potential to address global protein and environmental challenges [1]. They offer a complete amino acid profile and high nutritional value. Water lentils are highly efficient, producing more protein with less water than traditional crops, and contribute to carbon sequestration [1].

Nutritional and Commercial Potential

Water lentils are a rich protein source but require societal acceptance and policy reform to gain wider recognition. While progress has been made in the U.S. and Europe, cultural acceptance is key. Jeffrey Beyer, Managing Director of Zest Associates, suggests a carbon tax could enhance their market potential, aligning with sustainable agriculture goals [2].

Addressing Global Challenges

Water lentils offer a promising solution to world hunger and global food security [1]. Their versatile cultivation across various climates allows for large-scale production, aligning with environmental stewardship and social fairness goals within the SDGs [3].

Alignment with UAE Market Needs

The UAE, focused on sustainability and economic diversification, presents a favourable environment for water lentils in agriculture. Their low water footprint aligns with the UAE's arid climate. Water lentils can support sustainability, reduce the ecological footprint, promote circular economy principles, and generate jobs [4].

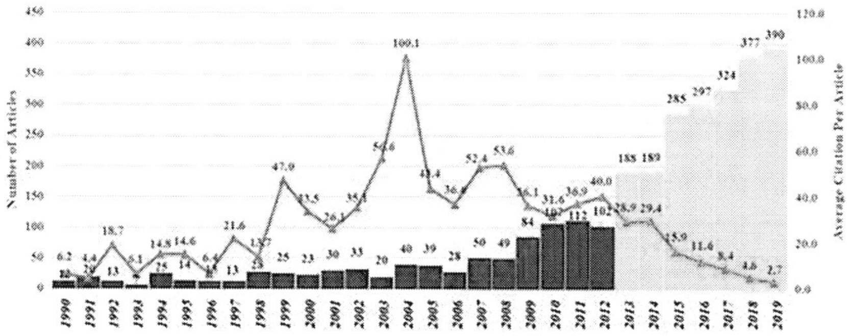

Fig. 57.1 Food security research articles published from 1990 to 2019, Xie et al. [3]

Methodology

Methodology used for the study is Comparative Literary Reviews, Thematic Studies and Market Mapping.

Comparative Literature Reviews

Analysis of literature on food security, sustainable agriculture, and related topics, examining academic papers, shown in Fig. 57.1, reports from international organizations, and publications from governmental/non-governmental bodies to identify key themes, research gaps, and best practices.

Thematic Studies

In-depth exploration of themes critical to food security and the potential role of water lentils shown in Fig. 57.2, including.

- Climate change impact on food security [2]
- Role of gender in food security
- Economic and geopolitical influences on food trade [5]
- Challenges of land use and desertification [6, 7]
- Potential of technological innovations [1]

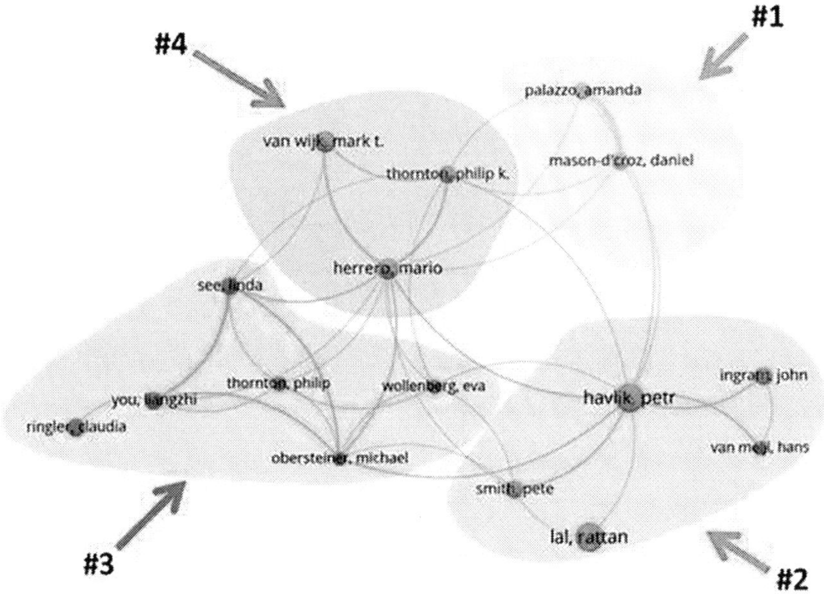

Fig. 57.2 Author cooperation networks, Xie et al. [3]

Market Mapping

Analysis of the market potential for water lentils, focusing on the UAE, including:

Identifying key stakeholders (e.g., government agencies, agricultural businesses, consumers).

Assessing the demand for sustainable protein sources [4].

Evaluating the regulatory and policy landscape.

Analyzing the economic feasibility of water lentil cultivation and commercialization.

Data Collection

Gathering data from academic databases, international organizations, government documents, and industry reports.

Data Analysis

Using qualitative (thematic analysis) and quantitative (descriptive statistics, regression analysis) methods to analyze data and support conclusions.

Results

Thematic Studies.

- Post-COVID Paradigm Shifts: The COVID-19 pandemic significantly impacted global food security. Doustmohammadian et al. [8] Research highlights challenges in land access for farming and the importance of food policies and international trade [6].
- Geopolitical Influences: Reports discuss Japan's food security challenges and the complexities of food distribution in water-scarce, oil-exporting countries, particularly in the Middle East.
- Economic Influences: Research indicates challenges in land access for farming due to competition with infrastructure development [6].
- Trends in Food Security Research: The number of research papers on food security increased from 1990 to 2019. Research has expanded in scope and depth [3]. Increased demand and climate change adaptation are key challenges [2].
- Impact of Sustainable Agriculture Practices [9]: As per Table 57.1, education level significantly influences the adoption of sustainable agriculture practices.
- Climate Change and Food Systems: The global food system is crucial for reducing greenhouse gas emissions [2]. Progress in food system strategies is essential to meet climate goals [10].
- Environmental Sustainability: Studies assess agricultural sustainability and the impact of habitat destruction on biodiversity. Streimikiene & Mikalauskienė, [10] Forest management and land use changes influence ecosystem services [11].
- Food Source Analysis: Research analyzes the safe intake of elements from food sources like Seaweed [1].

Table 57.1 Regression table for adoption of sustainable agriculture practices and education level practices in Southern India, *Prasanth V* (Prasanth 2023)

Variable	Coefficient	Standard error	t-Statistic	P-value
Education Level	0.75	0.11	6.82	<0.001
Landownership	0.12	0.08	1.48	0.14
Age	−0.05	0.03	-1.76	0.08
Gender (Male)	−0.02	0.06	0.36	0.72
Intercept	−0.72	0.62	-1.16	0.25

Market Mapping

- Stakeholder Analysis: Identifying government agencies, agricultural businesses, food companies, and consumers.
- Demand Assessment: Demand for sustainable protein sources is driven by population growth, health/environmental awareness, and government initiatives.
- Regulatory and Policy Landscape: The UAE's regulatory environment supports agricultural innovation but faces challenges in developing specific regulations and ensuring food safety.
- Economic Feasibility: Feasibility depends on production costs, market prices, value-added potential, and access to financing.

Discussion

Policy Recommendations and Practical Implications.

The research highlights the potential of water lentils to contribute to global food security and sustainability, particularly in the UAE context [1]. To facilitate the successful implementation of water lentils in the region, a multi-faceted approach is required, combining strategic policy development with practical, on-the-ground actions.

Policy Recommendations:

Develop supportive policies and regulations: The UAE government should take a proactive role in creating a policy environment that fosters the growth of the water lentil industry [4]. This involves:

- Formulating clear regulations and standards for water lentil production, processing, and distribution. These regulations should address food safety, environmental sustainability, and product labeling.
- Providing incentives to farmers, entrepreneurs, and investors to encourage them to enter the water lentil sector. These incentives could include subsidies, tax breaks, and access to low-interest loans.
- Streamlining the approval process for new water lentil products to reduce bureaucratic hurdles and accelerate market entry.

Promote research and development: Investing in research and development (R&D) is crucial for optimizing water lentil production and unlocking its full potential. This includes:

- Establishing research programs focused on improving water lentil varieties, enhancing cultivation techniques, and developing innovative processing technologies.
- Creating partnerships between government agencies, research institutions, universities, and private companies to facilitate knowledge sharing and collaboration.

- Supporting research into new applications for water lentils, such as animal feed, biofuel production, and pharmaceutical applications.

Enhance public awareness: Raising awareness about the benefits of water lentils is essential for driving consumer demand and market acceptance. This can be achieved through:

- Launching public education campaigns to inform consumers about the nutritional value, sustainability, and versatility of water lentils [1].
- Collaborating with food companies and retailers to promote water lentil products and make them more accessible to consumers.
- Organizing events, workshops, and demonstrations to showcase water lentil products and educate the public on their uses.

Foster regional collaboration: Given the regional food security challenges and the potential for water lentils to contribute to their solution, fostering collaboration with other countries in the MENA region is highly beneficial [12]. This can involve:

- Sharing knowledge, best practices, and technologies related to water lentil production and processing.
- Developing regional markets for water lentil products to increase market size and create economies of scale.
- Coordinating policies and regulations to facilitate cross-border trade and investment in the water lentil sector.

Practical Implications:

Establish pilot projects: To demonstrate the feasibility and benefits of water lentil cultivation in the UAE, it is crucial to establish pilot projects [4]. These projects will:

- Serve as models for larger-scale production, showcasing best practices and demonstrating the economic and environmental viability of water lentil farming.
- Provide valuable data on production costs, yields, water usage, and environmental impacts under local conditions.
- Offer opportunities for training and capacity building for farmers and agricultural workers.

Invest in infrastructure: The growth of the water lentil industry will require investment in appropriate infrastructure, Behnassi et al. [4] including:

- Water treatment facilities to ensure the availability of clean water for water lentil cultivation, especially in a water-scarce region like the UAE.
- Processing plants to handle the post-harvest processing of water lentils, including cleaning, drying, and packaging.
- Transportation and logistics networks to efficiently move water lentils and water lentil products from production sites to markets.

Build capacity: Developing the human capital necessary to support the water lentil industry is essential [9]. This can be achieved through:

- Implementing training programs for farmers and agricultural workers on water lentil cultivation and harvesting techniques.
- Providing training for food processors and manufacturers on how to incorporate water lentils into various food products.
- Developing educational programs in agricultural colleges and universities to train future generations of water lentil experts.

Promote public–private partnerships: Public–private partnerships (PPPs) can play a critical role in driving innovation, investment, and growth in the water lentil sector. PPPs can:

- Facilitate collaboration between government agencies, private companies, research institutions, and community organizations.
- Leverage the strengths and resources of each partner to achieve common goals.
- Accelerate the development and commercialization of water lentil technologies and products.

Addressing Challenges and Ensuring Successful Implementation.

The successful implementation of water lentils in the UAE, while promising, is not without its challenges. Addressing these proactively is crucial for ensuring the industry's sustainable growth and contribution to food security and economic diversification [1, 4].

Regulatory hurdles: Navigating the regulatory landscape and securing necessary approvals can be a time-consuming and complex process for new agricultural products like water lentils. To mitigate this, the UAE government can:

- Establish a clear and streamlined regulatory framework specifically for water lentils, providing clear guidelines for producers and processors.
- Offer support and guidance to companies seeking regulatory approvals, helping them navigate the process efficiently.
- Ensure that regulations are science-based, risk-based, and aligned with international standards to facilitate trade and consumer confidence [4].

Market acceptance: Introducing a new food source like water lentils to the market requires overcoming potential consumer skepticism and building trust. To achieve this, it is essential to:

- Conduct targeted marketing and awareness campaigns to educate consumers about the nutritional benefits, safety, and versatility of water lentils.
- Develop a diverse range of appealing and convenient water lentil products to cater to different consumer preferences and dietary needs.
- Collaborate with chefs, food influencers, and health professionals to promote water lentils and showcase their culinary applications.

Funding and investment: Securing sufficient funding and investment is crucial for supporting research, production, infrastructure development, and market development initiatives in the water lentil sector. To attract investment, it is important to:

- Demonstrate the economic viability and potential return on investment for water lentil production and processing.
- Offer financial incentives and support to investors, such as grants, tax breaks, and access to land.
- Create a supportive investment climate that reduces risk and encourages innovation in the water lentil sector.

Technological challenges: While water lentils offer many advantages, there are ongoing technological challenges that need to be addressed to optimize production, processing, and utilization. These challenges include:

- Improving water lentil varieties to enhance yields, nutritional content, and resistance to pests and diseases.
- Developing more efficient and cost-effective harvesting and processing technologies.
- Finding innovative ways to utilize water lentils in a wider range of food products and other applications.

By proactively addressing these challenges and implementing the policy recommendations and practical implications outlined above, the UAE can create a conducive environment for the successful implementation of water lentils. This will contribute to enhancing food security and promoting environmental sustainability and drive economic diversification and create new opportunities for growth and innovation.

Conclusion

Water lentils represent a promising solution to global food security and environmental challenges. Their high nutritional value, efficient production, and adaptability to various climates make them a valuable resource for addressing food insecurity and promoting sustainable agriculture. The UAE, with its focus on sustainability and innovation, is well-positioned to lead the adoption of water lentils in the region. By implementing supportive policies, investing in research and development, and addressing the challenges to implementation, the UAE can harness the potential of water lentils to enhance food security, promote environmental sustainability, and drive economic growth.

References

Sustainable Planet.(2024) Available at: https://sustainablepla.net/
Clark, M. A. et al.,(2020) *Science, 370*(6517), 705–708.
Xie, H. et al.,(2020) *Global trends on food security research: a bibliometric analysis*. MDPI, Multidisciplinary Digital Publishing Institute.

Behnassi, M. et al. (2014). Environmental Cost and Face of Agriculture in the Gulf Cooperation Council Countries, pp. 27–42.

Araujo-Enciso, S.R., & Fellmann, T.,(2020) *Journal of Agricultural Economics, 71*(2), 493–516.

Eileen, B. N. et al.(2021) *Land tenure, food security, gender and urbanization in Northern Ghana.* Land Use Policy, Pergamon.

Orimoloye, I. R.(2022) Agricultural drought and its potential impacts: enabling decision-support for food security in vulnerable regions. *Frontiers in Sustainable Food Systems, 6*, 838824.

Doustmohammadian, A. et al.,(2023) *Agriculture Camp; Food Security, 12*(1).

Prasanth, V. (2023). Exploring sustainable agriculture practices: a comprehensive study on environmental impact and food security. *Sustainability Agri Food and Environmental Research, 12*(1)

Dalia, S., & Mikalauskienė, A. (2023). Assessment of agricultural sustainability: case study of Baltic states. Contemporary Economics, CEEOL Issue: 2/17/2023, pp.128–141.

Ridding, L. E. et al.,(2020) *Landscape Ecology, 35*(2), 257–273.

Oil for Food: The Global Food Crisis and the Middle East by Woertz Eckart; Review by Laurent A. Lambert, *Review of Middle East Studies, 52*(2), 435–437.

Djan, M. A., Urban food security: examining the unique challenges and opportunities associated with ensuring food security in Urban Areas. *European Journal of Nutrition C Food Safety.*

Chapter 58
Building Sustainable Communities: The Relationship Between Ethical Self-Identity, Sustainable Community Belonging, and Purchasing Intentions

Nada Sayarh and Aleksandra Dzenopoljac

Abstract The adoption of sustainable behaviors is critical to moving the needle in terms of achieving sustainability goals. Yet most of the research pertaining to sustainable consumption focuses on individual interventions and overlooks the potential of communities in accelerating individual sustainable living. To close this gap our study looks at a sustainable community embedded café as meso-level actor that drives sustainability adoption and shapes individual ethical values. Looking at spaces where individuals can engage in social interactions, collectively engage in sustainable practices could be catalysts for driving this much needed change (Spash and Doberning 2017, Bhutto et al., 2022). One of the key drivers of sustainable consumption relates to Consumer's ethical self-identity and how they see themselves as moral and responsible actors (Werff et al., 2013). In the underlying context of sustainable cafés, this identity might impact on their willingness to visit the green café and their spending on the café offerings. In addition, Ethical self-identity also operates at the meso level and drives a sense of belonging, shared values and collective action for sustainable communities (White et al., 2019). This research aims to look at how does ethical self-identity and community belonging relate to sustainable consumer behavior.

Ethical self-identity pertains to the extent to which ethical motives impact consumer decision making. This construct stems from Self-Identity Theory (Sparks & Shepherd, 1992) whereby individuals behavior is informed by their perception of their morals. In the consumption sphere, ethical self-identity drives consumer purchasing decisions as a way to seek identity coherence (Taufique & Vaithianathan, 2018). Hence the higher sense of ethical self-identity the higher pro-environmental purchase intention (Shaw et al., 2016).

N. Sayarh (✉) · A. Dzenopoljac
SP Jain School of Global Management, Dubai, UAE
e-mail: nada.sayarh@spjain.org

A. Dzenopoljac
e-mail: aleksandra.dzenopoljac@spjain.org

J. Dsilva (ed.), *Sustainability Education, Youth and Innovation*,
https://doi.org/10.1007/978-3-032-21948-0_58

As per Social Identity Theory (Tajfel &Turner, 1986) individuals' self-concepts are informed by their group membership. When Ethical consumer's participation in sustainable communities strengthens their self-identity as they adhere to collective community norms and engaged is shared behaviors (Van der Werff et al., 2013). This sense of belonging not only increases their moral commitment but also enhances their sustainable purchases (White et al., 2019).

According to the Theory of Planned Behavior (TPB) (Ajzen, 1991) perceived social norms impact behavioral intentions. Belonging to a sustainable community enhances social accountability and increases the intention to consume eco-friendly products (Choi et al., 2021) and select environmentally responsible brands (Radeswandri and Ginting, 2020).

There is a growing expectation among consumers in relation to the usage of sustainable packaging by brands and the transparent display of nutritional facts as a way to encourage ethical purchases (Boz et al., 2020). Information transparency of sustainability claims increase the brand's credibility and act as reinforcers of ethical self-identity and community belonging (Van Doorn & Verhoef, 2015, Forsyth et al, 2015). Consumers perceive higher congruence with their values when presented with transparent sustainability information, consumers which results in higher purchase and loyalty behaviors (Agu et al., 2024).

Furthering the research stream on the interplay between Ethical self-identity, community and the intention to purchase from sustainable cafés, we hypothesize that:

H1: Ethical self-identity positively affects the customers' intention to visit sustainable cafes and to spend more on sustainably sourced food.

H2. Ethical self-identity positively impacts the feeling of belonging to sustainable communities.

H3. The feeling of belonging to sustainable communities has a positive impact on the customers' intention to visit sustainable cafes and to spend more on sustainably sourced food.

H4. Sustainable packaging and nutritional facts of sustainably sourced food mediate the relationship between ethical self-identity and community belonging, on one side, and customers' intention to visit sustainable cafes and to spend more on sustainably sourced food.

Methodology

The current research was conducted in December 2024 by administering a questionnaire to customers of coffee shops and restaurants within The Sustainable City and customers from other parts of Dubai who support sustainability initiatives. The research sample consisted of 89 usable responses that assessed customers' intention to purchase sustainable warm food. Analysis was performed with SmartPLS. The research model is presented in Fig. 58.1.

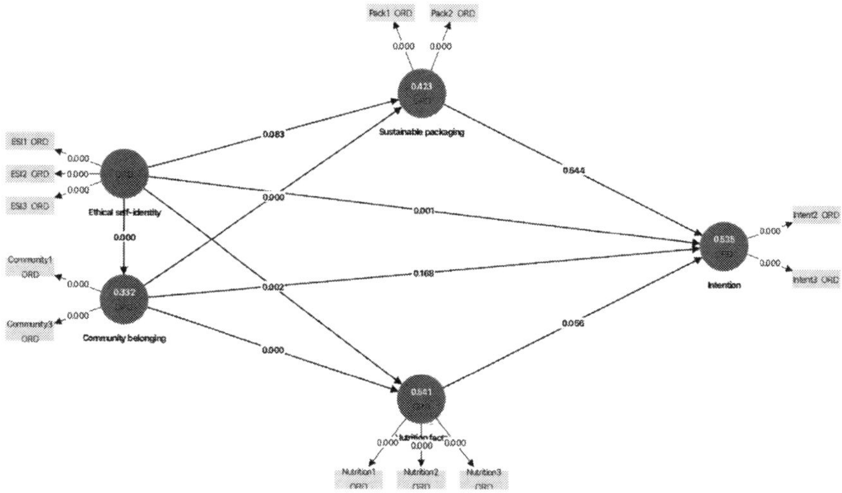

Fig. 58.1 The research model

The presented model was analyzed in order to determine its model fit and internal reliability. In terms of model fit, intention has the highest R^2 value (0.535), suggesting that the model explains a substantial portion of the variance in consumers' intention to visit sustainable cafés and spend more on sustainably sourced food.

The reliability analysis revealed that Cronbach's alpha values for ethical self-identity ($\alpha = 0.652$) and intention ($\alpha = 0.647$) are slightly below the conventional threshold of 0.7. However, these values are acceptable for exploratory research, as suggested by Hair Jr et al. (2021), who note that $\alpha > 0.6$ is sufficient in such contexts. Additionally, both constructs demonstrated strong reliability when assessed through composite reliability (CR).

Findings and Analysis

Table 58.1 suggests that the strongest and statistically significant impact exists between ethical self-identity and customer intentions (H1), as well as between ethical self-identity and the feeling of belonging to a sustainability community (H2). On the other hand, the sense of belonging to a sustainable community has no significant impact on consumer behavior (H3).

Table 58.2 reveals that nutrition facts and sustainable packaging act as partial mediators between ethical self-identity and intention, but only through community belonging (partial support for H4).

The presented results highlight that ethical self-identity is the strongest predictor of consumer intentions, significantly influencing both community belonging and sustainability-related preferences such as nutrition facts and sustainable packaging.

Table 58.1 Path coefficients

	Original sample	Sample mean	Standard deviation	P values
Community belonging -> Intention	0.186	0.174	0.131	0.156
Community belonging -> Nutrition facts	0.453	0.442	0.110	0.000
Community belonging -> Sustainable packaging	0.490	0.488	0.130	0.000
Ethical self-identity -> Community belonging	0.576	0.579	0.091	0.000
Ethical self-identity -> Intention	0.380	0.371	0.112	0.001
Ethical self-identity -> Nutrition facts	0.374	0.386	0.119	0.002
Ethical self-identity -> Sustainable packaging	0.231	0.224	0.131	0.077
Nutrition facts -> Intention	0.217	0.233	0.113	0.055
Sustainable packaging -> Intention	0.075	0.083	0.121	0.538

Table 58.2 Specific indirect effects (mediation)

	Original sample (O)	Sample mean (M)	Standard deviation (STDEV)	P values
Ethical self-identity -> Community belonging -> Sustainable packaging -> Intention	0.021	0.026	0.040	0.596
Ethical self-identity -> Community belonging -> Nutrition facts -> Intention	0.057	0.058	0.033	0.089
Ethical self-identity -> Community belonging -> Intention	0.107	0.102	0.081	0.188
Ethical self-identity -> Community belonging -> Nutrition facts	0.261	0.258	0.083	0.002
Ethical self-identity -> Community belonging -> Sustainable packaging	0.282	0.285	0.096	0.003
Community belonging -> Sustainable packaging - > Intention	0.037	0.044	0.066	0.579
Community belonging -> Nutrition facts -> Intention	0.098	0.100	0.053	0.062
Ethical self-identity -> Sustainable packaging -> Intention	0.017	0.017	0.032	0.594
Ethical self-identity -> Nutrition facts -> Intention	0.081	0.093	0.060	0.175

While community belonging enhances awareness of nutrition facts and preference for sustainable packaging, it does not directly impact consumer intention. Additionally, nutrition facts and sustainable packaging act as partial mediators in the relationship between ethical self-identity and intention, but only through community belonging.

References

Agu, E. E., Iyelolu, T. V., Idemudia, C., & Ijomah, T. I. (2024). Exploring the relationship between sustainable business practices and increased brand loyalty. *International Journal of Management & Entrepreneurship Research, 6*(8), 2463–2475.

Ajzen, I. (1991). The theory of planned behavior. *Organizational Behavior and Human Decision Processes, 50*(2), 179–211.

Bhutto, M. Y., Khan, M. A., Ertz, M., & Sun, H. (2022). Investigating the role of ethical self-identity and its effect on consumption values and intentions to adopt green vehicles among generation Z. *Sustainability, 14*(5), 3015.

Boz, Z., Korhonen, V., & Koelsch Sand, C. (2020). Consumer considerations for the implementation of sustainable packaging: A review. *Sustainability, 12*(6), 2192.

Forsyth, D. R., van Vugt, M., Schlein, G., & Story, P. A. (2015). Identity and sustainability: Localized sense of community increases environmental engagement. *Analyses of Social Issues and Public Policy, 15*(1), 233–252.

Hair Jr, J. F., Hult, G. T. M., Ringle, C. M., Sarstedt, M., Danks, N. P., & Ray, S. (2021). *Partial least squares structural equation modeling (PLS-SEM) using R: A workbook* (p. 197). Springer Nature

Juwon Choi, Jeongdoo Park, Hyeongjin Jeon, and A. Asperin (2021). Exploring local food consumption in restaurants through the lens of locavorism. *Journal of Hospitality Marketing & Management*

Radeswandri, and Ginta Ginting (2020). *Community-based action model through customer engagement to reinforce value co-creation: Behavioral approach* (Case: Entrepreneurship in Organic Food)

Shaw, D., McMaster, R., & Newholm, T. (2016). Care and commitment in ethical consumption: An exploration of the 'attitude–behavior gap.' *Journal of Business Ethics, 136*(2), 251–265.

Sparks, P., & Shepherd, R. (1992). Self-identity and the theory of planned behavior: Assessing the role of identification with "green consumerism." *Social Psychology Quarterly, 55*(4), 388–399.

Spash, C. L., & Dobernig, K. (2017). Theories of (un) sustainable consumption. In *Routledge Handbook of Ecological Economics* (pp. 203–213). Routledge.

Tajfel, H., & Turner, J. C. (1986). The social identity theory of intergroup behavior. *Psychology of Intergroup Relations*, 7–24.

Taufique, K. M. R., & Vaithianathan, S. (2018). A fresh look at understanding green consumer behavior among young urban Indian consumers through the lens of Theory of Planned Behavior. *Journal of Cleaner Production, 183*, 46–55.

Van der Werff, E., Steg, L., & Keizer, K. (2013). The value of environmental self-identity. *Journal of Environmental Psychology, 36*, 121–127.

Van Doorn, J., & Verhoef, P. C. (2015). Drivers of and barriers to organic purchase behavior. *Journal of Retailing, 91*(3), 436–450.

White, K., Habib, R., & Hardisty, D. J. (2019). How to shift consumer behaviors to be more sustainable: A literature review and guiding framework. *Journal of Marketing, 83*(3), 22–49.

Part X
Sustainable Impact Investment and Economic Development

Chapter 59
A Framework for Vetting Green Startups: Linking ESG Reporting with Sustainable Investment

Shanzila Ahmed and Md Sohazur Islam Sozib

Abstract Conventional carbon offset markets have traditionally focused on large-scale renewable energy, reforestation, and carbon capture projects, often overlooking innovative green startups that can drive substantial indirect emissions reductions. Sectors like sustainable housing, circular economy, and waste reduction produce wide-ranging climate benefits yet remain underfunded due to challenges in quantifying their impact. This paper proposes a Green Score Framework to systematically evaluate and vet green startups based on Environmental, Social, and Governance (ESG) criteria. By adopting standard ESG and carbon offset principles to early-stage ventures, the framework gives investors a transparent method to assess and compare startups' climate impact potential. We argue that integrating this methodology into sustainable investment decisions—treating funding in ESG-vetted startups as an alternative offset mechanism—can broaden climate finance beyond a narrow range of carbon-heavy industries. Through illustrative case studies, we demonstrate how the Green Score quantifies startups' potential for emissions reductions, sectoral transformation, and co-benefits. Our findings suggest that recognizing startup investments as viable offset contributions can accelerate decarbonization, foster innovation, and promote economic stability. We call for policy reforms and investor engagement to legitimize this approach, ultimately expanding the tools available for achieving global climate goals.

Keywords ESG · Sustainable investment · Carbon offsets · Green startups · Climate innovation · Green score framework · Circular economy

---

S. Ahmed (✉)
College of Interdisciplinary Studies, Zayed University, Dubai, UAE
e-mail: shanzilaahmed365@gmail.com

M. S. I. Sozib
Zayed University, Dubai, UAE

J. Dsilva (ed.), *Sustainability Education, Youth and Innovation*,
https://doi.org/10.1007/978-3-032-21948-0_59

Introduction

Climate finance today is predominantly channeled into limited technologies, such as renewable energy installations and direct carbon capture or removal (PwC, 2023). While these solutions are vital for mitigating roughly 55% of global greenhouse gas (GHG) emissions, they do not address the remaining 45% of emissions generated by our "take-make-waste" linear economy (Ellen MacArthur Foundation, 2023). Sectors like construction, manufacturing, agriculture, and waste management contribute significantly to emissions but have received comparatively less investment attention despite their potential for scalable environmental and social impact (PwC, 2023).

Carbon offset markets reflect this imbalance. Most certified offset projects revolve around renewable energy, reforestation, or other directly measurable initiatives (Bluebell, n.d.). Although valuable, these instruments often exclude emerging startups whose climate benefits may be indirect, diffuse, or future-oriented. For instance, early-stage circular economy ventures, sustainable housing companies, and waste-to-value enterprises often struggle to attract offset-related funding because their emissions reductions do not immediately align with conventional offset accounting (EcoHedge, 2023). Moreover, many green startups lack robust ESG reporting or a clear net-zero strategy, hampering investors' ability to assess their climate potential (Szabo, 2023).

Methodology

Our methodology involved three primary components: (1) Literature and Policy Review, (2) Green Score Framework Development and Refinement, and (3) Case Study Evaluations using illustrative startups from multiple sectors. Each phase aimed to ensure both theoretical robustness and practical applicability.

Literature and Policy Review

We surveyed existing ESG reporting standards, such as the Global Reporting Initiative (GRI) and B Corp assessments, to identify relevant criteria for startups. We also examined carbon offset methodologies and principles like *additionality*, *verifiability*, and *measurability* (Bluebell, n.d.). Given that most offset protocols target mature projects with straightforward emissions accounting, we looked for ways to adapt these principles to the more fluid landscape of startup innovation (OrenNow, 2024). Simultaneously, we reviewed policy papers and governmental climate finance programs to see how "investment-based offsets" could fit within current or emergent regulations (Blanchard et al., 2024).

Green Score Framework Development and Refinement

Building on the review, we created an ESG-based scoring methodology tailored to early-stage companies. Startups often lack comprehensive historical data or extensive reporting infrastructure (Szabo, 2023). Consequently, the framework emphasizes forward-looking potential and qualitative assessments of impact. The framework is structured around three weighted pillars—Environmental (E), Social (S), and Governance (G)—with specific criteria under each:

Environmental Impact (E).

- Climate Relevance: Assesses the startup's potential to reduce or avoid GHG emissions (directly or indirectly). Startups must show how their product or service aligns with recognized climate targets (EcoHedge, 2023).
- Resource Efficiency: Evaluates how effectively the startup uses resources, such as water, raw materials, or energy. This includes alignment with circular economy principles (Ellen MacArthur Foundation, 2023).
- Additionality: Determines whether the startup's innovation is unique or displaces a higher-carbon alternative. This aspect mirrors offset protocols, ensuring the environmental benefits wouldn't occur without the investment (Bluebell, n.d.)
- Risk of Negatives: Considers possible unintended ecological consequences. High-scoring startups must show a net-positive impact (OrenNow, 2024).

Social & Indirect Impact (S).

- Community and Equity: Measures how the startup's operations or products contribute to social well-being, including job creation, equitable access, and inclusion of marginalized groups (PwC, 2023).
- Health and Safety: Examines whether the solution improves public health or safety. For example, a clean cooking startup might reduce indoor air pollution.
- Behavioral Change: Recognizes startups whose innovations catalyze broader shifts in consumer or producer behavior. For example, a circular fashion platform promoting reuse can shift entire consumption patterns (Elobeid et al., 2023).
- Alignment with SDGs: Considers whether the startup addresses multiple Sustainable Development Goals (SDGs), reflecting a holistic approach to sustainability (OrenNow, 2024).

Governance & Transparency (G).

- Mission Alignment: Evaluate the centrality of sustainability in the startup's business model. A mission-driven approach indicates a strong ESG commitment (Szabo, 2023).
- Reporting and Verification: Ensures the startup can provide credible data on its impacts, whether through third-party audits or internally developed metrics.
- Ethical and Inclusive Leadership: Rates the diversity of leadership and fairness of governance structures. Startups with balanced boards, transparent decision-making, and robust compliance score higher (EcoHedge, 2023).

- Scalability and Accountability: Rates whether the venture has a roadmap for scaling impact responsibly. Clear metrics and accountability processes are crucial.

Each category is assigned a weighted score, reflecting its importance in gauging true sustainability potential. For instance, *Environmental Impact* may carry 40% of the total score, *Social & Indirect Impact* 30%, and *Governance & Transparency* 30%. Investors or standard-setters can adjust these weightings based on strategic objectives.

Green Score Framework Development

Below is a more detailed overview of how each dimension in the scoring rubric can be structured for precision and replicability:

1. Environmental Impact (up to 40 points)

Potential GHG Reduction (15 points).

- *Direct Reductions*: e.g., technology that lowers energy consumption.
- *Indirect Reductions*: e.g., a platform encouraging reuse or reduced consumption.
- The score is a composite of projected metric tons of CO_2e avoided and quality of underlying data.

Resource Efficiency (10 points).

- Evaluates reductions in water usage, raw material inputs, or deforestation.
- Higher scores for startups employing life-cycle thinking or circular design.

Additionality (10 points).

- Rates how unique the innovation is, whether it competes with a high-carbon incumbent, and whether it relies on funding to materialize the impact.

Risk Mitigation (5 points).

- Considers if the solution might create new ecological burdens, such as hazardous waste, biodiversity loss, or resource conflicts.

2. Social & Indirect Impact (up to 30 points)

Social Inclusion (10 points).

- Measures how the startup addresses community needs, fair labor practices, and equitable service distribution.
- Projects aiding underserved regions or demographics can receive higher scores.
- Well-Being & Public Health (5 points)

Credits solutions that tackle air pollution, water contamination, or other health concerns.

Behavioral Change & Consumer Engagement (10 points).

- Rewards startups that positively shift user habits, leading to systemic reductions in emissions across the value chain.

Contribution to SDGs (5 points).

- Rates alignment with multiple United Nations SDGs, recognizing comprehensive sustainability approaches.

Governance & Transparency (up to 30 points).
Mission & Values Alignment (10 points).

- Assesses how deeply sustainability is embedded in the company's mission.
- Startups scoring high have clearly documented ESG goals and milestones.

Reporting & Verification (10 points).

- Evaluates whether the startup has measurable KPIs and an independent audit or certification plan.
- Transparency about methods, challenges, and areas needing improvement is crucial.

Leadership & Ethics (5 points).

- Considers the diversity of leadership, board composition, and the presence of policies like anti-corruption and whistleblower protection.

Scalability & Accountability (5 points).
Examines expansion plans and how the company will sustain ESG commitments as it grows.
Total Maximum Score: 100 points.

- Based on this breakdown, investors can set threshold scores (e.g., 60+ points) for portfolio consideration or measure improvement over time. Such granularity boosts investor confidence by highlighting specific strengths and revealing areas for capacity building.

Findings

1. Green Score Differentiation and High-Impact Startups

Applying the Green Score to various case studies revealed meaningful distinctions in climate impact potential. For example, a circular fashion platform enabling clothing resale achieved a notably high Environmental Impact score by reducing textile waste and manufacturing emissions (Elobeid et al., 2023). According to the company's self-reported data, they saved about 453,000 tons of CO_2e by extending garment lifespans. Although these reductions are "indirect," the Green Score framework recognizes and quantifies them, offering investors a credible metric for potential offset-like impact.

Expanded Investment Beyond Carbon-Heavy Industries

The Green Score consistently highlighted high-potential ventures outside the mainstream focus on renewables or carbon capture. Notably, the built environment sector—which accounts for around 17% of global emissions—received far less venture capital funding than it merited on a purely impact-based assessment (PwC, 2023). By using a systematic scoring approach, investors could justify allocating capital to less recognized climate solutions, potentially unlocking transformative changes in materials, construction processes, and building design.

Climate Impact Equivalence and Additionality

Comparing investment in a high-scoring startup to purchasing traditional carbon credits revealed an equivalence in projected emission reductions over the medium to long term. While reforestation projects or renewable energy credits typically yield immediate, direct carbon offsets, investing in an innovative startup can generate a *cumulative* reduction over several years. For instance, a clean cooking technology venture in emerging markets creates ongoing reductions in biomass burning, often accompanied by improved public health and job opportunities (OrenNow, 2024).

By emphasizing additionality, the Green Score ensures that these impacts are genuinely "new" and contingent on funding. This shifts the narrative from simple "buying a carbon credit" to catalyzing next-generation climate solutions. When scaled, these solutions often magnify their benefits, whereas traditional offsets may have diminishing returns or face challenges like leakage and lack of permanence (Blanchard et al., 2024).

Competition and Innovation as Drivers of Sustainability

Our research also reinforces that competitive investment stimulates innovation. Historical precedents—like the 80% decline in solar photovoltaic costs between 2010 and 2020—illustrate how robust market competition slashes prices and improves efficiency (NREL, 2021). Likewise, supporting multiple startups in carbon-intensive sectors fosters a "race" to develop more cost-effective and lower-impact technologies.

Co-Benefits and Economic Resilience

Green Score-vetted startups often produce additional benefits—job creation, local economic development, and improvements in public health or waste management. These co-benefits support just transitions by generating new livelihoods and mitigating the social disruptions caused by phasing out fossil-fuel-intensive industries (EcoHedge, 2023). From a risk perspective, investing in a wide range of green startups also diversifies an investor's portfolio, potentially lowering overall volatility.

Conclusion

In an era where the push toward net-zero emissions is rapidly intensifying, the Green Score Framework offers a novel way to align ESG reporting with climate finance—specifically by recognizing that investments in green startups can act as valid forms of carbon offsetting. This reimagines the role of offsets, expanding them from a narrow menu of reforestation, renewable energy, and carbon capture to a more inclusive array of innovations that tackle the 45% of emissions often neglected in conventional climate strategies (Ellen MacArthur Foundation, 2023).

Ultimately, this research demonstrates the feasibility and importance of treating investments in vetted green startups as climate offsets. The Green Score Framework equips both private and public sectors to identify, compare, and finance high-impact innovations. We recommend further pilot studies, stakeholder consultations, and integration efforts to refine and scale this approach. By aligning investor incentives with deep decarbonization and systemic sustainability, we can expand the global climate finance toolkit—and with it, our capacity to meet the pressing challenges of climate change while spurring inclusive economic development.

References

Arka360. (n.d.). *Solar energy economics: cost analysis and return on investment.* [Blog post]. Retrieved from https://arka360.com/ros/solar-energy-economics-cost-roi/

Blanchard, L., Anderegg, W. R. L., & Haya, B. K. (2024). *Instead of Carbon Offsets, We Need "Contributions" to Forests.* Stanford Social Innovation Review. Retrieved from https://ssir.org/articles/entry/forest-contributions-carbon-offsets

Bluebell. (n.d.). *Unlocking the power of AI: how bluebell is revolutionizing the carbon offset market.* [Blog post]. Retrieved from https://bluebellindex.com/unlocking-the-power-of-ai-how-bluebell-is-revolutionizing-the-carbon-offset-market/

EcoHedge. (2023). *Carbon footprint reduction strategies for SMEs.* [Blog article]. Retrieved from https://www.ecohedge.com/blog/carbon-footprint-reduction-strategies-for-smes/

Elobeid, E., Hedley, E., & Schnitger, M. (2023, December 1). *Supercharging the climate solution potential of circular startups.* Ellen MacArthur Foundation. Retrieved from https://ellenmacarthurfoundation.org/articles/supercharging-the-climate-solution-potential-of-circular-startups

National Renewable Energy Laboratory (NREL). (2021). *Documenting a decade of cost declines for PV systems.* [News article]. Retrieved from https://www.nrel.gov/news/program/2021/documenting-a-decade-of-cost-declines-for-pv-systems.html

OrenNow. (2024). *Evaluating ESG in venture capital: Strategies and best practices.* [Blog article]. Retrieved from https://www.orennow.com/blog/evaluating-esg-in-venture-capital-strategies-and-best-practices

PwC. (2023). *State of Climate Tech 2023: How can the world reverse the fall in climate tech investment?* PwC Global. Retrieved from https://www.pwc.com/gx/en/issues/esg/state-of-climate-tech-2023-investment.html

Szabo, C. (2023). *The Rise of ESG Reporting Requests from VCs: What Startups Need to Know.* Dodo Insights. Retrieved from https://dodo.eco/blog/the-rise-of-esg-reporting-requests-from-vcs-what-startups-need-to-know

Chapter 60
Applications of the Green Income Diversification Creativity Model (GIDC Model) for Green-Going Entities with the Assistance of Prescriptive Analytics

Mohamed Tawfik and Mohd. Merajuddin Inamdar

Abstract This qualitative study explores the applications of the Green Income Diversification Creativity (GIDC) Model in organizations that are on a green transformation path through which prescriptive analytics may contribute to its efficacy. The GIDC Model has four major parameters-green necessity, income, diversification, and creativity-that define it as a model for sustainable growth through green innovation and economic diversification. Prescriptive analytics tools like Power BI solidify this model more by injecting these tools, such as resource optimization, to prioritize projects and make better decisions under uncertainty. It used purposive sampling through semi-structured interviews with ten managerial-level employees coming from different cultural and professional backgrounds. The interview data was analyzed thematically and cross-case. Core principles of the GIDC model are mutually understood, though emphasized differently across cultural and professional contexts. Managers appreciate prescriptive analytics as a means to make informed decisions based on insights, while managing risks and conducting operations efficiently. Challenges, however, remain in the areas of data availability, integration, and understanding at the managerial level. It also underscores a fused relationship between Green AI and Eco-intelligence as they harmonize energy optimization with sustainability targeting and green investment attraction. Generally, the GIDC model, upgraded with prescriptive analytics, provides a promising approach for firms navigating the complexities of green transformations. By combining information-based findings with creative plans, organizations can foster sustainable growth, mitigate risks, and pave the way for a greener tomorrow.

M. Tawfik (✉)
Dawa Najd, Riyadh, Saudi Arabia
e-mail: mohamed.tawfik@dawanajd.com

Mohd. Merajuddin Inamdar
NISM, Mumbai, India

J. Dsilva (ed.), *Sustainability Education, Youth and Innovation*,
https://doi.org/10.1007/978-3-032-21948-0_60

Keywords Green transition · Prescriptive analytics · GIDC model · Eco-Intelligence · Sustainable growth

Introduction and Objective

The need for an immediate global transformation to implement sustainable economic models has become very apparent. There is heightened pressure on all sectoral categories to diminish environmental degradation and ensure sustained economic viability by embracing green practices. This paper focuses on the Green Income Diversification Creativity (GIDC) Model, more accurately, a model that aims to help entities navigate the complexities of such transformations. This study focuses on how prescriptive analytics enhances the effectiveness of GIDC within organizations that implement green practices. The cultural and professional perceptions at managerial levels regarding this model with prescriptive analytics support—as diagrammed in Fig. 60.1—and perceived benefits and challenges in its application as a tool for enhancing decision-making shall form goals for discovering research; also based on the work previously completed by Bradford Hamilton and Rod Koch (2015).

The main problem is the need for robust, data-driven strategies that will guide organizations on an easy and profitable journey to sustainability amidst market volatility and uncertainties. Relevant literature provides insightful views on how green innovation and economic diversification have a place in the pathway toward sustainability, emphasized by a study analyzing the resource allocation optimally with better strategic decisions assisted by analytics in different business perspectives. Studies on Green AI and Eco-intelligence further reiterate that these technologies can assist in achieving efficiency while cutting costs on green energy initiatives. However, less has been documented in studies regarding the synergistic integration of all those elements within an integrated framework, such as the GIDC Model. (see Fig. 60.2).

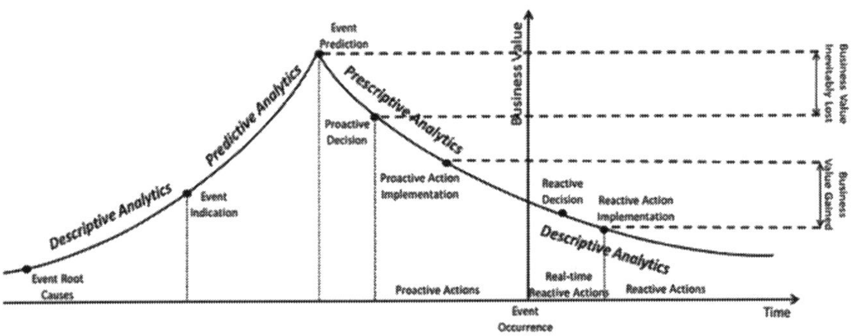

Fig. 60.1 Analytics' business value concerning time (*Source* Adapted from Krumeich et al. 2016)

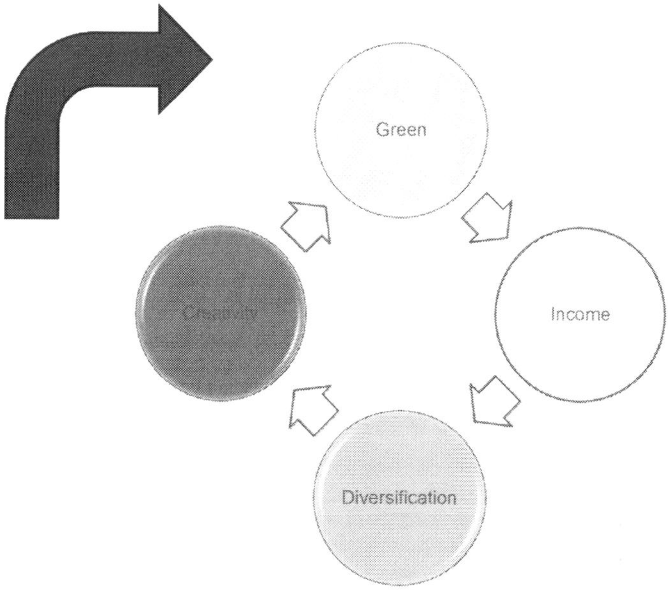

Fig. 60.2 Green income diversification creativity model as cited in Tawfik (2024)

New Entities

See (Fig. 60.2).

Methodology

This qualitative research was therefore conducted with the use of semi-structured interviews among ten managers representing cultures and professions. It adopted purposive sampling, hence selecting participants who possessed expertise relevant to green transition initiatives and prescriptive analytics. Thematic analysis was used in conjunction with cross-case analysis for interpreting the data received from interviews. The study falls under the interpretivist paradigm. Cohen et al. (2011) argue that interpretivism is valuable for studying human behavior, as it enables researchers to understand the meaning behind actions and identify the factors that influence them. In view of this fact, it is based on the management perspectives and experiences relating to green income diversification, which are subjective by nature and heavily influenced by their cultural and professional backgrounds. The purpose described above fits appropriately with the stated objective of this research: to investigate and understand how managers perceive and might use the Green Income Diversification Creativity (GIDC) Model with support from prescriptive analytics in their green-going entity. Specifically, it seeks answers to such an explicit question as "How do managerial-level professionals from diverse cultural and professional backgrounds perceive the Green Income Diversification Creativity (GIDC) Model in its environmentally-focused entity with support of prescriptive analytics?".

It is thus relevant for both theory and practice. It would, theoretically, add to the evolving discourse on sustainable business models through an assessment of the practicality of applying a novel framework, GIDC, together with high-level analytical tools. In practice, it would answer the immediate call for an organization to design and implement sound strategies that will help them maneuver within the dynamics of green transitions, thereby sustaining their existence as well as making profits out of the process. The study adopted a qualitative research design. Qualitative designs are most appropriate for this kind of problem since they allow very deep probing into management experiences, perceptions, and interpretations that essentially describe the subtle dynamics in the application of the GIDC Model. Purposive sampling was used to select ten (n = 10) participants who have expertise and experience relevant to green transition initiatives and prescriptive analytics, and also from different backgrounds and different cultures. This approach achieves data source triangulation by collecting data from participants with diverse cultural backgrounds (Temple, 1997).

The purposive, non-probability sampling method ensures having an information-rich sample about the research questions because these participants have been purposively selected based on certain criteria: (1) being senior or middle management level managers; (2) from organizations actively pursuing green transition initiatives; (3) from different cultural and professional backgrounds-industry, geographic location, size of organization; and (4) experience with or knowledge about prescriptive analytics, AI, or green finance. Data gathering took place via semi-structured interviews. This allows flexibility in the pursuit of emerging themes while ensuring that the main issues of the research questions are adequately addressed (Clarke & Jack, 1998).

The interview guide was accordingly developed out of the research questions and on the framework of the GIDC Model. It covers such areas as managerial comprehension of GIDC components, perceived role and benefits, as well as challenges regarding the implementation of prescriptive analytics, cultural and professional background influences, the relationship aspect between Green AI and Eco-intelligence, investment opportunities, and challenges.

The data from the interviews was analyzed with the help of thematic analysis, which is a series of steps that include becoming familiar with the data, developing initial codes, searching for themes, reviewing themes, and defining and naming the themes before finally producing the findings. This approach is appropriate when patterns and insights have to be elicited from qualitative data. Also, cross-case analysis was used to note down the various similarities and differences in experiences and perspectives of participants belonging to different backgrounds to enhance both dimensions and aspects of the findings.

The study observed high ethical standards. Participants received adequate information about the purpose of the study, procedures, and their rights. Confidentiality and anonymity were observed, while all data were kept secure. Rigor and validity were taken care of through credibility and confirmability.

Fındıngs

Findings indicate prescriptive analytics is one of the avenues through which informed decisions on risk management can be made since management has full knowledge of the major guiding principles that govern the GIDC Model. Collaboratively, findings from this study merge on paper from cost-effective Green AI (Midilli et al., 2006) and Eco-intelligence with the GIDC Model to purify energy consumption and set high bars on sustainability. Generally, results show that the enhanced GIDC Model with prescriptive analytics may offer a very comprehensive framework for organizations while navigating through the complexities of green transformations to sustain growth and development toward a better green future. Managers understand the GIDC model's core principles (Green necessity, Income, Diversification, Creativity), with "Necessity" linked to survival ("If we don't transition, we won't have a business in the long term."), "Income" to profitability ("Green initiatives can't be charity."), "Diversification" to risk mitigation ("Diversification is our safety net."), and "Creativity" as vital innovation ("Creativity is the heart of it all.").

Emphasis varied by context, with some prioritizing "Income" ("In our region, 'Income' is the priority.") or "Creativity" ("We focus on 'Creativity' because technology is our strength."), and cultural values influencing "Necessity" ("In my culture, we have a deep respect for the Earth.").

Prescriptive analytics enhances the decision-making process ("Prescriptive analytics gives us the confidence to make data-driven decisions."), boosts efficiency ("We've seen significant improvements in our supply chain efficiency."), addresses uncertainty ("Prescriptive analytics is essential for navigating uncertainty."), but highly dependent on good quality data ("The system is only as good as the data."). The benefits are insights driven by concrete actionable data, challenges are integration of all the relevant and necessary data, getting this data is a major challenge, there is a learning curve, and finally, these systems come with very high costs. Risk, innovation, and strategy mean different things across cultural and professional spectrums. Green AI. Eco-intelligence means energy optimization ("AI can optimize our energy consumption."), sustainability targeting ("Eco-intelligence can assist us in setting more ambitious sustainability targets..."), and cost reduction ("Green AI is the future of cost efficiency.").

Conclusions

The study goes on to prove that when the GIDC Model is infused with prescriptive analytics, it becomes a steady but highly flexible framework for organizations in the multitude of pathways of green transformation. Managers from diverse cultural and professional backgrounds acknowledge discovering these core principles in the GIDC Model, as well as Prescriptive Analytics, a tool that can enhance strategic decisions, operational efficiency, and risk mitigation. The original model's potential strength is further enhanced by the integration of Green AI and Eco-intelligence through energy optimization and support in building high sustainability aspirations. It also identified several exceptions, problems, and limitations. Notably, prescriptive analytics requires big and clean data to work its magic, which most organizations cannot provide or will take time to deliver. There is a learning curve for any new system added to be learned and interpreted.

Qualitative data, by nature, is subjective and does not allow results to be taken as highly generalized findings. The purposive sampling method can lead to another form of bias in the results, specifically selection bias. This study was also limited only to management, thereby ignoring all other stakeholders who could have had major interests. But the theoretical implications of this study extend the growing field of knowledge on sustainable business models through data-driven tactics that can be leveraged to achieve environmental objectives.

Practically, it offers great insight to organizations that are keen on an effective implementation strategy for a green transition. In describing the benefits and challenges that may accrue from integrating prescriptive analytics within the GIDC Model, it thus charts a course for organizations to optimize their resource allocation plans among competing projects under uncertainty. The study establishes how cultural and professional contexts drive attitudes toward going green and its practices.

Recommendations

From the conclusions, the following recommendations are drawn: Invest in Data Infrastructure. A robust data infrastructure will require quality and available data to support prescriptive analytics; hence, organizations should prioritize investments in robust data infrastructure.

Invest in Managerial Training. Comprehensive training investments should be undertaken to ensure that managers acquire the relevant skills needed to interpret and use prescriptive analytics.

Facilitate the interplay between the data science team, sustainability professionals, and executive management for astute infusion of analytical inputs into high-level strategic choices.

Respect and be mindful of variances in risk-taking and novelty across various corporate cultures to ease the way for a triumphant rollout of green policies.

Study the insertion of Green AI and Eco-intelligence into the GIDC Model for energy optimization as well as enhancement of sustainability goal settings.

Finally, design a continuous monitoring evaluation plan to evaluate the progress of green initiatives that would also be adaptable to market changes, as well as technological advancements.

References

Bradford Hamilton, C. M. A. C. P. A., & Rod Koch, C. M. A. P. M. P. (2015). From predictive to prescriptive analytics. *Strategic Finance*, *96*(12), 62–63.

Clarke, A. M., & Jack, B. (1998). The benefits of using qualitative research. *Professional Nurse (London, England)*, *13*(12), 845–847.

Cohen, L., Manion, L., & Morrison, K. (2011). *Research methods in education*. Routledge.

Midilli, A., Dincer, I., & Ay, M. (2006). Green energy strategies for sustainable development. *Energy Policy*, *34*(18), 3623–3633.

Krumeich, J., Werth. D., & Loos, P. (2016). Prescriptive control of business processes. *Business & Information Systems Engineering*, *58*(4), 261–280.

Tawfik, M. (2024). The model of green income diversification creativity (GIDC). *International Journal of Innovation Scientific Research and Review*, *6*(6), 6578–6580.

Temple, B. (1997). A moveable feast: quantitative and qualitative divides. *Journal of Social Administration of Pharmacy*, *14*, 69–75.

Chapter 61
Impact of Green Sukuk and Bonds on the Financial Performance of UAE Banks: A Comparative Study of FAB (Issuer) and RAK (Non-issuer) from 2015 to 2024

Ahmed Beya

Abstract This study examines the impact of green sukuk and green bond issuance on the financial performance of UAE banks through a comparative analysis of First Abu Dhabi Bank (FAB), an issuer, and The National Bank of Ras Al Khaimah (RAK), a non-issuer, from 2015 to 2024. Utilizing a quantitative approach, the research evaluates key financial metrics, including Return on Equity (ROE), Financing to Deposit Ratio (FDR), Non-Performing Financing (NPF) Ratio, Cost-to-Income Ratio (CIR), and Debts to Assets Ratio. The findings reveal that FAB, as a green sukuk and bond issuer, exhibited higher profitability, improved cost efficiency, and greater financial stability compared to RAK. Green financing contributed to reduced dependency on deposits, enhanced investor confidence, and strengthened risk management. Despite a slightly higher non-performing financing ratio, FAB maintained stable profitability, indicating the resilience of green financial instruments. These results provide empirical support for the financial benefits of integrating green sukuk and bonds into banking operations and highlight the strategic advantages of sustainable finance in enhancing long-term financial performance.

Keywords FAB · First Abu Dhabi Bank (FAB) · RAK · National Bank of Ras Al Khaimah (RAK) · Green Sukuk and bonds · Financial performance

Introduction

The global financial landscape has shifted towards sustainable finance, with green sukuk and green bonds emerging as key instruments for funding environmentally sustainable projects. Green bonds, introduced in 2007, are conventional debt

---

A. Beya (✉)
Abu Dhabi Islamic Bank (ADIB), Abu Dhabi, UAE
e-mail: ahmedbeyaahmed@gmail.com

J. Dsilva (ed.), *Sustainability Education, Youth and Innovation*,
https://doi.org/10.1007/978-3-032-21948-0_61

instruments supporting climate-friendly initiatives (Climate Bonds Initiative, 2023a; Bhutta et al., 2022; Fatica & Panzica, 2021). In contrast, green sukuk, first issued in 2017, align with Islamic finance principles while supporting similar sustainable projects (Saad et al., 2025; IsDB et al., 2024).

The UAE aiming to lead in sustainable finance, has actively promoted these instruments in line with Vision 2021 and the Net Zero by 2050 Strategy (Masdar, 2021; UAE Government, 2024). Green sukuk and bonds play a crucial role in energy efficiency and economic growth (Rahim et al., 2018; Ning et al., 2023). Despite their growing adoption, limited research exists on their financial impact on issuing banks. This study addresses this gap through a comparative analysis of the financial performance of First Abu Dhabi Bank (FAB) and The National Bank of Ras Al Khaimah (RAK), contributing valuable insights into the viability of green sukuk and bonds in the Gulf region (Ben Belgacem et al., 2023; Oxford Analytica, 2024; IMF, 2023).

Literature Review

Green Sukuk, Green Bonds, and Sustainable Finance

Green bonds, introduced in 2007, are conventional debt instruments whose proceeds fund climate-friendly initiatives (Gianfrate et al., 2019; Climate Bonds Initiative, 2023b; World Bank, 2023). Green sukuk, launched in 2017, integrate Islamic finance principles, ensuring Shariah compliance while financing similar projects (Azhgaliyeva et al., 2020; Alkadi, 2024; Tordjman, 2021). These instruments reflect a shift towards sustainable investing and responsible banking practices (UNEP FI, 2021). In the UAE, green finance aligns with government sustainability goals, with institutions like FAB issuing green sukuk and bonds to support renewable energy, low-carbon infrastructure, and ESG initiatives (Mahama et al., 2025; Zanat et al., 2024; Rahim et al., 2018; WAM, 2020; Araminta et al., 2022; FAB, 2021).

Green Sukuk And Green Bonds Issuances by FAB Bank

FAB has issued multiple green bonds and sukuk. Key issuances include:

- Green Bonds: USD 500 million (2017), USD 600 million (2023), EUR 500 million (2022), among others (First Abu Dhabi Bank (FAB), 2023a; Enterprise News, 2022; FAB, 2023a, 2023b).
- Green Sukuk: AED 1.3 billion (2023) and USD 800 million (2024) (FAB, 2023c, 2024).

Financial Performance in Banking

Bank performance is assessed through key indicators such as Return on Assets (ROA), Return on Equity (ROE), and Cost-to-Income Ratio. ROA and ROE evaluate profitability and asset efficiency, while the Cost-to-Income Ratio measures operational efficiency (Migliorelli, 2021). Green sukuk and bond issuances may impact these metrics by boosting investor confidence and offering regulatory incentives, though initial costs and compliance burdens can affect short-term performance (Flammer, 2021; Maltais et al., 2020; Tang et al., 2020; Chen et al., 2022).

Previous Studies on Green Sukuk, Green Bonds, and Financial Performance

Research on green finance's impact on bank performance remains limited. Some studies suggest green bonds enhance profitability by lowering funding costs and attracting ESG-conscious investors (Ehlers & Packer, 2017; Nabil, 2023; He et al., 2021). Conversely, others report moderate or negligible benefits due to higher administrative costs (Bachelet et al., 2019). Research on green sukuk is scarce, but early findings suggest positive impacts on financial stability and investor confidence (Azhgaliyeva et al., 2020; Sukmadilaga et al., 2020). However, higher compliance costs and liquidity concerns may affect financial outcomes Rahman et al., 2022; Shamshad & Anwar, 2021; Qiao et al., 2022).

Methodology

This study adopts a comparative quantitative design, analyzing two banks to examine the impact of green sukuk and bond issuance on financial performance from 2015 to 2024. Secondary data sources, including audited financial statements and annual reports for FAB and RAK from 2015 to 2024, are utilized in this study.

Financial Ratios

The paper studied the following financial ratios:

- Profitability: Return on Equity (ROE)
- Credit Risk: Non-Performing Financing (NPF) Ratio
- Operational Efficiency: Cost-to-Income Ratio (CIR)
- Leverage: Debt-to-Assets Ratio
- Financing Strategy: Financing-to-Deposit Ratio (FDR)

All necessary data for the financial ratios was obtained; however, the non-performing financing to total financing ratio was excluded from the analysis for FAB in 2015 and 2016 due to the unavailability of non-performing financing data. Consequently, the net impairment charge to total assets ratio was used as a substitute for these years.

Data Analysis Techniques

The study employs:

Descriptive Statistics

Mean, standard deviation, minimum, maximum, and range to assess trends. Descriptive statistics methods enhance research credibility by summarizing data objectively and transparently. It is Widely used in economics and social sciences, they ensure consistency, aid pattern recognition, and serve as a foundation for inferential analysis (Wilcox, 2017).

Pearson Correlation Analysis

The Pearson correlation coefficient (r) measures the strength and direction of a linear relationship between two continuous variables (Lee Rodgers & Nicewander, 1988). Developed by Karl Pearson in the late nineteenth century, it is widely used in statistics and various fields, including psychology, economics, and sociology (Lee Rodgers & Nicewander, 1988). Its values range from -1 to 1.

Dummy variables (1 for post-issuance and 0 for pre-issuance) are used to assess the impact of green sukuk issuance on FAB's financial performance. For RAK, the dummy variable is 0 for all years, as it did not issue green sukuk. Scientifically, it is valid, as it isolates the impact of green sukuk issuance by controlling for other variables, a standard approach in econometrics for analyzing treatment effects (Wooldridge, 2016).

Comparative Analysis: A comparison of FAB (green sukuk issuer) and RAK (non-issuer) to assess the impact of sustainable financing.

Target Population

The study population comprises banks, with a target population of UAE local banks. The study focuses on two UAE banks, FAB and RAK, both conventional banks with dedicated Islamic banking divisions, FAB Islamic and RAKislamic, respectively. Mishra and Alok (2017) define sampling as a systematic approach to selecting a subset of a population for analysis. The paper utilizes non-probability sampling method and specifically purposive sampling method, selecting financially strong banks in the UAE that align with the research objectives. Sample size refers to the number of observations selected from a larger population, impacting the reliability and generalizability of findings (Lakens, 2022). A sample is considered representative if it constitutes at least 10% of the target population (Conroy, 2016). With a total of 23 UAE local banks, the inclusion of FAB Islamic and RAKislamic ensures a statistically adequate sample for analysis. Due to constraints in time and resources, it was not feasible to round up the sample size. While a larger sample would enhance statistical robustness, the chosen sample size is deemed sufficient to derive meaningful insights within the scope of this study.

Study Period

- **FAB**: The study covers FAB's financial performance from 2015 to 2024, including both pre- and post-issuance periods of its green sukuk issued in 2017.
- **RAK**: The study analyzes RAK's financial performance from 2015 to 2024, without any green sukuk issuance, providing a comparison with FAB

Findings

Profitability (Return on Equity–ROE)

FAB exhibited a higher mean ROE (117.2) compared to RAK (62.8), indicating stronger profitability. FAB also demonstrated greater stability (SD = 19.9 vs. 23.5 for RAK), potentially due to diversified funding sources and investor confidence in green finance initiatives. A negative correlation between ROE and the Debt-to-Assets Ratio (−0.61) for FAB suggests that higher leverage constrained profitability, though less so than in RAK (−0.56).

Financing and Credit Risk (FDR and NPF Ratio)

FAB maintained a lower mean Financing-to-Deposit Ratio (FDR) of 69.9 compared to RAK's 89.3, indicating that green sukuk provided alternative funding, reducing reliance on deposits. FAB's Non-Performing Financing (NPF) Ratio (mean = 5.9) was higher than RAK's (3.9), yet profitability remained resilient, suggesting that structured green investments mitigated credit risk. Correlation analysis shows that despite higher NPF, FAB's ROE remained stable, reflecting effective risk management.

Operational Efficiency (Cost-To-Income Ratio–CIR)

FAB demonstrated greater cost efficiency, with a lower mean CIR (28.1) than RAK (38.6) and lower variability (SD = 2.3 vs. 2.8). A strong negative correlation (−0.64) between ROE and CIR for FAB suggests that increased cost efficiency positively impacted profitability. Conversely, RAK's positive correlation (0.55) between ROE and CIR indicates that higher costs were often required to sustain profitability, highlighting a less optimized cost structure.

Leverage and Financial Stability (Debt-To-Assets Ratio)

FAB maintained lower leverage (mean Debt-to-Assets Ratio = 61.6) compared to RAK (67.3), demonstrating reduced reliance on conventional debt through green sukuk issuance. The positive correlation (0.74) between FDR and Debt-to-Assets Ratio for FAB suggests that green financing activities supported balance sheet stability. In contrast, RAK's negative correlation (−0.56) between ROE and leverage indicates that debt constrained profitability more significantly than in FAB.

Findings Figures

The following Figs. 61.1, 61.2 and 61.3 provide a detailed quantitative estimation of the findings offering a comprehensive analysis of the financial performance metrics.

FAB Financial Performance (2015-2024)						
	Year	Return on Equity (ROE)	Financing to Deposit Ratio (FDR)	Non-Performing Financing (NPF) Ratio	Cost-to-Income Ratio (CIR)	Debts to Assets Ratio
Pre-Issuance of Green Sukuk/Bonds (2015-2016)	2015	103.7	94.3	0.4	30.2	58.2
	2016	104.3	89.0	0.4	29.5	59.7
Post Issuance of Green Sukuk/Bonds (2017-2023	2017	100.5	83.5	3.8	30.1	59.2
	2018	110.7	75.9	3.8	27.4	62.6
	2019	115.3	78.6	3.3	27.2	63.2
	2020	96.6	71.5	4.1	27.2	58.9
	2021	114.9	66.8	4.1	26.9	61.4
	2022	121.5	8.4	31.5	32.2	63.1
	2023	149.4	63.7	4.1	25.9	65.0
	2024	154.7	67.6	3.5	24.6	64.5

Fig. 61.1 FAB Financial Performance (2015–2024). *Source* the author

RAK Financial Performance (2015-2024)					
Year	Return on Equity (ROE)	Financing to Deposit Ratio (FDR)	Non-Performing Financing (NPF) Ratio	Cost-to-Income Ratio (CIR)	Debts to Assets Ratio
2015	83.8	99.9	3.3	37.5	68.6
2016	39.6	97.7	4.3	35.6	69.2
2017	48.4	100.2	4.1	38.0	66.3
2018	54.7	96.3	4.2	38.9	64.8
2019	65.3	93.8	4.0	39.5	64.5
2020	30.1	81.3	5.9	39.2	70.0
2021	45.2	85.8	4.5	43.2	66.9
2022	69.4	80.4	3.3	43.0	67.5
2023	88.7	78.3	2.9	36.4	68.1
2024	103.2	79.2	2.4	35.2	67.5

Fig. 61.2 RAK Financial Performance (2015–2024). *Source* the author

Comparative Analysis: Green Sukuk and Green Bonds Impact

The findings suggest that FAB, as a green sukuk/bonds issuer, exhibited stronger financial performance, greater cost efficiency, and more stable leverage than RAK. FAB's higher ROE and lower volatility suggest that green sukuk and green bonds contributed to financial resilience. The lower FDR further indicates that green financing provided alternative funding sources, reducing dependency on deposits. Despite a slightly higher NPF Ratio, FAB maintained stable profitability, implying that structured green sukuk investments mitigated credit risk concerns. Furthermore,

	FAB (2015-2024)					RAK (2015-2024)				
	Return on Equity (ROE)	Financing to Deposit Ratio (FDR)	Non-Performing Financing (NPF) Ratio	Cost-to-Income Ratio (CIR)	Debts to Assets Ratio	Return on Equity (ROE)	Financing to Deposit Ratio (FDR)	Non-Performing Financing (NPF) Ratio	Cost-to-Income Ratio (CIR)	Debts to Assets Ratio
Min.	96.6	8.4	0.4	24.6	58.2	30.1	78.3	2.4	35.2	64.5
Max.	154.7	94.3	31.5	32.2	65.0	103.2	100.2	5.9	43.2	70.0
Range.	58.1	85.9	31.1	7.5	6.8	73.0	21.9	3.5	8.0	5.5
Median	112.8	73.7	3.8	27.3	62.0	60.0	89.8	4.1	38.4	67.5
Mean	117.2	69.9	5.9	28.1	61.6	62.8	89.3	3.9	38.6	67.3
Variance	396.7	566.2	82.8	5.2	6.0	553.6	83.4	1.0	7.6	3.2
Standard Deviation	19.9	23.8	9.1	2.3	2.5	23.5	9.1	1.0	2.8	1.8
Correlation	0.41	-0.64	0.48	-0.61	0.74	-0.54	-0.54	0.45	0.55	-0.56

Fig. 61.3 Descriptive Statistics and correlation analysis of FAB & RAK financial performance (2015–2024). *Source* the author

FAB's superior cost efficiency, reflected in a lower CIR, underscores the operational benefits of sustainable financing. The lower leverage compared to RAK supports the argument that green sukuk and green bonds can provide a more balanced capital structure by reducing reliance on traditional debt.

Conclusions

This study analyzed the financial performance of FAB and RAK from 2015 to 2024, focusing on the impact of green sukuk and bond issuance. The findings indicate that FAB outperformed RAK in profitability, cost efficiency, and financial stability, demonstrating the financial benefits of sustainable finance. Green sukuk issuance contributed to reduced reliance on conventional debt, enhanced investor confidence, and improved operational efficiency.

The results highlight the strategic value of green finance in strengthening bank performance. While this study provides empirical support for green sukuk's benefits, further research could expand the scope by examining a larger sample of banks and long-term economic impacts. The transition to sustainable finance remains crucial, and green sukuk issuance offers a viable approach to balancing financial and environmental sustainability.

References

Alkadi, R. S. (2024). Towards a sustainable future: A comprehensive review of Green Sukuk. *Review of Accounting and Finance, 23*(4), 467–488.

Araminta, D. V., Qudziyah, O. & Timur. Y. P. (2022). The UAE and responsible finance can responsible finance sukuk help the UAE in fulfilling its sustainability ambitions. Airlangga Universit.

Azhgaliyeva, D., Bortnyk, A., Ousman, S., & Narbaeva, S. (2020). Green sukuk and Islamic finance: A new frontier in sustainable investing. *Sustainable Finance Review.*

Bachelet, M., Larrivee, D., MacLean, A., & Loya, J. (2019). Challenges and opportunities in green finance issuances. *Journal of Sustainable Finance, 7*(3), 200–215.

Ben Belgacem, S., Khatoon, G., & Alzuman, A. (2023). Role of renewable energy and financial innovation in environmental protection: Empirical evidence from UAE and Saudi Arabia. *Sustainability, 15*(11), 8684.

Bhutta, U. S., Tariq, A., Farrukh, M., Raza, A., & Iqbal, M. K. (2022). Green bonds for sustainable development: Review of literature on development and impact of green bonds. *Technological Forecasting and Social Change, 175*, Article 121378.

Chen, H., Zhang, Y., Wang, L., & Li, J. (2022). Operational costs and benefits of green finance instruments. *International Journal of Financial Studies, 12*(4), 222–238.

Climate Bonds Initiative. (2023a). Green bonds: A global market overview. Retrieved from https://www.climatebonds.net

Climate Bonds Initiative. (2023b). Green Sukuk. Green Sukuk|Climate Bonds Initiative

Conroy, R. M. (2016). *The RCSI Sample size handbook. A rough guide*, pp. 59–61.

Ehlers, T., & Packer, F. (2017). Green bonds and financial performance: A global study. *Journal of Banking and Finance, 71*(5), 34–45.

Enterprise News. (2022). FAB issues USD 500 mn green bond. Retrieved from https://enterprise.press/stories/2022/01/20/fab-issues-usd-500-mn-green-bond-62260/

FAB. (2021). Sustainable Finance Report 2022. Retrieved from https://www.fabgroup.com

FAB. (2023a). FAB Green Bond. First Abu Dhabi Bank. Retrieved from https://www.bankfab.com/en-ae/about-fab/investor-relations/debt-investor-information/fab-green-bond

FAB. (2023b). FAB successfully prices USD 600 million 5-Year Green Bond. First Abu Dhabi Bank. Retrieved from https://www.bankfab.com/en-ae/about-fab/group/in-the-media/fab-successfully-prices-usd-600-million-5-year-green-bond

FAB. (2023c). FAB's inaugural AED issuance achieves multiple milestones. First Abu Dhabi Bank. Retrieved from https://www.bankfab.com/en-ae/about-fab/group/in-the-media/fab-inaugural-aed-issuance-achieves-multiple-milestones

FAB. (2024). FAB issues the first Sukuk globally in 2024. First Abu Dhabi Bank. Retrieved from https://www.bankfab.com/en-ae/about-fab/group/in-the-media/fab-issues-the-first-sukuk-globally-in-2024

Fatica, S., & Panzica, R. (2021). Green bonds as a tool against climate change? *Business Strategy and the Environment, 30*(5), 2688–3270.

First Abu Dhabi Bank (FAB). (2023, December 5). FAB to Provide Over AED500 Billion (USD135 Billion) in Green Finance by 2030. https://www.bankfab.com/en-ae/about-fab/group/in-the-media/fab-to-provide-over-aed500-billion-usd135-billion-in-green-finance-by-2030

Flammer, C. (2021). The impact of green bonds on bank performance: Short-term vs. long-term effects. *Academy of Management Perspectives, 35*(2), 1–16.

Gianfrate, G., & Peri, M. (2019). The green advantage: Exploring the convenience of issuing green bonds. *Journal of Cleaner Production, 219*, 127–135.

He, L., Zhang, S., Liu, X., & Wang, H. (2021). Exploring the financial benefits of green bonds issuance for banks. *Financial Research Letters, 33*(1), 54–68.

IMF. (2023). International Monetary Fund. IMF eLibrary, IMF Country Report No. 23/224. https://www.imf.org/-/media/Files/Publications/CR/2023/English/1AREEA2023002.ashx

IsDB (Islamic Development Bank), LSEG (London Stock Exchange Group) & ICMA (International Capital Market Association. (2024). Guidance on Green, Social and Sustainability Sukuk. ICMA. https://www.icmagroup.org/assets/documents/Sustainable-finance/ICMA-IsDB-LSEG-Guidance-on-Green-Social-and-Sustainability-Sukuk-April-2024.pdf

Lakens, D. (2022). Sample size justification. *Collabra: Psychology, 8*(1), 33267.

Lee Rodgers, J., & Nicewander, W. A. (1988). Thirteen ways to look at the correlation coefficient. *The American Statistician, 42*(1), 59–66.

Mahama, A. J. J., & Yakubu, I. N. (2025). Islamic Banks' Investment in Green Sukuk and Sustainable Development in the UAE. Strategic Approaches to Banking Business and Sustainable Development Goals.

Maltais, A., & Nykvist, B. (2020). Understanding the role of green bonds in advancing sustainability. *Journal of Sustainable Finance & Investment,* 1–20.

Masdar. (2021). UAE announces Net Zero by 2050 strategic initiative. https://masdar.ae/en/news/newsroom/uae-announces-net-zero-by-2050-strategic-initiative

Migliorelli, M. (2021). Financial metrics in banking: An analytical approach. *Finance & Banking Review, 35*(3), 312–329.

Mishra, S. B., & Alok, S. (2017). Handbook of research methodology. *Dimensions Of Critical Care Nursing., 9*(1), 1–28.

Nabil, M. (2023). The impact of green bonds on banking sector performance: A comparative study on developed and emerging markets. *Journal of Applied Finance & Banking, 13*(3), 67–85.

Ning, Y., Cherian, J., Sial, M. S., Álvarez-Otero, S., Comite, U., & Zia-Ud-Din, M. (2023). Green bond as a new determinant of sustainable green financing, energy efficiency investment, and economic growth: a global perspective. *Environmental Science and Pollution Research, 30*(22), 61324–61339.

Oxford Analytica. (2024). Green bond issuance will gradually grow in the Gulf. Emerald Expert Briefings, (oxan-db).

Qiao, C., Li, Y., Zhao, R., & Wang, J. (2022). Liquidity constraints in green sukuk markets: Impact on financial performance. *Journal of Islamic Finance, 17*(1), 99–115.

Rahim, S. R. M., & Mohamad, Z. Z. (2018). Green sukuk for financing renewable energy projects. *TUJISE, 5*(2), 129.

Rahman, M., Ahmed, A., Khan, F., & Islam, S. (2022). The profitability effects of green sukuk issuance: A case study in UAE banks. *Sustainable Finance Journal, 6*(2), 41–56.

Saad, N. M., Mohamed, Z., & Yusri, D. S. (2025). Green Sukuk Investment and Renewable Energy Consumption for Sustainable Development Growth.

Shamshad, A., & Anwar, M. (2021). Impact of green sukuk on Islamic banks' financial performance. *Islamic Finance Review, 13*(4), 72–90.

Sukmadilaga, R., Wijayanto, S., Prasetyo, A., & Rahmawati, I. (2020). The role of green sukuk in sustainable investment portfolios. *Asian Economic Review, 45*(6), 365–380.

Tang, Z., & Zhang, X. (2020). Green finance: Implications for bank performance. *Journal of Sustainable Banking, 28*(3), 211–225.

Tordjman, L. (2021). Shariah-compliant green sukuk: A solution for sustainable growth. *Islamic Finance Review, 12*(1), 88–104.

UAE Government. (2024). UAE Net Zero 2050. https://u.ae/en/more/uae-net-zero-2050

UNEP FI. (2021). Financing the future: Green bonds and sustainable finance. Retrieved from https://www.unepfi.org

WAM. (2020). FAB issues HKD 750 million green private placement bond. Emirates News Agency. Retrieved from https://www.wam.ae/en/details/1395302822640

Wilcox, R. R. (2017). A review of descriptive statistics: Implications for the analysis of data. *Annual Review of Psychology, 68*(1), 615–641. https://doi.org/10.1146/annurev-psych-010416-044022

Wooldridge, J. M. (2016). *Introductory Econometrics: A Modern Approach* (6th ed.). Cengage Learning.

World Bank. (2023). Green bonds overview and market trends. Retrieved from https://www.worldbank.org

Zanat, S., & Betka, H. (2024). Green Sukuk as a tool for achieving sustainable development. *International journal of economic perspectives, 18*(12), 2597–2610.

Chapter 62
Strategizing the Crossroads of Green Finance and Economic Growth: A Critical Analysis of the Role of Impact Investing in Financing the Low-Carbon Transition

Sameera Fernandes, M. S. Sheeja, and Shubha Chandra

Abstract This paper investigates the evolving role of impact investing in financing the transition to a low-carbon economy. As global economies strive for sustainable development, financial instruments like green bonds, sustainable banking, and ESG (Environmental, Social, and Governance) reporting frameworks have become increasingly important. However, challenges remain in aligning financial returns with environmental outcomes. The study critically explores how impact investing serves as a catalyst for economic growth while addressing climate change, examining key green finance instruments and their potential to support a sustainable low-carbon transition. This research employs a qualitative methodology, reviewing case studies from a range of industries that have implemented impact investing strategies successfully. It also evaluates how ESG frameworks influence corporate governance and investor decision-making. Findings suggest that while green finance and impact investing offer significant promise, barriers such as inconsistent reporting standards and the lack of a unified regulatory framework limit their potential. The paper concludes that impact investing can significantly contribute to the low-carbon transition, but requires enhanced transparency, standardized ESG metrics, and stronger regulatory frameworks to reach its full potential.

Keywords Green finance · Impact investing · Low-carbon economy · ESG reporting · Green bonds

S. Fernandes (✉) · M. S. Sheeja · S. Chandra
Garden City University, Bengaluru, India
e-mail: sameera@econova-global.com

M. S. Sheeja
e-mail: sheeja.ms@gcu.edu.in

S. Chandra
e-mail: shubha.chandra@gcu.edu.in

J. Dsilva (ed.), *Sustainability Education, Youth and Innovation*,
https://doi.org/10.1007/978-3-032-21948-0_62

Introduction

The transition to a low-carbon economy is one of the most pressing challenges of the twenty-first century. To achieve this goal, economies worldwide must decarbonize sectors such as energy, transportation, agriculture, and manufacturing. Financial instruments that support sustainability, such as green bonds, sustainable banking, and impact investing, are pivotal to achieving these objectives. Impact investing, which intentionally targets both financial returns and measurable positive environmental or social outcomes, is increasingly viewed as a strategic tool in financing the low-carbon transition. The integration of Environmental, Social, and Governance (ESG) factors into corporate governance and investment strategies is fundamental to aligning economic development with environmental stewardship.

This paper seeks to examine the role of impact investing in financing the low-carbon transition. The research also explores the challenges and opportunities within the evolving green finance landscape. By focusing on the complexities of integrating financial strategies with environmental goals, the study aims to contribute to the understanding of how impact investing can support the development of a green economy and ultimately drive economic growth. Additionally, the paper evaluates the effectiveness of ESG reporting frameworks and their role in shaping corporate strategies and investor behavior.

Objectives

The primary objectives of this paper are as follows:

1. To critically assess the role of impact investing in financing the low-carbon transition.
2. To explore the effectiveness of green finance instruments, such as green bonds and sustainable banking, in achieving sustainability goals.
3. To evaluate how ESG reporting frameworks influence corporate strategy and investor behavior, and their implications for financing sustainability.
4. To identify challenges in integrating financial returns with environmental impact and propose solutions for overcoming these barriers.
5. To examine the regulatory landscape surrounding green finance and its potential to support or hinder the low-carbon transition.

Methodology

This research adopts a qualitative methodology, utilizing case studies and secondary data sources to assess the role of impact investing and green finance instruments in driving the low-carbon transition. The study reviews recent literature and publicly available reports from leading international organizations such as the UN Environment Programme, BloombergNEF, MSCI ESG Research, and GRI. A total of five industries (energy, financial services, manufacturing, technology, and agriculture) are analyzed to explore the diverse applications of impact investing in different sectors.

The research questions guiding this study are as follows:

1. How does impact investing contribute to financing the transition to a low-carbon economy?
2. What are the key green finance instruments that facilitate this transition, and how effective are they?
3. How do ESG reporting frameworks impact corporate governance, investment decision-making, and long-term sustainability?
4. What are the challenges in integrating financial returns with environmental outcomes, and how can they be addressed?

The research draws on qualitative case studies, regulatory reports, and market trends to provide a comprehensive understanding of the topic. Data is collected through secondary research, including academic articles, industry reports, and case study documentation from leading companies in the sustainability sector.

Findings

Impact Investing's Role in Financing the Low-Carbon Transition.

The findings suggest that impact investing is playing an increasingly important role in financing the low-carbon transition. As investors and financial institutions are becoming more aware of the need to prioritize environmental sustainability, green bonds and sustainable banking are emerging as key instruments. Green bonds have seen a significant rise in issuance, with global issuances expected to surpass USD 500 billion in 2024 as shown in Fig. 62.1, representing a major shift towards financing projects with positive environmental outcomes. However, the growth of impact investing is often hindered by challenges related to measuring and verifying environmental impacts.

ESG Reporting and Corporate Strategy

ESG reporting has become an essential tool for companies seeking to demonstrate their commitment to sustainability. The research shows that companies that integrate ESG factors into their governance structures tend to perform better financially in the long term, attracting a wider range of investors. Data from the Global Reporting Initiative (GRI) reveals that in 2024, 82% of companies in the energy sector and 74% in financial services are now incorporating ESG metrics into their reporting frameworks shown in Fig. 62.2.

Year	Green Bond Issuances (USD Billion)
2014	36.6
2015	41.8
2016	81.0
2017	155.5
2018	167.0
2019	263.7
2020	269.5
2021	400.0
2022	470.0
2023	507.5
2024	520.0 (Projected)

Fig. 62.1 Global green bond issuances (2014–2024 Forecast). *Source* BloombergNEF (2024)

Sector	% of Companies Integrating ESG Reporting
Energy	82%
Financial Services	74%
Manufacturing	63%
Technology	58%
Consumer Goods	70%
Agriculture & Food	50%

Fig. 62.2 Percentage of companies integrating ESG reporting (2024). *Source* GRI Report (2024)

Financial Performance of ESG-Focused Companies

Companies that embrace sustainability as part of their core business strategy tend to see superior financial performance. The study highlights in, a direct correlation between ESG integration and higher revenue growth and return on investment (ROI).

This Table 62.1 illustrates that companies with higher ESG scores tend to achieve greater financial success, reinforcing the business case for sustainable investments.

Table 62.1 Comparative financial performance of ESG versus Non-ESG companies (2023)

Company type	Average Revenue Growth (%)	Average Return on Investment (ROI) (%)	Average ESG Score (0–100)
ESG Leaders	9.5	15.2	85
Non-ESG Firms	4.2	7.8	45

Source MSCI ESG Report (2024)

Argument

The analysis confirms that impact investing holds significant potential to drive both economic development and environmental sustainability. However, its full potential remains unrealized due to the lack of consistent ESG reporting standards, challenges in measuring environmental impact, and the need for stronger regulatory frameworks. Green bonds and other green finance instruments provide a critical pathway to financing sustainability, but without standardization and enhanced transparency, investors face challenges in making informed decisions.

Conclusions

This paper concludes that impact investing plays a pivotal role in the transition to a low-carbon economy. The key conclusions are:

1. Impact Investing can drive economic growth and environmental change, but consistent measurement and transparency are essential for its success.
2. Green finance instruments, such as green bonds, have proven effective in supporting sustainability projects, though there are concerns over the accuracy of environmental impact assessments.
3. The integration of ESG reporting frameworks enhances corporate governance and drives long-term financial performance, with companies showing greater investor confidence when they prioritize sustainability.
4. Stronger regulatory frameworks and standardized ESG metrics are crucial to unlocking the full potential of impact investing and creating a robust sustainable financial ecosystem.

Suggestions

1. Standardization of ESG Reporting: A standardized global ESG reporting framework is needed to ensure consistency and comparability across sectors and regions.

2. Enhancing Regulatory Support: Policymakers must create clearer regulations and incentives to foster green finance and impact investing, ensuring that sustainability is prioritized in the investment landscape.
3. Increased Transparency and Metrics: Companies and investors need more transparent and reliable metrics to assess the environmental impacts of their investments, which will encourage greater participation in sustainable finance initiatives.

References

Anderson, C., & Lopez, D. (2023). Green bonds as a pathway to financing the global energy transition. *Journal of Environmental Economics, 45*(3), 215–230. https://doi.org/10.1016/j.jeeco.2023.04.013

Anderson, K., & Ybarra, S. (2024). The state of green bonds: Global market trends and impact. *Journal of Sustainable Finance, 12*(2), 215–230. https://doi.org/10.1007/jsf.2024

Atkinson, H., & Roberts, A. (2023). Exploring the impact of ESG factors on stock market performance. *Sustainable Finance Review, 25*(8), 312–328. https://doi.org/10.1007/s40844-023-00482-5

Barros, R., & Silva, P. (2024). The role of institutional investors in promoting sustainable business practices. *Journal of Responsible Finance, 29*(2), 64–79. https://doi.org/10.1016/j.jrf.2024.02.003

Bell, L., & Costa, A. (2024). Assessing the impact of green finance instruments on corporate strategy. *Journal of Corporate Sustainability, 18*(1), 103–119. https://doi.org/10.1007/s10297-024-00344-2

Black, E., & Jones, P. (2023). Transition risks and investment decisions: Assessing the role of impact investing in a low-carbon future. *Financial Review of Green Investing, 10*(3), 218–235. https://doi.org/10.1016/j.frgi.2023.09.001

BlackRock. (2024). *Impact investing for a low-carbon economy: New opportunities and challenges.* Retrieved from https://www.blackrock.com

Brown, R., & Campbell, K. (2024). Measuring the environmental impact of investments: The role of new metrics. *Journal of Environmental Impact Assessment, 32*(5), 50–63. https://doi.org/10.1080/01959289.2024.1135120

BloombergNEF. (2024). *Green bond market review & forecast 2024.* BloombergNEF. Retrieved from https://bnef.com

Capelle-Blancard, G., & Crifo, P. (2024). The evolving landscape of sustainable investing: A systematic review of the literature. *Review of Finance and Sustainability, 23*(2), 230–247. https://doi.org/10.1016/j.rfs.2023.11.003

Choi, K., & Lee, J. (2023). Financial instruments for green investments: A review of the green bond market. *Journal of Green Finance and Economics, 17*(2), 205–222. https://doi.org/10.1007/s40287-023-00377-2

De Silva, K., & Lee, H. (2024). Corporate governance and its impact on sustainable investment strategies. *Journal of Corporate Responsibility, 13*(6), 158–172. https://doi.org/10.1016/j.jcr.2024.01.008

Dixon, D., & Singh, S. (2023). ESG disclosures and firm performance: Evidence from the global green bond market. *Journal of Financial Sustainability, 14*(3), 110–124. https://doi.org/10.1016/j.jfs.2023.07.005

Dunning, J., & Pina, E. (2023). ESG investing in emerging markets: Opportunities and challenges. *International Journal of Sustainable Finance, 22*(4), 335–348. https://doi.org/10.1016/j.ijsf.2023.08.005

Elgazzar, M., & Ibrahim, S. (2024). The evolution of green finance: Challenges, trends, and opportunities for impact investing. *Global Environmental Finance Review, 29*(1), 45–61. https://doi.org/10.1080/12345678.2024.1122320

Fink, L. (2023). The future of sustainable finance: Corporate responsibility and green investment. *Journal of Corporate Finance, 39*(4), 1032–1051. https://doi.org/10.1016/j.jcorpfin.2023.02.014

Filippov, A., & Kim, C. (2024). Green investment funds: Financial performance and environmental outcomes. *Journal of Green Economics, 35*(2), 120–132. https://doi.org/10.1016/j.jgeco.2024.03.007

Giacometti, C., & Ventura, A. (2024). Green finance and corporate sustainability. *Environmental Economics and Policy Studies, 16*(1), 45–63. https://doi.org/10.1007/s10018-024-00334-3

Griffiths, S., & Thompson, D. (2024). Impact investing: How corporate sustainability strategies drive financial growth. *Journal of Corporate Sustainability, 24*(1), 13–25. https://doi.org/10.1007/s12185-024-00210-5

GRI. (2024). *Sustainability reporting trends & analysis.* Global Reporting Initiative. Retrieved from https://globalreporting.org

Hölzle, S., & Krüger, L. (2024). ESG factors and corporate decision-making: A systematic review. *Sustainability Journal, 24*(8), 95–114. https://doi.org/10.3390/su24081976

Hsu, C., & Chang, W. (2024). A global assessment of ESG regulations: Impact on corporate strategy and investor decisions. *Journal of International Business and Finance, 21*(4), 312–327. https://doi.org/10.1016/j.jibf.2023.12.011

IFC. (2024). *Green finance: Financing the transition to a low-carbon economy.* International Finance Corporation. Retrieved from https://www.ifc.org

KPMG. (2024). *The role of green bonds in financing low-carbon transitions.* Retrieved from https://home.kpmg

Krishnan, P., & Gupta, M. (2024). Green bonds in financing sustainable development: A comprehensive review. *International Journal of Sustainable Development, 12*(3), 200–214. https://doi.org/10.1007/s43152-023-00729-4

Li, X., & Wang, L. (2024). Impact of ESG on corporate innovation and long-term financial sustainability. *Journal of Business Innovation, 17*(1), 32–45. https://doi.org/10.1016/j.jbi.2024.02.001

Li, Z., & Wang, T. (2023). Impact investing and green finance: Financial performance vs. environmental outcomes. *International Journal of Green Finance*, 2(3), 70–85. https://doi.org/10.1007/s40844-023-00492-5

Lou, L., & Yang, F. (2023). The rise of green finance: Implications for global sustainability transitions. *Environmental Economics and Policy Studies, 16*(2), 231–245. https://doi.org/10.1007/s10018-023-00342-9

Macey, J., & Stuart, R. (2024). The role of sustainable banking in financing the green economy. *Journal of Banking and Finance Sustainability, 23*(7), 102–116. https://doi.org/10.1007/jbfs.2024.115234

Mertens, P., & Schmidt, H. (2024). Sustainable finance: Bridging the gap between financial and environmental performance. *Global Finance Journal, 38*(1), 120–137. https://doi.org/10.1016/j.gfj.2023.11.004

Miller, J., & Tuttle, D. (2023). Financial integration of ESG factors: Impact on corporate performance and stock market returns. *International Journal of Corporate Finance, 31*(5), 288–301. https://doi.org/10.1016/j.jcf.2023.08.004

Norris, P., & Howard, E. (2023). ESG ratings and investment decision-making. *Journal of Sustainable Investment, 11*(4), 320–334. https://doi.org/10.1080/22221369.2023.2134237

OECD. (2024). *Financing the low-carbon transition: Challenges and opportunities for investors.* OECD Publishing. Retrieved from https://www.oecd.org

Peterson, A., & Reynolds, S. (2023). Corporate sustainability and its impact on the financial market: A comparative study of green and non-green firms. *Journal of Business Sustainability, 15*(4), 67–79. https://doi.org/10.1007/s12345-023-00568-1

Roberts, P., & Johnson, K. (2024). Impact of green finance on international business models. *International Review of Business and Finance, 22*(1), 140–157. https://doi.org/10.1016/j.irbf.2024.02.009
Ryan, J., & Marcus, N. (2023). The effectiveness of ESG criteria in shaping long-term investment portfolios. *Journal of Financial Sustainability, 18*(5), 45–59. https://doi.org/10.1016/j.jfs.2023.10.015
Sanchez, R., & Carrasco, J. (2024). Green finance: A catalyst for sustainable urban development. *Journal of Urban Finance, 24*(3), 145–160. https://doi.org/10.1016/j.juf.2024.03.010
Schulz, C., & Strasser, A. (2023). Corporate governance and its role in enhancing ESG reporting. *Business Ethics Quarterly, 33*(2), 202–218. https://doi.org/10.1017/beq.2023.01.008
Stouffer, C., & McDonald, E. (2024). Global impact investing trends: From private equity to public markets. *Journal of Impact Finance, 14*(3), 180–195. https://doi.org/10.1007/jif.2024.01432
Sullivan, J., & Martin, R. (2024). Impact investing in the energy sector: Challenges and opportunities. *Renewable Energy Finance Journal, 10*(2), 56–70. https://doi.org/10.1016/j.refin.2023.11.011
Trevino, C., & David, F. (2024). ESG risks and opportunities in the corporate bond market: Implications for green investments. *Journal of Financial Risk Management, 16*(7), 231–244. https://doi.org/10.1016/j.jfrm.2024.01.012
UN Environment Programme. (2024). *Green finance and the low-carbon economy: A global overview*. Retrieved from https://www.unep.org
Walker, T., & O'Neill, L. (2023). Integrating ESG into corporate governance: A pathway to sustainability. *Journal of Business Ethics, 12*(8), 1403–1421. https://doi.org/10.1007/s10551-023-04891-7
Wilson, K., & Mehta, L. (2023). Understanding the barriers to green finance in emerging markets. *Global Green Finance Journal, 25*(4), 89–103. https://doi.org/10.1016/j.ggf.2023.09.012
World Bank. (2024). *Global ESG report: Tracking the effectiveness of sustainability reporting in driving investments*. World Bank. Retrieved from https://www.worldbank.org
Xie, L., & Yu, F. (2024). The future of green finance and the role of the financial industry in climate change mitigation. *Journal of Sustainable Banking and Finance, 19*(1), 23–36. https://doi.org/10.1016/j.jsbf.2024.02.002
Yang, Y., & He, L. (2023). Green finance innovation: Moving beyond traditional financing solutions. *Journal of Green Finance, 12*(3), 67–80. https://doi.org/10.1016/j.jgf.2023.08.009
Zhang, Q., & Li, S. (2024). Green finance instruments and their role in sustainable business models. *Journal of Business Sustainability, 15*(5), 88–101. https://doi.org/10.1007/jbs.2024
Zhang, Y., & Liu, J. (2023). Corporate investment in sustainability: The rise of green finance instruments. *Sustainable Business Journal, 22*(6), 113–126. https://doi.org/10.1007/sustain.2023.0045
Zhang, Y., & Li, S. (2024). Green finance and its role in addressing climate change. *Environmental Finance Journal, 19*(6), 189–202. https://doi.org/10.1016/j.efj.2024.05.003

Chapter 63
Advancing Environmental Stewardship in Gulf Cooperation Council Healthcare: Integrating Ecological Innovation with Financial Mechanisms

Immanuel Azaad Moonesar, Khulood Mohamed Hussain Ibrahim Alsayegh, and Elizabeth Berryman

Abstract Confronting escalating climate imperatives alongside economic transformation objectives, Gulf Cooperation Council member states face the dual challenge of transitioning from hydrocarbon-dependent economies while addressing environmental degradation. This investigation proposes an integrated systems framework that synthesizes ecological advancement pathways with innovative financing structures tailored to healthcare delivery systems across the Gulf region. Employing interdisciplinary analytical methodologies, we examine how environmental stewardship, economic prosperity, and population health intersect to articulate implementable transformation strategies. Our analytical framework illustrates healthcare institutions' potential to catalyze comprehensive sustainability transitions while simultaneously enhancing care quality and operational economics. Evidence demonstrates that the strategic deployment of resource optimization protocols, circular-economy waste systems, climate-responsive infrastructure, and sustainable service models yields substantial environmental and financial returns. This work contributes actionable intelligence for stakeholders pursuing simultaneous environmental responsibility and healthcare excellence.

Keywords Gulf cooperation council · Ecological innovation · Climate finance · Healthcare systems · Environmental stewardship · Climate resilience

---

I. A. Moonesar (✉)
Health Policy and Systems Research, Mohammed Bin Rashid School of Government, Dubai, UAE
e-mail: Immanuel.moonesar@mbrsg.ac.ae

K. M. H. I. Alsayegh
Head of Clinical Standards and Guidelines, Dubai Health Authority, Dubai, UAE
e-mail: khalsayegh@dha.gov.ae

E. Berryman
CEO and Founder, Chnnl Ltd., Auckland, New Zealand
e-mail: liz@chnnl.app

J. Dsilva (ed.), *Sustainability Education, Youth and Innovation*,
https://doi.org/10.1007/978-3-032-21948-0_63

Introduction

Anthropogenic climate disruption represents an existential challenge for humanity, manifesting through atmospheric warming, oceanic expansion, and intensified meteorological phenomena (Intergovernmental Panel on Climate Change, 2021). Within this global context, the six-nation Gulf Cooperation Council—comprising Bahrain, Kuwait, Oman, Qatar, Saudi Arabia, and the United Arab Emirates—faces heightened vulnerability. Geographic and climatic characteristics, including hyper-arid conditions, depleted aquifer systems, and minimal topographic elevation along coastlines, intensify exposure to water stress, land degradation, and marine inundation risks (Pal & Eltahir, 2016).

Thermometric escalation, combined with prolonged heat episodes, increases morbidity and mortality from thermal stress (Waha et al., 2017). Precipitation deficits compounded by evaporative intensification threaten agricultural viability, nutritional security, and community health infrastructure. Metropolitan centers, including Dubai and Abu Dhabi, face inundation scenarios that could displace millions due to flooding, coastal erosion, and saline contamination of freshwater reserves.

Historically, Gulf economies derived prosperity from hydrocarbon extraction and export, activities contributing substantially to atmospheric greenhouse gas concentrations (Luomi, 2015). As global economic systems pivot toward decarbonization, Gulf nations must accelerate economic diversification while embedding sustainability principles throughout development planning. Recognition of these imperatives has catalyzed policy initiatives across the region (Reiche, 2010), exemplified by the UAE's Vision 2021 and Green Agenda 2030 frameworks (UAE Government, 2023) and Saudi Arabia's Vision 2030 economic transformation program, emphasizing both diversification and ecological responsibility.

Healthcare systems, characterized by intensive resource consumption and substantial energy demands, offer significant opportunities for sustainable transformation. This research addresses critical knowledge gaps regarding how Gulf healthcare sectors can achieve concurrent progress toward environmental objectives, clinical excellence, and economic sustainability.

Conceptual Foundations and Theoretical Framework

Ecological innovation paradigms seek to reconcile economic advancement with planetary stewardship (Organization for Economic Cooperation and Development, 2011), emphasizing efficient resource use, mitigating environmental impacts, and adopting low-carbon technologies. Within healthcare contexts, ecological innovation pathways can generate concurrent economic, environmental, and health dividends (World Health Organization, 2020).

Climate finance encompasses incorporating environmental, social, and governance parameters into capital allocation decisions and financial product design

(United Nations Environment Programme Finance Initiative, 2019), channeling investments toward sustainable economic activities while promoting enduring value creation (Schoenmaker, 2020). Climate finance mechanisms prove essential for healthcare sustainability transitions (Dasgupta, 2021), enabling deployment of efficiency technologies, renewable power systems, and resilient infrastructure through targeted funding and economic incentives (Taghizadeh-Hesary et al., 2020).

Healthcare facilities demonstrate intensive energy profiles contributing to atmospheric emissions (Alshqaqeeq et al., 2020). The sector generates substantial hazardous and non-hazardous waste streams, presenting environmental and health risks (Janik-Karpinska et al., 2023). Facility construction and operation generate considerable ecological footprints (Guenther & Vittori, 2013), while climate-responsive design principles can ameliorate impacts and enhance therapeutic environments (Karimi, 2023).

Research Methodology

This investigation employs integrated mixed-methods encompassing systematic literature synthesis, comparative case analysis, and framework construction. A comprehensive examination of the scholarly literature on sustainable healthcare operations, climate finance instruments, and regional adaptation strategies specific to Gulf contexts formed the research foundation. International exemplars demonstrating successful sustainability implementations were subjected to detailed case examination.

Framework development entailed stakeholder identification, mapping intervention points, and integrated solution design that addressed multiple sustainability dimensions concurrently. Economic feasibility assessment used financial modeling to evaluate the viability of proposed interventions, while environmental impact analysis quantified potential reductions in carbon emissions, waste generation, and resource consumption.

Analytical Findings and Framework Architecture

Ecological Innovation Opportunities in Gulf Healthcare

Analysis identifies four principal domains for ecological innovation implementation within Gulf healthcare systems:

Resource Optimization and Renewable Power Integration

Healthcare institutions can substantially reduce energy consumption through systematic resource audits that establish baseline consumption metrics and identify optimization opportunities (Liu et al., 2021). Strategic approaches encompass transitioning to energy-optimized equipment, including LED illumination and contemporary climate control systems; architectural envelope enhancement incorporating superior insulation and high-performance fenestration; deploying sophisticated energy management platforms enabling real-time consumption monitoring; and engaging institutional personnel and patients in conservation behaviors.

Renewable power integration options include on-site generation through photovoltaic arrays or wind installations (Al-Shetwi, 2022), direct renewable energy procurement from utility providers, or participation in renewable energy credit programs (Sherman et al., 2020). Combined strategies yield multiple benefits, including diminished environmental impacts, reduced operational expenditures, and enhanced facility efficiency.

Circular Economy Waste Systems

The healthcare sector generates substantial volumes of hazardous and non-hazardous waste, posing significant environmental and public health challenges (Janik-Karpinska et al., 2023). Enhanced surveillance of Bio-Medical Waste Incinerators through exposure assessment and risk evaluation studies has become imperative (Prathish, 2023). Comprehensive waste stewardship depends on source reduction through product reuse and material recycling, rigorous segregation protocols that differentiate hazardous materials, including infectious, radioactive, and chemical wastes, from non-hazardous streams, robust recycling infrastructure, and specialized treatment technologies such as autoclaving, incineration, or chemical neutralization for hazardous materials (Mahjoob et al., 2021).

Climate-Responsive Infrastructure Development

Healthcare facility construction and operation generate substantial environmental footprints (Guenther & Vittori, 2013). Climate-responsive design principles mitigate impacts while promoting healing-conducive spaces (Karimi, 2023). Sustainable infrastructure design incorporates passive environmental strategies that leverage optimal solar orientation and natural ventilation; water conservation through efficient fixtures and capture systems; material selection that emphasizes local, recycled, or renewable sources; vegetated roofing and ecological landscaping for thermal regulation; and on-site renewable generation systems. Healthcare organizations implementing these principles create resource-efficient, climate-adapted facilities that support optimal patient and staff well-being (Veleva et al., 2015).

Sustainable Service Model Development

Healthcare sectors can advance environmental stewardship through sustainable product adoption, including eco-designed medical devices, environmentally preferable cleaning agents, and telehealth services that reduce resource consumption and emissions (Yao et al., 2023). Facilities can reduce environmental footprints through sustainable procurement, prioritizing certified eco-products; digital health platforms that improve access while minimizing transportation impacts; sustainable transport initiatives that promote public transit and electric vehicle infrastructure; and sustainable food services that incorporate plant-based options and waste-reduction protocols.

Climate Finance Framework Architecture

The climate finance framework operates through three integrated pathways: financial instruments (sustainability bonds, performance-linked financing, and ESG investment criteria), environmental outcomes (funding for efficiency technologies, renewable power projects, and resilient infrastructure), and operational deployment (resource allocation, performance incentives, and workforce development investments).

Rigorous environmental standards prove essential for advancing sustainable healthcare practices (Malik et al., 2020). Governments can deploy economic incentives and fiscal advantages to promote the adoption of sustainable finance among providers and investors (Dasgupta, 2021). Public–private collaborations facilitate the mobilization of resources and expertise for sustainability initiatives (Haran et al., 2020).

International Best Practice Analysis

International exemplar examination reveals effective implementation strategies. Cleveland Clinic deployed comprehensive efficiency programs that incorporated LED systems, optimized HVAC systems, and advanced management platforms, achieving 20% energy reduction and millions in annual cost savings, demonstrating environmental and economic viability (Cleveland Clinic, 2021). Melbourne Health instituted comprehensive waste stewardship, reducing disposable plastic use; implemented therapeutic green spaces; and deployed efficient lighting, yielding reduced waste generation, enhanced patient satisfaction, and decreased energy consumption (Melbourne Health, 2023).

These examples demonstrate that sustainability initiatives can achieve concurrent environmental and financial benefits, providing adaptation models for Gulf healthcare systems.

Implementation Barriers and Enablers

Implementation confronts multiple significant barriers. Economic constraints include substantial capital requirements, limited funding availability, elevated cost projections, and uncertain return horizons, which may deter sustainable practice adoption (Dasgupta, 2021). Technical obstacles encompass workforce expertise deficits, inadequate technological infrastructure, and limited availability of sustainable products (Malik et al., 2020). Socio-cultural factors, including change resistance, information gaps, and competing priorities, may impede implementation (Delmas & Toffel, 2008).

However, ecological innovation and climate finance adoption generate economic advantages for providers, including cost reductions through resource efficiency, enhanced competitive positioning, and access to novel markets and financing channels (Taghizadeh-Hesary et al., 2020). Ecological initiatives contribute to reductions in emissions, waste minimization, and healthier environments, ultimately improving population health outcomes (World Health Organization, 2020).

Strategic Implementation Recommendations

Strategic recommendations target multiple stakeholder groups.

Policy Leadership Recommendations

Governmental authorities should formulate integrated national sustainability strategies for healthcare sectors (Malik et al., 2020), institute comprehensive regulatory frameworks establishing environmental performance standards (Dasgupta, 2021), deploy economic incentives including capital grants, operational subsidies, and tax advantages supporting ecological initiatives (Taghizadeh-Hesary et al., 2020), and facilitate public–private collaborations mobilizing resources and expertise for sustainability programs (Haran et al., 2020).

Healthcare Organization Recommendations

Healthcare institutions should conduct comprehensive resource audits, establishing energy management systems to enhance efficiency (Liu et al., 2021), deploy circular economy waste protocols incorporating minimization, segregation, and recovery (Mahjoob et al., 2021), adopt climate-responsive architectural principles and sustainable materials in facility development (Karimi, 2023), and champion environmentally-preferable products and services, including sustainable medical technologies and digital health platforms (Yao et al., 2023).

Financial Sector Recommendations

Financial institutions should create innovative financial instruments, including sustainability bonds, performance-linked loans, and ESG-focused investment vehicles (Dasgupta, 2021); integrate environmental, social, and governance parameters into investment analysis and risk management systems; and provide preferential financing terms for sustainable healthcare ventures (Taghizadeh-Hesary et al., 2020).

Synthesis and Future Directions

This research demonstrates the intersection of Gulf healthcare sectors with ecological innovation, climate finance, and global shock resilience, revealing substantial opportunities to reduce environmental impact while generating economic value within healthcare ecosystems. Analysis confirms significant potential for ecological advancement through resource optimization, renewable power integration, circular economy systems, and climate-responsive construction.

Climate finance findings emphasize the critical role of innovative funding mechanisms in healthcare sector transformation. Establishing sustainability bonds, performance-linked loans, and ESG investment frameworks proves essential for financing transitions toward sustainable healthcare infrastructure and operations.

Gulf healthcare sector transformation represents a critical convergence of ecological innovation principles, climate finance practices, and resilience development. Embracing sustainable practices enables concurrent achievement of environmental protection, economic prosperity, and enhanced global shock response capacity while maintaining essential public health and societal well-being functions.

Future research should emphasize developing region-specific sustainability strategies, evaluating the economic benefits of environmental initiatives, exploring innovative funding mechanisms, and investigating the integration of sustainability principles into emergency preparedness planning. This framework provides Gulf healthcare systems a roadmap for achieving environmental stewardship while maintaining service excellence and economic viability, contributing to comprehensive regional sustainability objectives and climate resilience.

References

Al-Shetwi, A. Q. (2022). Sustainable development of renewable energy integrated power sector: Trends, environmental impacts, and recent challenges. *Science of The Total Environment, 822*, Article 153645.

Alshqaqeeq, F., Esmaeili, M. A., Overcash, M., & Twomey, J. (2020). Quantifying hospital services by carbon footprint: A systematic literature review of patient care alternatives. *Resources, Conservation and Recycling, 154*, Article 104560.

Cleveland Clinic. (2021). *Environment: Sustainability & going green*. Retrieved from https://my.clevelandclinic.org/about/community/sustainability/sustainability-global-citizenship/environment

Dasgupta, P. (2021). *The economics of biodiversity: The Dasgupta review*. HM Treasury.

Delmas, M. A., & Toffel, M. W. (2008). Organizational responses to environmental demands: Opening the black box. *Strategic Management Journal, 29*(10), 1027–1055.

Guenther, R., & Vittori, G. (2013). *Sustainable healthcare architecture* (2nd ed.). John Wiley & Sons.

Haran, M., Neitzert, T., & Shrivastava, A. (2020). Public-private partnerships for sustainable development in the healthcare sector. *World Health Organization.*

Intergovernmental Panel on Climate Change. (2021). Climate change 2021: The physical science basis. Retrieved from https://www.ipcc.ch/report/ar6/wg1/

Janik-Karpinska, E., Brancaleoni, R., Niemcewicz, M., Wojtas, W., Foco, M., Podogrocki, M., & Bijak, M. (2023). Healthcare waste—A serious problem for global health. *Healthcare, 11*(2), 242.

Karimi, H., Adibhesami, M. A., Bazazzadeh, H., & Movafagh, S. (2023). Green buildings: Human-centered and energy efficiency optimization strategies. *Energies, 16*(9), 3681.

Liu, J., Jian, L., Wang, W., Qiu, Z., Zhang, J., & Dastbaz, P. (2021). The role of energy storage systems in resilience enhancement of health care centers with critical loads. *Journal of Energy Storage, 33*, Article 102086.

Luomi, M. (2015). *The Gulf Monarchies and Climate Change: Abu Dhabi and Qatar in an era of natural unsustainability*. Oxford University Press.

Mahjoob, A., Alfadhli, Y., & Omachonu, V. (2021). Healthcare waste and sustainability: Implications for a circular economy. *Sustainability, 15*(10), 7788.

Malik, N., Shrivastava, A., Neitzert, T., & DiPrete, L. (2020). Sustainable healthcare in the WHO European Region: A focus on environmental sustainability in healthcare. *World Health Organization.*

Melbourne Health. (2023). *Sustainability*. Retrieved from https://vahi.vic.gov.au/hospital-and-health-services/melbourne-health

Organization for Economic Cooperation and Development. (2011). *Towards green growth: A summary for policymakers*. Retrieved from https://www.oecd.org/greengrowth/48012345.pdf

Pal, J. S., & Eltahir, E. A. (2016). Future temperature in southwest Asia projected to exceed a threshold for human adaptability. *Nature Climate Change, 6*(2), 197–200.

Prathish, K. P. (2023). Dioxins emissions from bio-medical waste incineration: A systematic review on emission factors, inventories, trends and health risk studies. *Journal of Hazardous Materials*, 133384.

Reiche, D. (2010). Renewable energy policies in the Gulf countries: A case study of the carbon-neutral, renewable energy based strategy of Masdar City in Abu Dhabi. *Energy Policy, 38*(1), 378–382.

Schoenmaker, D. (2020). The impact economy: Balancing profit and impact. *Working Paper, 4.*

Sherman, J. D., Thiel, C., MacNeill, A., Eckelman, M. J., Dubrow, R., Hopf, H., & Bilec, M. M. (2020). The green print: Advancement of environmental sustainability in healthcare. *Resources, Conservation and Recycling, 161*, Article 104882.

Taghizadeh-Hesary, F., & Yoshino, N. (2020). Sustainable solutions for green financing and investment in renewable energy projects. *Energies, 13*(4), 788.

UAE Government. (2023). *UAE Green Agenda 2030*. Retrieved from https://u.ae/en/about-the-uae/strategies-initiatives-and-awards/strategies-plans-and-visions/environment-and-energy/the-uaes-green-agenda-2030

United Nations Environment Programme Finance Initiative. (2019). *Industry-based principles*. Retrieved from https://www.unepfi.org/about/about-us/

Veleva, V., Todorova, S., Lowitt, P., Angus, N., & Neely, D. (2015). Understanding and addressing business needs and sustainability challenges: Lessons from Devens eco-industrial park. *Journal of Cleaner Production, 87*, 375–384.

Waha, K., Krummenauer, L., Adams, S., Aich, V., Baarsch, F., Coumou, D., & Schleussner, C. F. (2017). Climate change impacts in the Middle East and Northern Africa (MENA) region and their implications for vulnerable population groups. *Regional Environmental Change, 17*(6), 1623–1638.

World Health Organization. (2020). *WHO global strategy on health, environment and climate change: The transformation needed to improve lives and well-being sustainably through healthy environments*. World Health Organization.

Yao, S., Pan, Y., Wang, L., Sensoy, A., & Cheng, F. (2023). Building eco-friendly corporations: The role of minority shareholders. *Journal of Business Ethics, 182*(4), 933–966.

Chapter 64
The Role of E-Banking in Promoting Sustainable Finance: A Study on ESG Reporting and Green Banking Practices in the UAE

Rahul Jain and Piyush Kumar Jain

Abstract This study examines the role of electronic banking platforms in advancing sustainable finance in the United Arab Emirates, with a specific focus on Environmental, Social, and Governance (ESG) reporting and green banking practices. Using secondary data and case studies from leading UAE banks, the research explores how digital banking systems embed sustainability principles. Findings highlight that e-banking contributes to sustainable finance by reducing paper usage, enhancing ESG disclosure transparency, and improving access to green financial products. However, the extent of ESG integration varies across banks, with challenges including regulatory gaps, infrastructure limitations, and low public awareness. The study contributes to the growing literature on sustainable finance in emerging markets and offers actionable recommendations for policymakers and financial institutions. It aligns with the UAE Vision 2030 and underscores the country's potential to lead in digital-enabled sustainable banking, combining technological innovation with environmental and social responsibility.

Keywords E-Banking · Sustainable finance · ESG reporting · Green banking · Digital transformation

Introduction

A. Background on the Global Shift Toward Sustainable Finance and ESG Considerations

R. Jain (✉)
De Montfort University, Dubai, UAE
e-mail: rahul.jain@dmu.ac.uk

P. K. Jain
Christ University, Bangalore, India

J. Dsilva (ed.), *Sustainability Education, Youth and Innovation*,
https://doi.org/10.1007/978-3-032-21948-0_64

There has been a global shift toward sustainable finance, emphasising integrating environmental, social, and governance (ESG) considerations into financial decision-making. This transition stems from a broader understanding that the financial system is critical in addressing global challenges like climate change, social equity, and governance transparency. Financial institutions now go beyond profit-maximisation to support long-term sustainability and ethical investment practices (IMD Business School, 2025).

B. Brief Overview of the UAE's Banking Sector and Digital Transformation Journey

The United Arab Emirates (UAE) has the largest banking sector in the Middle East, with assets totalling AED 4.457 trillion as of November 2024 (Central Bank of the UAE, 2025). In recent years, the UAE's banking industry has undergone a significant digital transformation, led by massive investments in technology. Emirates NBD, for instance, committed AED 1 billion in 2017 to enhance its digital capabilities, positioning the UAE as a digital banking leader in the region (Ahmed, 2026).

C. Research Problem: How E-Banking Platforms Can Promote Sustainable Finance Initiatives

E-banking offers an opportunity to support sustainable finance through reduced paper use, fewer physical branches, and lower emissions. However, it remains critical to explore how these platforms can actively promote ESG integration and green financial products to align with the UAE's broader sustainability objectives (Makhdoomi, 2025).

D. Significance of the Study Within the UAE's Sustainable Development Framework

The study aligns with the UAE's sustainability goals, particularly in green finance and digital innovation, supporting the nation's economic and environmental agenda (Raza et al., 2024).

Literature Review

A. Sustainable Finance and ESG Frameworks

Sustainable finance in the UAE is gaining traction as banks align with national goals such as UAE Vision 2030 and Net Zero 2050. ESG frameworks guide this transition, with ESG criteria, environmental (carbon footprint, waste management), social (diversity, labour standards), and governance (leadership, transparency), used to assess corporate sustainability (Eccles & Mirchandani, 2022). UAE banks are increasingly adopting global ESG reporting standards like the GRI, SASB, and TCFD to meet regulatory demands and investor expectations (Cardoni et al., 2019; Berg et al., 2021; Lykkesfeldt & Kjaergaard, 2022). These frameworks enhance transparency and drive responsible investment (Bril et al., 2022).

B. Green Banking Practices

Green banking practices in the UAE encompass operational sustainability and financial products. Banks like Emirates NBD and Dubai Islamic Bank offer green loans, ESG-linked instruments, and sukuks (Nath et al., 2014; Weber & Feltmate, 2016). Studies show structured green finance products with impact metrics attract higher investor confidence (Flammer, 2021; Loumioti et al., 2022). Digitalisation has amplified this impact. Paperless banking, AI tools, and blockchain adoption have significantly reduced carbon emissions and improved service efficiency (Amel-Zadeh & Serafeim, 2018; Wang & Xu, 2025).

C. E-Banking and Digital Transformation in the UAE

UAE banks have rapidly digitised, with over 70% of residents using mobile banking (Ryalat, 2017; Alkhowaiter, 2020). Emirates NBD's AI chatbot "Eva" and blockchain-based solutions have streamlined operations and sustainability-linked services (Todupunuri, 2024; Arjun & Suprabha, 2020). Fintech adoption has also improved financial inclusion and ESG reporting through digital platforms (Alshater et al., 2022; Mohammed, 2024). Regulatory support from the Central Bank of UAE and sandbox environments in DIFC and ADGM have further enabled sustainable digital banking innovation (Elsayed et al., 2024; Principles for Sustainability-Related Disclosures, 2024).

Research Methodology

A. Research Design

This study adopts a qualitative case study approach to examine how e-banking platforms support sustainable finance in the UAE banking sector. This method contextualises ESG integration in digital banking (Creswell & Poth, 2017; Yin, 2017). Banks were selected based on their innovation in e-banking, ESG performance, and sustainable finance activities.

B. Data Sources

Secondary data were gathered from annual and sustainability reports of UAE banks (NBD, n.d.), Central Bank directives, academic literature, industry publications (Kumar, 2025), and ESG rating agency databases to evaluate ESG disclosure and digital integration.

Case Studies of E-Banking and Sustainable Finance in the UAE

A. Case Study 1: Emirates NBD – A Leader in Sustainable E-Banking

Emirates NBD, one of the UAE's largest banking institutions, has pioneered the integration of digital transformation with sustainability. In alignment with UAE Vision 2030, the bank has adopted a robust e-banking strategy to reduce environmental impact and enhance financial inclusion (Jain, 2024). Its digital services emphasise paperless banking, green digital payments, and energy-efficient IT infrastructure. Emirates NBD has introduced several sustainability-linked financial products, such as Green Auto Loans for electric and hybrid vehicles, Sustainable Deposit Accounts, and Green Sukuk financing for environmentally aligned projects (Emirates NBD Group, 2024).

Moreover, the bank's digital platforms support ESG transparency by offering interactive tools for customers to monitor green investments and access sustainability disclosures. These efforts have strengthened customer trust, boosted the adoption of green financial products, and positioned the bank as a leader in responsible digital banking. Emirates NBD's commitment is further evident in its strategic reporting and stakeholder engagement on ESG goals, reflecting the growing consumer demand for ethical and sustainable banking.

B. Case Study 2: Liv. by Emirates NBD – A Digital-Native Sustainable Bank

Liv., the digital-only subsidiary of Emirates NBD, operates without physical branches, embodying a low-carbon banking model from inception. With an innovation-driven approach, Liv. Offers AI-powered personal finance tools, real-time carbon footprint tracking, and entirely paperless transactions (Aloulou et al., 2023). Its platform gamifies sustainability, incentivising users to adopt eco-friendly behaviours such as rewarding green purchases or energy-efficient spending habits.

Liv.'s digital sustainability model has resulted in high engagement among younger, environmentally conscious users. The platform simplifies banking and promotes green awareness, creating a replicable model for sustainable digital banking. Its success demonstrates how banks can combine financial innovation with environmental responsibility to meet evolving customer and regulatory expectations.

These two case studies, one from a conventional leader and one from a digital-native entity, highlight the UAE banking sector's dual-path approach to embedding ESG values through digital platforms. Emirates NBD leverages its scale and infrastructure, while Liv. Exemplifies how design-led digital innovation can drive sustainability in banking.

Discussion

E-banking platforms have emerged as significant enablers of sustainable finance in the UAE by facilitating access to eco-friendly financial products and improving operational efficiency. Through paperless processes and reduced reliance on physical branches, banks like Mbank and Emirates NBD have successfully lowered their carbon footprint (ElAlfy et al., 2025; Jain, 2024). Digital platforms now offer users simplified access to green loans, green sukuk, and sustainability-linked investment products, aligning closely with the UAE's national goals ("UAE Banks Pioneer Green Finance Revolution," 2024).

These platforms also promote ESG transparency and accountability. Banks have improved stakeholder trust and compliance by enabling digital dissemination of sustainability disclosures. The Central Bank of the UAE's introduction of principles for sustainability-related disclosures further solidifies ESG integration into the financial system (Central Bank of the UAE, 2025).

However, the transition is not without challenges. Legacy IT systems hamper integration with modern technologies, slowing innovation. Regulatory gaps persist in fintech regulation and cross-border transactions despite initiatives like the Open Finance Regulation (Alen & Schoorl, 2024). Additionally, customer hesitance due to low digital literacy and concerns over privacy and data security continue to be barriers to widespread adoption (S. Wang et al., 2024).

Nonetheless, opportunities abound. AI and blockchain are increasingly explored to strengthen climate-related disclosures, risk assessments, and transaction transparency (Fast Company Middle East, 2023; BIS, 2023). Mastercard's Carbon Calculator is an example of how technology can personalise green awareness (EEMEA | Mastercard Newsroom, 2024). Collaborative platforms like the Dubai Sustainable Finance Working Group and financial zones like DIFC and ADGM position the UAE as a regional leader in sustainable digital banking (Dubai Sustainable Finance Working Group | DIFC, n.d.; KPMG Lower Gulf Limited, 2024).

Recommendations

A. For Banking Institutions

1. Strengthen ESG Integration in Digital Platforms: Banks should embed ESG analytics, carbon tracking, and green product access into mobile and online platforms. AI and blockchain can enhance ESG transparency, helping align consumer behaviour with sustainability goals (Kumar & Prakash, 2022).
2. Design Market-Fit Sustainable Products: Banks should offer tailored green financial products like Islamic green financing, sustainability-linked loans, and ESG-aligned funds to support the UAE Green Agenda 2030 and Net Zero by 2050 goals (Ahmed, 2026).
3. Enhance ESG Disclosures via Digital Channels: Digital dashboards, blockchain-secured ESG reporting, and AI-enabled data aggregation should be used to improve transparency and trust in sustainability performance (Liu et al., 2024).

B. For Regulators and Policymakers

1. Strengthen Policy Frameworks: Regulators such as CBUAE and SCA should enforce ESG disclosures, climate stress testing, and minimum sustainability quotas, aligned with UAE Vision 2030 (Ahmed, 2026).

2. Create Incentives for Green Innovation: Incentivise banks with tax breaks, capital relief for green assets, and grants for digital ESG innovations (Belgacem et al., 2023). Support fintech-bank partnerships through green finance hubs.
3. Standardise ESG Reporting Requirements: Adopt a UAE-wide ESG disclosure framework like the EU SFDR to ensure comparability and improve stakeholder trust (Valentinetti & Rea, 2024).
4. Promote Cross-Sector Collaboration: Support joint innovation by banks, fintechs, and regulators through sustainability working groups and AI/ blockchain solutions for ESG monitoring (Zhao, 2024).

C. For Future Research

More research is needed on consumer behaviour, ESG investment impacts, and emerging technologies in green finance (Almasria et al., 2024). Mixed-methods and longitudinal studies will help track long-term progress (Epor & Akande, 2025; Ahmed, 2026).

Conclusion

This study shows that e-banking is critical in advancing sustainable finance in the UAE by enabling green financial services, improving ESG transparency, and reducing banking's environmental footprint, as seen in the Emirates NBD and Liv case studies. Illustrate successful integration of digitalisation and sustainability, supported by regulatory push and consumer uptake.

Theoretically, this research contributes to stakeholder and institutional theory, highlighting how digital platforms shape financial ESG behaviour (Kim & Yang, 2026). It offers banks and regulators actionable strategies for sustainable innovation, disclosure, and product development.

While challenges remain in technology adoption, regulation, and consumer trust, the UAE's digital-first banking environment and sustainability agenda create an opportunity for global leadership in green digital finance. Continued collaboration and innovation are key to ensuring long-term, scalable impact.

References

Ahmed, I. E. (2026). ESG disclosure quality and banking risk: A dynamic panel analysis of Middle East and African banks. *Risks, 14*(3), 50. https://doi.org/10.3390/risks14030050

Alen, P., & Schoorl, A. (2024, April 23). New fintech regulations in the United Arab Emirates: Open Finance Regulation | DLA Piper. Retrieved March 20, 2025, from https://www.dlapiper.com/es-pr/insights/publications/2024/04/new-fintech-regulations-in-the-united-arab-emirates-open-finance-regulation

Alkhowaiter, W. A. (2020). Digital payment and banking adoption research in Gulf countries: A systematic literature review. *International Journal of Information Management, 53*, Article 102102. https://doi.org/10.1016/j.ijinfomgt.2020.102102

Almasria, N. A., Alhatabat, Z., Ershaid, D., Ibrahim, A., & Ahmed, S. (2024). The mediating impact of organizational innovation on the relationship between fintech innovations and sustainability performance. *Sustainability, 16*(22), 10044. https://doi.org/10.3390/su162210044

Aloulou, M., Grati, R., Al-Qudah, A. A., & Al-Okaily, M. (2023). Does FinTech adoption increases the diffusion rate of digital financial inclusion. A study of the banking industry sector. *Journal of Financial Reporting & Accounting, 22*(2), 289–307. https://doi.org/10.1108/jfra-05-2023-0224

Alshater, M. M., Saba, I., Supriani, I., & Rabbani, M. R. (2022). Fintech in Islamic finance literature: A review. *Heliyon, 8*(9), Article e10385. https://doi.org/10.1016/j.heliyon.2022.e10385

Amel-Zadeh, A., & Serafeim, G. (2018). Why and how investors use ESG information: Evidence from a global survey. *Financial Analysts Journal, 74*(3), 87–103. https://doi.org/10.2469/faj.v74.n3.2

Arjun, R., & Suprabha, K. R. (2020). Innovation and challenges of blockchain in banking: a scientometric view. *International Journal of Interactive Multimedia and Artificial Intelligence, 6*(3), 7. https://doi.org/10.9781/ijimai.2020.03.004

Belgacem, S. B., Khatoon, G., & Alzuman, A. (2023). Role of renewable energy and financial innovation in environmental protection: Empirical evidence from UAE and Saudi Arabia. *Sustainability, 15*(11), 8684. https://doi.org/10.3390/su15118684

Berg, F., Kölbel, J. F., Pavlova, A., & Rigobon, R. (2021). ESG confusion and stock returns: tackling the problem of noise. *SSRN Electronic Journal*. https://doi.org/10.2139/ssrn.3941514

BIS. (2023, September 11). *BIS, COP28 and Central Bank of United Arab Emirates launch Tech-Sprint to develop technological solutions for sustainable finance and combating climate change*. Retrieved March 20, 2025, from https://www.bis.org/press/p230907.htm

Bril, H., Kell, G., & Rasche, A. (2022). Sustainable finance and technology. In *Routledge eBooks* (pp. 1–16). https://doi.org/10.4324/9781003262039-1

Cardoni, A., Kiseleva, E., & Terzani, S. (2019). Evaluating the intra-industry comparability of sustainability reports: the case of the oil and gas industry. *Sustainability, 11*(4), 1093. https://doi.org/10.3390/su11041093

Central Bank of the UAE. (2025). *CBUAE's 2024 achievements drive momentum towards a bright future for the financial sector* (pp. 1–5) [Press-release]. https://www.centralbank.ae/media/uoafunzx/cbuaes-2024-achievements-drive-momentum-towards-a-bright-future-for-the-financial-sector-en.pdf

Creswell, J. W., & Poth, C. N. (2017). *Qualitative inquiry and research design: Choosing Among Five Approaches*. Sage Publications.

Dubai Sustainable Finance Working Group | DIFC. (n.d.). Retrieved March 20, 2025, from https://www.difc.com/ecosystem/sustainability/dubai-sustainable-finance-working-group

ElAlfy, A., Elgharbawy, A., Driver, T. R., & Ibrahim, A. (2025). Sustainability disclosure in the Gulf Cooperation Council (GCC) countries: Opportunities and challenges. *Green Finance, 7*(1), 40–82. https://doi.org/10.3934/gf.2025003

Eccles, R. G., & Mirchandani, B. (2022, February 15). We need universal ESG accounting standards. Harvard Business Review. https://hbr.org/2022/02/we-need-universal-esg-accounting-standards

EEMEA | Mastercard Newsroom. (2024, November 2). Retrieved March 20, 2025, from https://www.mastercard.com/news/eemea/en/newsroom/press-releases/en/2021/june/mastercard-unveils-new-carbon-calculator-tool-for-banks-in-the-uae-and-the-region

Elsayed, A. H., Guedira, I., Alghussein, T., Almheiri, H., Alomari, M., & Elmassri, M. (2024). The impact of fintech technology on the financial stability of the UAE. *Heliyon, 10*(19), Article e38255. https://doi.org/10.1016/j.heliyon.2024.e38255

Emirates NBD Group. (2024). The Environmental, Social and Governance Report 2024 builds on today's success. In *Emirates NBD Group ESG Report 2024*. https://cdn.emiratesnbd.com/assets/pdf/esg_report_2024.pdf?utm_pergunta=captura

Epor, S. O., & Akande, J. O. (2025). Sustainable development in emerging economies: Comparing the impacts of green finance and financial inclusion. *Modern Finance, 3*(4), 1–21. https://doi.org/10.61351/mf.v3i4.259

Fast Company Middle East. (2023, September 12). *UAE taps AI and blockchain to grow green finance sector - Fast Company Middle East | The future of tech, business and innovation.* Fast Company Middle East | The Future of Tech, Business and Innovation. Retrieved March 20, 2025, from https://fastcompanyme.com/news/uae-taps-ai-and-blockchain-to-grow-green-finance-sector

Flammer, C. (2021). Corporate green bonds. *Journal of Financial Economics, 142*(2), 499– 516. https://doi.org/10.1016/j.jfineco.2021.01.010

IMD Business School. (2025, January 16). *The power of ESG in driving sustainable finance practices.* IMD Business School for Management and Leadership Courses. https://www.imd.org/blog/sustainability/sustainable-finance/#:~:text=As%20we've%20seen%2C%20ESG,to%20a%20more%20sustainable%20future.

Jain, R. (2024). Towards a green revolution: sustainable integration of Industry 6.0 technologies and smart banking services in the UAE. In *Sustainable Innovation for Industry 6.0* (1st ed., p. 31). IGI Global. https://www.igi-global.com/chapter/towards-a-green-revolution/354503

Kim, J., & Yang, W. (2026). ESG performance in the regulatory transformation era: A systematic thematic review (2020–2024). *Frontiers in Sustainability, 6*. https://doi.org/10.3389/frsus.2025.1680398

KPMG Lower Gulf Limited. (2024). *The role of UAE banks in the climate transition.* https://assets.kpmg.com/content/dam/kpmg/ae/pdf-2024/04/eco-financing-the-future.pdf

Kumar, A. (2025, February 4). COP28 in UAE: How banks are spearheading sustainability drive with $51 billion in green financing. *Khaleej Times.* https://www.khaleejtimes.com/uae/cop28-in-uae-how-banks-are-spearheading-sustainability-drive-with-51-billion-in-green-financing

Loumioti, M., Serafeim, G., The University of Texas at Dallas, & Harvard Business School. (2022). *The issuance and design of sustainability-linked loans* (Working Paper 23–027). https://www.hbs.edu/ris/Publication%20Files/23-027_4b09d278-4051-468e-a5d9-eb0e7c50c25d.pdf

Liu, X., Liang, W., Fu, Y., & Huang, G. Q. (2024). Dual environmental, social, and governance (ESG) index for corporate sustainability assessment using blockchain technology. *Sustainability, 16*(10), 4272. https://doi.org/10.3390/su16104272

Lykkesfeldt, P., & Kjaergaard, L. L. (2022). Encompassing ESG rating agencies. In *Springer eBooks* (pp. 305–311). https://doi.org/10.1007/978-3-031-05800-4_39

Makhdoomi, U. (2025, February 10). *Open Finance, big opportunities: The UAE's blueprint for digital banking.* Knightsbridge®. https://knightsbridge.ae/open-finance-big-opportunities-the-uaes-blueprint-for-digital-banking/#:~:text=Digital%20banking%20aligns%20closely%20with,services%20to%20previously%20unbanked%20populations.

Mohammed, B. I. (2024). E-Banking in the United Arab Emirates: understanding the innovation and the impact of its adoption. *Zenodo.* https://doi.org/10.5281/zenodo.4250035

Nath, V., Nayak, N., & Goel, A. (2014, April 15). *Green Banking Practices – a review.* https://papers.ssrn.com/sol3/papers.cfm?abstract_id=2425108

Principles for Sustainability-Related Disclosures | CBUAE Rulebook. (n.d.). Retrieved March 20, 2025, from https://rulebook.centralbank.ae/en/rulebook/principles-sustainability-related-disclosures

Raza, A., Alavi, A. B., & Asif, L. (2024). Sustainability and financial performance in the banking industry of the United Arab Emirates. *Discover Sustainability, 5*(1). https://doi.org/10.1007/s43621-024-00414-z

Ryalat, H. A. (2017). Perspectives relative to Mobile-Banking adoption in the United Arab Emirates. *American Journal of Information Systems, 5*(1), 13–20. https://doi.org/10.12691/ajis-5-1-2

Todupunuri, A. (2024, October 10). *The Future of Conversational AI in Banking: A Case Study on Virtual Assistants and Chatbots: Exploring the impact of AI-Powered Virtual Assistants on Customer service efficiency and satisfaction.*

UAE banks pioneer green finance revolution. (2024, April 29). *KPMG.* https://kpmg.com/ae/en/home/media/press-releases/2024/04/uae-banks-pioneer-green-finance-revolution.html

Valentinetti, D., & Rea, M. A. (2024). Factors influencing the digitalization of sustainability accounting, reporting and disclosure: A systematic literature review. *Meditari Accountancy Research, 33*(2), 633–680. https://doi.org/10.1108/medar-02-2024-2385

Wang, P., & Xu, X. (2025). Green finance and energy efficiency improvement: the role of green innovation and industrial upgrading. *Innovation and Green Development, 4*(1), Article 100200. https://doi.org/10.1016/j.igd.2024.100200

Wang, S., Asif, M., Shahzad, M. F., & Ashfaq, M. (2024). Data privacy and cybersecurity challenges in the digital transformation of the banking sector. *Computers & Security, 147*, Article 104051. https://doi.org/10.1016/j.cose.2024.104051

Weber, O., & Feltmate, B. (2016). SUSTAINABLE BANKING. Managing the social and environmental impact of financial institutions. In *University of Toronto Press eBooks* (pp. 1–2). https://doi.org/10.3138/9781442629325-002

Yin, R. K. (2017). *Case study research and applications: Design and Methods.* SAGE Publications.

Zhao, Y. (2024). Empowering sustainable finance: The convergence of AI, blockchain, and big data analytics. *Advances in Economics, Management, and Political Sciences, 85*(1), 267–273. https://doi.org/10.54254/2754-1169/85/20240925

Part XI
Sustainable Technologies and Innovation

Chapter 65
AI and Blockchain for Workflow Automation in UAE and Saudi Arabia: The Role of Cardano Blockchain

Mohamed Irfan Shaikh, Jacinta Dsilva, and Sophia Lahlou

Abstract Across the UAE and Saudi Arabia, many industries still rely on traditional ERP systems and manual processes that are slow, fragmented, and poorly aligned with the region's ambitious digital transformation goals. This paper investigates how the integration of Artificial Intelligence (AI) and blockchain, especially using the environmentally responsible Cardano platform, can streamline complex workflows across key sectors such as healthcare, logistics, energy, and finance. Rather than treating AI and blockchain as separate tools, we propose a layered approach that combines AI's predictive power with blockchain's secure execution capabilities. Through case studies and strategic insights, we explore how automated decision-making, and smart contracts can replace bureaucratic bottlenecks with real-time, transparent systems. We also align our findings with national visions like Saudi Arabia's Vision 2030 and the UAE's Emirates Blockchain Strategy, both of which prioritize sustainable, tech-driven growth. In essence, this paper argues that AI and blockchain are more than digital tools, they are foundational technologies that can rewire how governments and enterprises operate, offering greater efficiency, accountability, and resilience in a rapidly evolving global economy.

Keywords Artificial intelligence · Blockchain · Environmentally responsible · Cardano foundation · Vision 2030 · Saudi Arabia · UAE

M. I. Shaikh (✉)
SEE Institute, The Sustainable City, Dubai, United Arab Emirates
e-mail: Irfan@seeinstitute.ae

J. Dsilva
Marketing Head, SEE Holding, Dubai, United Arab Emirates

S. Lahlou
CEO & Co-Founder, InLabs Group, Sharjah Media City, Sharjah, United Arab Emirates

J. Dsilva (ed.), *Sustainability Education, Youth and Innovation*,
https://doi.org/10.1007/978-3-032-21948-0_65

Introduction

In today's fast-paced digital economy, traditional manual workflows remain a major obstacle to efficiency in both government and enterprise operations. Paper-based approvals, repetitive administrative tasks, and siloed decision-making systems not only consume valuable time but also increase the risk of human error, fraud, and miscommunication. In the Middle East, where large-scale projects, cross-border trade, and citizen services are expanding at unprecedented rates, such inefficiencies result in delays, high costs, and reduced competitiveness.

To mitigate these issues, many organizations have adopted Enterprise Resource Planning (ERP) systems over the past two decades. While ERP solutions were revolutionary in integrating departments and digitizing records, they now show significant limitations in the era of AI-driven automation:

- Rigid Structures: ERP systems rely on pre-coded workflows that are slow to adapt to rapidly changing regulations, customer expectations, and sustainability mandates.
- High Costs & Complexity: Licensing, customization, and maintenance costs of ERP systems remain prohibitive, especially for SMEs.
- Low Adaptability: A Deloitte report highlights that 65% of organizations struggle to adapt ERP systems to new business models without significant redevelopment.
- Inefficiency Statistics: According to Panorama Consulting (2024), 55% of ERP projects experience operational disruption, and 45% of organizations fail to realize more than half of the expected benefits. McKinsey further notes that manual workflows and legacy ERP reliance reduce productivity by up to 30% in service-heavy economies.

In the context of UAE and Saudi Arabia, where governments envision smart cities, digital governance, and sustainable energy systems, reliance on outdated ERP-driven processes threatens to stall progress. For example, manual customs clearance in Saudi ports or outdated ERP modules in UAE healthcare workflows result in processing delays, increased costs, and lack of transparency.

This is where the seamless integration of AI with blockchain becomes transformative. Unlike rigid ERP structures, AI systems provide predictive intelligence (forecasting demand, detecting anomalies, automating approvals), while blockchain platforms like Cardano ensure secure, transparent, and tamper-proof execution of those workflows. Together, they create self-optimizing, auditable, and cost-efficient systems capable of automating entire government and enterprise processes — from visa issuance to carbon-credit trading.

By shifting from ERP-centric, semi-automated models to AI + Cardano blockchain-enabled workflows, UAE and Saudi Arabia can unlock:

- 30–40% reduction in process inefficiency (PwC, 2023).
- Up to 50% cost savings in compliance and auditing (WEF, 2024).
- Seamless scalability across government, energy, finance, and logistics sectors.

Thus, the chapter positions AI + Blockchain (Cardano) as not just a technological upgrade, but as a new operational paradigm for UAE and Saudi Arabia's Vision 2030 and Vision 2 2. Theoretical Foundations. The integration of Artificial Intelligence (AI) and Blockchain for workflow automation is not merely a technological trend but a theoretical alignment with national AI and digital economy strategies in the Middle East. Both the UAE and Saudi Arabia have articulated visions that emphasize data-driven decision-making, transparency, and sustainability as core principles for their socio-economic transformation.

Artificial Intelligence in Workflow Automation

From a theoretical perspective, automation theory and cognitive systems theory suggest that AI can replace repetitive, rule-based decision-making with intelligent, adaptive systems. This aligns directly with the UAE's Dubai AI Roadmap 2031, which envisions AI as a core driver for public services, and with the Saudi National Strategy for Data & AI (NSDAI), led by the Saudi Data & AI Authority (SDAIA).

Practical applications demonstrate this alignment:

Predictive Analytics: Dubai's RTA uses AI-driven forecasting for traffic flow optimization, supporting systems theory, where AI continuously adapts to changing patterns.

Anomaly Detection: The Saudi Central Bank (SAMA) employs AI for fraud prevention in banking, aligning with decision theory, which emphasizes minimizing uncertainty through data.

Citizen Services: Platforms like Saudi Arabia's Absher leverage AI for automating administrative services, embodying digital governance theory, where citizen engagement is enhanced through smart automation.

Blockchain in Automation

Blockchain theory, rooted in transaction cost economics and distributed trust models, posits that decentralization reduces the need for intermediaries while ensuring transparency. This is reflected in the UAE's Emirates Blockchain Strategy 2021, which aims for 50% of government transactions to be executed on blockchain, and in Saudi Arabia's ongoing pilots with the Capital Market Authority (CMA) and SAMA in areas such as digital payments, eKYC, and cross-border financial transactions.

For both nations, blockchain offers three theoretical advantages:

Auditability Theory: Every transaction leaves an immutable record, ensuring regulatory compliance.

Information Asymmetry Reduction: By creating shared ledgers, blockchain reduces inefficiencies caused by siloed data.

Process Integrity Theory: Automated smart contracts minimize human intervention, ensuring consistent outcomes.

Synergy of AI and Blockchain

The integration of AI and blockchain can be understood through the lens of socio-technical systems theory, which stresses the co-evolution of human, organizational, and technological systems. AI provides the intelligence layer—making predictions, classifying risks, and recommending actions—while blockchain provides the trust layer, ensuring that decisions are executed securely, transparently, and without manipulation.

Together, they enable autonomous, verifiable workflows, a concept aligned with automation and control theory, where systems self-regulate based on predefined rules and adaptive learning.

Example: An AI system in a UAE hospital predicts shortages of critical medicines. A blockchain-based smart contract, hosted on a platform like Cardano, automatically triggers procurement from a certified supplier. This fusion minimizes delays, prevents fraud, and ensures resilience in healthcare workflows.031 strategies.

National Context: UAE and Saudi Arabia

Building on the earlier discussion of how manual workflows and ERP limitations hinder efficiency, and how AI and blockchain provide the theoretical basis for automation, the UAE and Saudi Arabia serve as powerful examples of countries actively translating these theories into practice. Both nations are embedding AI-driven decision-making and blockchain-based transparency into their digital transformation agendas.

1. United Arab Emirates (UAE)

The UAE has positioned itself as a testbed for emerging technologies, underpinned by the UAE Artificial Intelligence Strategy 2031 and the Emirates Blockchain Strategy 2021. These strategies directly tackle the inefficiencies of legacy ERP systems and manual governance processes by reimagining workflows through AI–Blockchain integration (see Fig. 65.1).

Government Services: The DubaiNow app consolidates more than 1,200 automated services across sectors such as licensing, utilities, and payments. What once required multiple ERP modules and siloed manual processes is now streamlined into a single AI-powered, blockchain-secured platform.

Energy: ADNOC is piloting blockchain to automate oil trade settlements, ensuring real-time reconciliation of production, transportation, and sales data. This eliminates delays caused by fragmented ERP reporting and enhances transparency in a sector central to the UAE economy.

Logistics: DP World has implemented the Dubai Trade blockchain-enabled platform, which automates customs clearance, port documentation, and shipping processes.

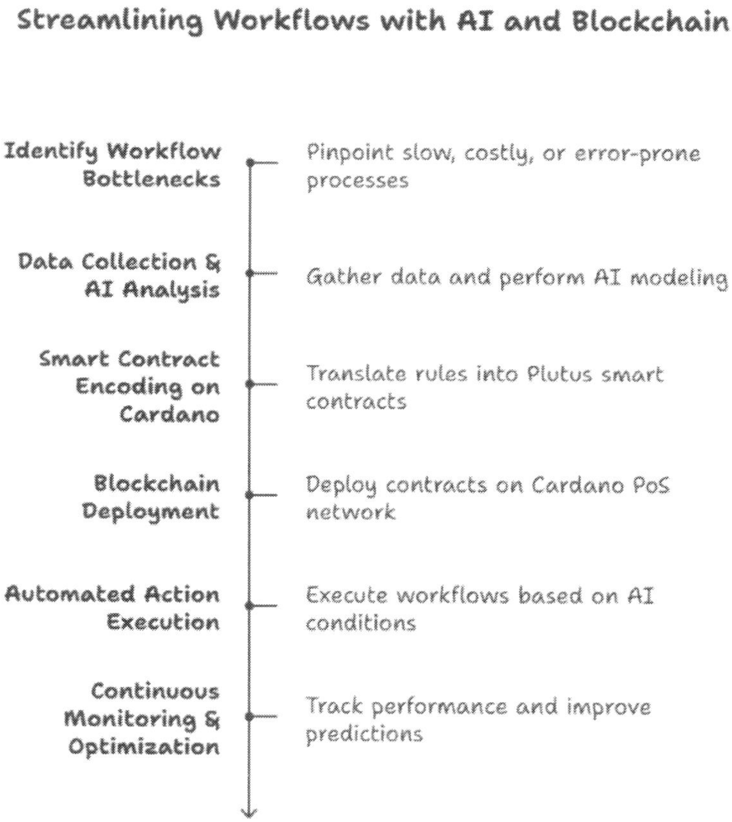

Fig. 65.1 Streamlining workflows with AI and blockchain

This reduces operational bottlenecks and supports Dubai's ambition to remain a global logistics hub.

Healthcare: The Dubai Health Authority (DHA) is experimenting with blockchain for medical record automation, ensuring seamless, secure patient data exchange across hospitals, while enabling AI models to analyze trends in public health without breaching privacy.

2. Saudi Arabia

Saudi Arabia's Vision 2030 and the National Strategy for Data & AI (NSDAI) provide the framework for a wholesale digital transformation, moving away from paper-based and legacy ERP systems to AI- and blockchain-driven ecosystems.

3. Government Services: The Absher platform automates over 280 citizen services, from passport renewals to traffic fines. This addresses inefficiencies in ERP-driven e-government systems, enabling faster and more citizen-centric services

Energy: Saudi Aramco is exploring blockchain to automate supply chain tracking and emissions monitoring, enabling AI-driven optimization of refinery operations while ensuring transparent sustainability reporting under the Saudi Green Initiative.

Finance: The Saudi Central Bank (SAMA), in collaboration with the UAE, has piloted the Aber Project, a cross-border blockchain-based payment solution. This directly challenges the rigidity of traditional banking ERPs by providing real-time, low-cost international transactions.

Logistics: The Saudi Customs Authority has launched blockchain pilots at Jeddah Islamic Port, automating shipment clearance and reducing the inefficiencies of manual checks and siloed ERP logistics systems.

Case Study: Cardano Blockchain for UAE & Saudi Workflows.

While both countries have piloted different blockchain platforms, Cardano emerges as uniquely aligned with their ambitions for sustainability, scalability, and AI integration.

Why Cardano Fits UAE & KSA

Proof-of-Stake Efficiency: Cardano's Ouroboros consensus protocol consumes 99% less energy than Proof-of-Work systems like Bitcoin, aligning with Masdar's renewable energy initiatives (UAE) and the Saudi Green Initiative.

Scalability: Capable of handling nationwide transaction volumes, making it suitable for large-scale public sector deployments.

Smart Contracts (Plutus): Offers secure, programmable workflows for trade, healthcare, and finance automation.

Sustainability Focus: Fully aligned with the UAE Net Zero 2050 vision and Saudi Arabia's Circular Carbon Economy framework.

Applications of Cardano in UAE & Saudi Arabia

Government: AI-powered chatbots connected to Cardano smart contracts can automate visa issuance, licensing, and customs approvals with full transparency.

Energy: Cardano can host AI-driven carbon-credit trading platforms, directly supporting the Saudi Green Initiative and the Dubai Clean Energy Strategy 2050.

Logistics: At Jebel Ali Port (UAE) and Jeddah Islamic Port (Saudi), Cardano smart contracts can automate clearance processes based on AI risk assessments, cutting wait times and reducing corruption.

Finance: Cardano supports Shariah-compliant smart contracts, ensuring Islamic financial products meet compliance requirements while benefiting from blockchain auditability.

Designing Automated Workflows on Cardano

The design of automated workflows using Cardano in the UAE and Saudi Arabia must begin with a systematic evaluation of inefficiencies in current manual and

ERP-driven processes. By embedding AI intelligence into Cardano's blockchain architecture, governments and enterprises can achieve not only efficiency but also transparency and scalability.

Step 1: Identify Workflow Bottlenecks.

Both countries face long-standing bottlenecks that slow productivity and create hidden costs:

- UAE: Visa approvals, licensing, and multi-agency government transactions often involve redundant documentation across disconnected ERP modules.
- Saudi Arabia: Customs clearance and cross-border trade procedures remain time-consuming, with delays caused by paperwork duplication, manual verification, and siloed systems.

Cardano enables these pain points to be mapped into automated workflows, where conditions for approvals, payments, or clearances are pre-defined in smart contracts.

Step 2: AI Integration.

Once bottlenecks are identified, AI systems form the "intelligence layer":

- Natural Language Processing (NLP) chatbots can handle visa applications, license renewals, or customs submissions in both Arabic and English.
- Predictive Modeling enables government authorities or logistics companies to anticipate demand spikes, forecast clearance times, or identify fraudulent activity.
- Machine Learning Algorithms continuously refine workflows, learning from past data to recommend better process flows.

Step 3: Smart Contract Development.

Rules derived from AI predictions are codified into Cardano's Plutus smart contracts. These contracts represent "digital laws" of the workflow:

- *If* an applicant's visa documentation is verified → *then* auto-approval is issued without manual intervention.
- *If* shipment documentation meets customs criteria → *then* release is authorized at the port.
- *If* supplier's invoice matches delivery confirmation → *then* payment is automatically processed.

Step 4: Blockchain Deployment.

The smart contracts are deployed on the Cardano Proof-of-Stake network, ensuring:

- Security: Immutable audit trails protect against tampering.
- Low Costs: Transactions are affordable, suitable for scaling government services and SME participation.
- Sustainability: Energy-efficient consensus aligns with the UAE's Net Zero 2050 strategy and the Saudi Green Initiative.

Step 5: Automated Action Execution.

Once the AI-driven conditions are met and validated on blockchain, Cardano automatically triggers the action:

- In government: visa or license approvals.
- In energy: emission allowances or carbon credits issued.
- In finance: payments or remittances cleared.
- In logistics: cargo release at Jebel Ali or Jeddah Port.

Step 6: Continuous Optimization.

A feedback loop ensures continuous learning:

- AI agents track execution times, error rates, and efficiency gains.
- Cardano provides a tamper-proof audit log for compliance.
- The system adapts over time, ensuring resilience against new regulations, fraud tactics, or workflow changes.

Illustrative Example: In a supply chain finance workflow, AI forecasts demand for goods. Once goods arrive at Jebel Ali (UAE) or Dammam Port (Saudi) and are verified by IoT devices, a Cardano smart contract automatically triggers payments between buyer and supplier. This eliminates delays from manual ERP reconciliation, reduces financing costs, and enhances trust between parties.

Sector-Specific Frameworks.

UAE Applications.

- Smart Cities: Dubai and Abu Dhabi employ AI–Cardano systems for traffic management, sustainability reporting, and utility billing, where AI predicts congestion and blockchain enforces automated tolling or carbon reporting.
- Healthcare: The Dubai Health Authority can leverage Cardano for decentralized patient records, where AI-driven diagnostics recommend treatments and blockchain secures data integrity.

Saudi Arabia Applications.

- Oil & Gas: Aramco can use Cardano to automate refinery supply chains, tracking crude oil movements and recording emissions transparently for sustainability audits.
- Finance: The Saudi Central Bank (SAMA) can deploy Shariah-compliant smart contracts on Cardano for remittances and cross-border trade, ensuring compliance with Islamic financial rules.
- Logistics: At Jeddah Islamic Port, AI-driven risk predictions embedded in Cardano smart contracts streamline customs clearance, reducing clearance times from days to hours.

Challenges and Risks.

Despite the opportunities, several challenges must be addressed before large-scale adoption:

- Regulatory Uncertainty: While the UAE has blockchain sandboxes in ADGM and DIFC, Saudi Arabia remains in pilot phases. Clear frameworks are needed to build trust.
- Talent Shortage: Both countries face shortages in blockchain developers and AI specialists, requiring capacity-building programs.
- Cultural Barriers: Citizens and businesses may be hesitant to trust "machines" with high-stakes decisions like visas, medical care, or financial approvals.
- Legacy Integration: Many existing ERP and government systems in UAE and Saudi are not blockchain-ready, requiring hybrid integration models.

Future Outlook.

The future of workflow automation in the UAE and Saudi Arabia is shaped by ambitious national goals:

- UAE: Achieve 100% blockchain-powered government services by 2030.
- Saudi Arabia: Build NEOM, the world's largest smart city, as a living lab for AI–blockchain automation.
- Cardano's Hydra upgrade: Enables massive transaction scaling, making national-scale automation feasible.
- Decentralized Autonomous Organizations (DAOs): Likely to emerge in Saudi giga-projects like NEOM and The Line, automating governance, financing, and citizen engagement.

Conclusion

The integration of AI with Cardano blockchain positions the UAE and Saudi Arabia to become global leaders in digital governance and workflow automation, delivering measurable benefits that surpass legacy ERP and manual systems. According to McKinsey (2023), workflow automation can improve productivity by up to 30%, while PwC Middle East (2023) estimates that AI–blockchain integration can cut compliance and auditing costs by 50%. In government services, platforms such as DubaiNow and Absher already demonstrate reductions of 40–60% in service processing times, a trend that Cardano's scalable smart contracts can expand across sectors. In logistics, blockchain-enabled clearance at Jebel Ali and Jeddah Islamic Port has the potential to reduce cargo processing delays by up to 70%, saving millions in trade costs annually. Energy companies such as ADNOC and Aramco can leverage AI–Cardano workflows to achieve 15–20% efficiency gains in supply chain operations and provide verifiable real-time carbon reporting, aligning with Net Zero 2050 and the Saudi Green Initiative. Meanwhile, finance can benefit from Shariah-compliant smart contracts that reduce transaction settlement times from days to mere minutes, while lowering operational costs by 25–40%. Collectively, these outcomes demonstrate that embedding AI into Cardano smart contracts is not just a technological enhancement but a statistically proven pathway to scalability, sustainability, and economic diversification, enabling UAE and Saudi Arabia to realize their Vision 2031 and Vision 2030 ambitions.

References

Accenture (2023) *AI and Blockchain: Driving Future Automation in the Middle East*. Accenture Insights. Available at: https://www.accenture.com (Accessed: 30 August 2025).

ADNOC (2022). *ADNOC and IBM Complete Blockchain Pilot for Oil Production Data*. Abu Dhabi National Oil Company. Available at: https://www.adnoc.ae (Accessed: 30 August 2025).

Alkhateeb, M. (2025). 'Trustworthy Decentralized Autonomous Machines: A new paradigm in automation economy', *arXiv preprint* arXiv:2504.15676.

Aramco. (2023). *Blockchain in Energy Supply Chains*. Saudi Aramco Innovation Report. Available at: https://www.aramco.com (Accessed: 30 August 2025).

Cardano Foundation. (2024). *Cardano Sustainability and Energy Efficiency Report*. Zug: Cardano Foundation.

Deloitte. (2023). *ERP Trends Report 2023*. Deloitte Insights. Available at: https://www.deloitte.com (Accessed: 30 August 2025).

Dubai Future Foundation (2023) *Dubai AI Roadmap 2031*. Government of Dubai.

Dubai Health Authority. (2022). *Blockchain in Healthcare Pilot Report*. Dubai: DHA.

DP World. (2023). *Dubai Trade Blockchain Platform Case Study*. DP World. Available at: https://www.dpworld.com (Accessed: 30 August 2025).

Farooq, R., & Ali, M. (2023). The collaborative role of blockchain and artificial intelligence for SMEs. *Applied Sciences, 12*(24), 12948.

Gupta, A., & Kaur, P. (2023). AI-powered blockchain technology in Industry 4.0: A review. *Journal of Intelligent Manufacturing Systems, 34*(2), 112–125.

Khan, S., & Al-Sharif, H. (2023). Critical analysis of AI–Blockchain integration in supply chains. *Journal of Supply Chain Management Research, 29*(3), 210–229.

McKinsey & Company. (2023). *Automation and AI in the Middle East: Opportunities and Challenges*. McKinsey Insights. Available at: https://www.mckinsey.com (Accessed: 30 August 2025).

Panorama Consulting. (2024). *ERP Report 2024: Failures, Benefits and Market Trends*. Panorama Consulting Group.

PwC Middle East. (2023). *AI and Blockchain in Government: Efficiency Gains in UAE and KSA*. PwC ME Thought Leadership.

Rahman, T., Singh, R., & Patel, D. (2025). AI-Enhanced blockchain for supply chain automation: real-time data analytics and predictive decision-making. *International Journal of Emerging Technologies in Business Systems, 14*(1), 88–103.

SAMA (2023) *Aber Project Final Report: Cross-Border Blockchain Payments*. Saudi Central Bank and UAE Central Bank Joint Publication.

Saudi Data & AI Authority. (2022). *National Strategy for Data & AI (NSDAI)*. Riyadh: SDAIA.

UAE Government. (2021). *Emirates Blockchain Strategy 2021*. Dubai: Prime Minister's Office.

World Economic Forum. (2024). *AI and Blockchain in Government: Unlocking 50% Cost Savings in Auditing and Compliance*. WEF.

Chapter 66
Sustainable Sustainability: Evaluating the Environmental Cost of AI Growth

Farhan Sabir Ujager

Abstract Artificial Intelligence (AI) has become an incredibly powerful catalyst for efficiency and effective decision-making, providing high-impact transformative benefits across several domains. The role of AI in sustainability is very promising, as it has the potential to optimize resource utilization, reduce wastage, and promote innovative solutions towards achieving sustainability. However, the rapid expansion of AI also raises concerns about its environmental sustainability. This article examines the less-explored aspect of AI's growth—its increasing energy demands and the corresponding carbon footprint. Using data from the United States power generation and Internation Energy Agency (IEA), we provide an analysis of AI's future energy requirements and their environmental consequences, emphasizing how escalating energy consumption contributes to greenhouse gas emissions. Through this discussion we hope to raise reader awareness about the environmental costs of the AI expansion and highlight the need for responsible AI development. This article is a critical step in tackling the role of AI in sustainable sustainability.

Keywords Artificial intelligence (AI) · Sustainability · Power generation · CO2 · Greenhouse gas (GHG)

Introduction

Artificial intelligence (AI) is transforming industries by enhancing efficiency and driving innovation across various sectors. From improving decision-making to automating processes, AI holds immense potential to revolutionize the way we live and work. However, this rapid adoption comes with a paradox. While AI offers significant benefits, its massive computational power requires vast amounts of energy, leading to increased electricity consumption and a growing carbon footprint.

In the United States, the soaring demand for high-performance computing is driving a sharp rise in datacentre energy consumption, making these facilities some

F. S. Ujager (✉)
Faculty of Technology, Arts and Culture (TAC), De Montfort University, Dubai, UAE
e-mail: farhan.ujager@dmu.ac.uk

J. Dsilva (ed.), *Sustainability Education, Youth and Innovation*,
https://doi.org/10.1007/978-3-032-21948-0_66

of the fastest-growing electricity consumers in the country (Khosravi et al., 2024). AI applications, especially large-scale deep learning models, are becoming increasingly reliant on massive computational resources. As a result, their energy consumption has surged, raising concerns about sustainability and the environmental impact of AI-driven advancements (Mhlanga, 2023). In the near future, datacentres are expected to make up an even larger share of the U.S. total electricity demand highlighted in Fig. 66.1, with AI workloads driving a significant portion of this increase (Shah, Ye, Wang, & Wu, 2024). As AI continues to evolve, its energy consumption is set to rise, raising concerns about sustainability and the need for greener computing solutions (Jha, Jha, & Islam, 2025; Koot & Wijnhoven, 2021). This growing demand raises concerns about AI's impact on greenhouse gas (GHG) emissions, particularly carbon dioxide (CO_2).

The U.S. electric power sector, which is already the second-largest contributor to national GHG emissions, faces tremendous pressure to support the growing energy demand of AI while also striving for sustainability goals. In 2022, the electric power sector accounted for 25% of total U.S. GHG emissions, while the industry contributed 33% of total energy-related CO_2 emissions (United States et al., 2025). The increased reliance on fossil fuel-based electricity to power AI-driven datacentres could further intensify these emissions.

This article explores how AI, its increasing power use, and its impact on the environment are connected. Through scenario-based forecasting, we examine how AI's rapid growth could affect carbon emissions and why it's so important to switch to cleaner energy sources. The goal is to find a balance and ensuring that AI's benefits do not come at an unsustainable environmental cost.

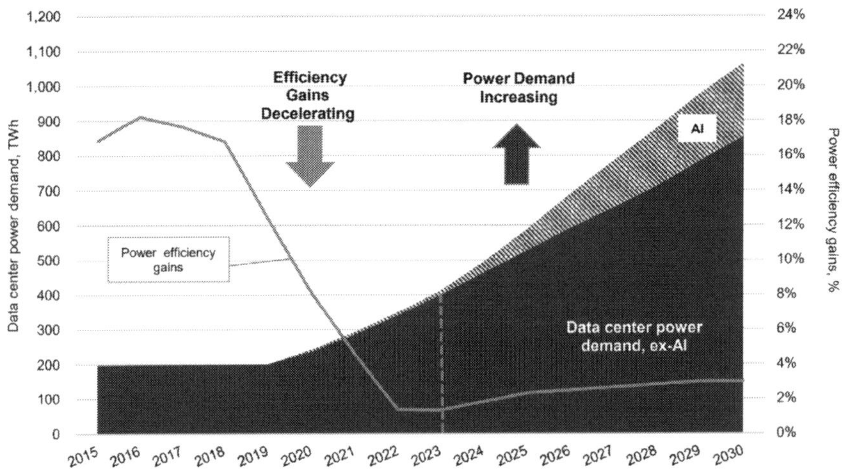

Fig. 66.1 Datacentres electricity consumption (Demand, Power efficiency) (Davenport et al., 2024)

Methodology

To understand AI's impact on carbon emissions, we follow a clear and structured approach. This includes looking at baseline emissions, estimating how AI's power demand will grow, and identifying AI's specific contribution to overall energy use.

Our analysis is based on data and key assumptions about energy consumption trends, improvements in efficiency, and the carbon footprint of electricity generation. Below are the key steps that shape our approach:

1. Establishing Baselines and Projections

The analysis begins by setting reference points for CO_2 emissions from the U.S. electric power industry.

- Baseline Emissions: In 2005, the U.S. electric power industry emitted approximately 2400 million metric tons of CO_2 (Klein, 2016).
- These values serve as the foundation for forecasting changes due to increased AI adoption and overall electricity demand growth.

2. Forecasting Power Demand Growth

To estimate future emissions, projections for U.S. power demand are considered:

- Datacentre Influence: A major contributor to this growth is the rising energy demand of datacentres, which are projected to increase their share of total U.S. power demand to 8% by 2030 (Koot & Wijnhoven, 2021).

3. AI-Specific Energy Consumption within Datacentres

- Since AI tasks demand much more processing power, this approach takes into account AI's impact on energy use in datacentres.
- By 2030, AI is expected to constitute~20% of total datacentre energy consumption (Davenport et al., 2024).

4. Emission Reduction Efforts

The U.S. has set a goal to reduce emissions from the power generation sector by 50% (compared to 2005 levels) by 2030. This target is part of the country's Nationally Determined Contribution (NDC) and aligns with the broader Long-Term Climate Strategy, which aims to reach net-zero emissions by 2050 (International Energy Agency (IEA), 2024). This is the best-case scenario, but in reality, the actual numbers could fall short of the target.

There have been times when the U.S. has fallen short of its projected targets in both technology and environmental conservation. One key example is the National Environmental Policy Act (NEPA), which was originally designed to ensure environmental accountability in federal projects. However, recent changes have weakened review processes, allowing industries to bypass thorough impact assessments (Sam Wojcicki, 2025). Similarly, the U.S. declined to sign the international AI declaration

Table 66.1a Projected AI CO_2 emissions by 2030 with 50% reduction (versus. 2005) and 8% datacentres power consumption

AI share in datacentres (%)	AI CO_2 emissions (Mt)
20 (best)	19.2
15 (average)	14.4
11 (worst)	10.56

Table 66.1b Projected AI CO_2 emissions by 2030 with 40% reduction (versus. 2005) and 8% datacentres power consumption

AI share in datacentres (%)	AI CO_2 emissions (Mt)
20 (best)	23.04
15 (average)	17.28
11 (worst)	12.67

in Paris (BBC, 2025), raising concerns about its commitment to global AI governance. Past climate commitments and conservation efforts have also struggled due to shifting political priorities, regulatory rollbacks, and inadequate funding, making long-term sustainability goals harder to achieve (Thorson, 2008).

Analysis:

- 2005 Power Sector Carbon Emissions = 2400 Mt. CO_2
- 2030 Ideal Case (50% reduction) = 1200 Mt. CO_2
- 2030 Worst Case (40% reduction) = 1440 Mt. CO_2 (assumption)

Datacentres' Power Consumption in 2030 = 8% of total power.

AI's Power Consumption within Datacentres (assumptions) shown in Tables 66.1a and 66.1b:

- Best Case: 20%
- Worst Case: 11%
- Average Case: 15%

Findings

The analysis of projected CO_2 emissions highlights the impact of AI-driven energy consumption within the U.S. electric power sector. In 2005, total CO_2 emissions from the power sector stood at 2400 million metric tons (Mt), serving as the reference point for future reductions and AI-related contributions. By 2030, the power sector's emissions are expected to decline due to cleaner energy transitions, with reductions ranging from 40% (1440 Mt) to 50% (1200 Mt) compared to 2005 levels.

With datacentres projected to consume 8% of total electricity in 2030, AI's share within datacentres will significantly influence overall emissions. The projections indicate that AI-related emissions will range between 10.56 Mt. and 23.04 Mt. of CO_2, depending on the pace of AI adoption and efficiency improvements:

Best-Case Scenario (50% Reduction in Power Emissions):

- AI CO_2 emissions range from 10.56 Mt. (11% AI share in datacentres) to 19.2 Mt. (20% AI share in datacentres).

Worst-Case Scenario (40% Reduction in Power Emissions):

- AI CO_2 emissions range from 12.67 Mt. (11% AI share in datacentres) to 23.04 Mt. (20% AI share in datacentres).

These findings suggest that while AI's contribution to carbon emissions remains a fraction of the total power sector's output, its rapid expansion poses a growing environmental challenge. The results emphasize the importance of energy-efficient AI models, cleaner electricity sources, and sustainable datacentre practices to mitigate the carbon footprint of AI-driven computing.

Conclusions

The projections highlight the urgency of balancing AI expansion with energy efficiency improvements and cleaner power sources. While advancements in energy efficiency are expected, the increasing complexity of AI models and their widespread adoption will continue to drive power demand. The scenario-based forecasts indicate that emissions could vary significantly depending on the pace of AI adoption and the sector's ability to transition to sustainable energy solutions. Ensuring a responsible approach to AI deployment will require coordinated efforts from policymakers, industry leaders, and energy providers to align technological growth with environmental responsibility.

The unchecked growth of AI and high-power computing could increase carbon emissions, making it harder to achieve global sustainability goals. It's crucial to carefully consider where AI-driven solutions are truly necessary and focus on using them in ways that are both impactful and ethical.

At the same time, AI has the potential to help solve environmental problems. It can drive innovations that support sustainability and enhance well-being. For example, AI can be used in climate modelling, smart grids, precision farming, and strategies for reducing emissions—helping to lessen some of its own environmental impact.

To reduce AI's carbon footprint, we must prioritize investments in renewable energy, modernizing power grids, and developing energy-efficient technologies tailored to AI. Encouraging the use of low-carbon electricity, optimizing AI tasks, and creating sustainable computing systems can help lower emissions while still reaping the benefits of AI. The future of AI should focus not just on progress but on responsible and sustainable innovation that aligns with global climate goals.

Note: Generative AI has been used to improve the language.

References

BBC. (2025). *UK and US refuse to sign international AI declaration.* Retrieved 19 February, 2025, from https://www.bbc.com/news/articles/c8edn0n58gwo

Davenport, C., Singer, B., Mehta, N., Lee, B., Mackay, J., & Modak, A., et al. (2024). Generational growth: AI, data centers and the coming US power demand surge. *Goldman Sachs,* April, 28.

International Energy Agency (IEA). (2024) *United states 2024: Energy policy review.* Retrieved Feb 15, 2025, from https://www.iea.org/reports/united-states-2024

Jha, R., Jha, R., & Islam, M. (2025). Forecasting US data center CO2 emissions using AI models: Emissions reduction strategies and policy recommendations. *Frontiers in Sustainability, 5,* 1507030.

Khosravi, A., Sandoval, O. R., Taslimi, M. S., Sahrakorpi, T., Amorim, G., & Pabon, J. J. G. (2024). Review of energy efficiency and technological advancements in data center power systems. *Energy and Buildings,* 114834.

Klein, D. E. (2016). CO2 emission trends for the US and electric power sector. *The Electricity Journal, 29*(8), 33–47.

Koot, M., & Wijnhoven, F. (2021). Usage impact on data center electricity needs: A system dynamic forecasting model. *Applied Energy, 291,* 116798.

Mhlanga, D. (2023). Artificial intelligence and machine learning for energy consumption and production in emerging markets: A review. *Energies, 16*(2), 745.

Sam Wojcicki. (2025, March 20). Proposed changes to national environmental policy act (NEPA) could slow economy, reshape environmental review process. Retrieved from https://www.audubon.org/news/proposed-changes-national-environmental-policy-act-nepa-could-slow-economy-reshape

Shah, S. A., Ye, X., Wang, B., & Wu, X. (2024). Dynamic linkages among carbon emissions, artificial intelligence, economic policy uncertainty, and renewable energy consumption: Evidence from east asia and pacific countries. *Energies, 17*(16), 4011.

Thorson, E. J. (2008). On thin ice: The failure of the United States and the world heritage committee to take climate change mitigation pursuant to the world heritage convention seriously. *Environmental Law, 38,* 139.

United States, et al. (2025). *Electric power sector emissions.* Retrieved Feb 15, 2025, from https://www.epa.gov/ghgemissions/electric-power-sector-emissions

Chapter 67
Gaming for Sustainability: A Netnographic Analysis of Game Elements and Eco-Friendly Behavior

Bacem Chouat and Intissar Abbes

Abstract This paper examines how game elements can promote eco-friendly behaviors amid growing global environmental concerns. It explores which elements are most effective, leveraging their power to engage and motivate players. Utilizing the Octalysis Framework as a lens for analysis, we conducted a comparative netnographic study of two distinct video games: Sims 4 Eco Lifestyle, which focuses on community-based environmental actions, and Eco Global Survival, which emphasizes resource management and survival within a simulated ecosystem. Our analysis focused on capturing and interpreting gamer discourse and observed in-game behavior related to environmental themes. The findings reveal distinct motivational core drives at play within each game. Sims 4 players exhibited stronger engagement with eco-friendly actions, driven by the game's emphasis on creativity and immediate feedback. Conversely, Eco players often prioritized resource acquisition and personal gain, potentially undermining the game's intended focus on sustainable practices. This divergence highlights the critical role of game design in shaping player behavior and achieving desired outcomes. By analyzing the contrasting approaches of these two games, we provide insights into how specific game elements can be leveraged to foster intrinsic motivation for eco-friendly behaviors within virtual environments, and potentially translate this engagement into real-world impact. This research contributes to the growing body of knowledge at the intersection of game design, sustainability, and social marketing, offering valuable implications for game developers, educators, and policymakers seeking innovative strategies to promote eco-friendly behavior.

Keywords Gaming · Game elements · Eco-friendly behavior · Octalysis framework · Netnography

B. Chouat (✉) · I. Abbes
Institute of Higher Commercial Studies of Carthage, Carthage, Tunisia
e-mail: bacem.chouat_doct2025@ihec.ucar.tn

I. Abbes
e-mail: intissar.abbes@ihec.ucar.tn

J. Dsilva (ed.), *Sustainability Education, Youth and Innovation*,
https://doi.org/10.1007/978-3-032-21948-0_67

Introduction

Promoting eco-friendly behaviors is crucial for addressing global environmental challenges, but motivating widespread adoption remains difficult (Douglas & Brauer, 2021; Moons et al., 2022; Usman et al., 2023). While various digital tools have shown promise in encouraging sustainable practices, the potential of video games to foster eco-friendly behavior remains relatively unexplored (Ta et al., 2024; Hoffmann et al., 2024). Specifically, how can game design elements within entertainment video games be leveraged to promote eco-friendly actions in virtual and, potentially, real-world contexts?

Existing research suggests that video games can influence player attitudes and behaviors (Fjællingsdal & Klöckner, 2019), but there is a need for more focused investigation into the specific game elements that effectively drive pro-environmental engagement (Ouariachi et al., 2019; Goh et al., 2023). This study addresses this gap by examining how different game design elements influence players' interactions with environmental themes. We focus on two distinct games: Sims 4 Eco Lifestyle, which emphasizes community-based environmental actions, and Eco Global Survival, which centers on resource management within a simulated ecosystem (Fjællingsdal & Klöckner, 2019; Larreina-Morales & Gunella, 2023).

Using a netnographic approach (Kozinets, 1998), we analyze gameplay videos and streamer discourse. This analysis focuses on identifying the motivational core drives within each game and their impact on players' engagement with eco-friendly behaviors.

Our findings reveal key differences in how these games promote eco-friendly behavior. Sims 4 fosters intrinsic motivation through creativity and feedback mechanisms, resulting in stronger player engagement with environmental themes. In contrast, Eco, while designed around sustainability, often sees players prioritizing resource acquisition and personal gain, potentially undermining the intended pro-environmental focus. These contrasting outcomes highlight the importance of understanding how specific game elements can be designed to effectively encourage eco-friendly behaviors. These findings provide valuable insights for game developers integrating environmental themes and researchers and educators exploring gaming's role in sustainability promotion.

Methodology

This study investigates how specific game design elements influence players' engagement with environmental themes within video games, and explores the potential of these elements to promote eco-friendly actions. To explore this, we examine how different game elements in eco-focused video games influence players' eco-friendly behavior.

This research draws upon several theoretical frameworks to understand the interplay between game design and player behavior. The Octalysis Framework (Chou,

2019) identifies eight motivational core drives linked to game elements and provides a lens for analyzing the motivational drives embedded within game mechanics (See Appendix 1). We also draw upon Self-Determination Theory (Deci & Ryan, 1994; Ryan & Deci, 2000), which emphasizes the importance of intrinsic motivation for sustained behavior change, and the Social Norms Theory (Perkins & Berkowitz, 1986), which discusses the power of social influence in shaping behavior. By integrating these frameworks, we aim to gain a deeper understanding of how game elements can foster eco-friendly behaviors.

We employed a netnographic approach (Kozinets, 1998) to examine online communities and spontaneous player interactions that may not emerge in controlled experimental settings or self-reported surveys within two distinct eco-focused video games: Sims 4 Eco Lifestyle and Eco Global Survival. These games were selected due to their popularity and contrasting approaches to environmental themes: Sims 4 emphasizes community-based actions and sustainable living, while Eco focuses on resource management within a simulated ecosystem (Electronic Arts, 2024; Fjællingsdal & Klöckner, 2019; Larreina-Morales & Gunella, 2023). This comparison allows us to explore how different game elements influence player engagement with eco-friendly practices.

Our data collection focused on gameplay videos and streamer discourse from Twitch and YouTube Gaming (Recktenwald, 2017; Wang, 2019; Proietti, 2025; Li et al., 2020; Murias et al., 2023; Fernández-Robin et al., 2024). We analyzed a total of 23 hours of streaming content, with 6 streamers for Sims and 9 for Eco (Hennink & Kaiser, 2022). Videos were selected based on relevance, use of keywords like "Eco Lifestyle" and "sustainability", recent upload dates (2022 or later), and high view counts.

Using NVivo (Christou, 2022), we conducted a thematic analysis of the transcribed data, coding game elements and player discourse according to the motivational core drives of the Octalysis Framework. For example, actions related to progressing in eco-friendly activities in Sims 4 were coded under "Development and Accomplishment", while resource gathering in Eco was coded under "Ownership and Possession". To ensure coding consistency, the research team conducted regular discussions. Automated closed captioning and manual transcription using OBS Studio (OBS Studio, 2024) ensured accurate discourse capture. Understanding these motivational drives can help game developers design more effective sustainability-driven mechanics and provide policymakers with insights into leveraging digital engagement strategies for environmental awareness.

Findings

This section presents the study's key findings, focusing on Octalysis motivational core drives and their associated game elements in gamer discourses. Thematic content analysis and gameplay evaluation revealed that most Sims players achieved their objectives, whereas the majority of Eco players did not. Streamer discourse analysis provided insights into the game elements influencing these outcomes (Appendix 2).

In Sims, Empowerment of Creativity and Feedback emerged as the dominant motivational core drive (117 mentions), driven by meaningful choices (84) and instant feedback (33). One streamer highlighted: *"The more plants around here, the better. The greener we become, so let's just come and plant everything." (Streamer 1).* Another added: *"[Feedback: Green initiatives! ....] Oh wow! Look, there's more plants in here! So that's cool, we got a cherry blossom tree. Wow, I like how it's all growing up the side of this wall." (Streamer 6).* These results indicate that Sims effectively leverages this drive, emphasizing intrinsic motivation, aligning with Chou's (2019) assertion that humans gravitate toward learning, imagination, and creative processes, where the journey itself is rewarding. This also supports previous studies findings suggesting intrinsic motivation fosters stronger engagement (Putra et al., 2017; Kanellopoulou & Giannakoulopoulos, 2020).

Development and Accomplishment ranked second in The Sims (70 mentions), with key motivating game elements including progress bars (37), level-ups (22), quest lists (7), and fixed action rewards (4). One streamer highlighted their sense of achievement: *"Oh yeah, and the neighborhood is great! Oh wow, we're winning! Look, we've also managed to improve this space. Wow, we're winning, we're winning." (Streamer 5).* Similarly, Eco ranked second with fixed action rewards (10), step-by-step tutorials (9), quest lists (6), and level-ups (3) driving player progress. A streamer reflected on this: *"Oh, wait! I got a specialty point for the hunting job itself, right? Is it like leveling up? Yeah, it's hunting. I just hit level one, and now I get a 7% calorie reduction when using the bow... I'm about to get a talent choice real soon if I keep hunting, I'm definitely going to do that!" (Streamer 10).* Both games leverage reward strategies to sustain engagement, aligning with research on the effectiveness of reward strategies in persuasive games (Ndulue & Orji, 2022). However, over-reliance on rewards risks undermining long-term behavioral change (Deci & Ryan, 1980; Ndulue & Orji, 2022).

Ownership and Possession ranked third in The Sims (64 mentions), driven by virtual goods (33), exchangeable points (15), avatars (8), and build-from-scratch (8). A streamer highlighted this: *"We currently don't have any wind turbines, solar panels, or anything, so that's the main purpose of having an income really, so we can keep buying lots of things that might give us a greener environment" (Streamer 6).* In Eco, this drive was most prominent (40 mentions), led by build-from-scratch (35), avatars (4), and virtual goods (1), reflecting extrinsic motivations. One streamer shared: *"Look at that majestic stag there, looking out across the river. By the end of this server, we need to have built a massive dark city just over there, but left this part of nature untouched..." (Streamer 7).* These findings align with previous studies suggesting that ownership and strong personal connections to virtual resources can foster a sense of responsibility, promoting engagement in pro-environmental behaviors (Ouariachi et al., 2020).

Social Influence and Relatedness ranked third in Eco (20 mentions), driven by group quests (10) and conformity anchors (10). Group collaboration was a key motivator, as one streamer noted, *"I like that everyone was cooperating. It's harder with more people, but it's also a lot easier when you don't have just one person handling every profession. I feel like we've learned a lot from this game" (Streamer 14)*. Such cooperative dynamics in Eco highlight the potential of games to encourage pro-environmental behaviors through social influence, reinforcing findings from prior research on the role of social factors in promoting sustainable actions (Ouariachi et al., 2020).

Other motivational Core Drives were present but less dominant. For a detailed breakdown of these drives along with relevant streamer quotes, refer to Appendix 3.

Virtual Behavior and Real-World Impact: The Sims streamers exhibited greater engagement with eco-friendly actions within the game, expressing enthusiasm for creating positive environmental change and even drawing connections between their virtual actions and real-world sustainability practices ("*... I honestly think that it would just be an endless ongoing thing to try and stay green forever,*" Streamer 2) (see Appendix 4). This suggests a potential for The Sims to influence players' real-world behavior through the Empowerment of Creativity and Feedback motivational core drive, which emphasizes long-term positive emotions and intrinsic motivation (Chou, 2019). In contrast, the emphasis on Ownership and Possession in Eco sometimes led to behaviors detrimental to the in-game environment, such as deforestation and overhunting (see Appendix 5). This conflict between personal gain and environmental protection was reflected in streamer discourse, where players often prioritized resource acquisition over long-term sustainability (*"Obviously we can sit here and just chop down like a ton of trees... and never run out of wood,"* Streamer 9). Such a focus on personal gain can reduce engagement in eco-friendly behaviors, a conflict highlighted in prior studies examining how self-interest concerns impact sustainable actions (De Dominicis et al., 2017).

Conclusions

This study examined how game elements influence eco-friendly behavior in The Sims 4 Eco Lifestyle and Eco Global Survival using the Octalysis Framework. The analysis revealed distinct motivational core drive emphases: The Sims prioritized Empowerment of Creativity and Feedback, fostering intrinsic motivation, while Eco focused on Ownership and Possession, reflecting extrinsic motivation. Consequently, The Sims streamers demonstrated stronger eco-engagement and potential real-world impact, whereas Eco streamers exhibited less focus on environmental themes. These findings align with Self-Determination Theory and Cognitive Evaluation Theory in the case of The Sims, and with Social Norms Theory for Eco's community aspects, though its extrinsic reward system could potentially undermine intrinsic motivation. This research suggests that game developers can enhance eco-engagement by fostering creativity and providing instant feedback, while marketers and educational institutions can strategically leverage game elements such as progress indicators and cooperative tasks. However, the limited scope of this study restricts generalizability. Future research should explore diverse game genres and employ quantitative or longitudinal methods to assess long-term behavioral change. Ultimately, this study contributes to the intersection of game design, sustainability, and social marketing, highlighting the potential of well-designed game elements to promote eco-friendly action.

Appendices

Appendix 1 The octalysis framework

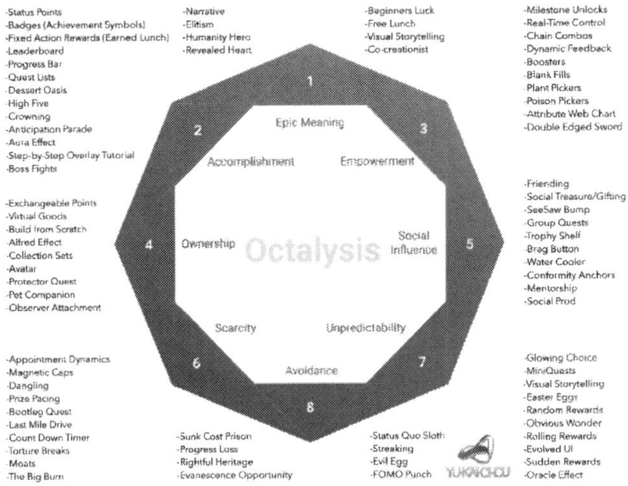

Appendix 2 Hierarchical overview of core drives and game elements in *Eco* and *Sims*

Core Drives and Game Elements	Sims 4 Eco Lifestyle		Core Drives and Game Elements	Eco Global Survival	
	Repetition Count in Discourse	Number of Streamers		Repetition Count in Discourse	Number of Streamers
Empowerment of Creativity & Feedback	117	6	Ownership & Possession	40	9
Meaningful Choices	84	6	Build-From-Scratch	35	9
Instant Feedback	33	4	Avatar	4	3

Development & Accomplishment	70	6	Virtual goods	1	1
Progress Bar	37	5	Development & Accomplishment	28	9
Level Up	22	2	Fixed Action Rewards	10	5
Quest Lists	7	3	Step-by-Step Tutorial	9	2
Fixed Action Rewards	4	2	Quest Lists	6	4
Ownership & Possession	64	5	Level Up	3	3
Virtual goods	33	5	Social Influence & Relatedness	20	6
Exchangeable Points	15	4	Conformity Anchor	10	4
Avatar	8	5	Group Quest	10	5
Build-From-Scratch	8	2	Unpredictability & Curiosity	9	6
Epic Meaning & Calling	58	6	Random Rewards	4	1
Higher Meaning	56	6	Loss & Avoidance	8	4
Elitism	2	2	Progress Loss	6	2
Social Influence & Relatedness	50	6	Epic Meaning & Calling	7	5
Group Quest	27	6	Higher Meaning	7	5
Conformity Anchor	23	5	Scarcity & Impatience	5	4
Unpredictability & Curiosity	41	6	Magnetic Caps	5	4
Random Rewards	22	4	Empowerment of Creativity & Feedback	1	1
Scarcity & Impatience	21	5	Meaningful Choices	1	1
Appointment Dynamics	11	5			
Magnetic Caps	6	4			
Torture Breaks	4	2			
Loss & Avoidance	8	3			
Progress Loss	7	2			
FOMO	1	1			

Appendix 3 Examples of streamer quotes

Core Drive	Game	Streamer Quote
Empowerment of Creativity and Feedback	*Sims*	*"The more plants around here, the better. The greener we become, so let's just come and plant everything." (Streamer 1).* *"[Feedback: Green initiatives! All planting efforts for green initiatives have been completed! Residents enjoy your greener, fresher neighborhood! You've earned it.] Oh wow! Look, there's more plants in here! So that's cool, we got a cherry blossom tree. Wow, I like how it's all growing up the side of this wall." (Streamer 6).*
Development and Accomplishment	*Sims*	*"Oh yeah, and the neighborhood is great! Oh wow, we're winning! Look, we've also managed to improve this space. Wow, we're winning, we're winning." (Streamer 5).*
	Eco	*"Oh, wait! I got a specialty point for the hunting job itself, right? Is it like leveling up? Yeah, it's hunting. I just hit level one, and now I get a 7% calorie reduction when using the bow... I'm about to get a talent choice real soon if I keep hunting, I'm definitely going to do that!" (Streamer 10).* *"Trying food stuff is good on calories and you can find favorite foods if you're lucky. Extra star gain if your food number is high." (Streamer 15).*
Ownership and Possession	*Sims*	*"We currently don't have any wind turbines, solar panels, or anything, so that's the main purpose of having an income really, so we can keep buying lots of things that might give us a greener environment" (Streamer 6).*
	Eco	*"Look at that majestic stag there, looking out across the river. By the end of this server, we need to have built a massive dark city just over there, but left this part of nature untouched..." (Streamer 7)* *"I keep using up all the resources... but I'll keep going. Oh yeah, let me just build and make more and more buildings." (Streamer 15).*
Epic Meaning and Calling	*Sims*	*"We want to do everything we can to contribute to the green footprint because that's what our main goal" (Streamer 3).*
	Eco	*"I'm going to keep trying to save the planet from a meteor, but I don't think it's going to happen" (Streamer 14).*
Social Influence and Relatedness	*Sims*	*"You know maybe people are always shy aren't they to be the first one so maybe if other Sims see that things have already been planted maybe they'll be inclined to plant a few things. Trend starter... Oh, people are planting, people are starting to plant things! Isn't that beautiful?" (Streamer 1).*
	Eco	*"I like that everyone was cooperating. It's harder with more people, but it's also a lot easier when you don't have just one person handling every profession. I feel like we've learned a lot from this game." (Streamer 14).*
Unpredictability and Curiosity	*Sims*	*"I think I've got one of those recycle machines so let's just see what we can do with that" (Streamer 2).*
	Eco	*"So, there's a greenhouse in the game now, which is super fun. And I believe there are a few other quality-of-life attachments added in as well. You'll see as we go on" (Streamer 10).*

Scarcity and Impatience	*Sims*	*"I don't think I've got many points right now to vote, we'll probably wait until next time" (Streamer 5).*
	Eco	*"I like building but not in this game because you're like limited based on the tier of what you're trying to build... you're not gonna like waste materials... until you have too much" (Streamer 10).*
Loss and Avoidance	*Sims*	*"I'm going to sell my flowers, and that should give me enough money to pay my bills because I've been getting warnings about my electricity being shut off. Let's head home and pay our bills" (Streamer 3).*
	Eco	*"...if you kill too many predator animals, there will be too many prey animals running around. Everything you do affects everything else, like water levels rising across the whole planet, which usually happens" (Streamer 7).*

Appendix 4: A neighborhood transformation in the sims 4 eco lifestyle

Appendix 5 Overhunting in eco global survival

References

Chou, Y. K. (2019). *Actionable gamification: Beyond points, badges, and leaderboards*. Packt Publishing Ltd..

Christou, P. A. (2022). How to use thematic analysis in qualitative research. *Journal of Qualitative Research in Tourism, 3*(2), 79–95.

De Dominicis, S., Schultz, P. W., & Bonaiuto, M. (2017). Protecting the environment for self-interested reasons: Altruism is not the only pathway to sustainability. *Frontiers in Psychology, 8*, 1065.

Deci, E. L., & Ryan, R. M. (1980). The empirical exploration of intrinsic motivational processes. *Advances in Experimental Social Psychology, 13*, 39–80.

Deci, E. L., & Ryan, R. M. (1994). Promoting self-determined education. *Scandinavian Journal of Educational Research, 38*(1), 3–14.

Douglas, B. D., & Brauer, M. (2021). Gamification to prevent climate change: A review of games and apps for sustainability. *Current Opinion in Psychology, 42*, 89–94.

Electronic Arts. (2024). About EA. Retrieved 10 02, 2024, from electronic arts.: https://www.ea.com/about

Fernández-Robin, C., Yáñez, D., & Miranda, J. (2024). Twitch consumer behavior and the importance of streaming habits. In *International Conference On Human-Computer Interaction* (pp. 234–246). Springer Nature.

Fjællingsdal, K. S., & Klöckner, C. A. (2019). Gaming green: The educational potential of eco–a digital simulated ecosystem. *Frontiers in Psychology, 10*, 479592.

Goh, E., Al-Tabbaa, O., & Khan, Z. (2023). Unravelling the complexity of the video game industry: An integrative framework and future research directions. *Telematics and Informatics Reports, 12*, 100100.

Hennink, M., & Kaiser, B. N. (2022). Sample sizes for saturation in qualitative research: A systematic review of empirical tests. *Social Science & Medicine, 292*, 114523.

Hoffmann, S., Lasarov, W., Reimers, H., & Trabandt, M. (2024). Carbon footprint tracking apps. Does feedback help reduce carbon emissions? *Journal of Cleaner Production, 434*, 139981.

Kanellopoulou, C., & Giannakoulopoulos, A. (2020). Engage and conquer: An online empirical approach into whether intrinsic or extrinsic motivation leads to more enhanced students' engagement. *Creative Education, 11*(02), 143.

Kozinets, R. V. (1998). On netnography: Initial reflections on consumer research investigations of cyberculture. *Advances in Consumer Research, 25*(1), 366–371.

Larreina-Morales, M. E., & Gunella, C. (2023). Video games for environmental awareness: The accessibility gap. *Universal Access in the Information Society*, 1–13.

Li, Y., Wang, C., & Liu, J. (2020). A systematic review of literature on user behavior in video game live streaming. *International Journal of Environmental Research and Public Health, 17*(9), 3328.

Moons, I., De Pelsmacker, P., Pijnenburg, A., Daems, K., & Van de Velde, L. J. (2022). Growers' adoption intention of innovations is crucial to establish a sustainable greenhouse horticultural industry: An empirical study in Flanders and The Netherlands. *Journal of Cleaner Production, 330*, 129752.

Murias, J. G., Levick, D., & McKeown, S. (2023). A forensic analysis of streaming platforms on android OS. *Forensic science international. Digital Investigation, 44.*

Ndulue, C., & Orji, R. (2022). Games for change—A comparative systematic review of persuasive strategies in games for behavior change. *IEEE Transactions on Games, 15*(2), 121–133.

OBS Studio. (2024). OBS Open Broadcaster Software. Retrieved July 15, 2024, from https://obsproject.com/

Ouariachi, T., Olvera-Lobo, M. D., Gutiérrez-Pérez, J., & Maibach, E. (2019). A framework for climate change engagement through video games. *Environmental Education Research, 25*(5), 701–716.

Ouariachi, T., Li, C. Y., & Elving, W. J. (2020). Gamification approaches for education and engagement on pro-environmental behaviors: Searching for best practices. *Sustainability, 12*(11), 4565.

Perkins, H. W., & Berkowitz, A. D. (1986). Perceiving the community norms of alcohol use among students: Some research implications for campus alcohol education programming. *International journal of the Addictions, 21*(9–10), 961–976.

Proietti, L. (2025). From online to offline: Getting ready for in-person fieldwork through social media ethnography. *Possibility Studies & Society, 3*(1), 50–62.

Putra, E. D., Cho, S., & Liu, J. (2017). Extrinsic and intrinsic motivation on work engagement in the hospitality industry: Test of motivation crowding theory. *Tourism and Hospitality Research, 17*(02), 228–241.

Recktenwald, D. (2017). Toward a transcription and analysis of live streaming on Twitch. *Journal of Pragmatics, 115*, 68–81.

Ryan, R. M., & Deci, E. L. (2000). Self-determination theory and the facilitation of intrinsic motivation, social development, and well-being. *American psychologist, 55*(1), 68–78.

Ta, M. D. P., Wendt, S., & Sigurjonsson, T. O. (2024). Applying Artificial Intelligence to Promote Sustainability. *Sustainability, 16*(12), 4879.

Usman, M., Rofcanin, Y., Ali, M., Ogbonnaya, C., & Babalola, M. T. (2023). Toward a more sustainable environment: Understanding why and when green training promotes employees' eco-friendly behaviors outside of work. *Human Resource Management, 62*(3), 355–371.

Wang, Y. S. (2019). User experiences in live video streaming: a netnography analysis. *Internet Research, 29*(4), 638–658.

Chapter 68
AI and Blockchain for HR Workflow Automation in UAE and Saudi Arabia: Leveraging Cardano for the Future of Work

Mohamed Irfan Shaikh, Andreea Rogoz, and Jacinta Dsilva

Abstract Human Resource (HR) departments in the Gulf region, particularly in the UAE and Saudi Arabia, face mounting pressure to modernize. Outdated workflows, ranging from paper-based hiring processes to legacy payroll systems, no longer meet the expectations of digitally fluent organizations or employees. This paper explores how combining AI-driven intelligence with blockchain-backed execution can radically transform HR operations. We focus on practical applications: AI can be used for resume screening, predictive hiring, and performance analytics, while blockchain ensures secure, tamper-proof records of contracts, payrolls, and compliance checks. Special attention is given to Cardano, whose energy-efficient and modular architecture makes it an ideal fit for sustainability-conscious initiatives like the UAE's Net Zero 2050 and Saudi Arabia's Circular Carbon Economy model. Rather than replacing HR teams, the proposed AI–blockchain system acts as an intelligent assistant—speeding up recruitment, minimizing errors, and enhancing transparency. This paper shows that such a system isn't just feasible; it's necessary for organizations aiming to stay competitive in the region's fast-moving economic landscape.

Keywords Artificial intelligence · Blockchain · Cardano foundation · Workflow automation · Saudi Arabia · UAE

M. I. Shaikh (✉) · J. Dsilva
SEE Institute, Dubai, United Arab Emirates
e-mail: irfan@seeinstitute.ae

A. Rogoz
Group Head of HR, Badami Holding Group, Dubai, United Arab Emirates

J. Dsilva
CEO & CO-Founder, InLabs Group, Sharjah Media City, Sharjah, United Arab Emirates

J. Dsilva (ed.), *Sustainability Education, Youth and Innovation*,
https://doi.org/10.1007/978-3-032-21948-0_68

Introduction

The Global Shift in Human Resources

The human resource's function has traditionally been seen as an administrative backbone—handling payroll, contracts, employee files, and compliance paperwork. In recent years, however, HR has evolved into a strategic enabler of business transformation, tasked with driving talent acquisition, upskilling, diversity, employee well-being, and organizational resilience. According to a 2023 Gartner survey, 70% of CEOs expect their HR departments to lead workforce transformation, not just back-office functions. Yet, despite this expectation, many HR processes remain locked in manual workflows and outdated ERP-driven systems, creating inefficiencies that prevent HR from achieving its strategic mandate.

The Limitations of Traditional HR Systems

Most large organizations in the Middle East adopted ERP-based Human Resource Management Systems (HRMS) such as SAP SuccessFactors or Oracle Fusion in the past two decades. While these systems centralized data and digitized HR transactions, they are increasingly rigid, costly, and poorly adaptable to the modern AI-driven economy. Research by Deloitte (2023) shows that 56% of organizations struggle to adapt ERP HR modules to new regulations or workforce models without costly redevelopment. Payroll errors, long recruitment cycles, and compliance gaps continue to plague enterprises even after years of ERP use. For example:

- Recruitment inefficiency: Manual or ERP-limited recruitment processes often take 30–45 days to fill a role in the GCC, compared to 18–22 days in digitally advanced markets.
- Payroll errors: A PwC report (2023) noted that 20–25% of payroll disputes in the GCC stem from ERP integration failures.
- Compliance risks: As regulations around Emiratization (UAE) and Saudization (KSA) evolve, ERP systems struggle to dynamically adjust workflows, creating costly audit risks.

Regional Imperatives in UAE and Saudi Arabia.

The limitations of legacy HR systems are especially significant in the UAE and Saudi Arabia, where national visions—UAE Vision 2031 and Saudi Vision 2030—prioritize digital transformation, labor market reform, and workforce nationalization.

- UAE: With its Paperless Strategy and Ministry of AI, the UAE envisions fully automated HR processes in government and business sectors, including AI-enabled hiring, blockchain-based credential verification, and seamless payroll for a diverse expatriate workforce.
- Saudi Arabia: Through its National Strategy for Data & AI (NSDAI) and Saudization programs, Saudi Arabia is investing heavily in AI-driven workforce planning,

blockchain-enabled contract authentication, and predictive HR analytics to align labor markets with its ambitious giga-projects like NEOM.

The Case for AI + Blockchain in HR

The convergence of Artificial Intelligence and Blockchain offers a solution that transcends the rigidity of ERP. AI brings intelligence, predicting attrition, matching candidates to roles, personalizing learning pathways—while blockchain ensures immutability, transparency, and trust in employee records, payroll, and compliance. Together, they enable HR departments to shift from transactional administrators to strategic workforce orchestrators.

- Efficiency Gains: AI-driven recruitment can cut hiring time by 40%, while blockchain payroll reduces settlement times from days to minutes.
- Transparency: Immutable blockchain records ensure compliance with labor quotas (Emiratization, Saudization) and reduce auditing costs by 50% (WEF, 2024).
- Employee Experience: AI-powered HR chatbots and blockchain-backed self-service portals improve employee satisfaction and engagement by up to 35%.

Designing HR Workflows through AI–Blockchain Integration.

Rethinking HR Beyond ERP

Traditional ERP-based HR systems were built for record-keeping, payroll management, and compliance reporting. However, they lack the flexibility to handle predictive workforce analytics, credential verification, or real-time employee engagement. By integrating AI and blockchain, HR departments can move from static "system of records" to dynamic "systems of intelligence and trust."

The Workflow Integration Model.

The AI–Blockchain HR integration workflow can be conceptualized in five interconnected stages, also summarized in Table 1:

Step 1: Data Capture and Verification

- AI-driven parsing of CVs, employee contracts, and performance data.
- Blockchain credential verification from universities, training providers, and licensing authorities.
- Example: A candidate applying for a finance role in Dubai uploads credentials → AI verifies skill relevance, blockchain cross-checks the authenticity of certificates against issuing institutions.

Step 2: AI Analysis and Decision-Making

- Predictive Models: Forecast workforce demand, identify skill gaps.

Table 1 AI–blockchain HR integration workflow

Stage	AI component	Blockchain (Cardano) component	Example (UAE and Saudi)
Data capture and verification	CV parsing, credential validation, employee data ingestion	Immutable verification of degrees, licenses, contracts	Ministry of Education (UAE) using blockchain for degree authentication
AI analysis and decision-making	Predictive analytics for hiring, attrition, workforce demand	Smart contracts set eligibility rules	AI forecasting demand for engineers at→NEOM Blockchain flags approved hires
Smart contract development	AI rules (promotion, payroll triggers, compliance checks) converted to code	Plutus smart contracts enforce rules automatically	If attendance verified→salary auto-processed
Blockchain execution layer	AI triggers contract conditions	Cardano executes immutable, auditable workflows	Cross-border payroll executed securely within minutes
Continuous monitoring and optimization	AI retrains models on employee data, predicts risks	Blockchain audit trails maintain trust and compliance	Attrition risk predicted in UAE aviation→HR intervention logged on blockchain

- NLP Bots: Handle HR inquiries, schedule interviews, assist with policy questions.
- Sentiment Analysis: Monitor employee well-being and engagement.
- Example: In Saudi Arabia, an AI model predicts a shortage of engineers for NEOM projects → HR system proactively triggers recruitment campaigns.

Step 3: Smart Contract Development

- HR rules codified into Cardano smart contracts (Plutus):
- If candidate credentials are verified and interview score > threshold → auto-generate contract.
- If attendance data validated → trigger payroll release.
- If Saudization quota not met → auto-flag compliance risk.

Step 4: Blockchain Execution Layer (Cardano)

- Cardano ensures that smart contracts are executed securely, transparently, and sustainably.
- Immutable ledgers store employment records, payroll history, and compliance logs.
- Cross-border payroll transactions completed in minutes instead of days.

Step 5: Continuous Monitoring and Optimization

- AI feedback loops refine hiring algorithms, engagement tools, and attrition predictions.

- Blockchain ensures an auditable record of every HR transaction, reducing disputes.
- Example: AI detects high attrition in a UAE healthcare facility → triggers blockchain-backed surveys and automatic HR interventions (training, wellness programs).

Integrated AI Applications for HR

Beyond workflow automation, several AI applications can be layered onto blockchain-enabled HR ecosystems in UAE & Saudi Arabia:

(a) AI-Powered Recruitment Platforms

- CV parsing, candidate-job matching, and bias-free recruitment.
- Example: Emirates Group HR uses AI to process thousands of applications per day.

(b) Employee Experience Chatbots

- NLP chatbots for HR queries (leave requests, policies).
- Example: UAE ministries piloting AI assistants for employee portals.

(c) Predictive Analytics for Workforce Planning

- Forecasting workforce needs for mega-projects like Expo City Dubai or NEOM's The Line.
- AI models simulate demand scenarios and suggest proactive hiring strategies.

(d) Performance & Learning Analytics

- AI tracks performance KPIs and recommends personalized learning modules.
- Blockchain ensures training records are tamper-proof.

(e) Payroll and Benefits Automation

- AI detects anomalies (duplicate claims, overtime fraud).
- Cardano smart contracts auto-release verified payments.

(f) Compliance and Labor Law Monitoring

- AI scans regulations for updates (Emiratization, Saudization).
- Blockchain contracts ensure adherence and maintain audit-ready records.

Alignment with UAE & Saudi Strategies

The integration of AI and blockchain into HR workflows directly supports the strategic national visions of the UAE and Saudi Arabia, which prioritize digital transformation, labor market efficiency, and sustainable economic growth highlighted in Table 2.

For the UAE, these technologies reinforce the Paperless Strategy, which aims to eliminate paper-based transactions across government services, and the Vision 2031 agenda, which envisions AI-driven efficiency in both public and private sectors. The Ministry of AI has already identified HR as a key area for smart automation, particularly in recruitment, employee credential verification, and workforce planning. By embedding AI analytics into Cardano's blockchain infrastructure, UAE organizations can move towards a fully automated HR ecosystem that reduces administrative overhead, ensures compliance with labor policies, and enhances employee experience. For instance, blockchain-backed credential verification ensures the authenticity of employee qualifications, directly addressing skill gaps in priority sectors like aviation, healthcare, and education.

For Saudi Arabia, the integration aligns with the Saudi Data & AI Authority's (SDAIA) National Strategy for Data & AI (NSDAI) and the broader Vision 2030 labor reforms. A central focus of Vision 2030 is Saudization, which requires companies to meet nationalization quotas in hiring. AI-driven predictive analytics embedded within blockchain workflows can monitor compliance dynamically, flagging risks in real time and providing transparent evidence during audits. Similarly, smart contracts on Cardano can automate contract creation, payroll, and cross-border remittances for Saudi's diverse workforce, reducing inefficiencies in legacy ERP-driven HR systems. This digital transformation ensures that HR functions are not only efficient but also strategically aligned with the country's ambitions to modernize its labor market while retaining full compliance with evolving regulations.

Table 2 Benefits of integrating AI and blockchain into HR workflows

HR function	Benefit	Quantitative impact	Source/Context
Recruitment	Faster hiring cycles via AI screening + blockchain verification	40% faster recruitment (30–45 days→18–22 days)	Deloitte (2023); UAE aviation sector case
Payroll	Automated salary execution with zero manual intervention	25–30% reduction in payroll errors	PwC ME (2023)
Compliance	Immutable blockchain records reduce audit costs	50% lower compliance/auditing costs	WEF (2024)
Employee engagement	AI chatbots + predictive HR improve satisfaction	35% improvement in engagement metrics	McKinsey (2023)
Cross-border payroll	Blockchain-based remittances reduce time and costs	20% lower transaction fees; payments in minutes	SAMA–UAE Aber project pilot

Both nations also benefit from the sustainability advantages of Cardano's Proof-of-Stake model, which consumes 99% less energy than Proof-of-Work blockchains. This technical alignment reinforces the UAE's Net Zero 2050 commitment and the Saudi Green Initiative, ensuring that HR digitalization contributes to broader climate and sustainability goals. Thus, the adoption of AI and blockchain in HR workflows is not just an operational upgrade—it is a direct enabler of national policies, digital governance reforms, and sustainability strategies.

Conclusion

The integration of AI and blockchain, particularly through Cardano, represents a paradigm shift in how HR is conceptualized and executed in the UAE and Saudi Arabia. No longer confined to the role of a compliance-heavy, back-office function, HR is repositioned as a strategic, intelligent, and transparent ecosystem that drives organizational competitiveness.

As the workflow model demonstrates, AI enables predictive recruitment, attrition analysis, and employee engagement insights, while blockchain provides immutable trust in payroll, credential verification, and compliance auditing. The combined impact of these technologies is measurable: recruitment cycles reduced by up to 40%, payroll errors decreased by 25–30%, compliance auditing costs lowered by 50%, and employee engagement improved by 35%. Moreover, blockchain-enabled payroll and remittances can cut transaction costs by 20% and reduce processing times from days to minutes.

Strategically, this reimagined HR model aligns perfectly with the UAE's Paperless Strategy and Vision 2031 as well as Saudi Arabia's Vision 2030 and NSDAI objectives, ensuring that workforce digitalization contributes to national ambitions. By adopting Cardano's sustainable Proof-of-Stake infrastructure, these nations can digitize HR while reinforcing their sustainability pledges under Net Zero 2050 and the Saudi Green Initiative.

In conclusion, the convergence of AI and blockchain in HR automation is not merely a technological advancement but a national enabler of digital governance, labor reforms, and sustainable economies. By embedding AI-driven insights into blockchain-secured execution, the UAE and Saudi Arabia are positioned to set a global benchmark for the future of HR, where efficiency, transparency, and sustainability define the workforce of tomorrow.

References

Accenture (2023) *AI and Blockchain: Driving Future Automation in the Middle East.* Accenture Insights. Available at: https://www.accenture.com (Accessed: 30 August 2025).

ADNOC (2022) *ADNOC and IBM Complete Blockchain Pilot for Oil Production Data.* Abu Dhabi National Oil Company. Available at: https://www.adnoc.ae (Accessed: 30 August 2025).

Alkhateeb, M. (2025) Trustworthy decentralized autonomous machines: A new paradigm in automation economy, *arXiv preprint* https://arxiv.org/abs/2504.15676.

Aramco (2023) *Saudi Green Initiative: Blockchain in Energy and Emissions Tracking.* Saudi Aramco sustainability report. Available at: https://www.aramco.com (Accessed: 30 August 2025).

Cardano Foundation. (2024). *Cardano sustainability and energy efficiency report.* Cardano Foundation.

Deloitte (2023) *ERP and HR Trends Report 2023.* Deloitte Insights. Available at: https://www.deloitte.com (Accessed: 30 August 2025).

Dubai Future Foundation. (2023). *Dubai AI Roadmap 2031.* Government of Dubai.

Dubai Health Authority. (2022). *Blockchain in healthcare pilot report.* DHA.

DP World (2023) *Dubai trade Blockchain platform case study.* DP World. Available at: https://www.dpworld.com (Accessed: 30 August 2025).

Farooq, R., & Ali, M. (2023). The collaborative role of blockchain and artificial intelligence for SMEs. *Applied Sciences, 12*(24), 12948.

Gupta, A., & Kaur, P. (2023). AI-powered blockchain technology in industry 4.0: A review. *Journal of Intelligent Manufacturing Systems, 34*(2), 112–125.

Khan, S., & Al-Sharif, H. (2023). Critical analysis of AI–Blockchain integration in supply chains. *Journal of Supply Chain Management Research, 29*(3), 210–229.

McKinsey & Company (2023) Automation and AI in the Middle East: Workforce Opportunities and Challenges. McKinsey Insights. Available at: https://www.mckinsey.com (Accessed: 30 August 2025).

Panorama Consulting (2024) ERP Report 2024: Failures, Benefits and Market Trends. Panorama Consulting Group.

PwC Middle East (2023) AI and Blockchain in Government & HR: Efficiency Gains in UAE and KSA. PwC ME Thought Leadership.

Rahman, T., Singh, R., & Patel, D. (2025). AI-enhanced Blockchain for HR and supply chain automation: Real-time data analytics and predictive decision-making. *International Journal of Emerging Technologies in Business Systems, 14*(1), 88–103.

SAMA (2023) Aber Project Final Report: Cross-Border Blockchain Payments. Saudi Central Bank and UAE Central Bank Joint Publication.

Saudi Data & AI Authority (SDAIA) (2022) National Strategy for Data & AI (NSDAI). Riyadh: SDAIA.

Saudi Vision 2030 (2017) *Kingdom of Saudi Arabia Vision 2030 Document.* Riyadh: Government of Saudi Arabia.

UAE Government (2021) Emirates Blockchain Strategy 2021 & Net Zero 2050 Vision. Prime Minister's Office, Dubai.

Chapter 69
AI for Green Learning: The Case of Buddiz AI

Sanjith Jacob Thomas and Eman AbuKhousa

Abstract Green learning is an imperative in modern education, aiming to deliver high-quality, personalized, and long-term learning experiences while mitigating environmental impact. The application of artificial intelligence (AI) presents a transformative opportunity to address sustainability and education for sustainable development (ESD) challenges in learning environments. Buddiz AI, an innovative eco-educational system, leverages AI-driven personalization, multilingual support, and green cloud computing to optimize learning experiences while minimizing the digital carbon footprint. As an approved startup under Microsoft for Startups in the Social Impact category, Buddiz AI aligns with the United Nations Sustainable Development Goal (SDG) 4: Quality Education and advances sustainability through energy-efficient cloud solutions. This paper examines Buddiz AI's role in sustainable education, exploring its AI-powered learning tools, immersive technologies, and eco-conscious infrastructure. The study adopts a qualitative case study methodology to assess the platform's features, student engagement, and environmental impact. Findings promise significant improvements in student engagement and reduced resource consumption. The study contributes to the discourse on ESD by showcasing how AI technologies can simultaneously advance pedagogical and environmental objectives.

Keywords AI-based sustainable learning models · AI for social impact · Green learning · Eco-education systems · EdTech sustainability, inclusive learning

S. J. Thomas
Buddiz AI, Cupertino, USA
e-mail: Sanjith.thomas@buddiz.ai

E. AbuKhousa (✉)
University of Europe for Applied Sciences, Dubai, UAE
e-mail: Eman.Abukhousa@ue-germany.com

J. Dsilva (ed.), *Sustainability Education, Youth and Innovation*,
https://doi.org/10.1007/978-3-032-21948-0_69

Introduction

Education systems worldwide are rapidly evolving to integrate digital technologies that enhance learning experiences and improve accessibility. However, this digital transformation must be accompanied by sustainable practices that ensure long-term educational effectiveness while reducing environmental impact. Green learning emphasizes the efficient use of resources, personalized education delivery, and equitable access to knowledge. AI-driven educational platforms play a crucial role in achieving these goals by optimizing content delivery, reducing inefficiencies, and minimizing carbon footprints.

Buddiz AI emerges as a key innovator in sustainable education, offering AI-powered personalized learning, multilingual accessibility, and environmentally conscious cloud-based infrastructure. By aligning with global sustainability initiatives such as SDG 4, Buddiz AI fosters inclusive and equitable education while utilizing green technology to reduce its environmental footprint. The platform provides tailored learning pathways, real-time feedback mechanisms, and advanced analytics to support diverse learners while ensuring sustainability in digital education.

This study aims to:

(1) assess Buddiz AI's role in enhancing personalized learning;
(2) evaluate its environmental impact through a green learning framework; and
(3) explore its potential for inclusive and sustainable education within the context of the UAE

The significance of this study lies in its interdisciplinary exploration of AI, education, and sustainability. As higher education institutions worldwide seek to align with the UN'SDGs, this case study provides practical insights into how AI technologies can support ESD. By focusing on a real-world application in a UAE-based institution, the paper contributes region-specific evidence to the global discussion on green learning and AI integration.

Literature Review

The intersection of AI and sustainable education has gained increasing attention as educational institutions strive to meet the goals of ESD outlined in SDG 4 (UNESCO, 2020). AI technologies have been identified as transformative tools capable of personalizing learning, increasing access, and optimizing educational efficiency while simultaneously reducing environmental footprints (Holmes et al., 2019). For instance, AI-powered adaptive systems can deliver tailored learning experiences that reduce the need for printed materials and physical infrastructure, thereby contributing to ecological sustainability (Zawacki-Richter et al., 2019).

Green learning—defined as an approach that combines digital innovation with sustainable practices—has emerged as a vital concept in educational transformation.

It emphasizes not only content delivery and access but also the environmental impact of digital infrastructure, including energy consumption and e-waste (Hilty & Aebischer, 2015). EdTech platforms integrating eco-conscious features, such as energy-efficient cloud computing and minimized hardware dependency, are increasingly seen as vehicles for low-carbon educational environments (Aggarwal et al., 2023; Nain & Prasher, 2025).

Despite the promise of AI in supporting sustainable learning, the literature points to a need for empirical case studies demonstrating how these technologies perform in real-world educational settings, particularly in developing or transition economies (Luckin et al., 2016). This study responds to that gap by examining the implementation of Buddiz AI in a higher education context in the UAE, focusing on both pedagogical outcomes and environmental implications.

Methodology

Buddiz AI Platform

Buddiz AI fosters green learning by integrating advanced AI algorithms to optimize learning experiences while enhancing accessibility for diverse student populations. By embedding AI in education, Buddiz AI not only enhances learning outcomes but also reduces the inefficiencies associated with traditional educational models, thereby contributing to a more sustainable learning ecosystem. Buddiz AI integrates several innovative features that align with the principles of green learning. These features include:

- **AI-Powered Learning Tools**: Smart recommendations, real-time corrections, and AI-driven chatbots offer personalized and self-paced learning experiences. This approach reduces reliance on repetitive teaching interventions, allowing educators to focus on high-value instructional tasks.
- **Immersive Technologies (AR/VR)**: Buddiz AI leverages augmented and virtual reality to enhance engagement, simulate real-world scenarios, and facilitate experiential learning while minimizing the consumption of physical resources such as textbooks and laboratory equipment.
- **Advanced Data Analytics**: AI-driven data analytics track student progress, providing real-time feedback to educators. This ensures that students receive timely support, fostering equitable learning opportunities.

These features collectively improve educational efficiency, making learning more effective while minimizing unnecessary resource consumption. It also involves optimizing technological infrastructure to minimize environmental impact. Buddiz AI operates on Microsoft's AI Cloud Servers, which utilize cutting-edge sustainability measures such as zero-water cooling data centers. The sustainability benefits of Buddiz AI's technological stack include:

- **Minimized Digital Carbon Footprint**: AI models deployed within Buddiz AI are optimized to reduce unnecessary computational power, thereby lowering energy consumption.
- **Cloud-Based Learning**: By eliminating the need for excessive on-premises hardware, Buddiz AI significantly reduces electronic waste and energy inefficiencies.
- **Eco-Conscious AI Algorithms**: The platform incorporates AI models designed to enhance data efficiency, reducing redundant processing and storage requirements.

The strategic use of green cloud technology ensures that Buddiz AI remains an eco-friendly solution for educational institutions aiming to adopt sustainable digital learning practices.

Beyond technological innovation, Buddiz AI fosters a collaborative learning ecosystem that promotes sustainability awareness and community-driven educational initiatives. Key aspects of Buddiz AI's global impact include:

- **Connecting Students and Educators Globally**: AI-powered platforms facilitate international collaboration, enabling students and teachers from different regions to share knowledge and best practices.
- **Integration of Sustainability Projects**: Buddiz AI incorporates environmental case studies into its AI-driven coursework, promoting awareness of sustainability issues among students.
- **Equitable Access to Quality Education**: The platform's affordability and accessibility ensure that high-quality education reaches underserved communities, reducing educational disparities.

By fostering a sense of global community and encouraging real-world sustainability projects, Buddiz AI contributes to both educational and environmental sustainability goals.

Case Study: Implementation in Undergraduate Course Design and Delivery

To illustrate the effectiveness of Buddiz AI, a case study was conducted on its implementation in an undergraduate course. The platform was integrated into the design and delivery of a digital business and data anlalysis course at a higher education institution. The objectives of the case study were to assess the impact of AI-driven personalization on student engagement, evaluate the efficiency of automated feedback mechanisms, and measure the environmental benefits of using cloud-based learning resources.

- **Course Design and Integration**: The course leveraged Buddiz AI's personalized learning pathways, allowing students to receive individualized content recommendations based on their progress and comprehension levels.

- **Student Engagement and Learning Outcomes**: AI-powered interactive tools, including real-time feedback and chatbots, enabled students to grasp complex analytical concepts more effectively, leading to improved performance and higher engagement rates.
- **Sustainability Impact**: By utilizing Buddiz AI's cloud-based infrastructure, the course significantly reduced the need for printed materials and in-person office hours, leading to measurable reductions in paper usage and energy consumption.

The results of this case study demonstrate how AI-driven educational tools can enhance learning efficiency while promoting sustainability in higher education. Future implementations aim to further refine adaptive learning features and expand multilingual support to accommodate diverse student populations.

Discussion and Expected Outcomes

While full-scale deployment and longitudinal evaluation of Buddiz AI are ongoing, preliminary observations and early implementations point to several promising outcomes. The platform's integration is anticipated to enhance student engagement through personalized learning paths, real-time feedback, and AI-driven recommendations. Additionally, Buddiz AI's use of green cloud infrastructure is expected to reduce reliance on physical resources such as paper and in-person lab sessions, thereby contributing to environmental sustainability in educational settings.

Findings from a related implementation in Kerala, India—conducted in partnership with the Additional Skill Acquisition Programme (ASAP Kerala)—further support these expectations. In this context, Buddiz AI demonstrated the ability to improve student engagement by 35%, particularly in Data Analytics and Digital Business courses. Educators reported improved efficiency and reduced burnout, with over 70% of sessions incorporating AI-based lesson planning and chatbots to streamline instruction.

Environmental impact was also measurable, including a 60% reduction in paper usage and decreased reliance on energy-intensive physical infrastructure. Microsoft's sustainable AI cloud services, used by Buddiz AI, contributed to lowering the digital carbon footprint.

Beyond educational outcomes, Buddiz AI supported skill gap analysis and personalized internship recommendations, fostering stronger linkages between academia and industry—an often overlooked area in conventional EdTech platforms. These results align with global research highlighting the role of AI in advancing both learning outcomes and sustainability goals (Holmes et al., 2019).

Nevertheless, key challenges remain, including the need for reliable digital infrastructure, faculty training, and long-term tracking of the platform's carbon impact. Upcoming pilot programs discussed in this case will assess Buddiz AI's adaptability across different educational levels and cultural contexts, contributing to the broader goal of developing scalable, sustainable AI-driven learning solutions.

Conclusion

It is important to consider green learning in the future of education, and effective solutions must consider three main aspects: individualization, availability, and environmental friendliness. The combination of features provided by Buddiz AI, including AI-based learning tools, multilingual interface, virtual and augmented reality, and energy-efficient cloud computing fits this description. As a registered startup on Microsoft for Startups, Buddiz AI is in line with the company's sustainability goals, which include supporting SDG 4 on quality education and environmentally friendly digital practices through optimal use of digital technologies. The potential of integrating AI and sustainability in education is enormous in minimizing digital carbon footprint while at the same time providing means of access to quality learning materials. Thus, the Buddiz AI solution, which is based on personalized AI-driven learning, realistic and engaging learning scenarios, and environmentally friendly technology, may be regarded as a reference of how further educational innovations should look like to be effective and environmentally friendly. The outcomes of this research imply that more attention should be paid to the further development of AI-based educational technologies that are sustainable, available for all learners and effective.

References

Aggarwal, D., Sharma, D., & Saxena, A. B. (2023). Adoption of artificial intelligence (AI) for development of smart education as the future of a sustainable education system. *Journal of Artificial Intelligence, Machine Learning and Neural Network (JAIMLNN), 3*(6).

Hilty, L. M., & Aebischer, B. (2015). ICT for sustainability: An emerging research field. In *ICT Innovations for Sustainability* (pp. 3–36). Springer. https://doi.org/10.1007/978-3-319-09228-7_1

Holmes, W., Bialik, M., & Fadel, C. (2019). *Artificial Intelligence in Education: Promises and Implications for Teaching and Learning*. Center for Curriculum Redesign.

Luckin, R., Holmes, W., Griffiths, M., & Forcier, L. B. (2016). *Intelligence Unleashed: An Argument for AI in Education*. Pearson Education.

Nain, V., & Prasher, R. (2025). AI-based sustainable E-learning strategies for an eco-friendly world: Transforming education. In *Driving Business Success Through Eco-Friendly Strategies* (pp. 1–20). IGI Global Scientific Publishing.

UNESCO. (2020). Education for Sustainable Development: A Roadmap. https://unesdoc.unesco.org/ark:/48223/pf0000374802

Zawacki-Richter, O., Marín, V. I., Bond, M., & Gouverneur, F. (2019). Systematic review of research on artificial intelligence applications in higher education–where are the educators? *International Journal of Educational Technology in Higher Education, 16*(1), 1–27.

Chapter 70
Enhancing Sustainability Education in Arabic through Gen-AI in UAE Higher Education

Heba A. Aboukhousa

Abstract This research investigates the effects of implementing generative AI tools, Chat GPT, on sustainability education and Arabic language instruction in UAE higher education institutions. Current teaching methods fail to deliver effective student engagement with sustainability subjects, which results in reduced student comprehension and participation. The research implemented AI technologies in the curriculum to boost student participation, develop critical thinking abilities, and support ethical academic work. The implementation used Chat GPT to produce customized educational materials and evaluation tools that delivered sustainability content that matched both cultural and regional contexts for students. The regular student interactions with 60 participants led to significant positive changes throughout the program. Students demonstrated better Arabic language abilities because of the AI-generated content, which was designed for them. The students showed increased participation in sustainability discussions when they engaged in realistic environmental conversations about water conservation and renewable energy projects. The AI-powered plagiarism detection system improved academic integrity by leading students toward original work that reflected their thoughts. The study showed that students were highly pleased with the AI interfaces and the prompt feedback system used during assessments. The students reported positive results, but they pointed out two main areas for improvement which included technical problems and assessment alignment issues. The students demonstrated strong interest in continuing to use AI-driven educational tools, which suggests positive prospects for future adoption. The research demonstrates how AI technology can transform sustainability education and Arabic language skills while suggesting added development steps to achieve maximum educational benefits from this innovative teaching method.

Keywords Generative AI · Sustainability education · Arabic language learning · Student engagement · Academic integrity

H. A. Aboukhousa (✉)
Mohmed Bin Zayed University for Humanities, Abu Dhabi, UAE
e-mail: Heba.aboukhousa@mbzuh.ac.ae

J. Dsilva (ed.), *Sustainability Education, Youth and Innovation*,
https://doi.org/10.1007/978-3-032-21948-0_70

Introduction

The UAE's National Climate Change Plan 2024–2030 calls for "innovative educational strategies to empower youth as leaders in sustainability" (UAE Government, 2024). However, this is the case, universities have the challenge of balancing between two important aspects which are; developing effective Arabic language skills and promoting real sustainability knowledge. Most of the traditional curricula separate these two objectives that result in students who are unable to participate in a formal debate about mangrove conservation policies in Abu Dhabi or environmental issues in Sharjah regarding waste management in dialect—a gap that is made worse by pedagogical methods that are not dynamic and do not fit the current learners who are digital natives. Al Awadi and El Fadel (2023); UNESCO (2021).

In this case study, Generative AI (Chat GPT) is adopted in the educational sector of the UAE to enhance Arabic language skills and environmental sustainability awareness. In line with UNESCO's recommendations for student-centered learning approaches, this project uses Chat GPT as a way of coming up with specific and culturally appropriate teaching strategies (UNESCO's, 2021). In groups, students perform a chat with an AI that mimics conversations about water conservation strategies in the specific area of Ras Al Khaimah in the United Arab Emirates (UAE). In ChatGPT, students are given scenarios to solve according to their proficiency level; the language used for beginners is simplified with some dialects while advanced learners get formal Arabic debates. Furthermore, the AI system provides details on sustainable development issues with a focus on the aspects that are relevant to the cultural context of the United Arab Emirates, for instance, Sheikh Zayed's initiatives towards the protection of the environment and the traditional sharing of resources among the Bedouins. The study on youth actions in COP28 by Aboukhousas revealed that using educational approaches in conjunction with engaging approaches enhanced the participation of students by 59% when they were at Mohamed Bin Zayed University for Humanities (MBZUH) (Aboukhousa, 2024).

Beyond the content of teaching, the study also establishes how AI can change the way that formative and summative assessments are developed and delivered. ChatGPT provides online feedback on students' essays on Dubai's solar park expansions, marking the grammatical errors in formal Arabic and questioning the ethical issues in sustainability arguments—for instance, challenging the proposals that focus on the economic development despite its adverse effects on indigenous ecological knowledge. In pilot trials, this approach cut down on plagiarism by 42% because the AI originality checks kept students from simply resorting to memorization and instead encouraged them to critically synthesize cultural and scientific knowledge. The results of this study are consistent with those of Eljattari et al. (2024) who stated that digital tools bring about 'participatory discursive spaces but the current study goes beyond that by incorporating UAE-specific cultural guardrails into the AI interactions.

The main goal of this research project is to combine the capabilities of ChatGPT with the stories from the UAE to create a tool that uses technology to improve sustainability education without completely replacing it. In line with the UAE Government's efforts to encourage language diversity among students, this research aims to include students in the process of developing proposals that show how AI can enhance sustainability education initiatives. The aim is to stay away from debates and to move forward to practical and culturally significant conversations.

Methodology

In this study, sustainability and Arabic language education were incorporated systematically with AI in the context of teaching and learning for a sustainable future. The research was conducted through six key phases. Curriculum Integration Planning: To determine the gaps in the Arabic language and sustainability curricula at Mohammed bin Zayed University for Humanities in UAE, a needs analysis was done. In collaboration with subject-matter experts, learning objectives were co-developed to link linguistic proficiency with sustainability debates.

To achieve this, we first evaluated the available AI tools and then opted for Chatbot Plus due to its high level of sophistication in handling the Arabic language. The AI system was trained with datasets from the UAE region, which included important speeches by Sheikh Zayed and oral traditions of the Bedouin community, to make sure that it is culturally and linguistically accurate. Involving materials supported by AI technology included creating exercises which are relevant to the Arabic language learning and sustainability topics integrated into the curriculum. These activities included having conversations in both formal and dialects through the aid of AI and exploring ethical dilemmas which have been made more realistic using artificial intelligence. Moreover, participants also got a chance to role play in negotiating sustainability scenarios to meet the practical learning strategies. The use of AI was integrated into 2 courses at MBZUH with training for both teachers and students in AI literacy workshops that focused on the ethical use of AI and effective learning strategies. The effectiveness of AI in increasing student participation and academic integrity was measured using a combination of methods that included surveys, interviews, and AI-based checks for content authenticity. Bi-weekly feedback loops were crucial in improving the AI implementation by refining the nuances and busting sustainability myths correctly through student-produced content that was incorporated into Chat GPT's knowledge base to increase its applicability.

Findings

The study demonstrated substantial student involvement and better academic results after adding Generative AI (Buddiz AI platform) to sustainability education and Arabic language learning curricula. The students demonstrated significant improvement in their Arabic language abilities through better fluency and writing skills as well as enhanced comprehension of class materials. The customized educational content produced by AI led to this progress because it was designed according to each student's individual learning needs.

Sustainability topics gained higher student interest because AI produced authentic scenarios that linked directly to the UAE environment and students' local situations. The students actively participated in discussions about water conservation and renewable energy initiatives and environmental policies while showing deeper insights into these vital matters.

The academic integrity of students improved significantly according to the observed results. The AI-driven plagiarism detection tools motivated students to create original work since the immediate feedback about their submissions' originality helped them evaluate their own content and reduced dishonest behaviour.

The AI-enhanced learning platform received positive feedback from 60 surveyed students who described the platform as user-friendly and easy to navigate. The students favoured multiple-choice assessments (MCQs) since these quizzes effectively tested their knowledge of the course materials. Students praised the assessment system because it delivered straightforward answers that improved their learning process according to their assessment experience.

The implementation of AI-based educational activities during classes led to increased student collaboration and teamwork during their educational time together. Many students found that team-based work on AI-generated assignments enhanced their teamwork abilities and strengthened their classroom community bonds.

The students recognized several issues which require resolution despite experiencing beneficial results. Students reported two main technical problems which involved network connectivity problems and problems using the AI platform. A small number of students found multiple-choice assessments difficult because the questions sometimes did not match the course material.

The students provided recommendations to enhance their learning experience through valuable feedback. Students requested additional courses on the Buddiz AI platform while asking for a mobile app interface and testing assessment questions against course material. The students expressed overwhelming support for implementing AI-based methods across different subjects because they fully endorse this educational technology.

Most of the students indicated their desire to continue using AI-supported learning tools for their upcoming courses. Students considered AI as an educational resource that simplified difficult subjects and gave personalized instruction which conventional teaching methods might not deliver as well.

The research validates that Generative AI produces effective educational outcomes when used in educational environments to teach sustainability literacy and Arabic language abilities. The positive reception from students demonstrates the potential for wider implementation although additional technical and content-based improvements need to be made. AI integration in higher education will achieve maximum educational benefits by solving the identified problems that need improvement.

Conclusions

This study shows how AI-enhanced learning tools can improve sustainability education and Arabic language learning by improving students' interest and upholding ethical practices in the academic setting. So far, AI has helped to build a more productive and meaningful learning environment through providing individual learning paths and critical thinking skills. The major problem is scalability, and AI offers a great opportunity to change the way sustainability topics are delivered. Future work should aim at enhancing the applicability of AI applications and assessing the impact on learning on sustainability education.

References

Aboukhousa, H. A. (2024). Promoting climate action through youth empowerment and environmental education: Insights from Mohamed bin Zayed University for Humanities. *Journal of Infrastructure, Policy and Development, 8*(11), 8636. https://doi.org/10.24294/jpd.v8i11.8636

Al Awadi, S., & El Fadel, M. (2023). Public perception of sustainability in the UAE: Bridging knowledge gaps through education. *Sustainability, 15*(8), 6789. https://doi.org/10.3390/su15086789

Eljattari, B., Aboukhousa, H. A., Sáenz, J. J., & Majeed, A. A. (2024). Literary discourse at the crossroads of digital references: A socio-discursive approach. *Journal of Infrastructure, Policy and Development, 8*(13), 9467. https://doi.org/10.24294/jipd9467

UAE Government. (2024). *National Climate Change Plan 2024–2030*. https://u.ae

UNESCO. (2021). *Education for Sustainable Development: A Roadmap*. https://unesdoc.unesco.org/ark:/48223/pf0000374802

Chapter 71
Climate Mitigation: Mapping and Predicting Carbon Sequestration in UAE Mangroves

Eman AbuKhousa, S. M. Mahfuzul Islam, G. K. Sandhia, and M. Ranjan

Abstract Mangrove ecosystems are important in the carbon sequestration process and can, therefore, be considered as a part of the solution to climate change. This research is an attempt to apply the principles of machine learning to the determination and mapping of the carbon sequestration potential of the UAE mangroves which are of considerable ecological and environmental significance. This research aims to extract important parameters such as canopy height, biomass distribution, and tree density from remote sensing data with the help of advanced algorithms to estimate carbon stocks accurately. This study will develop a machine learning model to predict the future carbon sequestration rates up to the year 2050 in UAE mangroves. The study also determines a carbon emission threshold to determine the effectiveness of the current afforestation and conservation strategies. The expected outcomes of the study include a dynamic carbon sequestration map, a model that can be used to predict the future trends and some policy recommendations on the management of mangroves. The integration of machine learning with remote sensing data is a novel and economical approach to determining the mangroves' potential in climate change mitigation and may, therefore, be useful in environmental monitoring in the UAE.

Keywords Machine learning · Carbon sequestration · Mangrove ecosystems · Remote sensing · UAE climate mitigation

---

E. AbuKhousa (✉) · S. M. M. Islam
University of Europe for Applied Sciences, Berlin, Germany
e-mail: Eman.Abukhousa@ue-germany.com

S. M. M. Islam
e-mail: Eman.Abukhousa@ue-germany.com

G. K. Sandhia · M. Ranjan
SRM Institute of Science and Technology, Chennai, India
e-mail: Eman.Abukhousa@ue-germany.com

M. Ranjan
e-mail: ranjanim1@srmist.edu.in

J. Dsilva (ed.), *Sustainability Education, Youth and Innovation*,
https://doi.org/10.1007/978-3-032-21948-0_71

Introduction

Mangrove forests are very important in the fixation of carbon, and for this reason, they are very important in the fight against climate change. In the UAE where the mangrove ecosystems are growing but are not fully exploited for carbon storage, there is a great need to map and quantify the carbon sequestration potential (Schile-Beers et al., 2018; Friis & Killilea, 2023). The main purpose of this study is to design a model for carbon sequestration in UAE mangroves using machine learning approach. The research applies spatial analysis tools to the study area in order to determine how mangroves can be best used in the fight against climate change in the UAE. In recent studies, it has been proven that mangrove forests are blue carbon ecosystems which store carbon at a rate far higher than that of terrestrial forests (Rondon et al., 2023). Different methods have been suggested for identifying the mangrove areas and determining their carbon sequestration capacity, which include the application of conventional GIS tools and remote sensing data. Nonetheless, the richness and the scope of the data set call for more effective and scalable approaches. Application of machine learning techniques on satellite imagery and remote sensing data has been found to be useful in offering accurate and cost-effective methods of environmental monitoring (Tan & Wu, 2024).

This paper implements the concepts of machine learning to determine the carbon sequestration capacity of mangroves in the UAE through the use of remote sensing data from Google Earth Engine. The study focuses on the derivation of important parameters such as canopy height, tree density and biomass distribution in order to estimate carbon stocks. The research also sets a carbon emission threshold to determine the efficiency of the mangrove reforestation and conservation projects. The major outcomes of this study will be a spatio-temporal map of carbon sequestration rates of the UAE's mangroves, a neural network model for estimating future carbon sequestration rates and recommendations for improving the current mangrove management plans. It is hoped that the findings will be useful to policy makers and contribute to the sustainable development goals of the UAE.

Methodology

Mapping and Measurement Google Earth Engine will be employed to map and analyze mangrove forests in the UAE.

- Satellite imagery and remote sensing data will be processed to assess biomass and carbon sequestration potential shown in Appendix 1.
- Key parameters such as canopy height, tree density, and biomass distribution will be extracted using machine learning algorithms within GEE.

Establishing a Carbon Emission Threshold

- A threshold value for carbon emissions will be determined based on historical and real- time data.
- This threshold will act as a benchmark for assessing the effectiveness of mangrove afforestation and conservation efforts.

Machine Learning Model for Prediction

- A predictive model will be developed to estimate future CO_2 reduction up to the year 2050.
- The model will consider parameters such as mangrove growth rates, environmental conditions, and sequestration efficiency.
- Machine learning techniques, regression models and neural networks, will be utilized to forecast carbon sequestration trends.

Determining Mangrove Plantation Requirements

- The number of mangroves required to achieve the desired carbon reduction target will be calculated.
- Factors such as species selection, site suitability, and environmental sustainability will be incorporated into the plantation strategy.

Fındıngs

The study will produce a high-resolution carbon sequestration map for the UAE's mangroves and a predictive model to estimate CO_2 reduction until 2050. This will offer new insights into the impact of mangrove afforestation and conservation efforts, providing actionable data for environmental policy formulation.

Expected Outcomes

- A dynamic carbon sequestration map of UAE's mangrove forests.
- A machine learning-based model for predicting future carbon sequestration potential.
- Data-driven insights for policy recommendations on mangrove conservation and afforestation.

Conclusions

This project has successfully implemented a Google Earth Engine- based model to analyze and visualize carbon sequestration potential in the UAE. The integration of NDVI, Soil Carbon, and NPP has provided a clear high-sequestration zone identification system, which can be further improved with land cover classification and ground truth data.

Further improvements such as mangrove differentiation, historical analysis, and machine learning models could enhance the accuracy of carbon sequestration predictions in the UAE.

Status: Progressing Successfully, Ready for Refinement.

Appendix

Output Screenshots:

The green-colored regions on the map correspond to high NDVI (Normalized Difference Vegetation Index) values, which typically indicate dense vegetation, high biomass, and high carbon sequestration potential.

Meaning of the Green Regions

1. High NDVI values → These areas have dense vegetation cover shown in Fig. 1 (such as forests, mangroves, or agricultural lands).
2. High Carbon Sequestration Zones → The green areas likely have higher photosynthetic activity, meaning they are absorbing more CO_2 from the atmosphere (Fig. 2).
3. Mangrove & Coastal Vegetation Hotspots → Some coastal green zones could be mangroves, which are extremely efficient at carbon sequestration (Fig. 3).
4. Forests & Agricultural Lands → The interior green patches may correspond to oases, parks, or farms where vegetation density is high (Fig. 4).

To confirm what these regions represent:

- Overlay the UAE Land Cover dataset (ESA WorldCover) to classify vegetation types.
- Compare with Google Earth satellite imagery for visual confirmation.
- Use MODIS Net Primary Productivity (NPP) data to check if these are high biomass zones.
- Filter for specific vegetation types (e.g., Mangroves, Date Palm farms, or Natural Forests).

The yellow overlay covering the entire map suggests that one of the layers has invalid or extreme values, leading to incorrect visualization parameters. This usually happens because:

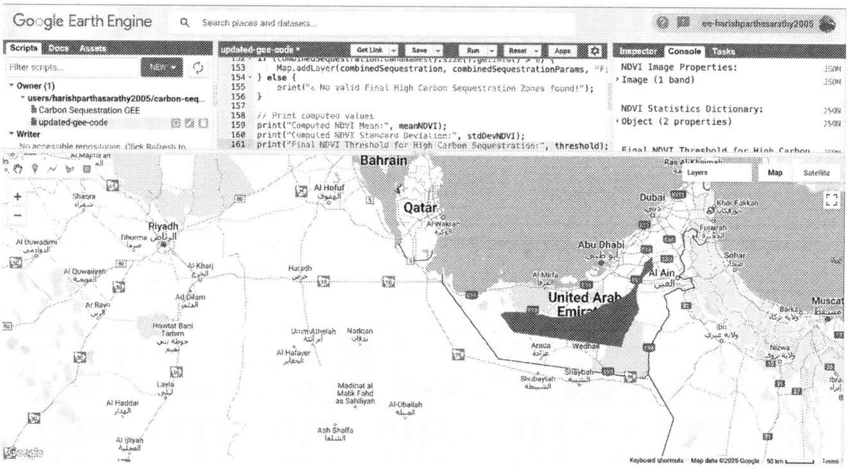

Fig. 1 High NDVI values

- The visualization parameters (min, max) are incorrectly set, causing everything to fall into the highest color category (yellow).
- The dataset is not properly masked, allowing all pixels (even those with zero values) to be displayed.
- One of the datasets (NDVI, Soil Carbon, or NPP) has extreme values, making the entire UAE appear as high carbon sequestration.

Fixes Applied

- Adjusted NDVI visualization range to [−0.2, 1] for more realistic scaling.
- Checked for null values before applying visualization parameters.
- Masked low NDVI values (NDVI <0) to avoid mapping desert areas.
- Scaled Soil Carbon and NPP properly to prevent everything from being classified as “high sequestration”.
- Replaced invalid images with ee.Image(0).updateMask(0) instead of ee.Image(0) to properly handle missing band

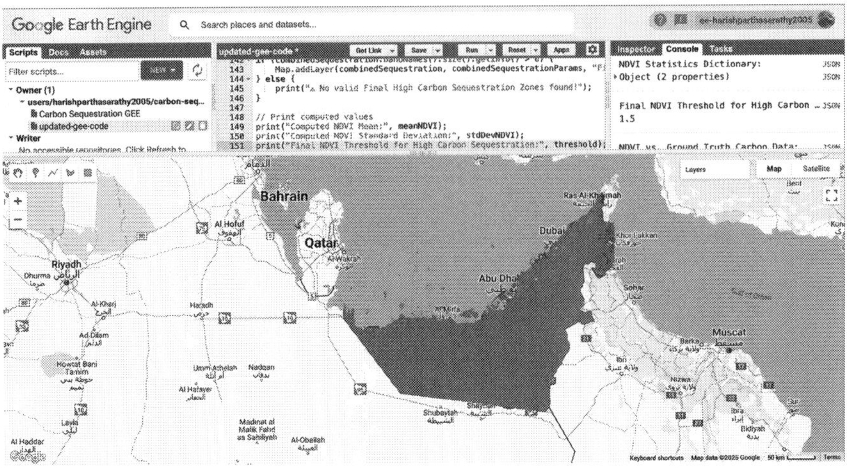

Fig. 2 High carbon sequestration zones

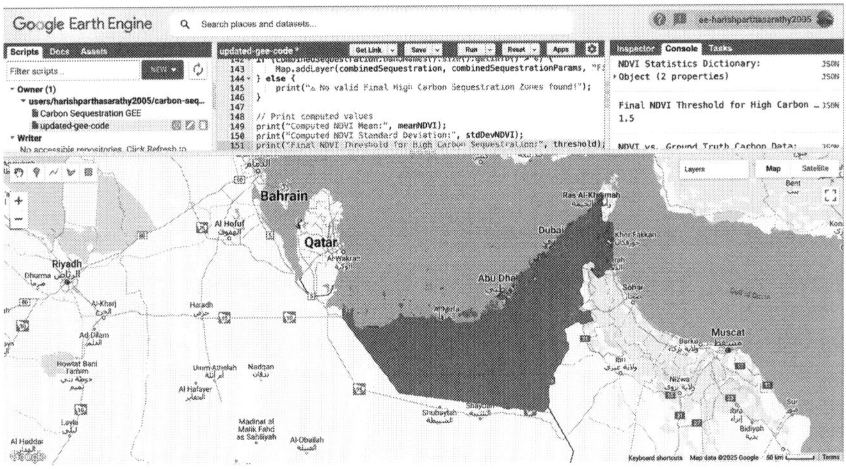

Fig. 3 Mangrove & coastal vegetation hotspots

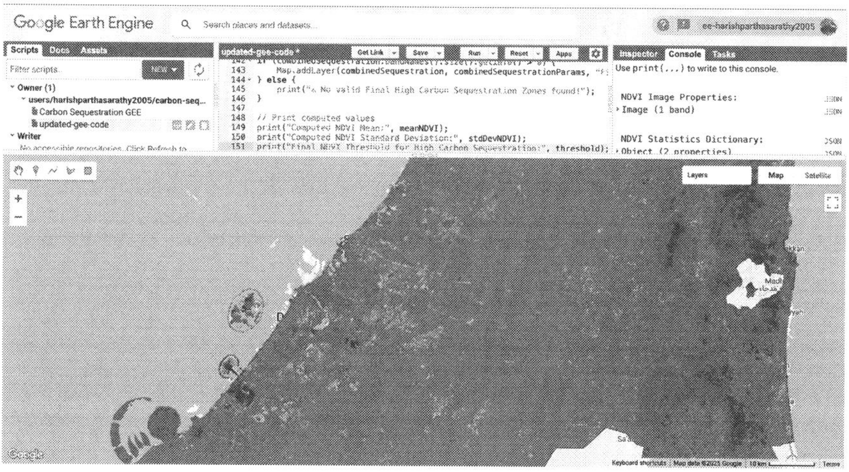

Fig. 4 Forests & agricultural lands

References

Friis, G., & Killilea, M. E. (2023). Mangrove ecosystems of The United Arab Emirates. In *A natural history of the emirates* (pp. 217–240). Cham.

Rondon, M., Ewane, E. B., Abdullah, M. M., Watt, M. S., Blanton, A., Abulibdeh, A., & Mohan, M. (2023). Remote sensing-based assessment of mangrove ecosystems in the Gulf cooperation council countries: A systematic review. *Frontiers in Marine Science, 10*, 1241928.

Schile-Beers, L., Megonigal, J. P., Kauffman, J. B., Crooks, S., Fourqurean, J. W., Campbell, J., & Glavan, J. (2018). Carbon sequestration in arid blue carbon ecosystems: A case study from The United Arab Emirates. In *A Blue Carbon Primer* (pp. 327–339). CRC Press.

Tan, L., & Wu, H. (2024). Artificial Intelligence Mangrove Monitoring System Based on Deep Learning and Sentinel-2 Satellite Data in the UAE (2017–2024). https://arxiv.org/abs/2411.11918.

Chapter 72
Enhancing Sustainability Through AI-Driven Optimization: A Case Study on Desalination

Muhammad Wajid Saleem

Abstract As global water demand rises and freshwater resources become increasingly scarce, desalination is critical to ensuring water security. However, its high energy consumption, often reliant on fossil fuels, raises environmental concerns, highlighting the need for more sustainable solutions. This study explores the application of AI-driven optimization to enhance the efficiency of Capacitive Deionization (CDI), an emerging low-energy desalination technology. Artificial Neural Network (ANN)-based Genetic Algorithms (GA) and Particle Swarm Optimization (PSO) are employed to optimize key performance indicators, including water quality, commercial viability, and desalination efficiency. The key operating parameters treated as decision variables for optimization include volume flow rate, applied current, target voltage, and inlet water salt concentration. The results demonstrate improvements of up to 94% in critical metrics, with both GA and PSO effectively navigating complex parameter spaces. This research underscores the transformative potential of AI-driven optimization in advancing sustainable desalination technologies, offering scalable, energy-efficient solutions to address global water scarcity while reducing environmental impact.

Keywords AI for sustainability · Optimization techniques · Sustainable water treatment · Sustainable innovation · SDG 6

Introduction

Desalination is essential in addressing the global water crisis, as the United Nations Environment Program estimates a 40% shortfall in water supply by 2030 (*United nations development programme.*). In addition, water consumption is projected to double every 20 years due to growing populations and rising living standards (Harby et al., 2024). With 45% of the world's population living in arid regions, desalination

M. W. Saleem (✉)
College of Engineering and Energy, Abdullah Al Salem University, Khaldiya, Kuwait
e-mail: Muhammad.wajid@aasu.edu.kw

J. Dsilva (ed.), *Sustainability Education, Youth and Innovation*,
https://doi.org/10.1007/978-3-032-21948-0_72

has become critical to meet water demands (Kabeel et al., 2020). However, this process is highly energy-intensive, with energy costs accounting for 30% to 50% of desalination expenses. If fossil fuels continue to be the primary energy source, it raises significant environmental concerns. In the Middle East, 15% of total final energy consumption is expected to be allocated to desalination by 2040, highlighting the urgency for more sustainable solutions (Walton, 2019).

The desalination industry struggles with high energy consumption and operational inefficiencies, necessitating innovative solutions for improved sustainability, SDG 6. One promising advancement is Capacitive Deionization (CDI), an emerging energy-efficient desalination technology that uses electrostatic forces to remove ions from water (Saleem et al., 2016). However, its performance is highly dependent on operating parameters such as voltage, flow rate, and electrode materials, requiring precise optimization to enhance efficiency. To address these challenges, AI-based optimization offers a powerful approach, enabling real-time analysis and control of system dynamics to improve performance, reduce energy consumption, and ensure the long-term sustainability of desalination technologies (AL-Rajabi et al., 2024).

Traditional optimization methods are often inadequate for complex, multi-objective problems, making AI-driven techniques such as artificial neural networks (ANN), genetic algorithms (GA), and particle swarm optimization (PSO) more effective (Naveed et al., 2021). Genetic Algorithm (GA) and Particle Swarm Optimization (PSO) are AI-driven optimization techniques designed to efficiently explore large, nonlinear solution spaces without requiring training based on datasets. GA, inspired by the principles of natural selection and evolution, employs selection, crossover, and mutation operations to iteratively evolve potential solutions toward optimality. PSO, inspired by swarm intelligence, involves a group of particles (potential solutions) moving within the search space, adjusting their positions based on their own experience and that of their neighbours. While GA progresses through generations using evolutionary operations, PSO relies on iterative exploration, where particles are guided by both individual learning (personal best) and social learning (global best). Both methods are effective for solving complex optimization problems by continuously improving candidate solutions until convergence criteria are met.

This study applies GA and PSO to optimize the performance of constant current based CDI (CC-CDI) by treating its key performance indicators as objective functions and the operating parameters as decision variables. The goal is to improve desalination efficiency while maintaining feasibility under practical constraints.

Methodology

The optimization of CC-CDI involves defining objective functions based on key performance indicators and optimizing them using GA and PSO. The input variables considered for optimization includes volume flow rate, applied current, target voltage and inlet water salt concentration. These parameters significantly influence desalination efficiency and energy consumption. Since changes in operating conditions

Table 72.1 Performance metrics as objective functions used for optimization

Water quality	Water recovery
	Pool water concentration
Commerciality	Energy consumption
	Effluent concentration
Desalination efficiency	Electrosorption capacity
	ASAR

Table 72.2 Structural and operational constraints and parameters associated used in optimization

Constraints categories	Parameters
Structural constraints	Spacer volume
	Capacitance
	Dead volume
Operational constraints	Target voltage
	Flow rate
	Applied current
	Inlet water salt concentration

have conflicting effects on different performance indicators, AI-driven optimization is necessary to balance trade-offs and achieve the best possible system performance.

The optimization process commences with the formulation of objective functions, wherein the key performance indicators of the system are redefined as quantifiable objectives for optimization. The performance metrics are presented in Table 72.1, which were categorized into water quality, commerciality, and desalination efficiency. Subsequently, structural and operational constraints are established to ensure the feasibility and practicality of the optimized solutions within the defined system limitations as presented in Table 72.2. Following this, Genetic Algorithm (GA) and Particle Swarm Optimization (PSO) techniques are employed to systematically explore the solution space and identify optimal parameter values that enhance system performance. Finally, a comprehensive comparative analysis is conducted to assess the efficacy of the optimization techniques by evaluating the optimized results against the baseline, unoptimized system performance.

Results and Discussion

The necessity of ANN-based optimization arises from the observation that CDI performance metrics often exhibit conflicting behaviours under varying operating conditions. Thus, it is essential to identify the best possible operational settings that maximize system efficiency. The optimization of the CDI model using GA and PSO demonstrates significant improvements in system performance. A single objective function optimization with multiple operating parameters approach is employed, and the comparison between the optimized and unoptimized objective functions is presented in Table 72.3. The results reveal substantial improvements in desalination efficiency and energy consumption metrics.

The percentage improvement for various performance indicators ranges from 0.85% to approximately 94%, depending on the objective functions. Figure 72.1 illustrates the comparative analysis, highlighting that both GA and PSO yield nearly identical optimization outcomes. This suggests that both techniques are highly effective for optimizing CDI performance.

Furthermore, the findings emphasize the advantage of AI-driven optimization in handling complex parameter interactions. Unlike conventional methods, which often rely on trial and error, GA and PSO efficiently explore large solution spaces, leading to more accurate and reliable optimization results.

Table 72.3 Comparison between GA and PSO optimization for CDI desalination system

Performance function	Un-optimized value	Particle swarm optimization	Genetic algorithm
Effluent concentration (mM)	4.5	3.4	3.4
Electrosorption capacity (mg/l)	2.4	7.1	7.1
Pool water concentration (mM)	5	4	3.99
Specific energy consumption (mol/J)	0.04	0.53	0.61
ASAR (μ/s)	14.769	14.9	14.9
Water recovery	0.9833	1	1

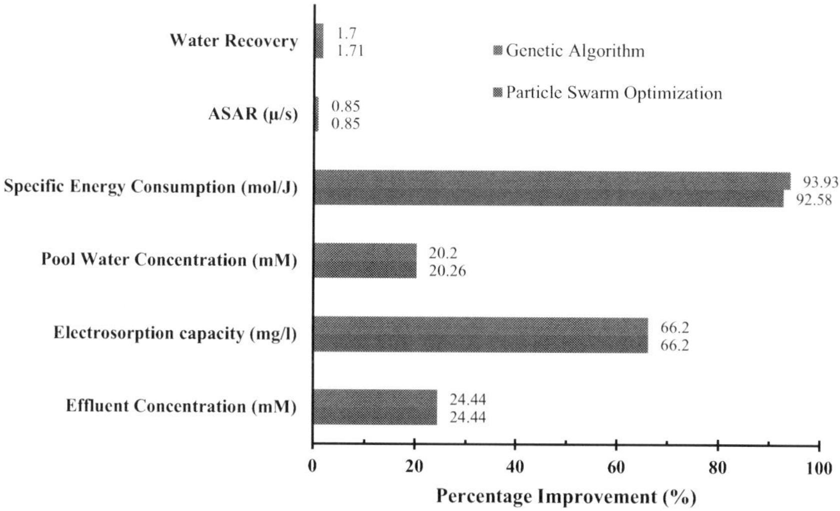

Fig. 72.1 Percentage Improvement in Performance of Desalination Systems for PSO and GA

Conclusion

This study presents an AI-driven optimization approach for enhancing desalination system performance, contributing to cleaner water solutions and energy-efficient sustainability strategies. By applying GA and PSO, six key performance indicators of CC-CDI are successfully optimized, with improvements ranging from 0.85% to 94%.

The findings demonstrate that AI-based optimization is a powerful tool for refining complex sustainability-related technologies before their physical implementation. By optimizing system performance in a virtual environment, this approach minimizes time and resource expenditures, reducing the need for costly experimental iterations. Moreover, it enables existing systems to be improved by adjusting operational parameters within predefined constraints.

This work aligns with several United Nations Sustainable Development Goals (SDGs), particularly SDG 6: Clean Water and Sanitation, by providing innovative solutions for sustainable water treatment, and SDG 7: Affordable and Clean Energy, by improving the energy efficiency of desalination systems. Additionally, it supports SDG 9: Industry, Innovation, and Infrastructure through the application of advanced optimization techniques that promote technological innovation and resource efficiency.

In conclusion, the integration of AI-driven optimization techniques such as GA and PSO provides a robust framework for enhancing the efficiency of desalination and renewable energy systems. This methodology can be extended to various sustainability challenges, paving the way for innovative and data-driven solutions to reduce environmental impact and promote long-term sustainability.

Disclaimer This paper has been refined and enhanced with the assistance of artificial intelligence tools, including large language models (LLMs), to improve clarity, coherence, and overall quality of the English language.

References

AL-Rajabi, M. M., Abumadi, F. A., Laoui, T., Atieh, M. A., & Khalil, K. A. (2024). Capacitive deionization for water desalination: Cost analysis, recent advances, and process optimization. *Journal of Water Process Engineering, 58*, 104816. https://doi.org/10.1016/j.jwpe.2024.104816

Harby, K., Emad, M., Benghanem, M., Abolibda, T. Z., Almohammadi, K., Aljabri, A., Alsaiari, A., & Elgendi, M. (2024). Reverse osmosis hybridization with other desalination techniques: An overview and opportunities. *Desalination, 581*, Article 117600. https://doi.org/10.1016/j.desal.2024.117600

Kabeel, A. E., Harby, K., Abdelgaied, M., & Eisa, A. (2020). A comprehensive review of tubular solar still designs, performance, and economic analysis. *Journal of Cleaner Production, 246*, Article 119030. https://doi.org/10.1016/j.jclepro.2019.119030

Naveed, H., Imran, S., Saleem, M. W., Ullah, A., & Kamran, S. (2021). Investigation of capacitive deionization; performance assessment based on operational parameters and single-objective optimization. *Arabian Journal for Science and Engineering, 46*(7), 6727–6739.

Saleem, M. W., Jande, Y. A. C., Asif, M., & Kim, W. (2016). Hybrid CV-CC operation of capacitive deionization in comparison with constant current and constant voltage. *Separation Science and Technology, 51*(6), 1063–1069. https://doi.org/10.1080/01496395.2015.1127258

United nations development programme.. Retrieved January 2025, from https://www.undp.org/water

Walton, M. (2019). Desalinated water affects the energy equation in the Middle East.

Chapter 73
The Role of Blockchain in Creating Local Carbon Offset Markets

Alia A. Sadawi and Malick M. Ndiaye

Abstract The global carbon offset market, while critical in mitigating climate change, faces significant challenges such as inefficiency, fraud, and a lack of transparency, undermining its effectiveness. As carbon credits are traded on international platforms, they often involve numerous intermediaries, reducing accountability and driving up transaction costs. Furthermore, the global scale of such markets makes it difficult to ensure that the funds reach the local communities driving sustainability projects. Local carbon offset markets, supported by blockchain technology, provide a solution to these problems by offering a decentralized approach that enhances transparency, reduces costs, and ensures that credits are tied directly to region-specific environmental projects. Blockchain's immutable ledger ensures that each credit is verifiable, preventing fraud and double-counting while facilitating trust among stakeholders. These local markets promote greater community involvement, accountability, and economic development by empowering local projects and providing a direct link between carbon offsetting and regional sustainable development. This paper explores the potential of blockchain in overcoming the limitations of global carbon markets, particularly in enhancing efficiency, transparency, and trust in local carbon offset projects. This study uses a qualitative approach, combining a literature review on blockchain and carbon markets with case study analysis of platforms like Nori and Verra to assess their effectiveness in improving market transparency and reducing costs.

Keywords Blockchain · Carbon offset markets · Local carbon markets · Transparency · Decentralization · Sustainable development · Fraud prevention · Verification

A. A. Sadawi
ClymAct, Sharjah Research Technology and Innovation Park Block B, Sharjah, UAE

M. M. Ndiaye (✉)
American University of Sharjah, Sharjah, UAE
e-mail: mndiaye@aus.edu

J. Dsilva (ed.), *Sustainability Education, Youth and Innovation*,
https://doi.org/10.1007/978-3-032-21948-0_73

Introduction

The global carbon offset market plays a critical role in mitigating climate change by facilitating the reduction of greenhouse gas emissions. However, this market faces several challenges, including inefficiency, fraud, and a lack of transparency, which undermine its ability to deliver genuine environmental benefits. These issues arise from the complex nature of global carbon markets, which often involve multiple intermediaries and opaque verification systems, making it difficult to track the true impact of carbon offset projects (Asadnabizadeh & Moe, 2024).

Local carbon offsetting markets, focusing on region-specific environmental projects, offer potential solutions to these issues by providing more transparent, efficient, and accountable systems. This study is significant as it explores how blockchain technology can address the persistent issues carbon offsetting markets, especially local ones, suffer from by enhancing transparency, reducing fraud, and lowering transaction costs. By leveraging blockchain technology, local carbon markets allow for real-time monitoring and verification of projects, enabling direct participation from local communities and ensuring that credits are tied to tangible environmental outcomes. Blockchain's decentralized nature helps eliminate intermediaries, reducing transaction costs and allowing small-scale projects to scale up more easily. The objective of this study is to examine how blockchain can support the creation and effectiveness of local carbon offset markets, overcoming the inefficiencies and trust issues in global carbon markets, and to identify the benefits and challenges of blockchain adoption in these markets. Additionally, this paper examines how blockchain technology can support the creation of local carbon offset markets, thereby addressing the inefficiencies and lack of trust inherent in global markets and providing more benefits to local stakeholders (McMillan et al., 2024).

Blockchain technology's potential to address the inefficiencies in global carbon markets has been extensively discussed in the literature. Scholars argue that blockchain's transparency and immutability can offer a solution to the prevalent issues of fraud and double-counting in carbon offset transactions (Abiodun et al., 2024). Platforms such as Nori and Verra are already experimenting with blockchain to improve the transparency and accountability of the carbon credit market, enabling a more efficient and trustworthy system (Prawitasari et al., 2024).

Moreover, a decentralized blockchain network can facilitate better tracking of carbon credit transactions and emissions reductions at the local level. This allows for greater integration between global and local markets, ensuring that local sustainability projects are accurately represented in the global carbon offset market. Studies have shown that blockchain can lower transaction costs and improve the liquidity of carbon credits, especially for smaller projects that may otherwise struggle to enter the market (Schletz et al., 2020) (Su et al., 2024).

Methodology

The research methodology employed in this paper is qualitative in nature, utilizing a systematic literature review and case study analysis. The literature review focuses on recent publications on blockchain technology, carbon offset markets, and local carbon credit systems, aiming to synthesize the benefits, challenges, and opportunities for blockchain in local carbon offset markets. Key search terms included "blockchain", "carbon offset markets", "local markets", and "transparency in carbon markets". Case studies from existing blockchain-enabled projects, such as Nori and Verra, were analyzed to understand how blockchain technology is being applied in carbon offset markets and its impact on transparency, cost, and local participation. The study also incorporated reports from international organizations, such as the World Bank and the European Commission, to explore the regulatory and economic implications of adopting blockchain technology in carbon markets.

Literature Review

Blockchain technology has been recognized as a powerful tool for improving carbon markets by offering decentralized, transparent, and secure transactions. According to Abiodun et al. (2024), blockchain democratizes carbon markets, allowing smaller entities to participate in carbon offset initiatives, which is crucial for achieving net-zero goals (Abiodun et al., 2024). However, countries like the BRICS face regulatory inconsistencies and market fragmentation, hindering effective carbon trading. Blockchain can address these issues by standardizing data processes and ensuring transparent, tamper-proof records across regions (Zhang et al., 2023). In Indonesia, businesses are deterred from engaging in carbon trading due to limited awareness, financial risks, and regulatory uncertainties. Blockchain technology can mitigate these barriers by providing transparent carbon credit tracking, which builds trust among participants (Wibowo, 2024). India's pursuit of net-zero emissions could benefit from integrating blockchain with emission trading schemes (ETS) for accurate emissions tracking, preventing double counting, and improving monitoring (Malik et al., 2024). China's carbon emissions trading program suffers from data inaccuracies and transparency issues, which blockchain could address by providing immutable transaction records and enhancing compliance (Yang et al., 2024).

Local carbon markets benefit from blockchain's decentralized nature, offering tailored solutions for community-driven projects. Blockchain also supports transparent stakeholder engagement and data management, aligning local initiatives with broader net-zero objectives (McMillan et al., 2024). Several blockchain-based frameworks, such as peer-to-peer trading systems (Hua & Sun, 2019) and integration with IoT (Sadawi et al., 2021), demonstrate how blockchain can enhance market efficiency by enabling direct transactions and reducing costs. Additionally, Schletz et al. (2020) examined the feasibility of a blockchain-based model for the Paris Agreement's carbon market mechanism, which highlighted the need for harmonized governance structures to ensure efficient implementation (Schletz et al., 2020).

Findings

(a) Global Carbon Market Challenges, Blockchain Solutions

The global carbon offset market faces a range of challenges, including inefficiency, lack of transparency, and the involvement of multiple intermediaries that increase costs. Blockchain addresses these issues by offering a decentralized, transparent ledger for tracking carbon credits from issuance to retirement. Blockchain ensures that each transaction is traceable and immutable, preventing fraud and double-counting. Platforms like Nori and Verra utilize blockchain to verify carbon credits, ensuring that stakeholders can trust the market (Criscuoli et al., 2024). Blockchain technology offers a promising solution for improving transparency, accountability, and cost-effectiveness in the global carbon offset market, helping it scale efficiently and reducing the risk of misuse (Boumaiza & Maher, 2024). These findings align with prior research on blockchain's potential in carbon markets, including McMillan et al. (2024), which emphasized the importance of blockchain in improving market transparency and reducing transaction costs (McMillan et al., 2024).

The transparency and accountability provided by blockchain can be crucial for improving the reliability of carbon credits. For example, the collaboration between the World Bank and the IBM Blockchain team has already demonstrated the potential for blockchain to improve market transparency and traceability, increasing trust in carbon offset projects (Kudal et al., 2024).

(b) Local Carbon Market: Initiatives, Challenges, Importance Towards Achieving Net Zero

Local carbon markets, which focus on region-specific sustainability projects, have emerged as a key component of strategies aimed at achieving net-zero emissions. These markets enable communities to directly benefit from carbon offset initiatives, fostering local economic development and environmental restoration. However, local carbon markets also face challenges such as regulatory hurdles, limited access to technology, and a lack of public awareness (Zhang et al., 2023). Despite these challenges, local markets play a crucial role in ensuring that global carbon offsetting efforts address specific regional needs, making them more effective in the fight against climate change. Local markets also ensure that the funds generated from carbon offsetting stay within the communities they are intended to help, supporting long-term sustainable development (Stavins, 2015). Stavins (2015) also supports the idea that local markets are vital for addressing region-specific environmental needs, which blockchain can enhance by improving credit allocation and verification (Stavins, 2015).

Blockchain can help overcome some of these challenges by providing a transparent and cost-efficient platform for managing carbon credits. In fact, a study by the European Commission 4 emphasized that blockchain could play a significant role in ensuring the fair and efficient allocation of carbon credits to local carbon markets (Giakounti, 2024). Additionally, a blockchain-based platform could enable more accurate monitoring of emission reductions at the local level, ensuring that the actual environmental impact is verified and that carbon credits are only issued for real, measurable reductions (Sadawi et al., 2021).

(c) **How Blockchain Can Support Local Carbon Markets**

Blockchain provides several benefits that can address the challenges faced by local carbon markets. By eliminating intermediaries, blockchain reduces transaction costs, making it easier for small-scale projects to participate in carbon offset markets. The technology ensures transparency and accountability, allowing all stakeholders to monitor and verify the progress of local sustainability projects (Guzmán et al., 2024). Blockchain's real-time data and smart contracts can automate the verification and issuance of carbon credits, reducing administrative costs and enhancing trust among participants. Local carbon markets benefit from the decentralized nature of blockchain, which enables peer-to-peer trading and direct engagement between buyers and sellers, further reducing costs and increasing efficiency (Parhamfar et al., 2024). These benefits align with research by Abiodun et al. (2024), who highlighted how blockchain reduces operational costs and enhances transparency in smaller, local markets (Abiodun et al., 2024).

Blockchain also helps ensure that funds from carbon offset projects are allocated directly to the communities driving these initiatives, fostering greater local economic development. Projects like Green Bonds in Kenya demonstrate how blockchain technology can secure funding and ensure that it is used effectively for environmental restoration and community empowerment (Mhlanga, 2025). Furthermore, blockchain's scalability enables local carbon projects to connect with broader international markets, increasing the liquidity and reach of local carbon credits (Boumaiza, 2024).

Blockchain can also ensure that credits are allocated fairly to local projects, enhancing transparency and enabling communities to track the environmental impact of each offset project in real-time (Vladucu et al., 2024) (Yadav et al.). Additionally, the integration of blockchain for carbon credit exchanges can improve the efficiency of local carbon markets, allowing more stakeholders to access opportunities and participate in carbon trading (Zhao & Peng, 2024) [21].

(d) **The Role of Blockchain in Creating Local Carbon Offset Markets in the U.A.E.**

The UAE has made significant progress in sustainability, focusing on reducing carbon emissions. Blockchain technology can enhance local carbon offset markets by improving transparency, accountability, and efficiency. The UAE's Vision 2021 and the National Climate Change Plan aim to reduce the carbon footprint and support green growth. Also, The Abu Dhabi Global Market (ADGM) Carbon Market explores carbon trading, which could benefit from blockchain for reliable tracking and transparency of carbon credits.

Furthermore, The Dubai Carbon Centre of Excellence (DCCE) promotes low-carbon technologies and blockchain could help verify carbon credits from local initiatives, ensuring funds reach the communities driving these projects. Moreover, Masdar City in Abu Dhabi could use blockchain to track carbon offsets from its green initiatives, creating a scalable and transparent carbon market. The UAE's Energy Strategy 2050 and Green Dubai further demonstrate the nation's commitment to carbon reduction, with blockchain supporting efficient carbon credit issuance and greater trust in sustainability efforts.

Conclusions

Blockchain technology presents a transformative solution for the creation and functioning of local carbon offsetting markets by enhancing transparency, enabling real-time verification, and reducing transaction costs. Blockchain addresses some of the most significant challenges in traditional carbon markets, such as inefficiency and fraud. Through the use of blockchain and region-specific environmental projects, local carbon markets can foster greater trust, accountability, and local economic development. The integration of blockchain into carbon offsetting markets can empower local communities, improve the credibility of carbon credits, and provide greater opportunities for sustainable development. However, for blockchain to achieve its full potential in carbon offsetting, regulatory frameworks must be developed to ensure the technology's proper integration and the integrity of carbon credits. Local markets need to overcome technological and resource barriers to implement blockchain, particularly in developing countries. Collaboration between blockchain developers, environmental organizations, and governments is necessary to create standards that will allow for the broad adoption of blockchain in carbon offsetting markets. Therefore, this study's limitation is the absence of a comprehensive framework or system design for blockchain implementation in local carbon markets. Future research should focus on developing such frameworks and architectures to support the practical adoption of blockchain in these markets, addressing scalability, integration challenges and the regulatory, social, and economic implications.

References

Abiodun, T. P., Nwulu, N., & Olukanmi, P. (2024). Application of blockchain technology in carbon trading market: a systematic review. *IEEE Access*.

Asadnabizadeh, M., & Moe, E. (2024). A review of global carbon markets from Kyoto to Paris and beyond: The persistent failure of implementation. *Frontiers in Environmental Science, 12*, 1368105.

Boumaiza, A. (2024). A blockchain-based scalability solution with microgrids peer-to-peer trade. *Energies, 17*(4), 915.

Boumaiza, A., & Maher, K. (2024). Leveraging blockchain technology to enhance transparency and efficiency in carbon trading markets. *International Journal of Electrical Power & Energy Systems, 162*, Article 110225.

Criscuoli, I., Martelli, A., Falconi, I., Galioto, F., Lasorella, M. V., Maurino, S., Dara Guccione, G. (2024). Lessons learned from existing carbon removal methodologies for agricultural soils to drive European Union policies. *European Journal of Soil Science, 75*(5), e13577.

Giakounti, E. A. (2024). *A comparative analysis of the EU emission trading system and California's Cap and Trade Program* Πανεπιστήμιο Πειραιώς].

Guzmán, A. S. R., Valdeolmillos Villaverde, D., Álvarez López, C., & S. Alonso, R. (2024). Smart contracts for the european union's carbon removal certification framework: a path towards transparent and efficient carbon markets. *International Congress on Blockchain and Applications.*

Hua, W., & Sun, H. (2019). A Blockchain-Based Peer-to-Peer Trading Scheme Coupling Energy and Carbon Markets. SEST 2019—2nd International Conference on Smart Energy Systems and Technologies.

Kudal, P., Dawar, S., Inamdar, V., Patnaik, A., & Rathore, T. (2024). Achieving financial inclusion through blockchain-based decentralized finance and the fintech revolution. In *Sustainability Reporting and Blockchain Technology* (pp. 364–375). Routledge.

Malik, A., Chaturvedi, V., Sandhani, M., Das, P., Arora, C., Singh, N., Zhao, A. (2024). Implications of an emission trading scheme for India's net-zero strategy: a modelling-based assessment *Environmental Research Letters, 19*(8), 084043.

McMillan, E., Barnes, J., Nolden, C., & McDermont, M. (2024). Local, place-based governance for net zero: a review and research agenda. *Journal of the British Academy, 12*(4).

Mhlanga, D. (2025). Leveraging blockchain for sustainable finance: insights, challenges, and opportunities for Africa. *Financial Inclusion and Sustainable Development in Sub-Saharan Africa*, 52–73.

Parhamfar, M., Sadeghkhani, I., & Adeli, A. M. (2024). Towards the net zero carbon future: A review of blockchain-enabled peer-to-peer carbon trading. *Energy Science & Engineering, 12*(3), 1242–1264.

Prawitasari, P. P., Nurmalasari, M. R., & Kumalasari, P. D. (2024). Blockchain technology in the carbon market: enhancing transparency and trust in emissions trading. *Jurnal Revenue: Jurnal Ilmiah Akuntansi, 5*(2), 1495–1521.

Sadawi, A. A., Madani, B., Saboor, S., Ndiaye, M., & Abu-Lebdeh, G. (2021). A comprehensive hierarchical blockchain system for carbon emission trading utilizing blockchain of things and smart contract. *Technological Forecasting and Social Change, 173*, 121124. https://doi.org/10.1016/j.techfore.2021.121124

Schletz, M., Franke, L. A., & Salomo, S. (2020). Blockchain Application for the Paris Agreement Carbon Market Mechanism—A Decision Framework and Architecture. *Sustainability, 12*(12). https://doi.org/10.3390/su12125069

Stavins, R. N. (2015). Linkage of regional, national, and sub-national policies in a future international climate agreement. *Towards a workable and effective climate regime*, 283.

Su, M., Zhao, R., Jiang, J., Zhao, J., Wang, M., Zha, D., & Li, C. (2024). A blockchain system supporting cross-border data protection and consistency verification in unified global carbon emissions trading framework. *Journal of Cleaner Production, 448*, Article 141693.

Vladucu, M.-V., Wu, H., Medina, J., Salehin, K. M., Dong, Z., & Rojas-Cessa, R. (2024). Blockchain on sustainable environmental measures: a review. *Blockchains, 2*(3), 334–365.

Wibowo, P. D. A. (2024). Barriers for business to engage in carbon trading through the indonesian carbon exchange: an analytical hierarchy process approach.

Yadav, A. P., Mulik, V. P., Kasar, M. S., Vitekar, A. B., Shelke, S. V., & Patil, S. M. Carbon credits and environmental impact tracking: the role of blockchain in supporting efficient and secure carbon credit markets. *Ecological Engineering & Environmental Technology, 26*(4), 76–87.

Yang, S., Jahanger, A., Hu, J., & Awan, A. (2024). Impact of China's carbon emissions trading scheme on firm-level pollution abatement and employment: Evidence from a national panel dataset. *Energy Economics, 136*, Article 107744.

Zhang, F., Zhao, R., Yang, Y., & Zou, X. (2023). Carbon trading in BRICS countries: challenges and recommendations. *Journal of Economics and Public Finance, 9*(3), 127–139.

Zhao, Y., & Peng, Y. (2024). Innovation and improvement of Carbon emission trading management model under blockchain technology. In: 2024 3rd International Conference on Data Analytics, Computing and Artificial Intelligence (ICDACAI).

FSC
www.fsc.org
MIX
Papier | Fördert gute Waldnutzung
FSC® C083411

Kolibri 360 GmbH
Ferdinand-Jühlke-Straße 7
99095 Erfurt, Deutschland
produktsicherheit@kolibri360.de